Mathematics: Advanced Concepts and Applications

Mathematics: Advanced Concepts and Applications

Editor: Alison Chapman

NY RESEARCH PRESS
New York

Published by NY Research Press
118-35 Queens Blvd., Suite 400,
Forest Hills, NY 11375, USA
www.nyresearchpress.com

Mathematics: Advanced Concepts and Applications
Edited by Alison Chapman

International Standard Book Number: 978-1-63238-662-5 (Hardback)

Cataloging-in-Publication Data

Mathematics : advanced concepts and applications / edited by Alison Chapman.
p. cm.
Includes bibliographical references and index.
ISBN 978-1-63238-662-5
1. Mathematics. I. Chapman, Alison.
QA37.3 .M38 2019
510--dc23

Contents

Preface

Mathematics is a fundamental science that is concerned with the study of quantity, structure, space and change through the use of abstraction and logic. These studies are approached from the disciplines of arithmetic, algebra, geometry and analysis. Logic, set theory, uncertainty, number theory, Riemann surfaces, etc. are significant sub-domains of mathematics. Applied mathematical domains of statistics, decision sciences and computational mathematics are of strategic importance in science, engineering, business and industry. This book strives to provide a fair idea about mathematics and to help develop a better understanding of the latest advances within this field. It covers in detail some existing theories and innovative concepts revolving around advanced mathematics and its applications. In this book, using case studies and examples, constant effort has been made to make the understanding of the difficult concepts of mathematics as easy and informative as possible for the readers.

All of the data presented henceforth, was collaborated in the wake of recent advancements in the field. The aim of this book is to present the diversified developments from across the globe in a comprehensible manner. The opinions expressed in each chapter belong solely to the contributing authors. Their interpretations of the topics are the integral part of this book, which I have carefully compiled for a better understanding of the readers.

At the end, I would like to thank all those who dedicated their time and efforts for the successful completion of this book. I also wish to convey my gratitude towards my friends and family who supported me at every step.

Editor

1

Approximation of common solutions for system of equilibrium problems and fixed-point problems

Yekini Shehu

Abstract In this paper, we complement the result of Zhu and Chang (J Ineq Appl 2013:146, 2013) by proving strong convergence theorems for approximation of a fixed point of a left Bregman strongly relatively nonexpansive mapping which is also a solution to a finite system of equilibrium problems in the framework of reflexive real Banach spaces using the Halpern–Mann's iterations used in Zhu and Chang (J Ineq Appl 2013:146, 2013). We also discuss the approximation of a common fixed point of a family of left Bregman strongly nonexpansive mappings which is also solution to a finite system of equilibrium problems in reflexive real Banach spaces. Our results complement many known recent results in the literature.

Keywords Left Bregman strongly relatively nonexpansive mapping · Left Bregman projection · Equilibrium problem · Banach spaces

Mathematics Subject Classification (2000) 47H06 · 47H09 · 47J05 · 47J25

Introduction

In this paper, let C be a nonempty, closed and convex subset of a real reflexive Banach space E with the dual space E^*. The norm and the dual pair between E and E^* are denoted by $\|.\|$ and $\langle .,. \rangle$, respectively. Let $T : C \to C$ be a nonlinear mapping. Denote by $F(T) := \{x \in C : Tx = x\}$ the set of fixed points of T. A mapping T is said to be *nonexpansive* if $||Tx - Ty|| \leq ||x - y||, \quad \forall x, y \in C$.

In 1994, Blum and Oettli [8] firstly studied the equilibrium problem: finding $x \in C$ such that

$$g(x, y) \geq 0, \quad \forall y \in C, \tag{1}$$

where $g : C \times C \to \mathbb{R}$ is a functional. Denote the set of solutions of the problem (1) by $EP(g)$. Since then, various equilibrium problems have been investigated. It is well known that equilibrium problems and their generalizations have been important tools for solving problems arising in the fields of linear or nonlinear programming, variational inequalities, complementary problems, optimization problems, fixed-point problems and have been widely applied to physics, structural analysis, management science and economics etc (see, for example [8, 26, 27]). One of the most important and interesting topics in the theory of equilibria is to develop efficient and implementable algorithms for solving equilibrium problems and their generalizations (see, e.g., [8, 26, 27, 47] and the references therein). Since the equilibrium problems have very close connections with both the fixed-point problems and the variational inequalities problems, finding the common elements of these problems has drawn many people's attention and has become one of the hot topics in the related fields in the past few years (see, e.g., [7, 17, 21, 29, 30, 31, 39, 42, 43, 44, 48] and the references therein).

In 1967, Bregman [11] discovered an elegant and effective technique for using of the so-called Bregman distance function D_f (see "Preliminaries", Definition 2.1) in the process of designing and analyzing feasibility and optimization algorithms. This opened a growing area of research in which Bregman's technique has been applied in various ways to design and analyze not only iterative algorithms for solving feasibility and optimization

Y. Shehu (✉)
Department of Mathematics, University of Nigeria, Nsukka, Nigeria
e-mail: deltanougt2006@yahoo.com

problems, but also algorithms for solving variational inequalities, for approximating equilibria, for computing fixed points of nonlinear mappings and so on (see, e.g., [3, 17, 41, 42, 43] and the references therein). In 2005, Butnariu and Resmerita [12] presented Bregman-type iterative algorithms and studied the convergence of the Bregman-type iterative method of solving some nonlinear operator equations.

Recently, using the Bregman projection, Reich and Sabach [35] presented the following algorithms for finding common zeroes of maximal monotone operators $A_i : E \to 2^{E^*}, (i = 1, 2, \ldots, N)$ in a reflexive Banach space E, respectively:

$$\begin{cases} x_0 \in E, \\ y_n^i = Res^f_{\lambda_n^i}(x_n + e_n^i), \\ C_n^i = \{z \in E : D_f(z, y_n^i) \le D_f(z, x_n + e_n^i)\}, \\ C_n = \cap_{i=1}^N C_n^i, \\ Q_n = \{z \in E : \langle \nabla f(x_0) - \nabla f(x_n), z - x_n \rangle \le 0\}, \\ x_{n+1} = proj^f_{C_n \cap Q_n} x_0, \quad n \ge 0 \end{cases} \tag{2}$$

and

$$\begin{cases} x_0 \in E, \\ \eta_n^i = \xi_n^i + \frac{1}{\lambda_n^i}(\nabla f(y_n^i) - \nabla f(x_n)), \quad \xi_n^i \in A_i y_n^i, \\ \omega_n^i = \nabla f^*(\lambda_n^i \eta_n^i + \nabla f(x_n)), \\ C_n^i = \{z \in E : D_f(z, y_n^i) \le D_f(z, x_n + e_n^i)\}, \\ C_n = \cap_{i=1}^N C_n^i, \\ Q_n = \{z \in E : \langle \nabla f(x_0) - \nabla f(x_n), z - x_n \rangle \le 0\}, \\ x_{n+1} = proj^f_{C_n \cap Q_n} x_0, \quad n \ge 0 \end{cases} \tag{3}$$

where $\{\lambda_n^i\}_{i=1}^N \subset (0, +\infty), \{e_n^i\}_{i=1}^N$ is an error sequence in E with $e_n^i \to 0$ and, $proj_C^f$ is the Bregman projection with respect to f from E onto a closed and convex subset C. Further, under some suitable conditions, they obtained two strong convergence theorems of maximal monotone operators in a reflexive Banach space. Reich and Sabach [36] also studied the convergence of two iterative algorithms for finitely many Bregman strongly nonexpansive operators in a Banach space. In [37], Reich and Sabach proposed the following algorithms for finding common fixed points of finitely many Bregman firmly nonexpansive operators $T_i : C \to C \ (i = 1, 2, \ldots, N)$ in a reflexive Banach space E if $\cap_{i=1}^N F(T_i) \neq \emptyset$:

$$\begin{cases} x_0 \in E, \\ Q_0^i = E, \quad i = 1, 2, \ldots, N, \\ y_n^i = T_i(x_n + e_n^i), \\ Q_{n+1}^i = \{z \in Q_n^i : \langle \nabla f(x_n + e_n^i) - \nabla f(y_n^i), z - y_n^i \rangle \le 0\}, \\ C_n = \cap_{i=1}^N C_n^i, \\ x_{n+1} = proj^f_{C_{n+1}} x_0, \quad n \ge 0. \end{cases} \tag{4}$$

Under some suitable conditions, they proved that the sequence $\{x_n\}$ generated by (4) converges strongly to $\cap_{i=1}^N F(T_i)$ and applied the result to the solution of convex feasibility and equilibrium problems.

In 2011, Chen et al. [18] introduced the concept of weak Bregman relatively nonexpansive mappings in a reflexive Banach space and gave an example to illustrate the existence of a weak Bregman relatively nonexpansive mapping and the difference between a weak Bregman relatively nonexpansive mapping and a Bregman relatively nonexpansive mapping. They also proved the strong convergence of the sequences generated by the constructed algorithms with errors for finding a fixed point of weak Bregman relatively nonexpansive mappings and Bregman relatively nonexpansive mappings under some suitable conditions.

Recently, Suantai et al. [40] considered strong convergence results for Bregman strongly nonexpansive mappings in reflexive Banach spaces by Halpern's iteration.In particular, they proved the following theorem.

Theorem 1.1 *Let E be a real reflexive Banach space and $f : E \to \mathbb{R}$ a strongly coercive Legendre function which is bounded, uniformly Fréchet differentiable and totally convex on bounded subsets of E. Let T be a Bregman strongly nonexpansive mapping on E such that $F(T) = \widehat{F}(T) \neq \emptyset$ Suppose that $u \in E$ and define the sequence $\{x_n\}$ as follows: $x_1 \in E$ and*

$$x_{n+1} = \nabla f^*(\alpha_n \nabla f(u) + (1 - \alpha_n)\nabla f(Tx_n)), \quad n \ge 1, \tag{5}$$

where $\{\alpha_n\} \subset (0,1)$ satisfying $\lim_{n\to\infty} \alpha_n = 0$ and $\sum_{n=1}^\infty \alpha_n = \infty$. Then $\{x_n\}$ converges strongly to $P_C^f(u)$, where $P^f_{F(T)}$ is the Bregman projection of E onto F(T).

Furthermore, using the Theorem 1.1, Suantai et al. [40] obtained some convergence theorems for a family of Bregman strongly nonexpansive mappings and gave some applications concerning the problems of finding zeroes of maximal monotone operators and equilibrium problems.

Very recently, Zhu and Chang [49] considered strong convergence results for Bregman strongly nonexpansive mappings in reflexive Banach spaces by modifying Halpern and Mann's iterations. Furthermore, they gave some applications concerning the problems of finding zeros of maximal monotone operators and equilibrium problems. In particular, they proved the following theorem.

Theorem 1.2 *Let E be a real reflexive Banach space and $f : E \to \mathbb{R}$ a strongly coercive Legendre function which is bounded, uniformly Fréchet differentiable and totally convex on bounded subsets of E. Let T be a Bregman strongly nonexpansive mapping on E such that $F(T) = \widehat{F}(T) \neq \emptyset$. Suppose that $u \in E$ and define the sequence $\{x_n\}$ as follows: $x_1 \in E$ and*

$$x_{n+1} = \nabla f^*(\alpha_n \nabla f(u) + (1 - \alpha_n)(\beta_n \nabla f(x_n) + (1 - \beta_n)\nabla f(Tx_n))), \quad n \geq 1, \tag{6}$$

where $\{\alpha_n\}$ *and* $\{\beta_n\}$ are sequences in (0,1) *satisfying*

$(C1)$ $\lim_{n\to\infty} \alpha_n = 0$;

$(C2)$ $\sum_{n=1}^{\infty} \alpha_n = \infty$;

$(C3)$ $0 < \liminf_{n\to\infty} \beta_n \leq \limsup_{n\to\infty} \beta_n < 1$.

Then $\{x_n\}$ *converges strongly to* $P^f_{F(T)}(u)$, *where* $P^f_{F(T)}$ *is the Bregman projection of E onto F(T).*

Motivated by the results of Suantai et al. [40] and Zhu and Chang [49], the purpose of this paper is to prove strong convergence theorems for approximation of a fixed point of a left Bregman strongly relatively nonexpansive mapping which is also a solution to a finite system of equilibrium problems in the framework of reflexive real Banach spaces. We also discuss the approximation of a common fixed point of a family of left Bregman strongly nonexpansive mappings which is also solution to a finite system of equilibrium problems in reflexive real Banach spaces. Our results complement many known recent results in the literature.

Preliminaries

In this section, we present the basic notions and facts that are needed in the sequel. The norms of E and E^*, its dual space, are denoted by $\|.\|$ and $\|.\|_*$, respectively. The pairing $\langle \xi, x\rangle$ is defined by the action of $\xi \in E^*$ at $x \in E$, that is, $\langle \xi, x\rangle := \xi(x)$. The *domain* of a convex function $f : E \to \mathbb{R}$ is defined to be

$$\text{dom} f := \{x \in E : f(x) < +\infty\}.$$

When dom $f \neq \emptyset$, we say that f is proper. The *Fenchel conjugate* function of f is the convex function $f^* : E \to \mathbb{R}$ defined by

$$f^*(\xi) = sup\{\langle \xi, x\rangle - f(x) : x \in E\}.$$

It is not difficult to check that when f is proper and lower semicontinuous, so is f^*. The function f is said to be *cofinite* if dom $f^* = E^*$.

Let $x \in$ int dom$f$, that is, let x belong to the interior of the domain of the convex function $f : E \to (-\infty, +\infty]$. For any $y \in E$, we define the *directional derivative* of f at x by

$$f^o(x,y) := \lim_{t\to 0^+} \frac{f(x+ty) - f(x)}{t}. \tag{7}$$

If the limit as $t \to 0^+$ in (7) exists for each y, then the function f is said to be *Gâteaux differentiable* at x. In this case, the gradient of f at x is the linear function $\nabla f(x)$, which is defined by $\langle \nabla f(x), y\rangle := f^o(x,y)$ for all $y \in E$ [19, Definition 1.3, page 3]. The function f is said to be Gâteaux differentiable if it is Gâteaux differentiable at each $x \in$ int dom$f$. When the limit as $t \to 0$ in (7) is attained uniformly for any $y \in E$ with $\|y\| = 1$, we say that f is *Fréchet differentiable* at x. Throughout this paper, $f : E \to (-\infty, +\infty]$ is always an *admissible function*, that is, a proper, lower semicontinuous, convex and Gâteaux differentiable function. Under these conditions, we know that f is continuous in int dom f (see [3], Fact 2.3, page 619).

The function f is said to be *Legendre* if it satisfies the following two conditions.

- $(L1)$ int dom $f \neq \emptyset$ and the subdifferential ∂f are single-valued on its domain.
- $(L2)$ int dom $f^* \neq \emptyset$ and ∂f^* are single-valued on its domain.

The class of Legendre functions in infinite-dimensional Banach spaces was first introduced and studied by Bauschke, Borwein and Combettes in [3]. Their definition is equivalent to conditions $(L1)$ and $(L2)$ because the space E is assumed to be reflexive (see [3], Theorems 5.4 and 5.6, page 634). It is well known that in reflexive spaces $\nabla f = (\nabla f^*)^{-1}$ (see [9], page 83). When this fact is combined with conditions $(L1)$ and $(L2)$, we obtain

$$\text{ran}\nabla f = \text{dom}\,\nabla f^* = \text{int dom} f^* \text{ and ran}\,\nabla f^* = \text{dom}\,\nabla f = \text{int dom} f.$$

It also follows that f is Legendre if and only if f^* is Legendre (see [3], Corollary 5.5, page 634) and that the functions f and f^* are Gateaux differentiable and strictly convex in the interior of their respective domains. When the Banach space E is smooth and strictly convex, in particular, a Hilbert space, the function $(\frac{1}{p})\|.\|^p$ with $p \in (1, \infty)$ is Legendre (cf. [3], Lemma 6.2, page 639). For examples and more information regarding Legendre functions, see, for instance, [3, 4].

Definition 2.1 The bifunction D_f : dom $f \times$ int dom $f \to [0, +\infty)$, which is defined by

$$D_f(y,x) := f(y) - f(x) - \langle \nabla f(x), y - x\rangle, \tag{8}$$

is called the *Bregman distance* (cf. [11, 15]).

The Bregman distance does not satisfy the well-known properties of a metric, but it does have the following important property, which is called the three point identity: for any $x \in$ dom$f$ and $y, z \in$ int dom$f$,

$$D_f(x,y) + D_f(y,z) - D_f(x,z) = \langle \nabla f(z) - \nabla f(y), x - y\rangle. \tag{9}$$

According to [13], Section 1.2, page 17 (see also [14]), the *modulus of total convexity* of f is the bifunction υ_f : int dom$f \times [0,+\infty) \to [0,+\infty]$ which is defined by

$$\upsilon_f(x,t) := \inf\{D_f(y,x) : y \in \text{dom} f, ||y-x|| = t\}.$$

The function f is said to be *totally convex at a point* $x \in$ int dom$f$ if $\upsilon_f(x,t) > 0$ whenever $t > 0$. The function f is said to be *totally convex* when it is totally convex at every point $x \in$ int dom$f$. This property is less stringent than uniform convexity (see [13], Section 2.3, page 92).

Examples of totally convex functions can be found, for instance, in [10, 12, 13]. We remark in passing that f is totally convex on bounded subsets if and only if f is uniformly convex on bounded subsets (see [12], Theorem 2.10, page 9).

The *Bregman projection* (cf. [11]) with respect to f of $x \in$ int dom$f$ onto a nonempty, closed and convex set $C \subset$ int dom$f$ is defined as the necessarily unique vector $proj_C^f(x) \in C$, which satisfies

$$D_f(proj_C^f(x), x) = \inf\{D_f(y,x) : y \in C\}. \tag{10}$$

Similarly to the metric projection in Hilbert spaces, the Bregman projection with respect to totally convex and Gâteaux differentiable functions has a variational characterization (cf. [12], Corollary 4.4, page 23).

Proposition 2.2 (Characterization of Bregman Projections). *Suppose that* $f : E \to (-\infty, +\infty]$ *is totally convex and Gâteaux differentiable in int dom f. Let* $x \in$ int dom$f$ *and let* $C \subset$ int dom$f$ *be a nonempty, closed and convex set. If* $\hat{x} \in C$, *then the following conditions are equivalent.*

1. *The vector* $\hat{x}$ *is the Bregman projection of* x *onto* C *with respect to f.*
2. *The vector* $\hat{x}$ *is the unique solution of the variational inequality*

$$\langle \nabla f(x) - \nabla f(z), z - y\rangle \geq 0 \quad \forall y \in C.$$

3. *The vector* $\hat{x}$ *is the unique solution of the inequality* $D_f(y,z) + D_f(z,x) \leq D_f(y,x) \forall y \in C.$

Recall that the function f is said to be *sequentially consistent* [5] if, for any two sequences $\{x_n\}$ and $\{y_n\}$ in E such that the first is bounded,

$$\lim_{n\to\infty} D_f(x_n, y_n) = 0 \Leftrightarrow \lim_{n\to\infty} ||x_n - y_n|| = 0. \tag{11}$$

Let C be a nonempty, closed and convex subset of E and $g : C \times C \to \mathbb{R}$ a bifunction that satisfies the following conditions:

A1. $g(x,x) = 0$ for all $x \in C$;
A2. g is monotone, i.e., $g(x,y) + g(y,x) \leq 0$ for all $x, y, \in C$;
A3. for each $x, y \in C$, $\lim_{t\to 0} g(tz + (1-t)x, y) \leq g(x,y)$;
A4. for each $x \in C$, $y \mapsto g(x,y)$ is convex and lower semicontinuous.

The resolvent of a bifunction $g : C \times C \to \mathbb{R}$ [19] is the operator $Res_g^f : E \to 2^C$ denoted by

$$Res_g^f(x) = \{z \in C : g(z,y) + \langle \nabla f(z) - \nabla f(x), y - z\rangle \geq 0 \quad \forall y \in C\}. \tag{12}$$

For any $x \in E$, there exists $z \in C$ such that $z = Res_C^f(x)$; see [36].

Let C be a convex subset of int dom$f$ and let T be a self-mapping of C. A point $p \in C$ is said to be an *asymptotic fixed point* of T if C contains a sequence $\{x_n\}_{n=0}^{\infty}$ which converges weakly to p and $\lim_{n\to\infty} ||x_n - Tx_n|| = 0$. The set of asymptotic fixed points of T is denoted by $\widehat{F}(T)$.

Recalling that the Bregman distance is not symmetric, we define the following operators.

Definition 2.3 A mapping T with a nonempty asymptotic fixed point set is said to be:

1. *left Bregman strongly nonexpansive* (see [5, 6]) with respect to a nonempty $\widehat{F}(T)$ if

$$D_f(p, Tx) \leq D_f(p,x), \quad \forall x \in C, \quad p \in \widehat{F}(T)$$

and if whenever $\{x_n\} \subset C$ is bounded, $p \in \widehat{F}(T)$ and

$$\lim_{n\to\infty} (D_f(p,x_n) - D_f(p, Tx_n)) = 0,$$

it follows that

$$\lim_{n\to\infty} D_f(Tx_n, x_n) = 0.$$

According to Martin-Marquez et al. [23], a left Bregman strongly nonexpansive mapping T with respect to a nonempty $\widehat{F}(T)$ is called *strictly left Bregman strongly nonexpansive mapping*.

2. An operator $T : C \to$ int dom$f$ is said to be: *left Bregman firmly nonexpansive* (L-BFNE) if

$$\langle \nabla f(Tx) - \nabla f(Ty), Tx - Ty\rangle \leq \langle \nabla f(x) - \nabla f(y), Tx - Ty\rangle$$

for any $x, y \in C$, or equivalently,

$$D_f(Tx, Ty) + D_f(Ty, Tx) + D_f(Tx, x) + D_f(Ty, y) \leq D_f(Tx, y) + D_f(Ty, x).$$

See [5, 10, 33] for more information and examples of L-BFNE operators (operators in this class are also called D_f-firm and BFNE). For two recent studies of the existence and approximation of fixed points of left Bregman firmly nonexpansive operators, see [24, 33]. It is also known that if T is left Bregman firmly nonexpansive and f is Legendre function which is bounded, uniformly Fréchet differentiable and totally convex on bounded subsets of E, then $F(T) = \widehat{F}(T)$ and $F(T)$ is closed and convex (see [33]). It also follows that every left Bregman firmly nonexpansive mapping is left Bregman strongly nonexpansive with respect to $F(T) = \widehat{F}(T)$.

Martin-Marquez et al. [23] called the Bregman projection defined in (10) and chracterized by Proposition 2.2 above as the *left Bregman projection* and they denoted the left Bregman projection by $\overleftarrow{Proj}^f_C$.

Let $f : E \to \mathbb{R}$ be a convex, Legendre and Gâteaux differentiable function. Following [1] and [15], we make use of the function $V_f : E \times E^* \to [0, +\infty)$ associated with f, which is defined by

$$V_f(x, x^*) = f(x) - \langle x^*, x\rangle + f^*(x^*), \forall x \in E, x^* \in E^*.$$

Then V_f is nonnegative and $V_f(x, x^*) = D_f(x, \nabla f^*(x^*))$ for all $x \in E$ and $x^* \in E^*$. Moreover, by the subdifferential inequality,

$$V_f(x, x^*) + \langle y^*, \nabla f^*(x^*) - x\rangle \leq V_f(x, x^* + y^*) \tag{13}$$

for all $x \in E$ and $x^*, y^* \in E^*$ (see also [20], Lemmas 3.2 and 3.3). In addition, if $f : E \to (-\infty, +\infty]$ is a proper lower semi-continuous function, then $f^* : E^* \to (-\infty, +\infty]$ is a proper weak* lower semi-continuous and convex function (see [28]). Hence V_f is convex in the second variable. Thus, for all $z \in E$,

$$D_f\Big(z, \nabla f^*\Big(\sum_{i=1}^{N} t_i \nabla f(x_i)\Big)\Big) \leq \sum_{i=1}^{N} t_i D_f(z, x_i), \tag{14}$$

where $\{x_i\}_{i=1}^N \subset E$ and $\{t_i\}_{i=1}^N \subset (0, 1)$ with $\sum_{i=1}^N t_i = 1$.

Finally, we state some lemmas that will be used in the proof of main results in next section.

Lemma 2.4 (Reich and Sabach [34]) *If $f : E \to \mathbb{R}$ is uniformly Fréchet differentiable and bounded on bounded subsets of E, then ∇f is uniformly continuous on bounded subsets of E from the strong topology of E to the strong topology of E*.*

Lemma 2.5 (Butnariu and Iusem [13]) *The function f is totally convex on bounded sets if and only if it is sequentially consistent.*

Lemma 2.6 (Reich and Sabach [35]) *Let $f : E \to \mathbb{R}$ be a Gâteaux differentiable and totally convex function. If $x_0 \in E$ and the sequence $\{D_f(x_n, x_0\}_{n=1}^\infty$ is bounded, then the sequence $\{x_n\}_{n=1}^\infty$ is also bounded.*

Lemma 2.7 (Reich and Sabach [36]) *Let $f : E \to (-\infty, +\infty)$ be a coercive Legendre function. Let C be a closed and convex subset of E. If the bifunction $g : C \times C \to \mathbb{R}$ satisfies conditions A1–A4, then*

1. *Res^f_C is single-valued;*
2. *Res^f_g is a Bregman firmly nonexpansive mapping;*
3. *$F(Res^f_g) = EP(g)$;*
4. *$EP(g)$ is a closed and convex subset of C;*
5. *for all $x \in E$ and $q \in F(Res^f_g)$,*

$$D_f(q, Res^f_g(x)) + D_f(Res^f_g(x), x) \leq D_f(q, x).$$

Lemma 2.8 (Xu [45]) *Let $\{a_n\}$ be a sequence of nonnegative real numbers satisfying the following relation:*

$$a_{n+1} \leq (1 - \alpha_n)a_n + \alpha_n\sigma_n + \gamma_n, n \geq 0,$$

where, (1) $\{\alpha_n\} \subset [0, 1]$, $\sum \alpha_n = \infty$; (2) $\limsup \sigma_n \leq 0$; (3) $\gamma_n \geq 0$; $(n \geq 0)$, $\sum \gamma_n < \infty$. *Then, $a_n \to 0$ as $n \to \infty$.*

Lemma 2.9 (Mainge [22]) *Let $\{a_n\}$ be a sequence of real numbers such that there exists a subsequence $\{n_i\}$ of $\{n\}$ such that $a_{ni} < a_{ni} + 1$ for all $i \in \mathbb{N}$. Then there exists a nondecreasing sequence $\{m_k\} \subset \mathbb{N}$ such that $m_k \to \infty$ and the following properties are satisfied by all (sufficiently large) numbers $k \in \mathbb{N}$:*

$$a_{m_k} \leq a_{m_k+1} \quad \text{and} \quad a_k \leq a_{m_k+1}.$$

In fact, $m_k = max\ \{j \leq k : a_j < a_{j+1}\}$.

Lemma 2.10 (Suantai et al. [40]) *Let E be a reflexive real Banach space. Let C be a nonempty, closed and convex subset of E. Let $f : E \to \mathbb{R}$ be a Gâteaux differentiable and totally convex function. Suppose T is a left Bregman strongly nonexpansive mappings of C into E such that $F(T) = \widehat{F}(T) \neq \emptyset$. If $\{x_n\}_{n=0}^\infty$ is a bounded sequence such that $x_n - Tx_n \to 0$ and $z := P^f_\Omega(u)$, then*

$$\limsup_{n\to\infty}\langle x_n - z, \nabla f(u) - \nabla f(z)\rangle \leq 0.$$

Main results

We first prove the following lemma.

Lemma 3.1 *Let E be a reflexive real Banach space and C a nonempty, closed and convex subset of E. For each $k = 1, 2, ..., N$, let g_k be a bifunction from $C \times C$ satisfying (A1) – (A4). Let $f : E \to \mathbb{R}$ a strongly coercive Legendre function which is bounded, uniformly Fréchet differentiable and totally convex on bounded subsets of E. Let T be a left Bregman strongly nonexpansive mapping of C into E such that $F(T) = \widehat{F}(T)$ and $\Omega := F(T) \cap (\cap_{k=1}^N EP(g_k)) \neq \emptyset$. Let $\{\alpha_n\}$ and $\{\beta_n\}$ be sequences in (0,1). Suppose $\{x_n\}_{n=1}^\infty$ is iteratively generated by u, $u_1 \in E$,*

$$\begin{cases} x_n = Res^f_{g_N} Res^f_{g_{N-1}} \ldots Res^f_{g_2} Res^f_{g_1} u_n, \\ u_{n+1} = \nabla f^*(\alpha_n \nabla f(u) + (1 - \alpha_n)(\beta_n \nabla f(x_n) + (1 - \beta_n)\nabla f(Tx_n))), \quad n \geq 1. \end{cases} \tag{15}$$

Then, $\{x_n\}_{n=0}^{\infty}$ *is bounded.*

Proof Let $x^* \in \Omega$. By taking $\theta_k^f = Res_{g_k}^f Res_{g_{k-1}}^f \ldots Res_{g_2}^f Res_{g_1}^f$, $k = 1, 2, \ldots, N$ and $\theta_0^f = I$, we obtain $x_n = \theta_N^f u_n$. Using the fact that Res_C^f, $k = 1, 2, \ldots, N$ is a strictly left quasi-Bregman nonexpansive mapping, we obtain from (15) that

$$\begin{aligned} D_f(x^*, x_{n+1}) =& D_f(x^*, \theta_N^f u_{n+1}) \leq D_f(x^*, u_{n+1}) \\ =& D_f(x^*, \nabla f^*(\alpha_n \nabla f(u) + \beta_n(1-\alpha_n)\nabla f(x_n) \\ &+ (1-\alpha_n)(1-\beta_n)\nabla f(Tx_n)) \\ \leq& \alpha_n D_f(x^*, u) + \beta_n(1-\alpha_n) D_f(x^*, x_n) \\ &+ (1-\alpha_n)(1-\beta_n) D_f(x^*, Tx_n) \\ \leq& \alpha_n D_f(x^*, u) + \beta_n(1-\alpha_n) D_f(x^*, x_n) \\ &+ (1-\alpha_n)(1-\beta_n) D_f(x^*, x_n) \\ =& \alpha_n D_f(x^*, u) + (1-\alpha_n) D_f(x^*, x_n) \\ \leq& \max\{D_f(x^*, u), D_f(x^*, x_n)\} \\ &\vdots \\ \leq& \max\{D_f(x^*, u), D_f(x^*, x_1)\}. \end{aligned} \tag{16}$$

Hence, $\{D_f(x^*, x_n)\}_{n=1}^{\infty}$ is bounded. We next show that the sequence $\{x_n\}$ is also bounded. Since $\{D_f(x^*, x_n)\}_{n=1}^{\infty}$ is bounded, there exists $M > 0$ such that

$$f(x^*) - \langle \nabla f(x_n), x^* \rangle + f^*(\nabla f(x_n)) = V_f(x^*, \nabla f(x_n)) = D_f(x^*, x_n) \leq M.$$

Hence, $\{\nabla f(x_n)\}$ is contained in the sublevel set $lev_{\leq}^{\psi}(M - f(x^*))$, where $\psi = f^* - \langle ., x^* \rangle$. Since f is lower semicontinuous, f^* is weak* lower semicontinuous. Hence, the function ψ is coercive by Moreau–Rockafellar Theorem (see [38], Theorem 7A and [25]). This shows that $\{\nabla f(x_n)\}$ is bounded. Since f is strongly accretive, f^* is bounded on bounded sets (see [46], Lemma 3.6.1 and [3], Theorem 3.3). Hence ∇f^* is also bounded on bounded subsets of E. (see [13], Proposition 1.1.11). Since f is a Legendre function, it follows that $x_n = \nabla f^*(\nabla f(x_n))$ is bounded for all $n \geq 0$. Therefore $\{x_n\}$ is bounded. So is $\{\nabla f(Tx_n)\}$. Indeed, since f is bounded on bounded subsets of E, ∇f is also bounded on bounded subsets of E (see [13], Proposition 1.1.11). Therefore $\{\nabla f(Tx_n)\}$ is bounded.

Now, following the method of proof in Suantai et al. [40], Zhu and Chang [49] and Mainge [22], we prove the following main theorem.

Theorem 3.2 *Let E be a reflexive real Banach space. Let C be a nonempty, closed and convex subset of E. For each $j = 1, 2, \ldots, N$, let g_j be a bifunction from $C \times C$ satisfying (A1) – (A4). Let $f : E \to \mathbb{R}$ a strongly coercive Legendre function which is bounded, uniformly Fréchet differentiable and totally convex on bounded subsets of E. Let T be a left Bregman strongly nonexpansive mapping of C into E such that $F(T) = \widehat{F}(T)$ and $\Omega := F(T) \cap (\cap_{j=1}^N EP(g_j)) \neq \emptyset$. Let $\{\alpha_n\}$ and $\{\beta_n\}$ be sequences in (0,1). Suppose $\{x_n\}_{n=1}^{\infty}$ is iteratively generated by (15) with the conditions*

1. $\lim_{n\to\infty} \alpha_n = 0$;
2. $\sum_{n=1}^{\infty} \alpha_n = \infty$.

Then, $\{x_n\}_{n=1}^{\infty}$ converges strongly to $\overrightarrow{Proj}_{\Omega}^f u$, where $\overrightarrow{Proj}_{\Omega}^f$ is the left Bregman projection of E onto Ω.

Proof Let $z_n := \nabla f^*(\alpha_n \nabla f(u) + \beta_n(1-\alpha_n)\nabla f(x_n) + (1-\alpha_n)(1-\beta_n)\nabla f(Tx_n))$, $n \geq 1$. Furthermore,

$$\begin{aligned} D_f(x^*, x_{n+1}) \leq& D_f(x^*, u_{n+1}) \\ =& V_f(x^*, \alpha_n \nabla f(u) + \beta_n(1-\alpha_n)\nabla f(x_n) \\ &+ (1-\alpha_n)(1-\beta_n)\nabla f(Tx_n)) \\ \leq& V_f(x^*, \alpha_n \nabla f(u) + \beta_n(1-\alpha_n)\nabla f(x_n) \\ &+ (1-\alpha_n)(1-\beta_n)\nabla f(Tx_n) - \alpha_n(\nabla f(u) \\ &- \nabla f(x^*)) \\ &- 2\langle \nabla f^*(\alpha_n \nabla f(u) + \beta_n(1-\alpha_n)\nabla f(x_n) \\ &+ (1-\alpha_n)(1-\beta_n)\nabla f(Tx_n)) - x^*, \\ &- \alpha_n(\nabla f(u) - \nabla f(x^*))\rangle \\ =& V_f(x^*, \alpha_n \nabla f(x^*) + \beta_n(1-\alpha_n)\nabla f(x_n) \\ &+ (1-\alpha_n)(1-\beta_n)\nabla f(Tx_n)) \\ &+ 2\alpha_n\langle z_n - x^*, \nabla f(u) - \nabla f(x^*)\rangle \\ =& D_f(x^*, \nabla f^*(\nabla f(x^*) + \beta_n(1-\alpha_n)\nabla f(x_n) \\ &+ (1-\alpha_n)(1-\beta_n)\nabla f(Tx_n)) \\ &+ 2\alpha_n\langle z_n - x^*, \nabla f(u) - \nabla f(x^*)\rangle \\ \leq& \alpha_n D_f(x^*, x^*) + \beta_n(1-\alpha_n) D_f(x^*, x_n) \\ &+ (1-\alpha_n)(1-\beta_n) D_f(x^*, Tx_n) \\ &+ 2\alpha_n\langle z_n - x^*, \nabla f(u) - \nabla f(x^*)\rangle \\ \leq& (1-\alpha_n) D_f(x^*, x_n) + 2\alpha_n\langle z_n - x^*, \nabla f(u) \\ &- \nabla f(x^*)\rangle. \end{aligned} \tag{17}$$

The rest of the proof will be divided into two parts.

Case 1 Suppose that there exists $n_0 \in \mathbb{N}$ such that $\{D_f(x^*, x_n)\}_{n=n_0}^{\infty}$ is nonincreasing. Then $\{D_f(x^*, x_n)\}_{n=0}^{\infty}$ converges and $D_f(x^*, x_{n+1}) - D_f(x^*, x_n) \to 0, \quad n \to \infty$. Observe that

$$D_f(x^*, x_{n+1}) \leq D_f(x^*, u_{n+1}) \leq \alpha_n D_f(x^*, u) + (1 - \alpha_n) D_f(x^*, x_n).$$

It then follows that

$$
\begin{aligned}
D_f(x^*,x_n)-D_f(x^*,Tx_n) =& D_f(x^*,x_n)-D_f(x^*,x_{n+1})\\
&+D_f(x^*,x_{n+1})-D_f(x^*,Tx_n)\\
\le& D_f(x^*,x_n)-D_f(x^*,x_{n+1})\\
&+\alpha_n(D_f(x^*,u)-D_f(x^*,Tx_n))\\
&\to 0,\quad n\to\infty.
\end{aligned}\tag{18}
$$

It then follows that

$$\lim_{n\to\infty} D_f(Tx_n,x_n)=0$$

Since $\{x_n\}$ is bounded, there exists a subsequence $\{x_{n\_j}\}$ of $\{x_n\}$ that converges weakly to p. Since $F(T)=\widehat{F}(T)$, we have $p\in F(T)$.

Next, we show that $p\in\cap_{k=1}^N EP(g_k)$. Now, using the fact that Res_C^f, $k=1, 2, ..., N$ is a strictly left quasi-Bregman nonexpansive mapping, we obtain

$$
\begin{aligned}
D_f(x^*,x_n) =& D_f(x^*,\theta_N^f u_n)\\
=& D_f(x^*,Res_{g_N}^f\theta_{N-1}^f u_n)\\
\le& D_f(x^*,\theta_{N-1}^f u_n)\le\ldots\le D_f(x^*,u_n).
\end{aligned}\tag{19}
$$

Since $x^*\in EP(g_N)=F(Res_{g_N}^f)$, it follows from Lemma 2.7, (19) and (17) that

$$
\begin{aligned}
D_f(x_n,\theta_{N-1}^f u_n) =& D_f(Res_{g_N}^f\theta_{N-1}^f u_n,\theta_{N-1}^f u_n)\\
\le& D_f(x^*,\theta_{N-1}^f u_n)-D_f(x^*,x_n)\\
\le& D_f(x^*,u_n)-D_f(x^*,x_n)\\
\le& \alpha_{n-1}M_1+D_f(x^*,x_{n-1})\\
&-D_f(x^*,x_n)\to 0,\quad n\to\infty,
\end{aligned}
$$

for some $M_1>0$. Thus, we obtain $\lim_{n\to\infty} D_f(\theta_N^f u_n,\theta_{N-1}^f u_n)=\lim_{n\to\infty} D_f(x_n,\theta_{N-1}^f u_n)=0$. From Lemma 2.5, we have

$$\lim_{n\to\infty}||\theta_N^f u_n-\theta_{N-1}^f u_n||=\lim_{n\to\infty}||x_n-\theta_{N-1}^f u_n||=0. \tag{20}$$

Since f is uniformly Fréchet differentiable, it follows from Lemma 2.4 and (20) that

$$\lim_{n\to\infty}||\nabla f(\theta_N^f u_n)-\nabla f(\theta_{N-1}^f u_n)||_*=0. \tag{21}$$

Again, since $x^*\in EP(g_{N-1})=F(Res_{g_{N-1}}^f)$, it follows from (19) and Lemma 2.7 that

$$
\begin{aligned}
D_f(\theta_{N-1}^f u_n,\theta_{N-2}^f u_n) =& D_f(Res_{g_{N-1}}^f\theta_{N-2}^f u_n,\theta_{N-2}^f u_n)\\
\le& D_f(x^*,\theta_{N-2}^f u_n)-D_f(x^*,\theta_{N-1}^f u_n)\\
\le& D_f(x^*,u_n)-D_f(x^*,x_n)\\
\le& \alpha_{n-1}M_1+D_f(x^*,x_{n-1})-D_f(x^*,x_n)\\
&\to 0,\quad n\to\infty.
\end{aligned}
$$

Again, we obtain $\lim_{n\to\infty} D_f(\theta_{N-1}^f u_n,\theta_{N-2}^f u_n)=0$. From Lemma 2.5, we have

$$\lim_{n\to\infty}||\theta_{N-1}^f u_n-\theta_{N-2}^f u_n||=0 \tag{22}$$

and hence,

$$\lim_{n\to\infty}||\nabla f(\theta_{N-1}^f u_n)-\nabla f(\theta_{N-2}^f u_n)||_*=0. \tag{23}$$

In a similar way, we can verify that

$$\lim_{n\to\infty}||\theta_{N-2}^f u_n-\theta_{N-3}^f u_n||=\cdots=\lim_{n\to\infty}||\theta_1^f u_n-u_n||=0. \tag{24}$$

From (20), (22) and (24), we can conclude that

$$\lim_{n\to\infty}||\theta_k^f u_n-\theta_{k-1}^f u_n||=0,\quad k=1,2,...,N \tag{25}$$

and

$$\lim_{n\to\infty}||x_n-u_n||=0.$$

Now, since $x_{n_j}\rightharpoonup p$ and $\lim_{n\to\infty}||x_n-u_n||=0$, we obtain that $u_{n_j}\rightharpoonup p$. Again, from (20), (22), (24) and $u_{n_j}\rightharpoonup p$, $n\to\infty$, we have that $\theta_k^f u_n\rightharpoonup p$, $j\to\infty$, for each $k=1, 2, ..., N$. Also, using (25), we obtain

$$\lim_{n\to\infty}||\nabla f(\theta_k^f u_n)-\nabla f(\theta_{k-1}^f u_n)||_*=0,\quad k=1,2,...,N. \tag{26}$$

By Lemma 2.7, we have that for each $k=1, 2, ..., N$

$$g_k(\theta_k^f u_{n_j},y)+\langle y-\theta_k^f u_{n_j},\nabla f(\theta_k^f u_{n_j})-\nabla f(\theta_{k-1}^f u_{n_j})\rangle\ge 0,\quad \forall y\in C.$$

Furthermore, using $(A2)$ we obtain

$$\langle y-\theta_k^f u_{n_j},\nabla f(\theta_k^f u_{n_j})-\nabla f(\theta_{k-1}^f u_{n_j})\rangle\ge g_k(y,\theta_k^f u_{n_j}). \tag{27}$$

By $(A4)$, (3.12) and $\theta_k^f u_{n_j}\rightharpoonup p$, we have for each $k=1, 2, ..., N$

$$g_k(y,p)\le 0,\quad \forall y\in C.$$

For fixed $y\in C$, let $z_{t,y}:=ty+(1-t)p$ for all $t\in(0,1)$. This implies that $z_t\in C$. This yields that $g_k(z_t,p)\le 0$. It follows from (A1) and (A4) that

$$
\begin{aligned}
0 =& g_k(z_t,z_t)\le tg_k(z_t,y)+(1-t)g_k(z_t,p)\\
\le& tg_k(z_t,y)
\end{aligned}
$$

and hence

$$0\le g_k(z_t,y).$$

From condition (A3), we obtain

$$g_k(p,y)\ge 0,\quad \forall y\in C.$$

This implies that $p\in EP(g_k)$, $k=1,2,...,N$. Thus, $p\in\cap_{k=1}^N EP(g_k)$. Hence, we have $p\in\Omega=F(T)\cap(\cap_{k=1}^N EP(g_k))$.

Let $y_n := \nabla f^*\left(\frac{\beta_n(1-\alpha_n)}{1-\alpha_n}\nabla f(x_n) + \frac{(1-\alpha_n)(1-\beta_n)}{1-\alpha_n}\nabla f(Tx_n)\right)$, $n \geq 1$, then

$$D_f(y_n, x_n) \leq \frac{\beta_n(1-\alpha_n)}{1-\alpha_n} D_f(x_n, x_n) + \frac{(1-\alpha_n)(1-\beta_n)}{1-\alpha_n} D_f(Tx_n, x_n) \to 0, \quad n \to \infty. \tag{28}$$

By Lemma 2.5, it follows that $||x_n - y_n|| \to 0, \quad n \to \infty$. Furthermore,

$$\begin{aligned} D_f(y_n, z_n) &= D_f(y_n, \nabla f^*(\alpha_n \nabla f(u) + (1-\alpha_n)\nabla f(y_n)) \\ &\leq \alpha_n D_f(y_n, u) + (1-\alpha_n) D_f(y_n, y_n) \\ &= \alpha_n D_f(y_n, u) \to 0, \quad n \to \infty. \end{aligned} \tag{29}$$

Again, by Lemma 2.5, it follows that $||y_n - z_n|| \to 0, \quad n \to \infty$. Then

$$||x_n - z_n|| \leq ||y_n - z_n|| + ||x_n - y_n|| \to 0, \quad n \to \infty. \tag{30}$$

Let $z := \overleftarrow{Proj}^f_\Omega u$. We next show that $\limsup_{n\to\infty}\langle y_n - z, \nabla f(u) - \nabla f(z)\rangle \leq 0$. To show the inequality $\limsup_{n\to\infty}\langle y_n - z, \nabla f(u) - \nabla f(z)\rangle \leq 0$, we choose a subsequence $\{x_{nj}\}$ of $\{x_n\}$ such that

$$\limsup_{n\to\infty}\langle x_n - z, \nabla f(u) - \nabla f(z)\rangle = \lim_{j\to\infty}\langle x_{n_j} - z, \nabla f(u) - \nabla f(z)\rangle.$$

By $||x_n - z_n|| \to 0, \quad n \to \infty$ and Lemma 2.10, we obtain

$$\limsup_{n\to\infty}\langle z_n - z, \nabla f(u) - \nabla f(z)\rangle = \limsup_{n\to\infty}\langle x_n - z, \nabla f(u) - \nabla f(z)\rangle \leq 0. \tag{31}$$

Now, using (31), (17) and Lemma 2.8, we obtain $D_f(z, x_n) \to 0, \quad n \to \infty$. Hence, by Lemma 2.5 we have that $x_n \to z, \quad n \to \infty$.

Case 2 Suppose there exists a subsequence $\{n_i\}$ of $\{n\}$ such that

$$D_f(x^*, x_{n_i}) < D_f(x^*, x_{n_i+1})$$

for all $i \in \mathbb{N}$. Then by Lemma 2.9, there exists a nondecreasing sequence $\{m_k\} \subset \mathbb{N}$ such that $m_k \to \infty$,

$$D_f(x^*, x_{m_k}) \leq D_f(x^*, x_{m_k+1}) \quad \text{and} \quad D_f(x^*, x_k) \leq D_f(x^*, x_{m_k+1})$$

for all $k \in \mathbb{N}$. Furthermore, we obtain

$$\begin{aligned} D_f(x^*, x_{m_k}) - D_f(x^*, Tx_{m_k}) &= D_f(x^*, x_{m_k}) - D_f(x^*, x_{m_k+1}) \\ &\quad + D_f(x^*, x_{m_k+1}) - D_f(x^*, Tx_{m_k}) \\ &\leq D_f(x^*, x_{m_k}) - D_f(x^*, x_{m_k+1}) \\ &\quad + \alpha_n(D_f(x^*, u) - D_f(x^*, x_{m_k})) \\ &\to 0, \quad k \to \infty. \end{aligned}$$

It then follows that

$$\lim_{k\to\infty} D_f(Tx_{m_k}, x_{m_k}) = 0.$$

By the same arguments as in Case 1, we obtain that

$$\limsup_{k\to\infty}\langle y_{m_k} - z, \nabla f(u) - \nabla f(z)\rangle \leq 0. \tag{32}$$

and

$$D_f(z, x_{m_k+1}) \leq (1-\alpha_{m_k})D_f(z, x_{m_k}) + 2\alpha_{m_k}\langle \nabla f(u) - \nabla f(z, y_{m_k} - x^*\rangle. \tag{33}$$

Since $D_f(z, x_{m\_k}) \leq D_f(z, x_{m\_k+1})$, we have

$$\alpha_{m_k} D_f(z, x_{m_k}) \leq D_f(z, x_{m_k}) - D_f(z, x_{m_k+1}) + 2\alpha_{m_k}\langle y_{m_k} - z, \nabla f(u) - \nabla f(z)\rangle \leq 2\alpha_{mk}\langle y_{m_k} - z, \nabla f(u) - \nabla f(z)\rangle.$$

In particular, since $\alpha_{mk} > 0$, we get

$$D_f(z, x_{m_k}) \leq 2\langle y_{m_k} - z, \nabla f(u) - \nabla f(z)\rangle. \tag{34}$$

It then follows from (32) that $D_f(z, x_{m_k}) \to 0, \quad k \to \infty$. From (34) and (33), we have

$$D_f(z, x_{m_k+1}) \to 0, \quad k \to \infty.$$

Since $D_f(z, x_k) \leq D_f(z, x_{mk+1})$ for all $k \in \mathbb{N}$, we conclude that $x_k \to z, \quad k \to \infty$. This implies that $x_n \to z, \quad n \to \infty$ which completes the proof.

Corollary 3.3 *Let E be a reflexive real Banach space. Let C be a nonempty, closed and convex subset of E. For each $j = 1, 2, ..., N$, let g_j be a bifunction from $C \times C$ satisfying (A1) – (A4). Let $f : E \to \mathbb{R}$ a strongly coercive Legendre function which is bounded, uniformly Fréchet differentiable and totally convex on bounded subsets of E. Let T be a left quasi-Bregman firmly nonexpansive mapping of C into E and $\Omega := F(T) \cap (\cap_{j=1}^N EP(g_j)) \neq \emptyset$. Let $\{\alpha_n\}$ and $\{\beta_n\}$ be sequences in (0,1). Suppose $\{x_n\}_{n=1}^\infty$ is iteratively generated by (15) with the conditions*

1. $\lim_{n\to\infty} \alpha_n = 0$;
2. $\sum_{n=1}^\infty \alpha_n = \infty$.

Then, $\{x_n\}_{n=1}^\infty$ converges strongly to $\overleftarrow{Proj}^f_\Omega u$, where $\overleftarrow{Proj}^f_\Omega$ is the left Bregman projection of E onto Ω.

Remark 3.4 Our Theorem 3.2 complements the results of Zhu and Chang [49] in the sense that it can be applied to the approximation of common solution of finite system of equilibrium problems and which is also a fixed point of left Bregman strongly nonexpansive mapping in a reflexive Banach space.

Convergence results concerning family of mappings

In this section, we present strong convergence theorems concerning approximation of common solution to a finite system of equilibrium problems which is also a common fixed point of a family of left Bregman strongly nonexpansive mappings in reflexive real Banach space.

Let C be a subset of a real Banach space E, $f : E \to \mathbb{R}$ a convex and Gâteaux differentiable function and $\{T_n\}_{n=1}^{\infty}$ a sequence of mappings of C such that $\cap_{n=1}^{\infty} F(T_n) \neq \emptyset$. Then $\{T_n\}_{n=1}^{\infty}$ is said to satisfy the AKTT condition [2] if, for any bounded subset B of C,

$$\sum_{n=1}^{\infty} \sup\{\|\nabla f(T_{n+1}z) - \nabla f(T_n z)\| : z \in B\} < \infty.$$

The following proposition is given in the results of Suantai et al. [40].

Proposition 4.1 *Let C be a nonempty, closed and convex subset of a real reflexive Banach space E. Let $f : E \to \mathbb{R}$ be a Legendre and Fréchet differentiable function. Let $\{T_n\}_{n=1}^{\infty}$ be a sequence of mappings from C into E such that $\cap_{n=1}^{\infty} F(T_n) \neq \emptyset$. Suppose that $\{T_n\}_{n=1}^{\infty}$ satisfies the AKTT condition. Then there exists the mapping $T : B \to E$ such that*

$$Tx = \lim_{n\to\infty} T_n x, \forall x \in B \tag{35}$$

and $\lim_{n\to\infty} \sup_{z\in B} \|\nabla f(Tz) - \nabla f(T_n z)\| = 0.$

In the sequel, we say that $(\{T_n\}, T)$ satisfies the AKTT condition if $\{T_n\}_{n=1}^{\infty}$ satisfies the AKTT condition and T is defined by (35) with $\cap_{n=1}^{\infty} F(T_n) = F(T)$.

By following the method of proof of Theorem 3.2, method of proof Theorem 4.2 of Suantai et al. [40] and Proposition 4.1, we prove the following theorem.

Theorem 4.2 *Let E be a reflexive real Banach space. Let C be a nonempty, closed and convex subset of E. For each $j = 1, 2, ..., N$, let G_j be a bifunction from $C \times C$ satisfying $(A1) - (A4)$. Let $f : E \to \mathbb{R}$ a strongly coercive Legendre function which is bounded, uniformly Fréchet differentiable and totally convex on bounded subsets of E. Let $\{T_n\}_{n=1}^{\infty}$ be a sequence of left Bregman strongly nonexpansive mappings on C such that $F(T_n) = \widehat{F}(T_n)$ for all $n \geq 0$ and $\Omega := (\cap_{n=1}^{\infty} F(T_n)) \cap (\cap_{j=1}^{N} EP(g_j)) \neq \emptyset$. Let $\{\alpha_n\}$ and $\{\beta_n\}$ be sequences in (0,1). Suppose $\{x_n\}_{n=0}^{\infty}$ is iteratively generated by by u, $u_0 \in E$,*

$$\begin{cases} x_n = Res_{g_N}^f Res_{g_{N-1}}^f \cdots Res_{g_2}^f Res_{g_1}^f u_n, \\ u_{n+1} = \nabla f^*(\alpha_n \nabla f(u) + (1-\alpha_n)(\beta_n \nabla f(x_n) \cdot \\ \qquad + (1-\beta_n)\nabla f(T_n x_n))), \quad n \geq 1, \end{cases} \tag{36}$$

with the conditions

1. $\lim_{n\to\infty} \alpha_n = 0;$
2. $\sum_{n=1}^{\infty} \alpha_n = \infty.$

If $(\{T_n\}, T)$ satisfies the AKTT condition, then $\{x_n\}_{n=1}^{\infty}$ converges strongly to $\overleftarrow{Proj}_{\Omega}^f u$, where $\overleftarrow{Proj}_{\Omega}^f$ is the left Bregman projection of E onto Ω.

Next, using the idea in [32], we consider the mapping $T : C \to C$ defined by $T = T_m \, T_{m-1} ... T_1$, where $T_i (i = 1, 2, \ldots, m)$ are left Bregman strongly nonexpansive mappings on E. Using Theorem 3.2 and Theorem 4.3 of Suantai et al. [40], we proof the following theorem.

Theorem 4.3 *Let E be a reflexive real Banach space. Let C be a nonempty, closed and convex subset of E. For each $j = 1, 2, ..., N$, let g_j be a bifunction from $C \times C$ satisfying $(A1) - (A4)$. Let $f : E \to \mathbb{R}$ a strongly coercive Legendre function which is bounded, uniformly Fréchet differentiable and totally convex on bounded subsets of E. Let $T_i (i = 1, 2, \ldots, m)$ be a sequence of left Bregman strongly nonexpansive mappings on C such that $F(T_i) = \widehat{F}(T_i)$ for all $n \geq 0$ and $\Omega := (\cap_{i=1}^{m} F(T_i)) \cap (\cap_{j=1}^{N} EP(g_j)) \neq \emptyset$. Let $\{\alpha_n\}$ and $\{\beta_n\}$ be sequences in (0,1). Suppose $\{x_n\}_{n=0}^{\infty}$ is iteratively generated by by u, $u_0 \in E$,*

$$\begin{cases} x_n = Res_{g_N}^f Res_{g_{N-1}}^f \cdots Res_{g_2}^f Res_{g_1}^f u_n, \\ u_{n+1} = \nabla f^*(\alpha_n \nabla f(u) + (1-\alpha_n)(\beta_n \nabla f(x_n) \\ \qquad + (1-\beta_n)\nabla f(T_m T_{m-1} ... T_1 x_n))), \quad n \geq 1, \end{cases} \tag{37}$$

with the conditions

1. $\lim_{n\to\infty} \alpha_n = 0;$
2. $\sum_{n=1}^{\infty} \alpha_n = \infty.$

Then $\{x_n\}_{n=1}^{\infty}$ converges strongly to $\overleftarrow{Proj}_{\Omega}^f u$, where $\overleftarrow{Proj}_{\Omega}^f$ is the left Bregman projection of E onto Ω.

References

1. Alber, Y.I.: Metric and generalized projection operator in Banach spaces: properties and applications, In: Theory and Applications of Nonlinear Operators of Accretive and Monotone Type, Lecture Notes in Pure and Applied Mathematics. vol. 78, pp. 15–50. Dekker, New York (1996)
2. Aoyama, K., Kimura, Y., Takahashi, W., Toyoda, M.: Approximation of common fixed points of a countable family of

nonexpansive mappings in a Banach space. Nonlinear Anal. **67**, 2350–2360 (2006)
3. Bauschke, H.H., Borwein, J.M., Combettes, P.L.: Essential smoothness, essential strict convexity, and Legendre functions in Banach spaces. Comm. Contemp. Math. **3**, 615–647 (2001)
4. Bauschke, H.H., Borwein, J.M.: Legendre functions and the method of random Bregman projections. J. Convex Anal. **4**, 27–67 (1997)
5. Bauschke, H.H., Borwein, J.M., Combettes, P.L.: Bregman monotone optimization algorithms. SIAM J. Control Optim. **42**, 596–636 (2003)
6. Bauschke, H.H., Wang, X. Yao, L.: General resolvents for monotone operators: characterization and extension, In: Proceedings of Biomedical mathematics: promising directions in imaging, therapy planning and inverse problems, pp. 57–74. Medical Physics Publishing, Madison (2009)
7. Bello Cruz, J.Y., Iusem, A.N.: An explicit algorithm for monotone variational inequalities. Optimization (2011). doi:10.1080/02331934.2010.536232.
8. Blum, E., Oettli, W.: From optimization and variational inequalities to equilibrium problems. Math. Stud. **63**, 123–145 (1994)
9. Bonnans, J.F., Shapiro, A.: Perturbation analysis of optimization problems. Springer, New York (2000)
10. Borwein, J.M., Reich, S., Sabach, S.: A characterization of Bregman firmly nonexpansive operators using a new monotonicity concept. J. Nonlinear Convex Anal. **12**, 161–184 (2011)
11. Bregman, L.M.: The relaxation method of finding the common point of convex sets and its application to the solution of problems in convex programming. USSR Comput. Math. Math. Phys. **7**, 200–217 (1967)
12. Butnariu, D., Resmerita, E.: Bregman distances, totally convex functions and a method for solving operator equations in Banach spaces. Abstr. Appl. Anal. **2006**, 1–39. Art. ID 84919 (2006)
13. Butnariu, D., Iusem, A.N.: Totally convex functions for fixed points computation and infinite dimensional optimization. Kluwer Academic Publishers, Dordrecht (2000)
14. Butnariu, D., Censor, Y., Reich, S.: Iterative averaging of entropic projections for solving stochastic convex feasibility problems. Comput. Optim. Appl. **8**, 21–39 (1997)
15. Censor, Y., Lent, A.: An iterative row-action method for interval convex programming. J. Optim. Theory Appl. **34**, 321–353 (1981)
16. Censor, Y., Reich, S.: Iterations of paracontractions and firmly nonexpansive operators with applications to feasibility and optimization. Optimization **37**, 323–339 (1996)
17. Chen, J.W., Cho, Y.J., Agarwal, R.P.: Strong convergence theorems for equilibrium problems and weak Bregman relatively nonexpansive mappings in Banach spaces. J. Ineq. Appl. **2013**, 119 (2013). doi:10.1186/1029-242X-2013-119
18. Chen, J.W., Wan, Z., Yuan, L.: Approximation of fixed points of weak Bregman relatively nonexpansive mappings in Banach spaces. Internet J. Math. Math. Sci. **2011**, 1–23, Art. ID 869684 (2010)
19. Combettes, P.L.:, Hirstoaga, S.A.: Equilibrium programming in Hilbert spaces. J. Nonlinear Convex Anal. **6**(2005), 117–136 (2002). ISBN 978-1-4020-0161-1
20. Kohsaka, F., Takahashi, W.: Proximal point algorithms with Bregman functions in Banach spaces. J. Nonlinear Convex Anal **6**, 505–523 (2005)
21. Kumam, P.: A new hybrid iterative method for solution of equilibrium problems and fixed point problems for an inverse strongly monotone operator and a nonexpansive mapping. J. Appl. Math. Comput. **29**, 263–280 (2009)
22. Maingé, P.E.: Strong convergence of projected subgradient methods for nonsmooth and nonstrictly convex minimization. Set-Valued Anal. **16**, 899–912 (2008)
23. Martin-Marquez, V., Reich, S., Sabach, S.: Right Bregman nonexpansive operators in Banach spaces. Nonlinear Anal. **75**, 5448–5465 (2012)
24. Martin-Marquez, V., Reich, S., Sabach, S.: Iterative methods for approximating fixed points of Bregman nonexpansive operators. Discret. Contin. Dyn. Syst. (in press)
25. Moreau J-J.: Sur la fonction polaire d'une fonction semi-continue superieurement. C. R. Acad. Sci. Paris **258**, 1128–1130 (1964)
26. Moudafi, A.: A partial complement method for approximating solutions of a primal dual fixed-point problem. Optim. Lett. **4**(3), 449–456 (2010)
27. Pardalos, P.M., Rassias, T.M., Khan, A.A.: Nonlinear analysis and variational problems. Springer (2010)
28. Phelps, R.P.: Convex functions, monotone operators, and differentiability, 2nd edn. In: Lecture Notes in Mathematics, vol. 1364, Springer, Berlin (1993)
29. Plubtieng, S., Punpaeng, R.: A new iterative method for equilibrium problems and fixed point problems of nonexpansive mappings and monotone mappings. Appl. Math. Comput. **197**, 548–558 (2008)
30. Qin, X., Cho, Y.J., Kang, S.M.: Convergence theorems of common elements for equilibrium problems and fixed point problems in Banach spaces. J. Comput. Appl. Math. **225**, 20–30 (2009)
31. Qin, X., Su, Y.: Strong convergence theorems for relatively nonexpansive mappings in a Banach space. Nonlinear Anal. **67**, 1958–1965 (2007)
32. Reich, S.A.: Weak convergence theorem for the alternating method with Bregman distances. In: Theory and applications of nonlinear operators of accretive and monotone type, pp. 313–318. Marcel Dekker, New York (1996)
33. Reich, S., Sabach, S.: Existence and approximation of fixed points of Bregman firmly nonexpansive operators in reflexive Banach spaces, In: Fixed-point algorithms for inverse problems in science and engineering, pptimization and its applications, vol. 49, pp. 301–316. Springer, New York (2011)
34. Reich, S., Sabach, S.: A strong convergence theorem for a proximal-type algorithm in reflexive Banach spaces. J. Nonlinear Convex Anal. **10**(3), 471–485 (2009)
35. Reich S, Sabach S (2010) Two strong convergence theorems for a proximal method in reflexive Banach spaces. Numer. Funct. Anal. Optim. **31**(1–3), 22–44
36. Reich, S., Sabach, S.: Two strong convergence theorems for Bregman strongly nonexpansive operators in reflexive Banach spaces. Nonlinear Anal. **73**(1), 122–135 (2010)
37. Reich, S., Sabach, S.A.: Projection method for solving nonlinear problems in reflexive Banach spaces. J. Fixed Point Theory Appl. doi:10.1007/s11784-010-0037-5.
38. Rockafellar, R.T.: Level sets and continuity of conjugate convex functions. Trans. Amer. Math. Soc. **123**, 46–63 (1966)
39. Shehu, Y.: A new iterative scheme for a countable family of relatively nonexpansive mappings and an equilibrium problem in Banach spaces. J. Glob. Optim. **54**, 519–535 (2012)
40. Suantai, S., Cho, Y.J., Cholamjiak, P.: Halpern's iteration for Bregman strongly nonexpansive mappings in reflexive Banach spaces. Comp. Math. Appl. **64**, 489–499 (2012)
41. Takahashi, S., Takahashi, W.: Viscosity approximation methods for equilibrium problems and fixed point problems in Hilbert spaces. J. Math. Anal. Appl. **331**, 506–518 (2007)
42. Takahashi, W., Zembayashi, K.: Strong convergence theorem by a new hybrid method for equilibrium problems and relatively nonexpansive mappings. Fixed Point Theory and and Applications. Art. ID 528476, 11 pages (2008)
43. Takahashi, W., Zembayashi, K.: Strong and weak convergence theorems for equilibrium problems and relatively nonexpansive mappings in Banach spaces. Nonlinear Anal. **70**, 45–57 (2000)

44. Wangkeeree, R.: An extragradient approximation method for equilibrium problems and fixed point problems of a countable family of nonexpansive mappings. Fixed Point Theory and Applications, vol. 2008, Art. ID 134148, 17 pages, (2008)
45. Xu, H.K.: Iterative algorithms for nonlinear operators. J. London Math. Soc. **66**(2), 240–256 (2002)
46. Zalinescu, C.: Convex analysis in general vector spaces. World Scientific Publishing Co., Inc., River Edge (2002)
47. Zegeye, H., Ofoedu, E.U., Shahzad, N.: Convergence theorems for equilibrium problems, variational inequality problem and countably infinite relatively nonexpansive mappings. Appl. Math.Comp. **216**, 3439–3449 (2010)
48. Zegeye, H., Shahzad, N.: A hybrid scheme for finite families of equilibrium, variational inequality and fixed point problems. Nonlinear Anal. **70**, 2707–2716 (2010)
49. Zhu, J.H., Chang, S.S.: Halpern-Mann's iterations for Bregman strongly nonexpansive mappings in reflexive Banach spaces with applications. J. Ineq. Appl. **2013**, 146 (2013) doi:10.1186/1029-242X-2013-146

2

Fixed point results for α-ψ-locally graphic contraction in dislocated qusai metric spaces

Muhammad Arshad · Fahimuddin · Abdullah Shoaib · Aftab Hussain

Abstract In this paper, we have obtained fixed point results for $\alpha - \psi$ -locally contractive type mappings in a closed ball in left K-sequentially complete and in right K-sequentially complete dislocated quasi metric spaces. Moreover the mappings under consideration are α-admissible with respect to η. We have used conditions weaker than those of Samet et al. [Nonlinear Anal. 75:2154–2165, (2012)]. As an application, we have derived some new fixed point theorems for ψ -graphic contractions defined on dislocated quasi metric space endowed with a graph as well as ordered dislocated metric space. Some comparative examples are constructed which illustrate the superiority of our results. In the process we have generalized several well known, recent and classical results from the literature.

Keywords Fixed point · Complete dislocated quasi metric space · Left K -sequentially complete · α-ψ contractive mappings · ψ -graphic contractions · Closed ball.

Introduction and preliminaries

Fixed point theory has wide and endless applications in many fields of engineering and science. Its core, the Banach contraction principle, has attracted many researchers who tried to generalize it in different aspects. Fixed point results of mappings satisfying certain contractive condition on the entire domain has been at the centre of rigorous research activities.

From the application point of view the situation is not yet completely satisfactory because it frequently happens that a mapping T is a contraction not on the entire space X but merely on a subset Y of X. However, we impose subtle restrictions to obtain fixed point results for such mapping. Recently Arshad et al. [7] proved a result concerning the existence of fixed points of a mapping satisfying a contractive conditions on closed ball in a complete dislocated metric space. Other results on closed ball can be seen in [6, 8–11, 27, 35, 36]. Recently, Karapınar et al. [24] introduced the concept of quasi-partial metric space. Zeyada et al. [37] introduced the concept of dislocated quasi metric which is basically the generalization of quasi-partial metric space.

Recently, Samet et al. [34] introduced the notions of α-ψ -contractive and α-admissible mapping in complete metric spaces. The existance of fixed points of α-ψ-contractive and α -admissible mapping in complete metric spaces has been studied by several researchers, (see [3, 4, 19, 20, 25, 26, 33]).

Consistent with [33, 34, 37], we give the following definitions which will be needed in the sequel.

Definition 1.1 [37] Let X be a nonempty set. Let $d_q : X \times X \to [0, \infty)$ be a function, called a dislocated qusai metric (or simply d_q-metric) if the following conditions hold for any $x, y, z \in X$:

(i) If $d_q(x, y) = d_q(y, x) = 0$, then $x = y$,
(ii) $d_q(x, y) \leq d_q(x, z) + d_q(z, y)$.

M. Arshad (✉) · Fahimuddin · A. Shoaib · A. Hussain
Department of Mathematics, International Islamic University, Islamabad 44000, Pakistan
e-mail: marshad_zia@iiu.edu.pk

Fahimuddin
e-mail: fahamiiu@gmail.com

A. Shoaib
e-mail: abdullahshoaib15@yahoo.com

A. Hussain
e-mail: aftabshh@gmail.com

The pair (X, d_q) is called a dislocated qusai metric space. It is clear that if $d_q(x, y) = d_q(y, x) = 0$, then from (i), $x = y$. But if $x = y$, $d_q(x, y)$ may not be 0. It is observed that if $d_q(x, y) = d_q(y, x)$ for all $x, y \in X$, then (X, d_q) becomes a dislocated metric space. We shall denote (X, d_l) for a dislocated metric space. The ball $\overline{B(x, \varepsilon)}$ where $\overline{B(x, \varepsilon)} = \{y \in X : d_q(x, y) \leq \varepsilon\}$ is a closed ball in dislocated qusai metric space, for some $x \in X$ and $\varepsilon > 0$. It is clear that any qusai-partial metric is a d_q-metric.

Example 1.2 If $X = R^+ \cup \{0\}$ then $d_q(x, y) = x + 2y$ defines a dislocated quasi metric d_q on X.

Example 1.3 If $X = R^+ \cup \{0\}$ then $d_q(x, y) = x + \max\{x, y\}$ defines a dislocated quasi metric d_q on X.

Let Ψ denote the family of all nondecreasing functions $\psi : [0, +\infty) \to [0, +\infty)$ such that $\sum_{n=1}^{+\infty} \psi^n(t) < +\infty$ for all $t > 0$, where ψ^n is the nth iterate of ψ.

Lemma 1.4 [33] *If* $\psi \in \Psi$, *then* $\psi(t) < t$ *for all* $t > 0$.

Definition 1.5 [34] Let (X, d) be a metric space and $S : X \to X$ be a given mapping. We say that S is α-ψ contractive mapping if there exist two functions $\alpha : X \times X \to [0, +\infty)$ and $\psi \in \Psi$ such that $\alpha(x, y)d(Sx, Sy) \leq \psi(d(x, y))$ for all $x, y \in X$.

Remark 1.6 [33] By definition, $\alpha(x, x) \neq 0$ for $x \in X$.

Definition 1.7 [33] Let $S : X \to X$ and $\alpha, \eta : X \times X \to [0, +\infty)$ be two functions. We say that S is α-admissible mapping with respect to η if $x, y \in X$ such that $\alpha(x, y) \geq \eta(x, y)$ then we have $\alpha(Sx, Sy) \geq \eta(Sx, Sy)$. Note that if we take $\eta(x, y) = 1$, then this definition reduces to Definition 1.1 of [33]. Also, if we take $\alpha(x, y) = 1$, then we say that T is an η-subadmissible mapping.

In this paper, we shall prove a theorem which is an extension of the results of Samet et al. [34].

Main results

Reilly et al. [31] introduced the notion of left (right) K-Cauchy sequence and left (right) K-sequentialy complete spaces(see [11, 16]). We use this concept to establish the following definition.

Definition 2.1 Let (X, d_q) be a dislocated quasi metric space.

(a) A sequence $\{x_n\}$ in (X, d_q) is called left (right) K-Cauchy if $\forall\ \varepsilon > 0$, $\exists\ n_0 \in N$ such that $\forall\ n > m \geq n_0$, $d_q(x_m, x_n) < \varepsilon$ (respectively $d_q(x_n, x_m) < \varepsilon$).

(b) A sequence $\{x_n\}$ dislocated quasi-converges (for short d_q -converges) to x if $\lim_{n\to\infty} d_q(x_n, x) = \lim_{n\to\infty} d_q(x, x_n) = 0$. In this case x is called a d_q-limit of $\{x_n\}$.

(c) (X, d_q) is called left (right) K-sequentially complete if every left (right) K-Cauchy sequence in X converges to a point $x \in X$ such that $d_q(x, x) = 0$.

One can easily observe that every complete dislocated quasi metric space is also left K-sequentially complete dislocated quasi metric space but the converse is not true in general.

Theorem 1 *Let* (X, d_q) *be a left K-sequentially complete dislocated quasi metric space. Suppose there exist two functions,* $\alpha, \eta : X \times X \to [0, +\infty)$. *Let* x_0 *be an arbitrary point in X and* $S : X \to X$ *be* α*-admissible with respect to* η *and* $\psi \in \Psi$. *Assume that,*

$$x, y \in \overline{B(x_0, r)}, \alpha(x, y) \geq \eta(x, y) \Longrightarrow d_q(Sx, Sy) \leq \psi(d_q(x, y)) \tag{1}$$

and

$$\sum_{i=0}^{j} \psi^i(d_q(x_0, Sx_0)) \leq r, \ \text{for all } j \in N \text{ and } r > 0. \tag{2}$$

Suppose that the following assertions hold:

(i) $\alpha(x_0, Sx_0) \geq \eta(x_0, Sx_0)$;

(ii) *for any sequence* $\{x_n\}$ *in* $\overline{B(x_0, r)}$ *such that* $\alpha(x_n, x_{n+1}) \geq \eta(x_n, x_{n+1})$ *for all* $n \in N \cup \{0\}$ *and* $x_n \to u \in \overline{B(x_0, r)}$ *as* $n \to +\infty$ *then* $\alpha(x_n, u) \geq \eta(x_n, u)$ *for all* $n \in N \cup \{0\}$.

Then, there exists a point x^* *in* $\overline{B(x_0, r)}$ *such that* $x^* = Sx^*$.

Proof Choose a point x_1 in X such that $x_1 = Sx_0$ let $x_2 = Sx_1$. Continuing this process, we construct a sequence x_n of points in X such that,

$$x_{i+1} = Sx_i, \quad \text{where i} = 0, 1, 2, \ldots$$

□

By assumption $\alpha(x_0, x_1) \geq \eta(x_0, x_1)$ and S is α-admissible with respect to η, we have, $\alpha(Sx_0, Sx_1) \geq \eta(Sx_0, Sx_1)$ we deduce that $\alpha(x_1, x_2) \geq \eta(x_1, x_2)$ which also implies that $\alpha(Sx_1, Sx_2) \geq \eta(Sx_1, Sx_2)$. Continuing in this way obtain $\alpha(x_n, x_{n+1}) \geq \eta(x_n, x_{n+1})$ for all $n \in N \cup \{0\}$. First we show that $x_n \in \overline{B(x_0, r)}$ for all $n \in N$. Using inequality (2), we have,

$$\sum_{i=0}^{n} \psi^i(d_q(x_0, Sx_0)) \leq r \quad \text{for all} j \in N.$$

It follows that,

$$x_1 \in \overline{B(x_0, r)}.$$

Let $x_2, \ldots, x_j \in \overline{B(x_0, r)}$ for some $j \in N$. so using inequality (1), we obtain,

$$\begin{aligned} d_q(x_j, x_{j+1}) &= d_q(Sx_{j-1}, Sx_j) \\ &\leq \psi(d_q(x_{i-1}, x_i)) \\ &\leq \psi^2(d_q(x_{j-2}, x_{j-1})) \\ &\leq \cdots \leq \psi^j(d_q(x_0, x_1)). \end{aligned}$$

Thus we have,

$$d_q(x_j, x_{j+1}) \leq \psi^j(d_q(x_0, x_1)). \tag{3}$$

Now,

$$\begin{aligned} d_q(x_0, x_{j+1}) &= d_q(x_0, x_1) + d_q(x_1, x_2) + d_q(x_2, x_3) + \ldots \\ &\quad + d_q(x_j, x_{j+1}) \\ &\leq \sum_{i=0}^{j} \psi^i(d_q(x_0, x_1)) \\ &\leq r. \end{aligned}$$

Thus $x_{j+1} \in \overline{B(x_0, r)}$. Hence $x_n \in \overline{B(x_0, r)}$ for all $n \in N$. Now inequality (3) can be written as

$$d_q(x_n, x_{n+1}) \leq \psi^n(d_q(x_0, x_1)), \quad \text{for all n} \in \text{N}. \tag{4}$$

Fix $\varepsilon > 0$ and let $n(\varepsilon) \in N$ such that $\sum \psi^n(d_q(x_0, x_1)) < \varepsilon$. let $n, m \in N$ with $m > n > n(\varepsilon)$ using the triange inequality, we obtain,

$$\begin{aligned} d_q(x_n, x_m) &\leq \sum_{k=n}^{m-1} d_q(x_k, x_{k+1}) \leq \sum_{k=n}^{m-1} \psi^k(d_q(x_0, x_1)) \\ &\leq \sum_{n \geq n(\varepsilon)} \psi^k(d_q(x_0, x_1)) < \varepsilon. \end{aligned}$$

Thus we have proved that $\{x_n\}$ is a left K-Cauchy sequence in $(\overline{B(x_0, r)}, d_q)$. As $\overline{B(x_0, r)}$ is closed, so it is left K-sequentially complete. Therefore, there exists a point $x^* \in \overline{B(x_0, r)}$ such that $x_n \to x^*$. Also

$$\lim_{n\to\infty} d_q(x_n, x^*) = 0. \tag{5}$$

On the other hand, from (ii), we have,

$$\alpha(x^*, x_n) \geq \eta(x^*, x_n) \quad \text{for all } n \in N \cup \{0\}. \tag{6}$$

Now using the triangle inequality, also by using (1) and (6), we get

$$d_q(Sx^*, x_{i+1}) \leq \psi(d_q(x^*, x_i)) < d_q(x^*, x_i). \tag{7}$$

Letting $i \to \infty$ and by using inequality (7), we obtain $d_q(Sx^*, x^*) < 0$. Hence $Sx^* = x^*$.

If $\eta(x, y) = 1$ for all $x, y \in X$ in Theorem 2.2, we obtain following result.

Corollary 2 *Let (X, d_q) be a left K-sequentially complete dislocated quasi metric space. Suppose there exist a function, $\alpha : X \times X \to [0, +\infty)$. Let x_0 be an arbitrary point in X and $S : X \to X$ be α-admissible and $\psi \in \Psi$. Assume that,*

$$x, y \in \overline{B(x_0, r)}, \alpha(x, y) \geq 1 \Longrightarrow d_q(Sx, Sy) \leq \psi(d_q(x, y))$$

and

$$\sum_{i=0}^{j} \psi^i(d_q(x_0, Sx_0)) \leq r, \quad \text{forall } j \in N \text{ and } r > 0.$$

Suppose that the following assertions hold:

(i) $\alpha(x_0, Sx_0) \geq 1$;

(ii) *for any sequence $\{x_n\}$ in $\overline{B(x_0, r)}$ such that $\alpha(x_n, x_{n+1}) \geq 1$ for all $n \in N \cup \{0\}$ and $x_n \to u \in \overline{B(x_0, r)}$ as $n \to +\infty$ then $\alpha(x_n, u) \geq 1$ for all $n \in N \cup \{0\}$.*

Then, there exists a point x^* in $\overline{B(x_0, r)}$ such that $x^* = Sx^*$.

Theorem 3 *Adding condition "if x and y are any fixed point in $\overline{B(x_0, r)}$ then $\alpha(x, y) \geq \eta(x, y)$" to the hypotheses of Theorem 2.2. Then S has a unique fixed point x^* and $d_q(x^*, x^*) = 0$.*

Proof Assume that x^* and y^* be two fixed point of S in $\overline{B(x_0, r)}$, then, by assumption, $\alpha(x^*, y^*) \geq \eta(x^*, y^*)$,

$$d_q(x^*, y^*) = d_q(Sx^*, Sy^*) \leq \psi(d_q(x^*, y^*))$$

A contradiction to the fact that for each $t > 0$, $\psi(t) < t$. So $x^* = y^*$. Hence S has no fixed point other than x^*. Now, $\alpha(x^*, x^*) \geq \eta(x^*, x^*)$, then,

$$d_q(x^*, x^*) = d_q(Sx^*, Sx^*) \leq \psi(d_q(x^*, x^*)).$$

This implies that,

$$d_q(x^*, x^*) = 0.$$

□

Example 2.2 Let $X = Q^+ \cup \{0\}$ and let $d_q : X \times X \to X$ be the complete ordered dislocated quasi metric on X defined by $d_q(x, y) = x + 2y$, endowed with usual order. Let $S : X \to X$ be defined by,

$$Sx = \begin{cases} \dfrac{x}{7} & \text{if} x \in [0, 1] \\ x - \dfrac{1}{2} & \text{if} x \in (1, \infty) \end{cases}$$

$x_0 = 1$, $r = 3$, $\overline{B(x_0, r)} = [0, 1]$ and $\alpha(x, y) = 3$ for all x, y. Clearly S is an α-ψ-contractive mapping, where $\psi(t) = \frac{t}{3}$.

$$d_q(1, S1) = d_q\left(1, \frac{1}{7}\right) = 1 + \frac{2}{7} = \frac{9}{7}$$

$$\sum_{i=1}^{n} \psi^n(d_q(x_0, Sx_0)) = \frac{9}{7}\sum_{i=1}^{n} \frac{1}{3^n} < \frac{3}{2}\left(\frac{9}{7}\right) = \frac{27}{14} < 3$$

If $x, y \in \overline{B(x_0, r)}$, then

$$\frac{3x}{7}+\frac{6y}{7}\leq x+2y$$
$$\frac{x}{7}+\frac{2y}{7}\leq\frac{x+2y}{3}$$
$$d_q(Sx,\,Sy)\leq\psi(d_q(x,\,y))$$

Also if $x,\,y\in(1,\,\infty)$, then

$$3x+6y-\frac{9}{2}>x+2y$$
$$x+2y-\frac{3}{2}>\frac{x+2y}{3}$$
$$\left(x-\frac{1}{2}\right)+2\left(y-\frac{1}{2}\right)>\psi(x+2y)$$
$$d_q(Sx,\,Sy)>\psi(d_q(x,\,y))$$

Then the contractive condition does not hold on X.

Therefore, all the conditions of corollary 2.3 are satisfied and 0 is the unique fixed point of S.

Corollary 4 *Let $(X,\,p_q)$ be a left K-sequentially complete partial quasi metric space. Suppose there exist two functions, $\alpha,\eta:X\times X\to[0,\,+\infty)$. Let x_0 be an arbitrary point in X and $S:X\to X$ be α-admissible with respect to η and $\psi\in\Psi$. Assume that,*

$$x,\,y\in\overline{B(x_0,\,r)},\alpha(x,\,y)\geq\eta(x,\,y)\Longrightarrow p_q(Sx,\,Sy)\leq\psi(p_q(x,\,y))$$

and

$$\sum_{i=0}^{j}\psi^i(p_q(x_0,Sx_0))\leq r+p(x_0,\,x_0)\quad\text{for all } j\in N \text{ and } r>0.$$

Suppose that, the following assertions hold:

(i) $\alpha(x_0,\,Sx_0)\geq\eta(x_0,\,Sx_0)$;

(ii) for any sequence $\{x_n\}$ in $\overline{B(x_0,\,r)}$ such that $\alpha(x_n,\,x_{n+1})\geq\eta(x_n,\,x_{n+1})$ for all $n\in N\cup\{0\}$ and $x_n\to u\in\overline{B(x_0,\,r)}$ as $n\to+\infty$ then $\alpha(x_n,\,u)\geq\eta(x_n,\,u)$ for all $n\in N\cup\{0\}$.

Then, there exists a point x^* in $\overline{B(x_0,\,r)}$ such that $x^*=Sx^*$.

Fixed point results for graphic contractions in dislocated quasi metric spaces

Consistent with Jachymski [23], let $(X,\,d_q)$ be a dislocated quasi metric space and Δ denotes the diagonal of the Cartesian product $X\times X$. Consider a directed graph G such that the set $V(G)$ of its vertices coincides with X, and the set $E(G)$ of its edges contains all loops, i.e., $E(G)\supseteq\Delta$. We assume G has no parallel edges, so we can identify G with the pair $(V(G),\,E(G))$. Moreover, we may treat G as a weighted graph (see [23]) by assigning to each edge the distance between its vertices. If x and y are vertices in a graph G, then a path in G from x to y of length m ($m\in\mathbb{N}$) is a sequence $\{x_i\}_{i=0}^{m}$ of $m+1$ vertices such that $x_0=x,\ x_m=y$ and $(x_{n-1},\,x_n)\in E(G)$ for $i=1,\dots,m$. A graph G is connected if there is a path between any two vertices. G is weakly connected if $\tilde{G}$ is connected (see for details [1, 13, 21, 23]).

Definition 3.1 [23] We say that a mapping $T:X\to X$ is a Banach G-contraction or simply G-contraction if T preserves edges of G, i.e.,

$$\forall x\,,y\in X((x,\,y)\in E(G)\Rightarrow(Tx,\,Ty)\in E(G))$$

and T decreases weights of edges of G in the following way:

$$\exists k\in(0,\,1),\forall x\,,y\in X((x,y)\in E(G)\Rightarrow d(Tx,Ty)\leq kd(x,y)).$$

Definition 3.2 Let $(X,\,d_q)$ be a dislocated quasi metric space endowed with a graph G and $S:X\to X$ be self-mapping. Assume that for $r>0$, $x_0\in X$ and $\psi\in\Psi$, following conditions hold,

$$\forall x,\,y\in\overline{B(x_0,\,r)}\ ((x,\,y)\in E(G)\Rightarrow(Sx,\,Sy)\in E(G).$$
$$\forall x,\,y\in\overline{B(x_0,\,r)},\ (x,\,y)\in E(G)\Rightarrow d_q(Sx,\,Sy)\leq\psi(d_q(x,\,y)).$$

Then the mapping S is called a ψ-graphic contractive mapping. If $\psi(t)=kt$ for some $k\in[0,\,1)$, then we say S is G -contractive mappings.

Theorem 5 *Let $(X,\,d_q)$ be a left K-sequentially complete dislocated quasi metric space endowed with a graph G and $S:X\to X$ be ψ-graphic contractive mapping. Suppose that the following assertions hold:*

Definition 3.3

(i) $(x_0,\,Sx_0)\in E(G)$ and $\sum_{i=0}^{j}\psi^i(d_q(x_0,\,Sx_0))\leq r$ for all $j\in N$ and $r>0$.

(ii) if $\{x_n\}$ is a sequence in $\overline{B(x_0,\,r)}$ such that $(x_n,\,x_{n+1})\in E(G)$ for all $n\in\mathbb{N}$ and $x_n\to x$ as $n\to+\infty$, then $(x_n,\,x)\in E(G)$ for all $n\in\mathbb{N}$.

Then S has a fixed point.

Proof Define, $\alpha:X^2\to(-\infty,\,+\infty)$ by $\alpha(x,\,y)=\begin{cases}1, & \text{if } (x,\,y)\in E(G)\\ 0, & \text{otherwise}\end{cases}$. At fist we prove that the mapping S is α-admissible. Let $x,\,y\in\overline{B(x_0,\,r)}$ with $\alpha(x,\,y)\geq1$, then $(x,\,y)\in E(G)$. As S is ψ-graphic contractive mappings, we have, $(Sx,\,Sy)\in E(G)$. That is, $\alpha(Sx,\,Sy)\geq1$. Thus S is α-admissible mapping. From (i) there exists x_0 such that $(x_0,\,Sx_0)\in E(G)$. That is, $\alpha(x_0,\,Sx_0)\geq1$. If $x,\,y\in\overline{B(x_0,\,r)}$ with $\alpha(x,\,y)\geq1$, then $(x,\,y)\in E(G)$. Now, since S, is ψ -graphic contractive mapping, so $d_q(Sx, Sy) \leq \psi(d_q(x, y))$. That is,

$$\alpha(x, y) \geq 1 \Longrightarrow d_q(Sx, Sy) \leq \psi(d_q(x, y)).$$

Let $\{x_n\} \subset \overline{B(x_0, r)}$ with $x_n \to x$ as $n \to \infty$ and $\alpha(x_n, x_{n+1}) \geq 1$ for all $n \in \mathbb{N}$. Then, $(x_n, x_{n+1}) \in E(G)$ for all $n \in \mathbb{N}$ and $x_n \to x$ as $n \to +\infty$. So by (ii) we have, $(x_n, x) \in E(G)$ for all $n \in \mathbb{N}$. That is, $\alpha(x_n, x) \geq 1$. Hence, all conditions of Theorem 1 are satisfied and S has a fixed point. □

Theorem 3.2 (2°) [23] and corollary 2.3(2)[14] are extended to ψ-graphic contractive defined on a dislocated quasi metric space as follows.

Corollary 6 *Let (X, d_q) be a left K-sequentially complete dislocated quasi metric space endowed with a graph G and $S : X \to X$ be ψ-graphic contractive mapping. Suppose that the following assertions hold:*

(i) $(x_0, Sx_0) \in E(G)$ and $\sum_{i=0}^{j} \psi^i(d_q(x_0, Sx_0)) \leq r$ for all $j \in N$ and $r > 0$.

(ii) $(x, z) \in E(G)$ and $(z, y) \in E(G)$ imply $(x, y) \in E(G)$ for all $x, y, z \in X$, that is, $E(G)$ is a quasi-order [23] and if $\{x_n\}$ is a sequence in $\overline{B(x_0, r)}$ such that $(x_n, x_{n+1}) \in E(G)$ for all $n \in \mathbb{N}$ and $x_n \to x$ as $n \to +\infty$, then there is a subsequence $\{x_{k_n}\}$ with $(x_{k_n}, x) \in E(G)$ for all $n \in \mathbb{N}$.

Then S has a fixed point.

Proof Condition (ii) implies that of *(ii)* in Theorem 5 (see Remark 3.1 [23]). Now the conclusion follows from Theorem 5. □

Corollary 7 *Let (X, d_q) be a left K-sequentially complete dislocated quasi metric space endowed with a graph G and $S : X \to X$ be a mapping. Suppose that the following assertions hold:*

(i) S is Banach G-contraction on $\overline{B(x_0, r)}$;

(ii) $(x_0, Sx_0) \in E(G)$ and $d_q(x_0, Sx_0)) \leq (1 - k)r$;

(iii) if $\{x_n\}$ is a sequence in $\overline{B(x_0, r)}$ such that $(x_n, x_{n+1}) \in E(G)$ for all $n \in \mathbb{N}$ and $x_n \to x$ as $n \to +\infty$, then $(x_n, x) \in E(G)$ for all $n \in \mathbb{N}$.

Then S has a fixed point.

Corollary 8 *Let (X, d_q) be a left K-sequentially complete dislocated quasi metric space endowed with a graph G and $S : X \to X$ be a mapping. Suppose that the following assertions hold:*

(i) S is Banach G-contraction on X and there is $x_0 \in X$ such that $(x_0, Sx_0) \in E(G)$;

(iii) if $\{x_n\}$ is a sequence in X such that $(x_n, x_{n+1}) \in E(G)$ for all $n \in \mathbb{N}$ and $x_n \to x$ as $n \to +\infty$, then $(x_n, x) \in E(G)$ for all $n \in \mathbb{N}$.

Then S has a fixed point.

The study of existence of fixed points in partially ordered sets has been initiated by Ran and Reurings [30] with applications to matrix equations. Agarwal, et al. [2], Bhaskar and Lakshmikantham [12], Ciric et al. [15] and Hussain et al. [22] presented some new results for nonlinear contractions in partially ordered metric spaces and noted that their theorems can be used to investigate a large class of problems. Here as an application of our results we deduce some new common fixed point results in partially ordered dislocated quasi metric spaces.

Recall that if $(X, \preceq)$ is a partially ordered set and $S : X \to X$ is such that for $x, y \in X$, with $x \preceq y$ implies $Sx \preceq Sy$, then the mapping S is said to be non-decreasing.

Let $(X, d_q, \preceq)$ be a partially ordered dislocated quasi metric space. Define the graph G by

$$E(G) := \{(x, y) \in X \times X : x \preceq y\}.$$

For this graph, first condition in Definition 3.2 means S is non-decreasing with respect to this order. We derive following important results in partially ordered dislocated quasi metric spaces.

Corollary 9 *Let $(X, \preceq, d_q)$ be a partially ordered left K-sequentially complete dislocated quasi metric space and $S : X \to X$ be a nondecreasing map. Suppose that the following assertions hold:*

(i) there exists $k \in [0, 1)$ such that $d_q(Sx, Sy) \leq kd_q(x, y)$ for all $x, y \in \overline{B(x_0, r)}$ with $x \preceq y$;

(ii) $x_0 \preceq Sx_0$ and $d_q(x_0, Sx_0) \leq (1 - k)r$;

(iii) if $\{x_n\}$ is a nondecreasing sequence in $\overline{B(x_0, r)}$ such that $x_n \to x \in \overline{B(x_0, r)}$ as $n \to +\infty$, then $x_n \preceq x$ for all n.

Then S has a fixed point.

Corollary 10 *Let $(X, \preceq, d_q)$ be a partially ordered left K-sequentially complete dislocated quasi metric space and $S : X \to X$ be a nondecreasing map. Suppose that the following assertions hold:*

(i) there exists $k \in [0, 1)$ such that $d_q(Sx, Sy) \leq kd_q(x, y)$ for all $x, y \in X$ with $x \preceq y$;

(ii) there exists $x_0 \in X$ such that $x_0 \preceq Sx_0$;

(iii) if $\{x_n\}$ is a nondecreasing sequence in X such that $x_n \to x \in X$ as $n \to +\infty$, then $x_n \preceq x$ for all n.

Then S has a fixed point.

Corollary 11 ([29]) *Let $(X, \preceq, d)$ be a partially ordered complete metric space and $S : X \to X$ be a nondecreasing mapping such that*

$$d(Sx, Sy) \le kd(x, y)$$

for all $x, y \in X$ with $x \preceq y$ where $0 \le k < 1$. Suppose that the following assertions hold:

(i) there exists $x_0 \in X$ such that $x_0 \preceq Sx_0$;
(ii) if $\{x_n\}$ is a sequence in X such that $x_n \preceq x_{n+1}$ for all $n \in \mathbb{N}$ and $x_n \to x$ as $n \to +\infty$, then $x_n \preceq x$ for all $n \in \mathbb{N}$.

Then S has a fixed point.

Remark The above results can easily be proved in right K-sequentially dislocated quasi metric space.

Acknowledgments The authors sincerely thank the learned referees.

Competing interests The authors declare that they have no competing interests.

References

1. Abbas, M., Nazir, T.: Common fixed point of a power graphic contraction pair in partial metric spaces endowed with a graph. Fixed Point Theory and Appl. **2013**, 20 (2013)
2. Agarwal, R.P., El-Gebeily, M.A., O'Regan, D.: Generalized contractions in partially ordered metric spaces. Appl. Anal. **87**, 109–116 (2008)
3. Ali, M. U., Kamran T., Shahzad N.: Best proximity point for α-ψ-proximal contractive multimaps, Abstr. Appl. Anal., p 7 (2014), Article ID 141489
4. Ali, M.U., Kamran, T., Karapinar, E.: Fixed point of α-ψ-contractive type mappings in uniform spaces. Fixed Point Theory Appl. **2014**, 150 (2014). doi:10.1186/1687-1812-2014-150
5. Arshad, M., Azam A., Vetro, P.: Some common fixed point results in cone metric spaces. Fixed Point Theory Appl., p 11 (2009) Article ID 493965
6. Arshad, M., Shoaib, A., Abbas, M., Azam, A.: Fixed points of a pair of Kannan type mappings on a closed ball in ordered partial metric spaces. Miskolc Math. Notes **14**(3), 769–784 (2013)
7. M. Arshad, A. Shoaib and I. Beg, Fixed point of a pair of contractive dominated mappings on a closed ball in an ordered complete dislocated metric space. Fixed Point Theory Appl. **2013**, 1–15 (2013)
8. M. Arshad, A. Shoaib, and P. Vetro, Common Fixed Points Of A Pair Of Hardy Rogers Type Mappings On A Closed Ball In Ordered Dislocated Metric Spaces, J.Funct Spaces Appl. **2013**, 9 (2013), article ID 638181
9. Arshad, M., Azam, A., Abbas, M., Shoaib, A.: Fixed point results of dominated mappings on a closed ball in ordered partial metric spaces without continuity. U.P.B. Sci. Bull., Series A, **76**(2), 123–134 (2014)
10. Azam, A., Hussain, S., Arshad, M.: Common fixed points of Chatterjea type fuzzy mappings on closed balls. Neural Computing & Applications **21**(Suppl 1), S313–S317 (2012)
11. Azam, A., Waseem, M., Rashid, M.: Fixed point theorems for fuzzy contractive mappings in quasi-pseudo-metric spaces. Fixed Point Theory Appl. **2013**, 27 (2013)
12. Bhaskar, T.G., Lakshmikantham, V.: Fixed point theorems in partially ordered metric spaces and applications. Nonlinear Anal. **65**, 1379–1393 (2006)
13. Bojor, F.: Fixed point theorems for Reich type contraction on metric spaces with a graph. Nonlinear Anal. **75**, 3895–3901 (2012)
14. Bojor, F.: Fixed point of φ-contraction in metric spaces endowed with a graph. Ann. Univ. Craiova Math. Comput. Sci. Ser. **37**(4), 85–92 (2010)
15. Ćirić, L., Abbas, M., Saadati, R., Hussain, N.: Common fixed points of almost generalized contractive mappings in ordered metric spaces. Appl. Math. Comput. **217**, 5784–5789 (2011)
16. Cobzas, S.: Functional analysis in Asymmeric Normed Spaces, Frontiers in Mathematics. Birkhauser, Basel (2013)
17. Haghi, R.H., Rezapour, Sh, Shahzad, N.: Some fixed point generalizations are not real generalizations. Nonlinear Anal. **74**, 1799–1803 (2011)
18. Hitzler, P., Seda, A. K.: Dislocated topologies, J. Electr. Eng. **51**(12/s), 3–7 (2000)
19. Hussain, N., Arshad, M., Shoaib A.: Fahimuddin, common fixed point results for α-ψ-contractions on a metric space endowed with graph. J. Inequal. Appl. **2014**:136 (2014)
20. Hussain, N.: E. karapinar, P. Salimi and F, Akbar α -admissible mappings and related fixed point theorems. J. Inequal. Appl.**2013**, 114 (2013)
21. Hussain, N., Al-Mezel S., SALIMI, Peyman.: Fixed points for α-ψ-graphic contractions with application to integral equations. Abstr. Appl. Anal., (2013) Article 575869
22. Hussain, N., Khan A.R., Agarwal, Ravi P.: Krasnosel'skii and Ky Fan type fixed point theorems in ordered Banach spaces, J. Nonlinear Convex Anal., **11**(3), (2010), 475–489
23. Jachymski, J.: The contraction principle for mappings on a metric space with a graph. Proc. Amer. Math. Soc. **1**(136), 1359–1373 (2008)
24. Karapınar, E., Erhan, İ.M., Öztürk, A.: Fixed point theorems on quasi-partial metric spaces. Math. Comp. Model. (2012). doi:10.1016/j.mcm.2012.06.036
25. Karapinar, E., Samet, B.: Generalized α-ψ -contractive type mappings and related fixed point theorems with applications, Abstr. Appl. Anal. (2012) Article ID 793486
26. Kutbi, M. A., Arshad, M., Hussain, A.: ON modified (α- η)-contractive mappings. p 7 (2014) Article ID 657858
27. Kutbi, M. A., Ahmad, J., Hussain N., Arshad, M.: Common fixed point results for mappings with rational expressions, Abstr. Appl. Anal. p 11 (2013) Article ID 549518
28. Matthews, S. G.: Partial metric topology. In: Proceedings 8th Summer Conference on General Topology and Applications, Ann. New York Acad. Sci., **728**, 183–197(1994)
29. Nieto, J.J., Rodríguez-López, R.: Contractive mapping theorems in partially ordered sets and applications to ordinary differential equetions. Order. **22**, 223–229 (2005)
30. Ran, A.C.M., Reurings, M.C.B.: A fixed point theorem in partially ordered sets and some applications to matrix equations. Proc. Amer. Math. Soc. **132**, 1435–1443 (2003)
31. Reilly, I.L., Subrahmanyam, P.V., Vamanamurthy, M.K.: Cauchy sequences in quasi-pseudo-metric spaces. Monatsh. Math. **93**, 127–140 (1982)
32. Ren, Y., Li, J., Yu, Y.:Common fixed point Theorems for nonlinear contractive mappings in dislocated metric spaces, Abstr. Appl. Anal., Vol. 2013, p 5 (2013), Article ID 483059
33. Salimi, P., Latif, A., Hussain, N.: Modified α-ψ -contractive mappings with applications. Fixed Point Theory and Appl. **2013**, 151 (2013)

34. Samet, B., Vetro, C., Vetro, P.: Fixed point theorems for α -ψ-contractive type mappings. Nonlinear Anal. **75**, 2154–2165 (2012)
35. Shoaib, A., Arshad M., Ahmad, J.: Fixed point results of locally contractive mappings in ordered quasi-partial metric spaces. Sci. World J. (2013), Article ID 194897, p 8
36. Shoaib, A., Arshad, M., Kutbi, M.A.: Common fixed points of a pair of Hardy Rogers Type Mappings on a closed ball in ordered partial metric spaces. J. Comput. Anal. Appl. **17**(2), 255–264 (2014)
37. Zeyada, F.M., Hassan, G.H., Ahmed, M.A.: A generalization of a fixed point theorem due to Hitzler and Seda in dislocated quasi-metric spaces. Arab J. Sci. Eng. A**31**(1), 111–114 (2006)

3

Fixed point approximation of Picard normal *S*-iteration process for generalized nonexpansive mappings in hyperbolic spaces

Mohammad Imdad[1] · **Samir Dashputre**[2]

Abstract In this paper, we establish strong and Δ-convergence theorems for a relatively new iteration process generated by generalized nonexpansive mappings in uniformly convex hyperbolic spaces. The theorems presented in this paper generalizes corresponding theorems for uniformly convex normed spaces of Kadioglu and Yildirim (Approximating fixed points of nonexpansive mappings by faster iteration process, arXiv:1402.6530v1 [math.FA], 2014) and CAT(0)-spaces of Abbas et al. (J Inequal Appl 2014:212, 2014) and many others in this direction.

Keywords Generalized nonexpansive mappings · Strong and Δ-convergence · Uniformly convex hyperbolic spaces · Picard normal *S*-iteration process

Introduction

In this paper, $\mathbb{N}$ denotes the set of all positive integers while $F(T)$ denotes the set of all fixed points of T, i.e., $F(T) = \{Tx = x; x \in C\}$.

Let C be a nonempty subset of normed space X and mapping $T : C \to C$ is said to be

(i) *nonexpansive*, if $\|Tx - Ty\| \le \|x - y\|$, for all $x, y \in C$,
(ii) *quasi-nonexpansive*, if $\|Tx - p\| \le \|x - p\|$, for all $x \in C$ and $p \in F(T)$.

Many nonlinear equations are naturally formulated as fixed point problems,

$$x = Tx, \tag{1.1}$$

where T, the fixed point mapping, may be nonlinear. A solution x^* of the problem (1.1) is called a fixed point of the mapping T. Consider a *fixed point iteration*, which is given by

$$x_{n+1} = Tx_n, \forall n \in \mathbb{N}. \tag{1.2}$$

The iterative method (1.2) is also known as *Picard iteration* or the method of successive substitution. For the Banach contraction mapping theorem, the Picard iteration converges unique fixed point of T, but it fails to approximate fixed point for nonexpansive mappings, even when the existence of a fixed point of T is guaranteed.

Example 1.1 Consider a self mapping T on [0, 1] defined by $Tx = 1 - x$ for $0 \le x \le 1$. Then T is nonexpansive with unique fixed point at $x = \frac{1}{2}$. If we choose a starting value $x = a \ne \frac{1}{2}$, then successive iteration of T yield the sequence $\{1 - a, a, 1 - a, \ldots\}$.

Thus, when a fixed point of nonexpansive mappings exists, other approximation techniques are needed to approximate it. In the last fifty years, the numerous numbers of researchers attracted in these direction and developed iterative process has been investigated to approximate fixed point for not only nonexpansive mapping, but also for some wider class of nonexpansive mappings (see e.g., Agarwal et al. [3], Ishikawa [9], Krasnosel'skiĭ [12], Mann

✉ Mohammad Imdad
mhimdad@gmail.com

Samir Dashputre
samir231973@gmail.com

1 Department of Mathematics, Aligarh Muslim University, Aligarh Uttar Pradesh, India

2 Department of Applied Mathematics, Shri Shankaracharya Technical Campus, Shri Shankaracharya Group of Institutions (F.E.T), Junwani, Bhilai 490020, India

[18], Noor [19], Schaefer [23]), and compare which one is faster.

Sahu [21] has introduced Normal S-iteration Process, whose rate of convergence similar to the Picard iteration process and faster than other fixed point iteration processes (see [21, Theorem 3.6]).

(NS) *Normal S-iteration process* (see Sahu [21]) defined as follows:

For C a convex subset of normed space X and a nonlinear mapping T of C into itself, for each $x_1 \in C$, the sequence $\{x_n\}$ in C is defined by

$$\begin{cases} x_{n+1} = Ty_n \\ y_n = (1-\alpha_n)x_n + \alpha_n Tx_n, \quad n \in \mathbb{N}, \end{cases} \tag{1.3}$$

where $\{\alpha_n\}$ is real sequences in (0, 1).

It brings a following natural question.

Question 1.1 *Does there exists an iteration process whose rate of convergence is faster than Normal S-iteration process for contraction mappings?*

The question have been resolved in affirmative way by Abbas et al. [2], Kadioglu and Yildirim [11, Theorem 5], Thakur et al. [26, Theorem 2.3], developed new iteration processes for approximating the fixed point, as earliest as possible compare Normal S-iteration process. The following iteration process developed by Kadioglu and Yildirim [11] for approximating the fixed point for nonexpansive mapping and establish some strong and weak convergence theorems in uniformly convex Banach spaces.

(PNS) *Picard normal S-iteration process* (see Kadioglu and Yildirim [11]) defined as follows: With C, X and T as in (NS), for each $x_1 \in C$, the sequence $\{x_n\}$ in C is defined by

$$\begin{cases} x_{n+1} = Ty_n \\ y_n = (1-\alpha_n)z_n + \alpha_n Tz_n \\ z_n = (1-\beta_n)x_n + \beta_n Tx_n, \quad n \in \mathbb{N}, \end{cases} \tag{1.4}$$

where $\{\alpha_n\}$ and $\{\beta_n\}$ are real sequences in (0, 1).

Remark 1.1 If $\beta_n = 0$ and $\alpha_n = \beta_n = 0$ in the process (1.4) then it reduces to Normal S-iteration process (1.3) and Picard iteration process (1.2) respectively.

The purpose of this paper is to establish strong and Δ-convergence theorems for a new iteration process generated by generalized nonexpansive mappings in uniformly convex hyperbolic spaces. The theorems presented in this paper generalizes corresponding theorems for uniformly convex normed spaces of Kadioglu and Yildirim [11] and CAT(0)-spaces of Abbas et al. [1] and many others in this directions (see Itoh [8], Kim et al. [14], Sahu [21] etc.).

Preliminaries

Let (X, d) be a metric space and C be a nonempty subset of X. Suzuki [24] introduced a class of single valued mappings called Suzuki-generalized nonexpansive mappings (or condition (C)), satisfying a condition

$$\frac{1}{2}d(x,Tx) \leq d(x,y) \implies d(Tx,Ty) \leq d(x,y),$$

which is weaker than nonexpansiveness and stronger than quasi nonexpansiveness. The following examples make obvious this fact.

Example 2.1 [24] Define a mapping T on [0, 3] by

$$Tx = \begin{cases} 0, & \text{if } x \neq 3, \\ 1, & \text{if } x = 3. \end{cases}$$

Then T satisfies condition (C), but T is not nonexpansive.

Example 2.2 [24] Define a mapping T on [0, 3] by

$$Tx = \begin{cases} 0, & \text{if } x \neq 3, \\ 2, & \text{if } x = 3. \end{cases}$$

Then $F(T) = \{0\} \neq \emptyset$ and T is quasi-nonexpansive, but T does not satisfy condition (C).

In [10], Karapinar and Tas introduced some new definitions which are modifications of Suzuki's-generalized nonexpansive mappings (or condition (C)) as follows.

Definition 2.1 Let C be a nonempty subset of a metric space X. The mapping $T : C \to C$ is said to be

(i) *Suzuki-Ciric mapping (SCC)* [10] if

$$\frac{1}{2}d(Tx,Ty) \leq d(x,y) \implies d(Tx,Ty) \leq M(x,y)$$

where

$$M(x,y) = \max\{d(x,y), d(x,Tx), d(y,Ty), d(x,Ty), d(y,Tx)\}$$

for all $x, y \in C$;

(ii) *Suzuki-KC mapping* (SKC) if

$$\frac{1}{2}d(Tx,Ty) \leq d(x,y) \implies d(Tx,Ty) \leq N(x,y)$$

where

$$N(x,y) = \max\left\{d(x,y), \frac{d(x,Tx)+d(y,Ty)}{2}, \frac{d(x,Ty)+d(y,Tx)}{2}\right\}$$

for all $x, y \in C$;

(iii) *Kannan Suzuki mapping* (KSC) if

$$\frac{1}{2}d(Tx,Ty)\le d(x,y) \implies d(Tx,Ty) \le \frac{d(x,Tx)+d(y,Ty)}{2}$$

for all $x,y\in C$;

(iv) *Chatterjea–Suzuki mappings* (CSC) if

$$\frac{1}{2}d(Tx,Ty)\le d(x,y) \implies d(Tx,Ty) \le \frac{d(y,Tx)+d(x,Ty)}{2}$$

for all $x,y\in C$;

Theorem 2.1 [10] *Let T be a mapping on a closed subset C of a metric space X and T satisfy condition SKC. Then* $d(x,Ty)\le 5d(Tx,x)+d(x,y)$ *holds for* $x,y\in C$.

Remark 2.1 Theorem 2.1 holds if one replaces condition SKC by one of the conditions KSC, SCC, and CSC.

Recently, García-Falset et al. [7] introduced two generalizations of nonexpansive mappings which in turn include Suzuki generalized nonexpansive mappings contained in [24].

Definition 2.2 Let T be a mapping defined on a subset C of metric space X and $\mu\ge 1$. Then T is said to satisfy *condition* (E_μ), if (for all $x,y\in C$)

$$d(x,Ty)\le \mu d(x,Tx)+d(x,y).$$

Often, T is said to satisfy *condition (E)* whenever T satisfies condition (E_μ) for some $\mu\ge 1$.

Remark 2.2 If T satisfies one of the conditions SKC, KSC, SCC, and CSC, then T satisfies condition E_μ for $\mu=5$.

Definition 2.3 Let T be a mapping defined on a subset C of a metric space X and $\lambda\in(0,1)$. Then T is said to satisfy the *condition* (C_λ) if for all $x,y\in C$

$$\lambda d(x,Tx)\le d(x,y) \implies d(Tx,Ty)\le d(x,y).$$

For $0<\lambda_1<\lambda_2<1$, the condition (C_{λ_1}) implies the condition (C_{λ_2}).

The following example shows that the class of mappings satisfying conditions (*E*) and (C_λ) for some $\lambda\in(0,1)$ is larger than the class of mappings satisfying the condition (*C*).

Example 2.3 [7] For a given $\lambda\in(0,1)$, define a mapping T on [0, 1] by

$$Tx=\begin{cases}\dfrac{x}{2}, & \text{if } x\ne 1,\\ \dfrac{1+\lambda}{2+\lambda}, & \text{if } x=1.\end{cases}$$

The mapping T satisfies the condition (C_λ) but it fails the condition (C_{λ_1}), whenever $0<\lambda_1<\lambda$. Moreover, T satisfies the condition (E_μ) for $\mu=\frac{2+\lambda}{2}$.

Throughout, this paper we work in the setting of hyperbolic spaces introduced by Kohlenbach [15].

A *hyperbolic space* (*X*, *d*, *W*) is a metric space (*X*, *d*) together with a convexity mapping $W: X^2\times[0,1]\to X$ satisfying

(W_1) $d(u,W(x,y,\alpha))\le \alpha d(u,x)+(1-\alpha)d(u,y)$;
(W_2) $d(W(x,y,\alpha),W(x,y,\beta))=|\alpha-\beta|d(x,y)$;
(W_3) $W(x,y,\alpha)=W(y,x,1-\alpha)$;
(W_4) $d(W(x,z,\alpha),W(y,w,\alpha))\le(1-\alpha)d(x,y)+\alpha d(z,w)$,
for all $x,y,z,w\in X$ and $\alpha,\beta\in[0,1]$.

A metric space is said to be a *convex metric space* in the sense of Takahashi [25], where a triple (*X*, *d*, *W*) satisfy only (W_1). The concept of hyperbolic spaces in [15] is more restrictive than the hyperbolic type introduced by Goebel and Kirk [5] since (W_1) and (W_2) together are equivalent to (*X*, *d*, *W*) being a space of hyperbolic type in [5]. But it is slightly more general than the hyperbolic space defined in Reich and Shafrir [20] (see [15]). This class of metric spaces in [15] covers all normed linear spaces, ℝ-trees in the sense of Tits, the Hilbert ball with the hyperbolic metric (see [6]), Cartesian products of Hilbert balls, Hadamard manifolds (see [20]), and CAT(0) spaces in the sense of Gromov (see [4]). A thorough discussion of hyperbolic spaces and a detailed treatment of examples can be found in [15] (see also [5, 6, 20]).

If $x,y\in X$ and $\lambda\in[0,1]$, then we use the notation $(1-\lambda)x\oplus\lambda y$ for $W(x,y,\lambda)$. The following holds even for the more general setting of convex metric space [25]: for all $x,y\in X$ and $\lambda\in[0,1]$,

$$d(x,(1-\lambda)x\oplus\lambda y)=\lambda d(x,y) \quad\text{and}$$
$$d(y,(1-\lambda)x\oplus\lambda y)=(1-\lambda)d(x,y).$$

As consequence,

$$1x\oplus 0y=x,\quad 0x\oplus 1y=y$$

and

$$(1-\lambda)x\oplus\lambda x=\lambda x\oplus(1-\lambda)x=x.$$

A hyperbolic space (*X*, *d*, *W*) is *uniformly convex* [16] if for any $r>0$ and $\varepsilon\in(0,2]$, there exists $\delta\in(0,1]$ such that for all $a,x,y\in X$,

$$d\left(\frac{1}{2}x \oplus \frac{1}{2}y, a\right) \leq (1-\delta)r.$$

provided $d(x,a) \leq r, d(y,a) \leq r$ and $d(x,y) \geq \varepsilon r$.

A mapping $\eta : (0,\infty) \times (0,2] \to (0,1]$, which providing such a $\delta = \eta(r,\varepsilon)$ for given $r > 0$ and $\varepsilon \in (0,2]$, is called as a *modulus of uniform convexity*. We call the function η is *monotone* if it decreases with r (for fixed ε), that is $\eta(r_2,\varepsilon) \leq \eta(r_1,\varepsilon), \forall r_2 \geq r_1 > 0$.

In [16], Leuştean proved that CAT(0) spaces are uniformly convex hyperbolic spaces with modulus of uniform convexity $\eta(r,\varepsilon) = \frac{\varepsilon^2}{8}$ quadratic in ε. Thus, the class of uniformly convex hyperbolic spaces are a natural generalization of both uniformly convex Banach spaces and CAT(0) spaces.

Now, we give the concept of Δ-convergence and some of its basic properties.

Let C be a nonempty subset of metric space (X, d) and $\{x_n\}$ be any bounded sequence in X while diam(C) denote the diameter of C. Consider a continuous functional $r_a(\cdot, \{x_n\}) : X \to \mathbb{R}^+$ defined by

$$r_a(x, \{x_n\}) = \limsup_{n\to\infty} d(x_n, x), \quad x \in X.$$

The infimum of $r_a(\cdot, \{x_n\})$ over C is said to be the *asymptotic radius* of $\{x_n\}$ with respect to C and is denoted by $r_a(C, \{x_n\})$.

A point $z \in C$ is said to be an *asymptotic center* of the sequence $\{x_n\}$ with respect to C if

$$r_a(z, \{x_n\}) = \inf\{r_a(x, \{x_n\}) : x \in C\},$$

the set of all asymptotic centers of $\{x_n\}$ with respect to C is denoted by $AC(C, \{x_n\})$. This set may be empty, a singleton, or certain infinitely many points.

If the asymptotic radius and the asymptotic center are taken with respect to X, then these are simply denoted by $r_a(X, \{x_n\}) = r_a(\{x_n\})$ and $AC(X, \{x_n\}) = AC(\{x_n\})$, respectively. We know that for $x \in X$, $r_a(x, \{x_n\}) = 0$ if and only if $\lim_{n\to\infty} x_n = x$. It is known that every bounded sequence has a unique asymptotic center with respect to each closed convex subset in uniformly convex Banach spaces and even CAT(0) spaces.

The following Lemma is due to Leuştean [17] and ensures that this property also holds in a complete uniformly convex hyperbolic space.

Lemma 2.1 [17, Proposition 3.3] *Let (X, d, W) be a complete uniformly convex hyperbolic space with monotone modulus of uniform convexity η. Then every bounded sequence $\{x_n\}$ in X has a unique asymptotic center with respect to any nonempty closed convex subset C of X.*

Recall that, a sequence $\{x_n\}$ in X is said to Δ-*converge* to $x \in X$, if x is the unique asymptotic center of $\{u_n\}$ for every subsequence $\{u_n\}$ of $\{x_n\}$. In this case, we write Δ-$\lim_n x_n = x$ and call x the Δ-limit of $\{x_n\}$.

Lemma 2.2 [13] *Let (X, d, W) be a uniformly convex hyperbolic space with monotone modulus of uniform convexity η. Let $x \in X$ and $\{t_n\}$ be a sequence in $[a, b]$ for some $a, b \in (0,1)$. If $\{x_n\}$ and $\{y_n\}$ are sequences in X such that*

$$\limsup_{n\to\infty} d(x_n, x) \leq c, \quad \limsup_{n\to\infty} d(y_n, x) \leq c,$$
$$\lim_{n\to\infty} d(W(x_n, y_n, t_n), x) = c,$$

for some $c \geq 0$, then $\lim_{n\to\infty} d(x_n, y_n) = 0$.

Lemma 2.3 *Let (X, d) be complete uniformly convex hyperbolic space with monotone modulus of convexity η, C be a nonempty closed convex subset of X and $T : C \to C$ be a mapping which satisfies conditions (C_λ) (for some $\lambda \in (0,1)$) and (E) on C. Suppose $\{x_n\}$ is bounded sequence in C such that*

$$\lim_{n\to\infty} d(x_n, Tx_n) = 0,$$

then T has a fixed point.

Proof Since $\{x_n\}$ is bounded sequence in X, then by Lemma 2.1, has unique asymptotic center in C, i.e., $AC(C, \{x_n\}) = \{x\}$ is singleton and $\lim_{n\to\infty} d(x_n, Tx_n) = 0$. Since T satisfies the condition (E_μ) on C, there exists $\mu > 1$ such that

$$d(x_n, Tx) \leq \mu d(x_n, Tx_n) + d(x_n, x).$$

Taking lim sup as $n \to \infty$ both the sides, we have

$$\begin{aligned} r_a(Tx, \{x_n\}) &= \limsup_{n\to\infty} d(x_n, Tx) \\ &\leq \limsup_{n\to\infty} [\mu d(x_n, Tx_n) + d(x_n, x)] \\ &\leq \limsup_{n\to\infty} d(x_n, x) = r_a(x, \{x_n\}). \end{aligned}$$

By using the uniqueness of asymptotic center, $Tx = x$, so x is fixed point of T. Hence, $F(T)$ is nonempty. □

Main results

We begin with the definition of Fejér monotone sequences:

Definition 3.1 Let C be a nonempty subset of hyperbolic space X and $\{x_n\}$ be a sequence in X. Then $\{x_n\}$ is *Fejér monotone* with respect to C if for all $x \in C$ and $n \in \mathbb{N}$,

$$d(x_{n+1}, x) \leq d(x_n, x).$$

Example 3.1 Let C be a nonempty subset of X, and $T : C \to C$ be a quasi-nonexpansive (in particular, nonexpansive) mapping such that $F(T) \neq \emptyset$ and $x_0 \in C$. Then the sequence $\{x_n\}$ of Picard iterates is Fejér monotone with respect to $F(T)$.

We can easily prove the following proposition.

Proposition 3.1 *Let $\{x_n\}$ be a sequence in X and C be a nonempty subset of X. Suppose that $\{x_n\}$ is Fejér monotone with respect to C, then we have the followings:*

(1) $\{x_n\}$ *is bounded.*
(2) *The sequence $\{d(x_n, p)\}$ is decreasing and converges for all $p \in F(T)$.*

We now define Picard Normal S-iteration process (PNS) in hyperbolic spaces:

(PNS) *Picard normal S-iteration process*: Let C be a nonempty closed convex subset of a hyperbolic space X and $T : C \to C$ be a mapping which satisfies the condition (C_λ) for some $\lambda \in (0, 1)$. For any $x_1 \in C$ the sequence $\{x_n\}$ is defined by

$$\begin{cases} x_{n+1} = W(Ty_n, 0, 0) \\ y_n = W(z_n, Tz_n, \alpha_n) \\ z_n = W(x_n, Tx_n, \beta_n), \quad n \in \mathbb{N}, \end{cases} \tag{3.1}$$

where $\{\alpha_n\}$ and $\{\beta_n\}$ are in $[\epsilon, 1 - \epsilon]$ for all $n \in \mathbb{N}$ and some $\epsilon \in (0, 1)$.

Lemma 3.1 *Let C be a nonempty closed convex subset of a hyperbolic space X and $T : C \to C$ be a mapping which satisfies the condition (C_λ) for some $\lambda \in (0, 1)$. If $\{x_n\}$ is a sequence defined by (3.1), then $\{x_n\}$ is Fejér monotone with respect to $F(T)$.*

Proof Since T satisfies the condition (C_λ) for some $\lambda \in (0, 1)$ and $p \in F(T)$, we have

$$\lambda d(p, Tp) = 0 \leq d(p, z_n)$$

$$\lambda d(p, Tp) = 0 \leq d(p, y_n)$$

and

$$\lambda d(p, Tp) = 0 \leq d(p, x_n),$$

for all $n \in \mathbb{N}$ so that, we have

$$d(Tp, Tz_n) \leq d(p, z_n)$$

$$d(Tp, Ty_n) \leq d(p, y_n)$$

and

$$d(Tp, Tx_n) \leq d(p, x_n).$$

Using (3.1), we have

$$\begin{aligned} d(z_n, p) &= d(W(x_n, Tx_n, \beta_n), p) \\ &= d((1 - \beta_n)x_n \oplus \beta_n Tx_n, p) \\ &\leq (1 - \beta_n)d(x_n, p) + \beta_n d(Tx_n, p) \\ &\leq d(x_n, p). \end{aligned} \tag{3.2}$$

From (3.1) and (3.2), we have

$$\begin{aligned} d(y_n, p) &= d(W(z_n, Tz_n, \alpha_n), p) \\ &= d((1 - \alpha_n)z_n \oplus \alpha_n Tz_n, p) \\ &\leq (1 - \alpha_n)d(z_n, p) + \alpha_n d(Tz_n, p) \\ &\leq (1 - \alpha_n)d(z_n, p) + \alpha_n d(z_n, p) \\ &\leq d(z_n, p) \\ &\leq d(x_n, p). \end{aligned} \tag{3.3}$$

Again, using (3.2) and (3.3), we have

$$\begin{aligned} d(x_{n+1}, p) &= d(W(Ty_n, 0, 0), p) \\ &= d(Ty_n, p) \\ &\leq d(y_n, p) \\ &\leq d(x_n, p), \end{aligned} \tag{3.4}$$

that is, $d(x_{n+1}, p) \leq d(x_n, p)$ for all $p \in F(T)$. Thus, $\{x_n\}$ is Fejér monotone with respect to $F(T)$. □

Lemma 3.2 *Let C be a nonempty closed convex subset of a complete uniformly convex hyperbolic space with monotone modulus of uniform convexity η and $T : C \to C$ be a mapping which satisfies the condition (C_λ) for some $\lambda \in (0, 1)$. If $\{x_n\}$ is a sequence defined by (3.1), then $F(T)$ is nonempty if and only if the sequence $\{x_n\}$ is bounded and $\lim_{n\to\infty} d(x_n, Tx_n) = 0$.*

Proof Suppose that the fixed point set $F(T)$ is nonempty and $p \in F(T)$. Then by Lemma 3.1, $\{x_n\}$ is Fejér monotone with respect to $F(T)$ and hence by Proposition 3.1, $\{x_n\}$ is bounded and $\lim_{n\to\infty} d(x_n, p)$ exists, let $\lim_{n\to\infty} d(x_n, p) = c \geq 0$.

(i) If $c = 0$, we obviously have

$$\begin{aligned} d(x_n, Tx_n) &\leq d(x_n, p) + d(Tx_n, p) \\ &\leq 2d(x_n, p), \end{aligned}$$

by taking lim as $n \to \infty$ on both the sides above inequality, we have

$$\lim_{n\to\infty} d(x_n, Tx_n) = 0.$$

(ii) If $c > 0$, since T satisfies the condition (C_λ) for some $\lambda \in (0, 1)$ and $p \in F(T)$, we have

$$d(Tx_n, p) \leq d(x_n, p),$$

taking lim sup as $n \to \infty$ both the sides, we get

$$\limsup_{n\to\infty} d(Tx_n, p) \leq c.$$

Taking lim sup as $n \to \infty$ both the sides in (3.2), we have

$$\limsup_{n\to\infty} d(z_n, p) \leq c. \tag{3.5}$$

Since

$$d(x_{n+1},p)\le d(z_n,p),$$

therefore, we take lim inf as $n\to\infty$ both the sides, we get

$$\begin{aligned}\liminf_{n\to\infty} d(x_{n+1},p) &\le \liminf_{n\to\infty} d(z_n,p)\\ c &\le \liminf_{n\to\infty} d(z_n,p)\end{aligned}\tag{3.6}$$

From (3.5) and (3.6), we have

$$\lim_{n\to\infty} d(z_n,p)=c,$$

it implies that

$$\begin{aligned}c &= \limsup_{n\to\infty} d(z_n,p)\\ &= \limsup_{n\to\infty}[d(W(x_n,Tx_n,\beta_n),p)]\\ &= \limsup_{n\to\infty}[d((1-\beta_n)x_n\oplus\beta_nTx_n,p)]\\ &\le \limsup_{n\to\infty}[(1-\beta_n)d(x_n,p)+\beta_nd(Tx_n,p)]\\ &\le (1-\beta_n)\limsup_{n\to\infty} d(x_n,p)+\beta_n\limsup_{n\to\infty} d(Tx_n,p)=c.\end{aligned}$$

Hence, it follows from Lemma 2.2, we have

$$\lim_{n\to\infty} d(x_n,Tx_n)=0.$$

Conversely, suppose that sequence $\{x_n\}$ is bounded and $\lim_{n\to\infty} d(x_n,Tx_n)=0$. Hence, it holds all the assumption of Lemma 2.3, so we have $Tx=x$, i.e., $F(T)$ is nonempty. □

Theorem 3.1 *Let C be a nonempty closed convex subset of a complete uniformly convex hyperbolic space X with monotone modulus of uniform convexity η and $T:C\to C$ be a mapping which satisfies conditions (C_λ) (for some $\lambda\in(0,1)$) and (E) on C with $F(T)\neq\varnothing$. If $\{x_n\}$ is the sequence defined by (3.1), then the sequence $\{x_n\}$ Δ-converges to a fixed point of T.*

Proof From Lemma 3.2, we observe that $\{x_n\}$ is a bounded sequence therefore, $\{x_n\}$ has a Δ-convergent subsequence. We now prove that every Δ-convergent subsequence of $\{x_n\}$ has unique Δ-limit $F(T)$. For this, let u and v Δ-limits of the subsequences $\{u_n\}$ and $\{v_n\}$ of $\{x_n\}$ respectively. By Lemma 2.1, $AC(C,\{u_n\})=\{u\}$ and $AC(C,\{v_n\})=\{v\}$. By Lemma 3.2, we have $\lim_{n\to\infty} d(u_n,Tu_n)=0$.

We claim that u and v are fixed points of T and it is unique.

By Lemma 2.3, u and v are fixed points of T. Now we show that $u=v$. If not, then by uniqueness of asymptotic center

$$\begin{aligned}\limsup_{n\to\infty} d(x_n,u) &= \limsup_{n\to\infty} d(u_n,u)\\ &< \limsup_{n\to\infty} d(u_n,v)\\ &= \limsup_{n\to\infty} d(x_n,v)\\ &= \limsup_{n\to\infty} d(v_n,v)\\ &< \limsup_{n\to\infty} d(v_n,u)\\ &= \limsup_{n\to\infty} d(x_n,u),\end{aligned}$$

which is a contradiction. Hence $u=v$, the sequence $\{x_n\}$ Δ-converges to a fixed point of T. □

Theorem 3.2 *Let C be a nonempty closed convex subset of a complete uniformly convex hyperbolic space X with monotone modulus of uniform convexity η and $T:C\to C$ be a mapping which satisfies conditions (C_λ) (for some $\lambda\in(0,1)$) and (E) on C with $F(T)\neq\varnothing$. Then the sequence $\{x_n\}$ which is defined by (3.1), converges strongly to some fixed point of T if and only if $\liminf_{n\to\infty} D(x_n,F(T))=0$, where $D(x_n,F(T))=\inf_{x\in F(T)} d(x_n,x)$.*

Proof Necessity is obvious, we have to prove only sufficient part. First, we show that the fixed point set $F(T)$ is closed, let $\{x_n\}$ be a sequence in $F(T)$ which converges to some point $z\in C$. As

$$\lambda d(x_n,Tx_n)=0\le d(x_n,z),$$

in view of the condition (C_λ), we have

$$d(x_n,Tz)=d(Tx_n,Tz)\le d(x_n,z).$$

By taking the limit of both sides we obtain

$$\lim_{n\to\infty} d(x_n,Tz)\le \lim_{n\to\infty} d(x_n,z)=0.$$

In view of the uniqueness of the limit, we have $z=Tz$, so that $F(T)$ is closed. Suppose

$$\liminf_{n\to\infty} D(x_n,F(T))=0.$$

From (3.4)

$$D(x_{n+1},F(T))\le D(x_n,F(T)),$$

it follows from Lemma 3.1 and Proposition 3.1 that $\lim_{n\to\infty} d(x_n,F(T))$ exists. Hence we know that $\lim_{n\to\infty} D(x_n,F(T))=0$.

Consider a subsequence $\{x_{n_k}\}$ of $\{x_n\}$ such that

$$d(x_{n_k},p_k)<\frac{1}{2^k},$$

for all $k\ge 1$ where $\{p_k\}$ is in $F(T)$. By Lemma 3.1, we have

$$d(x_{n_{k+1}}, p_k) \le d(x_{n_k}, p_k) < \frac{1}{2^k},$$

which implies that

$$\begin{aligned} d(p_{k+1}, p_k) &\le d(p_{k+1}, x_{n_{k+1}}) + d(x_{n_{k+1}}, p_k) \\ &< \frac{1}{2^{k+1}} + \frac{1}{2^k} \\ &< \frac{1}{2^{k-1}}. \end{aligned}$$

This shows that $\{p_k\}$ is a Cauchy sequence. Since *F*(*T*) is closed, $\{p_k\}$ is a convergent sequence. Let $\lim_{k\to\infty} p_k = p$. Then we know that $\{x_n\}$ converges to *p*. In fact, since

$$d(x_{n_k}, p) \le d(x_{n_k}, p_k) + d(p_k, p) \to 0, \text{as} k \to \infty,$$

we have $\lim_{k\to\infty} d(x_{n_k}, p) = 0$. Since $\lim_{n\to\infty} d(x_n, p)$ exists, the sequence $\{x_n\}$ is convergent to *p*.

We recall the definition of condition (*I*) due to Senter and Doston [22], define as follows:

Definition 3.2 [22] Let *C* be a nonempty subset of a metric space *X*. A mapping $T: C \to C$ with nonempty fixed point set *F*(*T*) in *C* is said to satisfy Condition (*I*) if there is a nondecreasing function $f: [0, \infty) \to [0, \infty)$ with $f(0) = 0, f(t) > 0$ for all $t \in (0, \infty)$, such that $d(x, Tx) \ge f(D(x, F(T)))$ for all $x \in C$, where $D(x, F(T))) = \inf\{d(x, p) : p \in F(T)\}$.

Theorem 3.3 *Let C be a nonempty closed convex subset of a complete uniformly convex hyperbolic space X with monotone modulus of uniform convexity* η *and* $T: C \to C$ *be a mapping which satisfies conditions* (C_λ) *(for some* $\lambda \in (0, 1)$*) and (E) on C. Moreover, T satisfies the condition (I) with* $F(T) \neq \emptyset$*. If* $\{x_n\}$ *is the sequence defined by* (3.1), *then the sequence* $\{x_n\}$ *converges strongly to some fixed point of T.*

Proof As in proof of Theorem 3.2, it can be shown that *F*(*T*) is closed. Observe that by Lemma 3.1, we have $\lim_{n\to\infty} d(x_n, Tx_n) = 0$. It follows from the condition (*I*) that

$$\lim_{n\to\infty} f(D(x_n, F(T)) \le \lim_{n\to\infty} d(x_n, Tx_n) = 0.$$

Therefore, we have

$$\lim_{n\to\infty} f(D(x_n, F(T))) = 0.$$

Since $f: [0, \infty] \to [0, \infty)$ is a nondecreasing mapping satisfying $f(0) = 0$ and $f(t) > 0$ for all $t \in (0, \infty)$, we have $\lim_{n\to\infty} d(x_n, F(T)) = 0$. Rest of the proof follows in lines of Theorem 3.2. □

In the view of the Remark 1.1 the following Corollaries are trivially true.

Corollary 3.1 *Let C be a nonempty closed convex subset of a complete uniformly convex hyperbolic space X with monotone modulus of uniform convexity* η *and* $T: C \to C$ *be a mapping which satisfies conditions* (C_λ) *(for some* $\lambda \in (0, 1)$*) and (E) on C with* $F(T) \neq \emptyset$*. If* $\{x_n\}$ *is the sequence defined by (for each* $x_1 \in C$*)*

$$\begin{cases} x_{n+1} = W(Ty_n, 0, 0) \\ y_n = W(x_n, Tx_n, \alpha_n), \quad n \in \mathbb{N}, \end{cases} \tag{3.7}$$

then the sequence $\{x_n\}$ Δ*-converges to a fixed point of T.*

Corollary 3.2 *Under the assumption of Corollary* 3.1 *with* $F(T) \neq \emptyset$*. The sequence* $\{x_n\}$ *which is defined by* (3.7), *converges strongly to some fixed point of T if and only if* $\lim_{n\to\infty} \inf D(x_n, F(T)) = 0$*, where* $D(x_n, F(T)) = \inf_{x\in F(T)} d(x_n, x)$.

Corollary 3.3 *Under the assumption of Corollary* 3.1 *with* $F(T) \neq \emptyset$ *and T satisfies the condition (I). The sequence* $\{x_n\}$ *which is defined by* (3.7), *converges strongly to some fixed point of T.*

In the view of the Remark 2.2, we have the following Corollaries:

Corollary 3.4 *Let C be a nonempty closed convex subset of a complete uniformly convex hyperbolic space X with monotone modulus of uniform convexity* η *and* $T: C \to C$ *be a SKC mapping with* $F(T) \neq \emptyset$*. The sequence* $\{x_n\}$ *defined by* (3.1), Δ*-converges to a fixed point of T.*

Corollary 3.5 *Under the assumption of Corollary* 3.4 *with* $F(T) \neq \emptyset$*. The sequence* $\{x_n\}$ *which is defined by* (3.1), *converges strongly to some fixed point of T if and only if* $\liminf_{n\to\infty} D(x_n, F(T)) = 0$*, where* $D(x_n, F(T)) = \inf_{x\in F(T)} d(x_n, x)$.

Corollary 3.6 *Under the assumption of Corollary* 3.4 *with* $F(T) \neq \emptyset$ *and T* satisfies the condition (*I*). The sequence $\{x_n\}$ *which is defined by* (3.1), *converges strongly to some fixed point of T.*

Corollary 3.7 *Let C be a nonempty closed convex subset of a complete uniformly convex hyperbolic space X with monotone modulus of uniform convexity* η *and* $T: C \to C$ *be a SKC mapping with* $F(T) \neq \emptyset$*. Then the sequence* $\{x_n\}$ *defined by* (3.7), Δ*-converges to a fixed point of T.*

Corollary 3.8 *Under the assumption of Corollary* 3.7 *with* $F(T) \neq \emptyset$*. The sequence* $\{x_n\}$ *which is defined by* (3.1), *converges strongly to some fixed point of T if and only if* $\liminf_{n\to\infty} D(x_n, F(T)) = 0$*, where* $D(x_n, F(T)) = \inf_{x\in F(T)} d(x_n, x)$.

Corollary 3.9 *Under the assumption of Corollary* 3.7 *with* $F(T) \neq \emptyset$ *and T satisfies the condition (I). The* *sequence* $\{x_n\}$ *which is defined by* (3.1), *converges strongly to some fixed point of* T.

References

1. Abbas, M., Khan, S.H., Postolache, M.: Existence and approximation results for SKC mappings in CAT(0) spaces. J. Inequal. Appl. **2014**, 212 (2014)
2. Abbas, M., Nazir, T.: A new faster iteration process applied to constrained minimization and feasibility problems. Mat. Vesn. **66**, 223–234 (2014)
3. Agarwal, R.P., O'Regan, D., Sahu, D.R.: Iterative construction of fixed points of nearly asymptotically nonexpansive mappings. J. Convex Anal. **8**(1), 61–79 (2007)
4. Bridson, N., Haefliger, A.: Metric Spaces of Non-Positive Curvature. Springer, Berlin (1999)
5. Goebel, K., Kirk, W.A.: Iteration processes for nonexpansive mappings. In: Singh, S.P., Thomeier, S., Watson, B. (eds) Topological Methods in Nonlinear Functional Analysis (Toronto, 1982), pp. 115–123. Contemporary Mathematics, vol 21. American Mathematical Society, New York (1983)
6. Goebel, K., Reich, S.: Uniformly Convexity, Hyperbolic Geometry, and Nonexpansive Mappings. Marcel Dekker Inc, New York (1984)
7. García-Falset, J., Llorens-Fuster, E., Suzuki, T.: Fixed point theory for a class of generalized nonexpansive mappings. J. Math. Anal. Appl. **375**, 185–195 (2011)
8. Itoh, S.: Some fixed point theorems in metric spaces. Fund. Math. **102**, 109–117 (1979)
9. Ishikawa, S.: Fixed points by new iteration method. Proc. Am. Math. Soc. **149**, 147–150 (1974)
10. Karapinar, E., Tas, K.: Generalized (C)-conditions and related fixed point theorems. Comput. Math. Appl. **61**, 3370–3380 (2011)
11. Kadioglu, N., Yildirim, I.: Approximating fixed points of nonexpansive mappings by faster iteration process. arXiv:1402.6530v1 [math.FA] (2014)
12. Krasnosel'skiĭ, M.A.: Two remarks on the method of successive approximations. Usp. Mat. Nauk. **10**, 123–127 (1955)
13. Khan, A.R., Fukhar-ud-din, H., Khan, M.A.: An implicit algorithm for two finite families of nonexpansive maps in hyperbolic spaces. Fixed Point Theory Appl. **2012**, 54 (2012)
14. Kim, J.K., Pathak, R.P., Dashputre, S., Diwan, S.D., Gupta, R.: Fixed point approximation of generalized nonexpansive mappings in hyperbolic spaces. Int. J. Math. Math. Sci. **2015**, 6 (2015)
15. Kohlenbach, U.: Some logical metathorems with applications in functional analysis. Trans. Am. Math. Soc. **357**(1), 89–128 (2005)
16. Leuştean, L.: A quadratic rate of asymptotic regularity for CAT(0) spaces. J. Math. Anal. Appl. **325**(1), 386–399 (2007)
17. Leuştean, L.: Nonexpansive iteration in uniformly convex W-hyperbolic spaces. In: Leizarowitz, A., Mordukhovich, B.S., Shafrir, I., Zaslavski, A. (eds) Nonlinear Analysis and Optimization I. Nonlinear Analysis. Contemporary Mathematics, vol. 513, pp. 193–210. Ramat Gan American Mathematical Society, Bar Ilan University, Providence (2010)
18. Mann, W.R.: Mean value methods in iteration. Proc. Am. Math. Soc. **4**, 506–510 (1953)
19. Noor, M.A.: New approximation schemes for general variational inequalities. J. Math. Anal. Appl. **251**, 217–229 (2000)
20. Reich, S., Shafrir, I.: Nonexpansive iterations in hyperbolic spaces. Nonlinear Anal. **15**, 537–558 (1990)
21. Sahu, D.R.: Application of the S-iteration process to constrained minimization problem and split feasibility problems. Fixed Point Theory **12**, 187–204 (2011)
22. Senter, H.F., Doston Jr., W.G.: Approximating fixed points of nonexpansive mappings. Proc. Am. Math. Soc. **44**(2), 375–380 (1974)
23. Schaefer, H.: Über die methode sukzessiver approximationen. Jber. Dtsch. Math. Ver. **59**, 131–140 (1957)
24. Suzuki, T.: Fixed point theorems and convergence theorems for some generalized nonexpansive mappings. J. Math. Anal. Appl. **340**, 1088–1095 (2008)
25. Takahashi, W.A.: A convexity in metric space and nonexpansive mappings I. Kodai Math. Sem. Rep. **22**, 142–149 (1970)
26. Thakur, D., Thakur, B.S., Postolache, M.: New iteration scheme for numerical reckoning fixed points of nonexpansive mappings. J. Inequal. Appl. **2014**, 328 (2014)

4

New contractive conditions of integral type on complete S-metric spaces

Nihal Yilmaz Özgür[1] · **Nihal Taş**[1]

Abstract An S-metric space is a three-dimensional generalization of a metric space. In this paper our aim is to examine some fixed-point theorems using new contractive conditions of integral type on a complete S-metric space. We give some illustrative examples to verify the obtained results. Our findings generalize some fixed-point results on a complete metric space and on a complete S-metric space. An application to the Fredholm integral equation is also obtained.

Keywords Integral-type contractive conditions · Fixed point · S-metric

Mathematics Subject Classification Primary 47H10 · Secondary 54H25

Introduction

Recently, the notion of an S-metric has been introduced and studied as a generalization of a metric. This notion has been defined by Sedghi et al. [13] as follows:

Definition 1.1 [13] Let $X \neq \emptyset$ be any set and $S : X \times X \times X \to [0, \infty)$ be a function satisfying the following conditions for all $u, v, z, a \in X$.

(S1) $S(u, v, z) = 0$ if and only if $u = v = z$.
(S2) $S(u, v, z) \leq S(u, u, a) + S(v, v, a) + S(z, z, a)$.

Then the function S is called an S-metric on X and the pair (X, S) is called an S-metric space.

Some fixed-point theorems have been given for self-mappings satisfying various contractive conditions on an S-metric space (see [4, 6, 8, 9, 13, 14]). One of the important results among these studies is the Banach's contraction principle on a complete S-metric space.

Theorem 1.2 [13] *Let (X, S) be a complete S-metric space, $h \in (0, 1)$ and $T : X \to X$ be a self-mapping of X such that*

$$S(Tu, Tu, Tv) \leq hS(u, u, v),$$

for all $u, v \in X$. Then T has a unique fixed point in X.

On the other hand some generalizations of the well-known Ćirić's and Nemytskii-Edelstein fixed-point theorems obtained on S-metric spaces via some new fixed point results (see [8, 9, 13, 14] for more details).

Later, different applications of some contractive conditions have been constructed on an S-metric space such as differential equations, complex valued functions etc. (see [5, 7, 10, 11]).

In recent years, fixed-point theory has been examined for various contractive conditions. For example, contractive conditions of integral type were adapted into some studied fixed-point results. So more general fixed-point theorems were obtained.

Through the whole paper we assume that $\varsigma : [0, \infty) \to [0, \infty)$ is a Lebesgue-integrable mapping which is summable (i.e., with finite integral) on each compact subset of $[0, \infty)$, nonnegative and such that for each $\varepsilon > 0$,

$$\int_0^{\varepsilon} \varsigma(t)dt > 0. \qquad (1)$$

✉ Nihal Yilmaz Özgür
nihal@balikesir.edu.tr

Nihal Taş
nihaltas@balikesir.edu.tr

[1] Department of Mathematics, Balıkesir University, 10145 Balıkesir, Turkey

Branciari [1] studied a fixed-point theorem for a general contractive condition of integral type on a complete metric space as seen in the following theorem.

Theorem 1.3 [1] *Let (X, ρ) be a complete metric space, $h \in (0,1)$, the function $\varsigma : [0,\infty) \to [0,\infty)$ be defined as in (1) and $T : X \to X$ be a self-mapping of X such that*

$$\int_0^{\rho(Tu,Tv)} \varsigma(t)\mathrm{d}t \leq h \int_0^{\rho(u,v)} \varsigma(t)\mathrm{d}t,$$

for all $u, v \in X$, then T has a unique fixed point $w \in X$ such that

$$\lim_{n\to\infty} T^n u = w,$$

for each $u \in X$.

After the study of Branciari, some researchers have investigated new generalized contractive conditions of integral type using different known inequalities on various metric spaces (see [2, 3, 12]).

The purpose of this paper is to give new contractive conditions of integral type satisfying some new generalized inequalities given in [6] on a complete S-metric space. Our results generalize some known fixed-point results on a complete metric space and on a complete S-metric space.

Fixed-point results under some contractive conditions of integral type

In this section we obtain new fixed-point theorems using some contractive conditions of integral type on a complete S-metric space. We construct three examples to show the validity of our results. At first we recall some basic results about S-metric spaces.

Lemma 2.1 [13] *Let (X, S) be an S-metric space. Then we have*

$$S(u,u,v) = S(v,v,u).$$

The above Lemma 2.1 can be considered as a symmetry condition on an S-metric space. The following definition is related to convergent sequences on an S-metric space.

Definition 2.2 [13] Let (X, S) be an S-metric space.

(1) A sequence $\{u_n\}$ in X converges to u if and only if $S(u_n, u_n, u) \to 0$ as $n \to \infty$. That is, there exists $n_0 \in \mathbb{N}$ such that for all $n \geq n_0$, $S(u_n, u_n, u) < \varepsilon$ for each $\varepsilon > 0$. We denote this by

$$\lim_{n\to\infty} u_n = u \text{or} \lim_{n\to\infty} \mathrm{S}(\mathrm{u_n}, \mathrm{u_n}, \mathrm{u}) = 0.$$

(2) A sequence $\{u_n\}$ in X is called a Cauchy sequence if $S(u_n, u_n, u_m) \to 0$ as $n, m \to \infty$. That is, there exists $n_0 \in \mathbb{N}$ such that for all $n, m \geq n_0$, $S(u_n, u_n, u_m) < \varepsilon$ for each $\varepsilon > 0$.

(3) The S-metric space (X, S) is called complete if every Cauchy sequence is convergent.

In the following lemma we see the relationship between a metric and an S-metric.

Lemma 2.3 [4] *Let (X, ρ) be a metric space. Then the following properties are satisfied :*

(1) $S_\rho(u, v, z) = \rho(u, z) + \rho(v, z)$ *for all $u, v, z \in X$ is an S-metric on X.*

(2) $u_n \to u$ *in* (X, ρ) *if and only if* $u_n \to u$ *in* (X, S_ρ).

(3) $\{u_n\}$ *is Cauchy in* (X, ρ) *if and only if* $\{u_n\}$ *is Cauchy in* (X, S_ρ).

(4) (X, ρ) *is complete if and only if* (X, S_ρ) *is complete.*

We call the function S_ρ defined in Lemma 2.3 (1) as the S-metric generated by the metric ρ. It can be found an example of an S-metric which is not generated by any metric in [4, 9].

Now we give the following theorem.

Theorem 2.4 *Let (X, S) be a complete S-metric space, $h \in (0,1)$, the function $\varsigma : [0,\infty) \to [0,\infty)$ be defined as in (1) and $T : X \to X$ be a self-mapping of X such that*

$$\int_0^{S(Tu,Tu,Tv)} \varsigma(t)\mathrm{d}t \leq h \int_0^{S(u,u,v)} \varsigma(t)\mathrm{d}t, \tag{2}$$

for all $u, v \in X$. Then T has a unique fixed point $w \in X$ and we have

$$\lim_{n\to\infty} T^n u = w,$$

for each $u \in X$.

Proof Let $u_0 \in X$ and the sequence $\{u_n\}$ be defined as

$$T^n u_0 = u_n.$$

Suppose that $u_n \neq u_{n+1}$ for all n. Using the inequality (2), we obtain

$$\int_0^{S(u_n,u_n,u_{n+1})} \varsigma(t)\mathrm{d}t \leq h \int_0^{S(u_{n-1},u_{n-1},u_n)} \varsigma(t)\mathrm{d}t \leq \cdots \leq h^n \int_0^{S(u_0,u_0,u_1)} \varsigma(t)\mathrm{d}t. \tag{3}$$

If we take limit for $n \to \infty$, using the inequality (3) we get

$$\lim_{n\to\infty} \int_0^{S(u_n,u_n,u_{n+1})} \varsigma(t)\mathrm{d}t = 0,$$

since $h \in (0,1)$. The condition (1) implies

$$\lim_{n\to\infty} S(u_n, u_n, u_{n+1}) = 0.$$

Now we show that the sequence $\{u_n\}$ is a Cauchy sequence. Assume that $\{u_n\}$ is not Cauchy. Then there exists an $\varepsilon > 0$ and subsequences $\{m_k\}$ and $\{n_k\}$ such that $m_k < n_k < m_{k+1}$ with

$$S(u_{m_k}, u_{m_k}, u_{n_k}) \geq \varepsilon \tag{4}$$

and

$$S(u_{m_k}, u_{m_k}, u_{n_k-1}) < \varepsilon.$$

Hence using Lemma 2.1, we have

$$\begin{aligned} S(u_{m_k-1}, u_{m_k-1}, u_{n_k-1}) &\leq 2S(u_{m_k-1}, u_{m_k-1}, u_{m_k}) \\ &\quad + S(u_{n_k-1}, u_{n_k-1}, u_{m_k}) \\ &< 2S(u_{m_k-1}, u_{m_k-1}, u_{m_k}) + \varepsilon \end{aligned}$$

and

$$\lim_{k\to\infty} \int_0^{S(u_{m_k-1}, u_{m_k-1}, u_{n_k-1})} \varsigma(t)\mathrm{d}t \leq \int_0^{\varepsilon} \varsigma(t)\mathrm{d}t. \tag{5}$$

Using the inequalities (2), (4) and (5) we obtain

$$\begin{aligned} \int_0^{\varepsilon} \varsigma(t)\mathrm{d}t \leq \int_0^{S(u_{m_k}, u_{m_k}, u_{n_k})} \varsigma(t)\mathrm{d}t \leq h \int_0^{S(u_{m_k-1}, u_{m_k-1}, u_{n_k-1})} \varsigma(t)\mathrm{d}t \\ \leq h \int_0^{\varepsilon} \varsigma(t)\mathrm{d}t, \end{aligned}$$

which is a contradiction with our assumption since $h \in (0,1)$. So the sequence $\{u_n\}$ is Cauchy. Using the completeness hypothesis, there exists $w \in X$ such that

$$\lim_{n\to\infty} T^n u_0 = w.$$

From the inequality (2) we find

$$\int_0^{S(Tw,Tw,u_{n+1})} \varsigma(t)\mathrm{d}t = \int_0^{S(Tw,Tw,Tu_n)} \varsigma(t)\mathrm{d}t \leq h \int_0^{S(w,w,u_n)} \varsigma(t)\mathrm{d}t.$$

If we take limit for $n \to \infty$, we get

$$\int_0^{S(Tw,Tw,w)} \varsigma(t)\mathrm{d}t = 0,$$

which implies $Tw = w$.

Now we show the uniqueness of the fixed point. Suppose that w_1 is another fixed point of T. Using the inequality (2) we have

$$\int_0^{S(w,w,w_1)} \varsigma(t)\mathrm{d}t = \int_0^{S(Tw,Tw,Tw_1)} \varsigma(t)\mathrm{d}t \leq h \int_0^{S(w,w,w_1)} \varsigma(t)\mathrm{d}t,$$

which implies

$$\int_0^{S(w,w,w_1)} \varsigma(t)\mathrm{d}t = 0,$$

since $h \in (0,1)$. Using the inequality (1) we get $w = w_1$. Consequently, the fixed point w is unique. □

Remark 2.5

(1) If we set the function $\varsigma : [0,\infty) \to [0,\infty)$ in Theorem 2.4 as

$$\varsigma(t) = 1,$$

for all $t \in [0,\infty)$, then we obtain the Banach's contraction principle on a complete S-metric space.

(2) Since an S-metric space is a generalization of a metric space, Theorem 2.4 is a generalization of the classical Banach's fixed-point theorem.

(3) If we set the S-metric as $S : X \times X \times X \to \mathbb{C}$ and take the function $\varsigma : [0,\infty) \to [0,\infty)$ as

$$\varsigma(t) = 1,$$

for all $t \in [0,\infty)$ in Theorem 2.4, then we get Theorem 3.1 in [10] and Corollary 2.5 in [5] for $n = 1$.

Example 2.6 Let $X = \mathbb{R}$, $k > 1$ be a fixed real number and the function $S : X \times X \times X \to [0,\infty)$ be defined as

$$S(u,v,z) = \frac{k}{k+1}(|v-z| + |v+z-2u|),$$

for all $u, v, z \in \mathbb{R}$. It can be easily seen that the function S is an S-metric. Now we show that this S-metric can not be generated by any metric ρ. On the contrary, we assume that there exists a metric ρ such that

$$S(u,v,z) = \rho(u,z) + \rho(v,z), \tag{6}$$

for all $u, v, z \in \mathbb{R}$. Hence we find

$$S(u,u,z) = 2\rho(u,z) = \frac{2k}{k+1}|u-z|$$

and

$$\rho(u,z) = \frac{k}{k+1}|u-z|. \tag{7}$$

Similarly, we get

$$S(v,v,z) = 2\rho(v,z) = \frac{2k}{k+1}|v-z|$$

and

$$\rho(v,z) = \frac{k}{k+1}|v-z|. \tag{8}$$

Using the equalities (6), (7) and (8), we obtain

$$\frac{k}{k+1}(|v-z|+|v+z-2u|) = \frac{k}{k+1}|u-z| + \frac{k}{k+1}|v-z|,$$

which is a contradiction. Consequently, S is not generated by any metric and $(\mathbb{R}, S)$ is a complete S-metric space.

Let us define the self-mapping $T : \mathbb{R} \to \mathbb{R}$ as

$$Tu = \frac{u}{6},$$

for all $u \in \mathbb{R}$ and the function $\varsigma : [0,\infty) \to [0,\infty)$ as

$$\varsigma(t) = 3t^2,$$

for all $t \in [0,\infty)$. Then we get

$$\int_0^{\varepsilon} \varsigma(t)dt = \int_0^{\varepsilon} 3t^2 dt = \varepsilon^3 > 0,$$

for each $\varepsilon > 0$. Therefore T satisfies the inequality (2) in Theorem 2.4 for $h = \frac{1}{2}$. Indeed, we have

$$\frac{k^3}{27(k+1)^3}|u-v|^3 \le \frac{4k^3}{(k+1)^3}|u-v|^3,$$

for all $u, v \in \mathbb{R}$. Consequently, T has a unique fixed point $u = 0$.

Now we give the first generalization of Theorem 2.4.

Theorem 2.7 *Let (X, S) be a complete S-metric space, the function* $\varsigma : [0,\infty) \to [0,\infty)$ *be defined as in (1) and* $T : X \to X$ *be a self-mapping of X such that*

$$\begin{aligned}\int_0^{S(Tu,Tu,Tv)} \varsigma(t)dt \le h_1 \int_0^{S(u,u,v)} \varsigma(t)dt + h_2 \int_0^{S(Tu,Tu,v)} \varsigma(t)dt \\ + h_3 \int_0^{S(Tv,Tv,u)} \varsigma(t)dt \\ + h_4 \int_0^{\max\{S(Tu,Tu,u),S(Tv,Tv,v)\}} \varsigma(t)dt,\end{aligned} \tag{9}$$

for all $u, v \in X$ *with nonnegative real numbers* h_i $(i \in \{1,2,3,4\})$ *satisfying* $\max\{h_1+3h_3+2h_4, h_1+h_2+h_3\} < 1$. *Then T has a unique fixed point* $w \in X$ *and we have*

$$\lim_{n\to\infty} T^n u = w,$$

for each $u \in X$.

Proof Let $u_0 \in X$ and the sequence $\{u_n\}$ be defined as

$$T^n u_0 = u_n.$$

Suppose that $u_n \neq u_{n+1}$ for all n. Using the inequality (9), the condition $(S2)$ and Lemma 2.1 we get

$$\begin{aligned}\int_0^{S(u_n,u_n,u_{n+1})} \varsigma(t)dt &= \int_0^{S(Tu_{n-1},Tu_{n-1},Tu_n)} \varsigma(t)dt \le h_1 \int_0^{S(u_{n-1},u_{n-1},u_n)} \varsigma(t)dt \\ &+ h_2 \int_0^{S(u_n,u_n,u_n)} \varsigma(t)dt + h_3 \int_0^{S(u_{n+1},u_{n+1},u_{n-1})} \varsigma(t)dt \\ &+ h_4 \int_0^{\max\{S(u_n,u_n,u_{n-1}),S(u_{n+1},u_{n+1},u_n)\}} \varsigma(t)dt \\ &= h_1 \int_0^{S(u_{n-1},u_{n-1},u_n)} \varsigma(t)dt + h_3 \int_0^{S(u_{n+1},u_{n+1},u_{n-1})} \varsigma(t)dt \\ &+ h_4 \int_0^{\max\{S(u_n,u_n,u_{n-1}),S(u_{n+1},u_{n+1},u_n)\}} \varsigma(t)dt \\ &\le h_1 \int_0^{S(u_{n-1},u_{n-1},u_n)} \varsigma(t)dt + h_3 \int_0^{2S(u_{n+1},u_{n+1},u_n)} \varsigma(t)dt \\ &+ h_3 \int_0^{S(u_{n-1},u_{n-1},u_n)} \varsigma(t)dt + h_4 \int_0^{S(u_n,u_n,u_{n-1})} \varsigma(t)dt \\ &+ h_4 \int_0^{S(u_{n+1},u_{n+1},u_n)} \varsigma(t)dt \\ &= (h_1+h_3+h_4) \int_0^{S(u_{n-1},u_{n-1},u_n)} \varsigma(t)dt \\ &+ (2h_3+h_4) \int_0^{S(u_n,u_n,u_{n+1})} \varsigma(t)dt,\end{aligned}$$

which implies

$$\int_0^{S(u_n,u_n,u_{n+1})} \varsigma(t)dt \le \left(\frac{h_1+h_3+h_4}{1-2h_3-h_4}\right) \int_0^{S(u_{n-1},u_{n-1},u_n)} \varsigma(t)dt. \tag{10}$$

If we put $h = \frac{h_1+h_3+h_4}{1-2h_3-h_4}$ then we find $h < 1$ since $h_1 + 3h_3 + 2h_4 < 1$. Using the inequality (10) we have

$$\int_0^{S(u_n,u_n,u_{n+1})} \varsigma(t)\mathrm{d}t \leq h^n \int_0^{S(u_0,u_0,u_1)} \varsigma(t)\mathrm{d}t. \qquad (11)$$

If we take limit for $n \to \infty$, using the inequality (11) we get

$$\lim_{n\to\infty} \int_0^{S(u_n,u_n,u_{n+1})} \varsigma(t)\mathrm{d}t = 0,$$

since $h \in (0,1)$. The condition (1) implies $\lim_{n\to\infty} S(u_n, u_n, u_{n+1}) = 0$.

By the similar arguments used in the proof of Theorem 2.4, we see that the sequence $\{u_n\}$ is Cauchy. Then there exists $w \in X$ such that

$$\lim_{n\to\infty} T^n u_0 = w,$$

since (X, S) is a complete S-metric space. From the inequality (9) we find

$$\begin{aligned}\int_0^{S(u_n,u_n,Tw)} \varsigma(t)\mathrm{d}t = \int_0^{S(Tu_{n-1},Tu_{n-1},Tw)} \varsigma(t)\mathrm{d}t &\leq h_1 \int_0^{S(u_{n-1},u_{n-1},w)} \varsigma(t)\mathrm{d}t \\ &+ h_2 \int_0^{S(u_n,u_n,w)} \varsigma(t)\mathrm{d}t + h_3 \int_0^{S(Tw,Tw,u_{n-1})} \varsigma(t)\mathrm{d}t \\ &+ h_4 \int_0^{\max\{S(u_n,u_n,u_{n-1}),S(Tw,Tw,w)\}} \varsigma(t)\mathrm{d}t.\end{aligned}$$

Taking limit for $n \to \infty$ and using Lemma 2.1 we get

$$\int_0^{S(Tw,Tw,w)} \varsigma(t)\mathrm{d}t \leq (h_3 + h_4) \int_0^{S(Tw,Tw,w)} \varsigma(t)\mathrm{d}t,$$

which implies $Tw = w$ since $h_3 + h_4 < 1$.

Now we show the uniqueness of the fixed point. Let w_1 be another fixed point of T. Using the inequality (9) and Lemma 2.1, we get

$$\begin{aligned}\int_0^{S(w,w,w_1)} \varsigma(t)\mathrm{d}t = \int_0^{S(Tw,Tw,Tw_1)} \varsigma(t)\mathrm{d}t &\leq h_1 \int_0^{S(w,w,w_1)} \varsigma(t)\mathrm{d}t \\ &+ h_2 \int_0^{S(w,w,w_1)} \varsigma(t)\mathrm{d}t + h_3 \int_0^{S(w_1,w_1,w)} \varsigma(t)\mathrm{d}t \\ &+ h_4 \int_0^{\max\{S(w,w,w),S(w_1,w_1,w_1)\}} \varsigma(t)\mathrm{d}t,\end{aligned}$$

which implies

$$\int_0^{S(w,w,w_1)} \varsigma(t)\mathrm{d}t \leq (h_1 + h_2 + h_3) \int_0^{S(w,w,w_1)} \varsigma(t)\mathrm{d}t.$$

Then we obtain

$$\int_0^{S(w,w,w_1)} \varsigma(t)\mathrm{d}t = 0,$$

that is, $w = w_1$ since $h_1 + h_2 + h_3 < 1$. Consequently, T has a unique fixed point $w \in X$. □

Remark 2.8

(1) If we set the function $\varsigma : [0,\infty) \to [0,\infty)$ in Theorem 2.7 as

$$\varsigma(t) = 1,$$

for all $t \in [0,\infty)$, then we obtain Theorem 3 in [6].

(2) Theorem 2.7 is a generalization of Theorem 2.4 on a complete S-metric space. Indeed, if we take $h_1 = h$ and $h_2 = h_3 = h_4 = 0$ in Theorem 2.7, then we get Theorem 2.4.

(3) Since Theorem 2.7 is a generalization of Theorem 2.4, Theorem 2.7 generalizes the classical Banach's fixed-point theorem.

(4) If we set the S-metric as $S : X \times X \times X \to \mathbb{C}$ and take the function $\varsigma : [0,\infty) \to [0,\infty)$ as

$$\varsigma(t) = 1,$$

for all $t \in [0,\infty)$ in Theorem 2.7, then we get Theorem 3.1 in [7].

Now we give the second generalization of Theorem 2.4.

Theorem 2.9 *Let (X, S) be a complete S-metric space, the function $\varsigma : [0,\infty) \to [0,\infty)$ be defined as in (1) and $T : X \to X$ be a self-mapping of X such that*

$$\begin{aligned}\int_0^{S(Tu,Tu,Tv)} \varsigma(t)\mathrm{d}t &\leq h_1 \int_0^{S(u,u,v)} \varsigma(t)\mathrm{d}t + h_2 \int_0^{S(Tu,Tu,u)} \varsigma(t)\mathrm{d}t \\ &+ h_3 \int_0^{S(Tu,Tu,v)} \varsigma(t)\mathrm{d}t \\ &+ h_4 \int_0^{S(Tv,Tv,u)} \varsigma(t)\mathrm{d}t + h_5 \int_0^{S(Tv,Tv,v)} \varsigma(t)\mathrm{d}t \\ &+ h_6 \int_0^{\max\{S(u,u,v),S(Tu,Tu,u),S(Tu,Tu,v),S(Tv,Tv,u),S(Tv,Tv,v)\}} \varsigma(t)\mathrm{d}t,\end{aligned} \qquad (12)$$

for all $u, v \in X$ *with nonnegative real numbers* h_i $(i \in \{1,2,3,4,5,6\})$ *satisfying* $\max\{h_1+h_2+3h_4+h_5+3h_6, h_1+h_3+h_4+h_6\}<1$. *Then* T *has a unique fixed point* $w \in X$ *and we have*

$$\lim_{n\to\infty} T^n u = w,$$

for each $u \in X$.

Proof Let $u_0 \in X$ and the sequence $\{u_n\}$ be defined as

$$T^n u_0 = u_n.$$

Suppose that $u_n \neq u_{n+1}$ for all n. Using the inequality (12), the condition $(S2)$ and Lemma 2.1 we get

$$\int_0^{S(u_n,u_n,u_{n+1})} \varsigma(t)\mathrm{d}t = \int_0^{S(Tu_{n-1},Tu_{n-1},Tu_n)} \varsigma(t)\mathrm{d}t \leq h_1 \int_0^{S(u_{n-1},u_{n-1},u_n)} \varsigma(t)\mathrm{d}t$$
$$+ h_2 \int_0^{S(u_n,u_n,u_{n-1})} \varsigma(t)\mathrm{d}t + h_3 \int_0^{S(u_n,u_n,u_n)} \varsigma(t)\mathrm{d}t$$
$$+ h_4 \int_0^{S(u_{n+1},u_{n+1},u_{n-1})} \varsigma(t)\mathrm{d}t + h_5 \int_0^{S(u_{n+1},u_{n+1},u_n)} \varsigma(t)\mathrm{d}t$$
$$+ h_6 \int_0^{\max\{S(u_{n-1},u_{n-1},u_n),S(u_n,u_n,u_{n-1}),S(u_n,u_n,u_n),S(u_{n+1},u_{n+1},u_{n-1}),S(u_{n+1},u_{n+1},u_n)\}} \varsigma(t)\mathrm{d}t$$
$$\leq (h_1+h_2+h_4+h_6) \int_0^{S(u_{n-1},u_{n-1},u_n)} \varsigma(t)\mathrm{d}t + (2h_4+h_5+2h_6) \int_0^{S(u_{n+1},u_{n+1},u_n)} \varsigma(t)\mathrm{d}t,$$

which implies

$$\int_0^{S(u_n,u_n,u_{n+1})} \varsigma(t)\mathrm{d}t \leq \left(\frac{h_1+h_2+h_4+h_6}{1-2h_4-h_5-2h_6}\right) \int_0^{S(u_{n-1},u_{n-1},u_n)} \varsigma(t)\mathrm{d}t. \tag{13}$$

If we put $h = \frac{h_1+h_2+h_4+h_6}{1-2h_4-h_5-2h_6}$ then we find $h<1$ since $h_1+h_2+3h_4+h_5+3h_6<1$. Using the inequality (13) we have

$$\int_0^{S(u_n,u_n,u_{n+1})} \varsigma(t)\mathrm{d}t \leq h^n \int_0^{S(u_0,u_0,u_1)} \varsigma(t)\mathrm{d}t. \tag{14}$$

If we take limit for $n \to \infty$, using the inequality (14) we get

$$\lim_{n\to\infty} \int_0^{S(u_n,u_n,u_{n+1})} \varsigma(t)\mathrm{d}t = 0,$$

since $h \in (0,1)$. The condition (1) implies

$$\lim_{n\to\infty} S(u_n,u_n,u_{n+1}) = 0.$$

By the similar arguments used in the proof of Theorem 2.4, we see that the sequence $\{u_n\}$ is Cauchy. Then there exists $w \in X$ such that

$$\lim_{n\to\infty} T^n u_0 = w,$$

since (X, S) is a complete S-metric space. From the inequality (12) we find

$$\int_0^{S(u_n,u_n,Tw)} \varsigma(t)\mathrm{d}t = \int_0^{S(Tu_{n-1},Tu_{n-1},Tw)} \varsigma(t)\mathrm{d}t \leq h_1 \int_0^{S(u_{n-1},u_{n-1},w)} \varsigma(t)\mathrm{d}t$$
$$+ h_2 \int_0^{S(u_n,u_n,u_{n-1})} \varsigma(t)\mathrm{d}t + h_3 \int_0^{S(u_n,u_n,w)} \varsigma(t)\mathrm{d}t$$
$$+ h_4 \int_0^{S(Tw,Tw,u_{n-1})} \varsigma(t)\mathrm{d}t + h_5 \int_0^{S(Tw,Tw,w)} \varsigma(t)\mathrm{d}t$$
$$+ h_6 \int_0^{\max\{S(u_{n-1},u_{n-1},w),S(u_n,u_n,u_{n-1}),S(u_n,u_n,w),S(Tw,Tw,u_{n-1}),S(Tw,Tw,w)\}} \varsigma(t)\mathrm{d}t.$$

If we take limit for $n \to \infty$, using Lemma 2.1 we get

$$\int_0^{S(Tw,Tw,w)} \varsigma(t)\mathrm{d}t \le (h_4+h_5+h_6)\int_0^{S(Tw,Tw,w)} \varsigma(t)\mathrm{d}t,$$

which implies $Tw = w$ since $h_4+h_5+h_6<1$.

Now we show the uniqueness of the fixed point. Let w_1 be another fixed point of T. Using the inequality (12) and Lemma 2.1, we get

$$\begin{aligned}\int_0^{S(w,w,w_1)} \varsigma(t)\mathrm{d}t = \int_0^{S(Tw,Tw,Tw_1)} \varsigma(t)\mathrm{d}t &\le h_1 \int_0^{S(w,w,w_1)} \varsigma(t)\mathrm{d}t \\ &+ h_2 \int_0^{S(w,w,w)} \varsigma(t)\mathrm{d}t + h_3 \int_0^{S(w,w,w_1)} \varsigma(t)\mathrm{d}t \\ &+ h_4 \int_0^{S(w_1,w_1,w)} \varsigma(t)\mathrm{d}t + h_5 \int_0^{S(w_1,w_1,w_1)} \varsigma(t)\mathrm{d}t \\ &+ h_6 \int_0^{\max\{S(w,w,w_1),S(w,w,w),S(w,w,w_1),S(w_1,w_1,w),S(w_1,w_1,w_1)\}} \varsigma(t)\mathrm{d}t,\end{aligned}$$

which implies

$$\int_0^{S(w,w,w_1)} \varsigma(t)\mathrm{d}t \le (h_1+h_3+h_4+h_6)\int_0^{S(w,w,w_1)} \varsigma(t)\mathrm{d}t.$$

Then we obtain

$$\int_0^{S(w,w,w_1)} \varsigma(t)\mathrm{d}t = 0,$$

that is, $w = w_1$ since $h_1+h_3+h_4+h_6<1$. Consequently, T has a unique fixed point $w \in X$. □

Remark 2.10

(1) In Theorem 2.9, if we set the function $\varsigma : [0,\infty) \to [0,\infty)$ as

$$\varsigma(t) = 1,$$

for all $t \in [0,\infty)$, then we obtain Theorem 4 in [6].

(2) Theorem 2.9 is a generalization of Theorem 2.4 on a complete S-metric space. Indeed, if we take $h_1 = h$ and $h_2 = h_3 = h_4 = h_5 = h_6 = 0$ in Theorem 2.9, then we get Theorem 2.4.

(3) Since Theorem 2.9 is another generalization of Theorem 2.4, Theorem 2.9 generalizes the classical Banach's fixed-point theorem.

(4) If we set the S-metric as $S : X \times X \times X \to \mathbb{C}$ and take the function $\varsigma : [0,\infty) \to [0,\infty)$ as

$$\varsigma(t) = 1,$$

for all $t \in [0,\infty)$ in Theorem 2.9, then we get Theorem 3.4 in [7].

In the following example we give a self-mapping satisfying the conditions of Theorems 2.7 and 2.9, respectively, but does not satisfy the condition of Theorem 2.4.

Example 2.11 Let $\mathbb{R}$ be the complete S-metric space with the S-metric defined in Example 1 given in [9]. Let us define the self-mapping $T : \mathbb{R} \to \mathbb{R}$ as

$$Tu = \begin{cases} u+80 & \text{if } \ u \in \{0,2\} \\ 75 & \text{if } \ \text{otherwise} \end{cases},$$

for all $u \in \mathbb{R}$ and the function $\varsigma : [0,\infty) \to [0,\infty)$ as

$$\varsigma(t) = 2t,$$

for all $t \in [0,\infty)$. Then we get

$$\int_0^{\varepsilon} \varsigma(t)\mathrm{d}t = \int_0^{\varepsilon} 2t\mathrm{d}t = \varepsilon^2 > 0,$$

for each $\varepsilon > 0$. Therefore T satisfies the inequality (9) in Theorem 2.7 for $h_1 = h_2 = h_3 = 0$, $h_4 = \frac{1}{2}$ and the inequality (12) in Theorem 2.9 for $h_1 = h_3 = h_4 = h_5 = 0$, $h_2 = h_6 = \frac{1}{3}$. Hence T has a unique fixed point $u = 75$. But T does not satisfy the inequality (2) in Theorem 2.4. Indeed, if we take $u = 0$ and $v = 1$, then we obtain

$$\int_0^{10} 2t\mathrm{d}t = 100 \le h \int_0^{2} 2t\mathrm{d}t = 4h,$$

which is a contradiction since $h \in (0,1)$.

Finally, we give another generalization of Theorem 2.4.

Theorem 2.12 Let (X, S) be a complete S-metric space, the function $\varsigma : [0,\infty) \to [0,\infty)$ be defined as in (1) and $T : X \to X$ be a self-mapping of X such that

$$\begin{aligned}\int_0^{S(Tu,Tu,Tv)} \varsigma(t)\mathrm{d}t &\le h_1 \int_0^{S(u,u,v)} \varsigma(t)\mathrm{d}t + h_2 \int_0^{S(Tu,Tu,u)} \varsigma(t)\mathrm{d}t \\ &+ h_3 \int_0^{S(Tv,Tv,v)} \varsigma(t)\mathrm{d}t \\ &+ h_4 \int_0^{\max\{S(Tu,Tu,v),S(Tv,Tv,u)\}} \varsigma(t)\mathrm{d}t,\end{aligned} \tag{15}$$

for all $u, v \in X$ with nonnegative real numbers h_i $(i \in \{1,2,3,4\})$ satisfying $h_1+h_2+h_3+3h_4<1$. Then T has a unique fixed point $w \in X$ and we have

$$\lim_{n\to\infty} T^n u = w,$$

for each $u \in X$.

Proof Let $u_0 \in X$ and the sequence $\{u_n\}$ be defined as $T^n u_0 = u_n$.

Suppose that $u_n \neq u_{n+1}$ for all n. Using the inequality (15), the condition $(S2)$ and Lemma 2.1 we get

$$\begin{aligned}\int_0^{S(u_n,u_n,u_{n+1})} \varsigma(t)dt &= \int_0^{S(Tu_{n-1},Tu_{n-1},Tu_n)} \varsigma(t)dt \leq h_1 \int_0^{S(u_{n-1},u_{n-1},u_n)} \varsigma(t)dt \\ &+ h_2 \int_0^{S(u_n,u_n,u_{n-1})} \varsigma(t)dt + h_3 \int_0^{S(u_{n+1},u_{n+1},u_n)} \varsigma(t)dt \\ &+ h_4 \int_0^{\max\{S(u_n,u_n,u_n),S(u_{n+1},u_{n+1},u_{n-1})\}} \varsigma(t)dt \\ &\leq h_1 \int_0^{S(u_{n-1},u_{n-1},u_n)} \varsigma(t)dt + h_2 \int_0^{S(u_{n-1},u_{n-1},u_n)} \varsigma(t)dt \\ &+ h_3 \int_0^{S(u_n,u_n,u_{n+1})} \varsigma(t)dt \\ &+ h_4 \int_0^{2S(u_n,u_n,u_{n+1})+S(u_{n-1},u_{n-1},u_n)} \varsigma(t)dt \\ &\leq (h_1+h_2+h_4) \int_0^{S(u_{n-1},u_{n-1},u_n)} \varsigma(t)dt \\ &+ (h_3+2h_4) \int_0^{S(u_n,u_n,u_{n+1})} \varsigma(t)dt,\end{aligned}$$

which implies

$$\int_0^{S(u_n,u_n,u_{n+1})} \varsigma(t)dt \leq \left(\frac{h_1+h_2+h_4}{1-h_3-2h_4}\right) \int_0^{S(u_{n-1},u_{n-1},u_n)} \varsigma(t)dt. \quad (16)$$

If we put $h = \frac{h_1+h_2+h_4}{1-h_3-2h_4}$ then we find $h<1$ since $h_1+h_2+h_3+3h_4<1$. Using the inequality (16) and mathematical induction, we have

$$\int_0^{S(u_n,u_n,u_{n+1})} \varsigma(t)dt \leq h^n \int_0^{S(u_0,u_0,u_1)} \varsigma(t)dt. \quad (17)$$

Taking limit for $n \to \infty$ and using the inequality (17) we find

$$\lim_{n\to\infty} \int_0^{S(u_n,u_n,u_{n+1})} \varsigma(t)dt = 0,$$

since $h \in (0,1)$. The condition (1) implies

$$\lim_{n\to\infty} S(u_n,u_n,u_{n+1}) = 0.$$

By the similar arguments used in the proof of Theorem 2.4, we see that the sequence $\{u_n\}$ is Cauchy. Then there exists $w \in X$ such that

$$\lim_{n\to\infty} T^n u_0 = w,$$

since (X, S) is a complete S-metric space. From the inequality (15) we find

$$\begin{aligned}\int_0^{S(u_n,u_n,Tw)} \varsigma(t)dt &= \int_0^{S(Tu_{n-1},Tu_{n-1},Tw)} \varsigma(t)dt \leq h_1 \int_0^{S(u_{n-1},u_{n-1},w)} \varsigma(t)dt \\ &+ h_2 \int_0^{S(u_n,u_n,u_{n-1})} \varsigma(t)dt + h_3 \int_0^{S(Tw,Tw,w)} \varsigma(t)dt \\ &+ h_4 \int_0^{\max\{S(u_n,u_n,w),S(Tw,Tw,u_{n-1})\}} \varsigma(t)dt.\end{aligned}$$

If we take limit for $n \to \infty$, using Lemma 2.1 we get

$$\int_0^{S(Tw,Tw,w)} \varsigma(t)dt \leq (h_3+h_4) \int_0^{S(Tw,Tw,w)} \varsigma(t)dt,$$

which implies $Tw = w$ since $h_3+h_4<1$.

Now we show the uniqueness of the fixed point. Let w_1 be another fixed point of T. Using the inequality (15) and Lemma 2.1, we get

$$\begin{aligned}\int_0^{S(w,w,w_1)} \varsigma(t)dt &= \int_0^{S(Tw,Tw,Tw_1)} \varsigma(t)dt \leq h_1 \int_0^{S(w,w,w_1)} \varsigma(t)dt \\ &+ h_2 \int_0^{S(w,w,w)} \varsigma(t)dt + h_3 \int_0^{S(w_1,w_1,w_1)} \varsigma(t)dt \\ &+ h_4 \int_0^{\max\{S(w,w,w_1),S(w_1,w_1,w)\}} \varsigma(t)dt,\end{aligned}$$

which implies

$$\int_0^{S(w,w,w_1)} \varsigma(t)dt \leq (h_1+h_4) \int_0^{S(w,w,w_1)} \varsigma(t)dt.$$

Then we obtain

$$\int_0^{S(w,w,w_1)} \varsigma(t)dt = 0,$$

that is, $w = w_1$ since $h_1 + h_4 < 1$. Consequently, T has a unique fixed point $w \in X$. □

Remark 2.13

(1) If we set the function $\varsigma : [0, \infty) \to [0, \infty)$ in Theorem 2.12 as

$$\varsigma(t) = 1,$$

for all $t \in [0, \infty)$, then we obtain Theorem 2 in [6].

(2) Theorem 2.12 is another generalization of Theorem 2.4 on a complete S-metric space. Indeed, if we take $h_1 = h$ and $h_2 = h_3 = h_4 = 0$ in Theorem 2.12, then we get Theorem 2.4.

(3) Since Theorem 2.12 is another generalization of Theorem 2.4, Theorem 2.12 generalizes the classical Banach's fixed-point theorem.

Let us consider the self-mapping $T : \mathbb{R} \to \mathbb{R}$ and the function $\varsigma : [0, \infty) \to [0, \infty)$ defined in Example 2.11. Then T satisfy the contractive condition (15) in Theorem 2.12 and so $u = 75$ is a unique fixed point of T. Notice that T does not satisfy the inequality (2) in Theorem 2.4.

An application to the Fredholm integral equation

In this section, we give an application of the contraction condition (2) to the Fredholm integral equation

$$y(u) = l(u) + \lambda \int_a^b k(u,t)y(t)\mathrm{d}t, \tag{18}$$

where $y : [a, b] \to \mathbb{R}$ with $-\infty < a < b < \infty$, $k(u, t)$ which is called the kernel of the integral equation (18) is continuous on the squared region $[a, b] \times [a, b]$ with $|k(u, t)| \leq M$ ($M > 0$) and $l(u)$ is continuous on $[a, b]$.

Let $C[a, b] = \{f \mid f : [a, b] \to \mathbb{R} \text{isacontinuousfunction}\}$. Now we define the function $S : C[a, b] \times C[a, b] \times C[a, b] \to [0, \infty)$ by

$$S(f, g, h) = \sup_{u\in[a,b]} |f(u) - h(u)| + \sup_{u\in[a,b]} |f(u) + h(u) - 2g(u)|, \tag{19}$$

for all $f, g, h \in C[a, b]$. Then the function S is an S-metric. Now we show that this S-metric can not be generated by any metric ρ. We assume that this S-metric is generated by any metric ρ, that is, there exists a metric ρ such that

$$S(f, g, h) = \rho(f, h) + \rho(g, h), \tag{20}$$

for all $f, g, h \in C[a, b]$. Then we get

$$S(f,f, h) = 2\rho(f, h) = 2 \sup_{u\in[a,b]} |f(u) - h(u)|$$

and

$$\rho(f, h) = \sup_{u\in[a,b]} |f(u) - h(u)|. \tag{21}$$

Similarly, we obtain

$$S(g, g, h) = 2\rho(g, h) = 2 \sup_{u\in[a,b]} |g(u) - h(u)|$$

and

$$\rho(g, h) = \sup_{u\in[a,b]} |g(u) - h(u)|. \tag{22}$$

Using the equalities (20), (21) and (22), we find

$$\sup_{u\in[a,b]} |f(u) - h(u)| + \sup_{u\in[a,b]} |f(u) + h(u) - 2g(u)|$$
$$= \sup_{u\in[a,b]} |f(u) - h(u)| + \sup_{u\in[a,b]} |g(u) - h(u)|,$$

which is a contradiction. Hence this S-metric is not generated by any metric ρ. Consequently, $(C[a, b], S)$ is a complete S-metric space.

Proposition 3.1 *Let $(C[a, b], S)$ be a complete S-metric space with the S-metric defined in* (19) *and λ be a real number with*

$$|\lambda| < \frac{1}{M(b - a)}.$$

Then the Fredholm integral equation (18) *has a unique solution $y : [a, b] \to \mathbb{R}$.*

Proof Let us define the function $T : C[a, b] \to C[a, b]$ as

$$Ty(u) = l(u) + \lambda \int_a^b k(u,t)y(t)\mathrm{d}t.$$

Now we show that T satisfies the contractive condition (2). We get

$$\begin{aligned}
S(Ty_1, Ty_1, Ty_2) &= 2 \sup_{u\in[a,b]} |Ty_1(u) - Ty_2(u)| \\
&= 2 \sup_{u\in[a,b]} \left| \lambda \int_a^b k(u,t)(y_1(u) - y_2(u))\mathrm{d}t \right| \\
&\leq 2|\lambda|M \sup_{u\in[a,b]} \left| \int_a^b (y_1(u) - y_2(u))\mathrm{d}t \right| \\
&\leq 2|\lambda|M \sup_{u\in[a,b]} \int_a^b |y_1(u) - y_2(u)|\mathrm{d}t \\
&\leq 2|\lambda|M \sup_{u\in[a,b]} |y_1(u) - y_2(u)| \left| \int_a^b \mathrm{d}t \right| \\
&\leq |\lambda|M(b - a)S(y_1, y_1, y_2) \\
&< S(y_1, y_1, y_2),
\end{aligned}$$

which implies

$$\int_0^{S(Ty_1,Ty_1,Ty_2)} \varsigma(t)dt < \int_0^{S(y_1,y_1,y_2)} \varsigma(t)dt.$$

Consequently, the contractive condition (2) is satisfied and the Fredholm integral equation (18) has a unique solution y. □

Now we give an example of Proposition 3.1.

Example 3.2 Let us consider the Fredholm integral equation defined as

$$y(u) = e + \lambda \int_1^e \frac{\ln u}{t} y(t)dt. \tag{23}$$

Now we find a solution of the Fredholm integral equation (23) with the initial condition $y_0(u) = 0$. We solve this equation for $|\lambda| < \frac{1}{e-1}$ since $\left|\frac{\ln u}{t}\right| < 1$ for all $1 \le u, t \le e$. We obtain

$$y_1(u) = e,$$

$$y_2(u) = e + \lambda \int_1^e \frac{\ln u}{t} e dt = e + \lambda e \ln u,$$

$$\begin{aligned} y_3(u) &= e + \lambda \int_1^e \frac{\ln u}{t}(e + \lambda e \ln t)dt \\ &= e + \lambda e \ln u + \frac{\lambda^2}{2} e \ln u, \end{aligned}$$

$$\begin{aligned} y_4(u) &= e + \lambda \int_1^e \frac{\ln u}{t}\left(e + \lambda e \ln t + \frac{\lambda^2}{2} e \ln t\right)dt \\ &= e + \lambda e \ln u + \frac{\lambda^2}{2} e \ln u + \frac{\lambda^3}{2} e \ln u, \end{aligned}$$

$$\dots$$

$$\begin{aligned} y_n(u) &= e + \lambda e \ln u \left[1 + \frac{\lambda}{2} + \frac{\lambda^2}{4} + \cdots + \frac{\lambda^n}{2^n}\right] \\ &\to e + \frac{2\lambda}{2 - \lambda} e \ln u. \end{aligned}$$

Consequently, this is a solution of the Fredholm integral equation (18) for $|\lambda| < \frac{1}{e-1} < 1$.

References

1. Branciari, A.: A fixed point theorem for mappings satisfying a general contractive condition of integral type. Int. J. Math. Math. Sci. **29**(9), 531–536 (2002)
2. Dey, D., Ganguly, A., Saha, M.: Fixed point theorems for mappings under general contractive condition of integral type. Bull. Math. Anal. Appl. **3**(1), 27–34 (2011)
3. Gu, F., Ye, H.: Common fixed point theorems of Altman integral type mappings in *G*-metric spaces, *Abstr. Appl. Anal.* vol. 2012, Article ID 630457, 13 pages. doi:10.1155/2012/630457
4. Hieu, N.T., Ly, N.T., Dung, N.V.: A generalization of Ciric quasi-contractions for maps on *S*-metric spaces. Thai J. Math. **13**(2), 369–380 (2015)
5. Mlaiki, N.M.: Common fixed points in complex *S*-metric space. Adv. Fixed Point Theory **4**(4), 509–524 (2014)
6. Özgür, N.Y., Taş, N.: Some Generalizations of Fixed Point Theorems on *S*-Metric Spaces, Essays in Mathematics and Its Applications in Honor of Vladimir Arnold. Springer, New York (2016)
7. Özgür, N.Y., Taş, N.: Some generalizations of the Banach's contraction principle on a complex valued *S*-metric space. J. New Theory **2**(14), 26–36 (2016)
8. Özgür, N.Y., Taş, N.: Some fixed point theorems on *S*-metric spaces. Mat. Vesnik **69**(1), 39–52 (2017)
9. Özgür, N.Y., Taş, N.: Some new contractive mappings on *S*-metric spaces and their relationships with the mapping **(S25)**. Math. Sci. **11**(7), 7–16 (2017). doi:10.1007/s40096-016-0199-4
10. Özgür, N.Y., Taş, N.: Common fixed point results on complex valued *S*-metric spaces. (submitted for publication)
11. Özgür, N.Y., Taş, N.: The Picard theorem on *S*-metric spaces. (submitted for publication)
12. Rahman, M.U., Sarwar, M., Rahman, M.U.: Fixed point results of Altman integral type mappings in *S*-metric spaces. Int. J. Anal. Appl. **10**(1), 58–63 (2016)
13. Sedghi, S., Shobe, N., Aliouche, A.: A generalization of fixed point theorems in *S*-metric spaces. Mat. Vesn. **64**(3), 258–266 (2012)
14. Sedghi, S., Dung, N.V.: Fixed point theorems on *S*-metric spaces. Mat. Vesn. **66**(1), 113–124 (2014)

5

A novel operational matrix method based on shifted Legendre polynomials for solving second-order boundary value problems involving singular, singularly perturbed and Bratu-type equations

W. M. Abd-Elhameed[1,2] · Y. H. Youssri[2] · E. H. Doha[2]

Abstract In this article, a new operational matrix method based on shifted Legendre polynomials is presented and analyzed for obtaining numerical spectral solutions of linear and nonlinear second-order boundary value problems. The method is novel and essentially based on reducing the differential equations with their boundary conditions to systems of linear or nonlinear algebraic equations in the expansion coefficients of the sought-for spectral solutions. Linear differential equations are treated by applying the Petrov–Galerkin method, while the nonlinear equations are treated by applying the collocation method. Convergence analysis and some specific illustrative examples include singular, singularly perturbed and Bratu-type equations are considered to ascertain the validity, wide applicability and efficiency of the proposed method. The obtained numerical results are compared favorably with the analytical solutions and are more accurate than those discussed by some other existing techniques in the literature.

Keywords Shifted Legendre polynomials · Second-order equations · Singular and singularly perturbed · Bratu equation · Petrov–Galerkin method · Collocation method

Mathematics Subject Classification 65M70 · 65N35 · 35C10 · 42C10

✉ Y. H. Youssri
youssri@sci.cu.edu.eg

1 Department of Mathematics, Faculty of Science, University of Jeddah, Jeddah, Saudi Arabia

2 Department of Mathematics, Faculty of Science, Cairo University, Giza, Egypt

Introduction

Spectral methods (see, for instance [1–6]) are one of the principal methods of discretization for the numerical solution of differential equations. The main advantage of these methods lies in their accuracy for a given number of unknowns. For smooth problems in simple geometries, they offer exponential rates of convergence/spectral accuracy. In contrast, finite difference and finite-element methods yield only algebraic convergence rates. The three spectral methods, namely, the Galerkin, collocation, and tau methods are used extensively in the literature. Collocation methods [7, 8] have become increasingly popular for solving differential equations, since they are very useful in providing highly accurate solutions to nonlinear differential equations. Petrov–Galerkin method is widely used for solving ordinary and partial differential equations; see for example [9–13]. The Petrov–Galerkin methods [14] have generally come to be known as "stablized" formulations, because they prevent the spatial oscillations and sometimes yield nodally exact solutions where the classical Galerkin method would fail badly. The difference between Galerkin and Petrov–Galerkin methods is that the test and trial functions in Galerkin method are the same, while in Petrov–Galerkin method, they are not.

The subject of nonlinear differential equations is a well-established part of mathematics and its systematic development goes back to the early days of the development of calculus. Many recent advances in mathematics, paralleled by a renewed and flourishing interaction between mathematics, the sciences, and engineering, have again shown that many phenomena in applied sciences, modeled by differential equations, will yield some mathematical explanation of these phenomena (at least in some approximate sense).

Even order differential equations have been extensively discussed by a large number of authors due to their great importance in various applications in many fields. For example, in the sequence of papers [12, 15–17], the authors dealt with such equations by the Galerkin method. They constructed suitable basis functions which satisfy the boundary conditions of the given differential equation. For this purpose, they used compact combinations of various orthogonal polynomials. The suggested algorithms in these articles are suitable for handling one- and two-dimensional linear high even-order boundary value problems. In this paper, we aim to give some algorithms for handling both of linear and nonlinear second-order boundary value problems based on introducing a new operational matrix of derivatives, and then applying Petrov–Galerkin method on linear equations and collocation method on nonlinear equations.

Of the important high-order differential equations are the singular and singular perturbed problems (SPPs) which arise in several branches of applied mathematics, such as quantum mechanics, fluid dynamics, elasticity, chemical reactor theory, and gas porous electrodes theory. The presence of a small parameter in these problems prevents one from obtaining satisfactory numerical solutions. It is a well-known fact that the solutions of SPPs have a multi-scale character, that is, there are thin layer(s) where the solution varies very rapidly, while away from the layer(s) the solution behaves regularly and varies slowly.

Also, among the second-order boundary value problems is the one-dimensional Bratu problem which has a long history. Bratu's own article appeared in 1914 [19]; generalizations are sometimes called the Liouville–Gelfand or Liouville–Gelfand–Bratu problem in honor of Gelfand [20] and the nineteenth century work of the great French mathematician Liouville. In recent years, it has been a popular testbed for numerical and perturbation methods [21–27].

Simplification of the solid fuel ignition model in thermal combustion theory yields an elliptic nonlinear partial differential equation, namely the Bratu problem. Also due to its use in a large variety of applications, many authors have contributed to the study of such problem. Some applications of Bratu problem are the model of thermal reaction process, the Chandrasekhar model of the expansion of the Universe, chemical reaction theory, nanotechnology and radiative heat transfer (see, [28–32]).

The Bratu problem is nonlinear (BVP) and extensively used as a benchmark problem to test the accuracy of many numerical methods. It is given by:

$$y''(x) + \lambda\, e^{y(x)} = 0, \quad y(0) = y(1) = 0, \quad 0 \leqslant x \leqslant 1, \tag{1}$$

where $\lambda > 0$. The Bratu problem has the following analytical solution:

$$y(x) = -2\ln\left[\frac{\cosh\left(\frac{\theta}{4}(2x-1)\right)}{\cosh\left(\frac{\theta}{4}\right)}\right], \tag{2}$$

where θ is the solution of the nonlinear equation $\theta = \sqrt{2\lambda}\cosh\theta$.

Our main objectives in the present paper are:

- Introducing a new operational matrix of derivatives based on using shifted Legendre polynomials and harmonic numbers.
- Using Petrov–Galerkin matrix method (PGMM) to solve linear second-order BVPs.
- Using collocation matrix method (CMM) to solve a class of nonlinear second-order BVPs, including singular, singularly perturbed and Bratu-type equations.

The outlines of the paper is as follows. In "Some properties and relations of Shifted Legendre polynomials and harmonic numbers", some relevant properties of shifted Legendre polynomials are given. Some properties and relations of harmonic numbers are also given in this section. In "A shifted Legendre matrix of derivatives", and with the aid of shifted Legendre polynomials polynomials, a new operational matrix of derivatives is given in terms of harmonic numbers. In "Solution of second-order linear two point BVPs", we use the introduced operational matrix for reducing a linear or a nonlinear second-order boundary value problems to a system of algebraic equations based on the application of Petrov–Galerkin and collocation methods, and also we state and prove a theorem for convergence. Some numerical examples are presented in "Numerical results and discussions" to show the efficiency and the applicability of the suggested algorithms. Some concluding remarks are given in "Concluding remarks".

Some properties and relations of shifted Legendre polynomials and harmonic numbers

Shifted Legendre polynomials

The shifted Legendre polynomials $L_k^*(x)$ are defined on $[a, b]$ as:

$$L_k^*(x) = L_k\left(\frac{2x-a-b}{b-a}\right), \qquad k = 0, 1, \ldots,$$

where $L_k(x)$ are the classical Legendre polynomials. They may be generated by using the recurrence relation

$$(k+1)\,L_{k+1}^*(x) - (2k+1)\left(\frac{2x-b-a}{b-a}\right)L_k^*(x) - k\,L_{k-1}^*(x), \quad k = 1, 2, \ldots, \tag{3}$$

with $L_0^*(x) = 1$, $L_1^*(x) = \dfrac{2x-b-a}{b-a}$.These polynomials are orthogonal on $[a, b]$, i.e.,

$$\int_a^b L_m^*(x)\, L_n^*(x)\, dx = \begin{cases} \dfrac{b-a}{2n+1}, & m=n, \\ 0, & m \neq n. \end{cases} \quad (4)$$

The polynomials $L_k^*(x)$ are eigenfunctions of the following singular Sturm–Liouville equation:

$$-D\big[(x-a)(x-b)\, D\, \phi_k(x)\big] + k(k+1)\, \phi_k(x) = 0,$$

where $D \equiv \frac{d}{dx}$.

Harmonic numbers

The nth harmonic number is the sum of the reciprocals of the first n natural numbers, i.e.,

$$H_n = \sum_{i=1}^{n} \frac{1}{i}. \quad (5)$$

The numbers H_n satisfy the recurrence relation

$$H_n - H_{n-1} = \frac{1}{n}, \quad n = 1, 2, \ldots,$$

and have the integral representation

$$H_n = \int_0^1 \frac{1-x^n}{1-x}\, dx.$$

The following Lemma is of fundamental importance in the sequel.

Lemma 1 The harmonic numbers satisfy the following three-term recurrence relation:

$$(2i-1)\, H_{i-1} - (i-1)\, H_{i-2} = i\, H_i, \qquad i \geq 2. \quad (6)$$

Proof The recurrence relation (6) can be easily proved with the aid of the relation (5). □.

A shifted Legendre matrix of derivatives

Consider the space (see, [33])

$$L_0^2[a,b] = \{\phi(x) \in L^2[a,b] : \phi(a) = \phi(b) = 0\},$$

and choose the following set of basis functions:

$$\phi_i(x) = (x-a)(b-x)\, L_i^*(x), \quad i = 0, 1, 2, \ldots. \quad (7)$$

It is not difficult to show that the set of polynomials $\{\phi_k(x) : k = 0, 1, 2, \ldots\}$ are linearly independent and orthogonal in the complete Hilbert space $L_0^2[a,b]$, with respect to the weight function $w(x) = \dfrac{1}{(x-a)^2\,(b-x)^2}$, i.e.,

$$\int_a^b \frac{\phi_i(x)\, \phi_j(x)\, dx}{(x-a)^2\,(b-x)^2} = \begin{cases} 0, & i \neq j, \\ \dfrac{b-a}{2i+1}, & i = j. \end{cases}$$

Any function $y(x) \in L_0^2[a,b]$ can be expanded as

$$y(x) = \sum_{i=0}^{\infty} c_i\, \phi_i(x), \quad (8)$$

where

$$c_i = \frac{2i+1}{b-a} \int_a^b \frac{y(x)\, \phi_i(x)}{(x-a)^2 (b-x)^2}\, dx = \Big(y(x), \phi_i(x)\Big)_{w(x)}.$$

If the series in Eq. (8) is approximated by the first $(N+1)$ terms, then

$$y(x) \simeq y_N(x) = \sum_{i=0}^{N} c_i\, \phi_i(x) = \boldsymbol{C}^T\, \boldsymbol{\Phi}(x), \quad (9)$$

where

$$\boldsymbol{C}^T = [c_0, c_1, \ldots, c_N], \quad \boldsymbol{\Phi}(x) = [\phi_0(x), \phi_1(x), \ldots, \phi_N(x)]^T. \quad (10)$$

Now, we state and prove the basic theorem, from which a new operational matrix can be intoduced.

Theorem 1 Let $\phi_i(x)$ be as chosen in (7). Then for all $i \geq 1$, one has

$$D\, \phi_i(x) = \frac{2}{b-a} \sum_{\substack{j=0 \\ (i+j)\ \text{odd}}}^{i-1} (2j+1)(1 + 2H_i - 2H_j)\, \phi_j(x) + \delta_i(x), \quad (11)$$

where $\delta_i(x)$ is given by

$$\delta_i(x) = \begin{cases} a+b-2x, & i \text{ even}, \\ a-b, & i \text{ odd}. \end{cases} \quad (12)$$

Proof We proceed by induction on i. For $i = 1$, it is clear that the left hand side of (11) is equal to its right-hand side, which is equal to: $a - b + \dfrac{6\,(x-a)(b-x)}{b-a}$. Assuming that relation (11) is valid for $(i-2)$ and $(i-1)$, we want to prove its validity for i. If we multiply both sides of (3) by $(x-a)(x-b)$ and make use of relation (7), we get

$$\phi_i(x) = \left(\frac{2\,i-1}{i}\right)\left(\frac{2\,x-b-a}{b-a}\right)\phi_{i-1}(x) - \left(\frac{i-1}{i}\right) \phi_{i-2}(x), \quad i = 2, 3, \ldots, \quad (13)$$

which immediately gives

$$D\phi_i(x) = \left(\frac{2\,i-1}{i(b-a)}\right)\Big[(2x-b-a)D\phi_{i-1}(x) + 2\,\phi_{i-1}(x)\Big] - \left(\frac{i-1}{i}\right) D\phi_{i-2}(x). \quad (14)$$

Now, application of the induction step on $D\phi_{i-1}(x)$ and $D\phi_{i-2}(x)$ in (14), yields

$$D\phi_i(x) = \frac{2(2i-1)(2x-b-a)}{i(b-a)^2} \sum_{\substack{j=0 \\ (i+j)\,\text{even}}}^{i-2} (2j+1)\left(1+2H_{i-1}-2H_j\right)\phi_j(x) - \frac{2(i-1)}{i(b-a)} \sum_{\substack{j=0 \\ (i+j)\,\text{odd}}}^{i-3} (2j+1)\left(1+2H_{i-2}-2H_j\right)\phi_j(x) + \frac{2(2i-1)}{i(b-a)}\phi_{i-1}(x) + \xi_i(x), \tag{15}$$

where

$$\xi_i(x) = \frac{(2i-1)(2x-b-a)}{i(b-a)}\,\delta_{i-1}(x) - \frac{i-1}{i}\,\delta_{i-2}(x) = \begin{cases} a+b-2x, & i\ \text{even}, \\ \dfrac{(2i-1)(2x-b-a)^2}{i(a-b)} - \dfrac{i-1}{i}(a-b), & i\ \text{odd}. \end{cases} \tag{16}$$

Substitution of the recurrence relation (13) in the form

$$\left(\frac{2x-a-b}{b-a}\right)\phi_j(x) = \frac{j+1}{2j+1}\left[\phi_{j+1}(x) + \frac{j}{j+1}\phi_{j-1}(x)\right],$$

into relation (15), and after performing some rather lenghthy algebraic manipulations, give

$$D\phi_i(x) = \sum_{\substack{j=1 \\ (i+j)\,\text{odd}}}^{i-3} m_{ij}\,\phi_j(x) + \frac{2(2i-1)}{(b-a)}\left[1+\frac{2(i-1)}{i}(H_{i-1}-H_{i-2})\right]\phi_{i-1}(x) - \frac{2c_i}{(b-a)}\left[\frac{2(i-1)}{i}H_{i-2} - \frac{2(2i-1)}{i}H_{i-1} + \frac{3i-2}{i}\right]\phi_0(x) + \xi_i(x), \tag{17}$$

where

$$m_{ij} = \frac{2(2j+1)}{(b-a)}\left[1 - \frac{2(2i-1)j}{i(2j+1)}H_{j-1} + \frac{2(i-1)}{i}H_j - \frac{2(2i-1)(j+1)}{i(2j+1)}H_{j+1}. \tag{18}$$
$$- \frac{2(i-1)}{i}H_{i-2} + \frac{2(2i-1)}{i}H_{i-1}\right],$$

$$c_i = \begin{cases} 1, & \text{i odd}, \\ 0, & \text{i even}. \end{cases}$$

Now, the elements m_{ij} in (18) can be written in the alternative form

$$m_{ij} = \frac{2(2j+1)}{(b-a)}\left[1 + \frac{2}{i}\{(2i-1)H_{i-1} - (i-1)H_{i-2}\} - \frac{2(2i-1)}{i(2j+1)}\{jH_{j-1} + (j+1)H_{j+1}\} + \frac{2(i-1)}{i}H_j\right], \tag{19}$$

which can be simplified with the aid of Lemma 1, to take the form

$$m_{ij} = \frac{2(2j+1)}{(b-a)}\left(1+2H_i-2H_j\right).$$

Repeated use of Lemma 1 in (17), and after performing some manipulation, leads to

$$D\phi_i(x) = \frac{2}{b-a}\left[\sum_{\substack{j=0 \\ (i+j)\,\text{odd}}}^{i-1} (2j+1)\left(1+2H_i-2H_j\right)\phi_j(x) - \frac{2(2i-1)}{i}c_i(x-a)(b-x)\right] + \xi_i(x), \tag{20}$$

and by noting that

$$\xi_i(x) - \frac{4(2i-1)}{i(b-a)}c_i(x-a)(b-x) = \delta_i(x),$$

then

$$D\phi_i(x) = \frac{2}{b-a} \sum_{\substack{j=0 \\ (i+j)\,\text{odd}}}^{i-1} (2j+1)\left(1+2H_i-2H_j\right)\phi_j(x) + \delta_i(x), \tag{21}$$

and this completes the proof of Theorem 1. □

Corollary 1 Let $x \in [-1,1] = [a,b]$, $\psi_i(x) = (1-x^2)L_i(x)$. Then for all $i \geqslant 1$, one has

$$D\psi_i(x) = \sum_{\substack{j=0 \\ (i+j)\,\text{odd}}}^{i-1} (2j+1)\left(1+2H_i-2H_j\right)\psi_j(x) + \gamma_i(x), \tag{22}$$

where

$$\gamma_i(x) = \begin{cases} -2x, & i\ \text{even}, \\ -2, & i\ \text{odd} \end{cases}$$

Now, and based on Theorem 1, it can be easily shown that the first derivative of the vector $\mathbf{\Phi}(x)$ defined in (10) can be expressed in the matrix form:

$$\frac{d\boldsymbol{\Phi}(x)}{dx} = M\,\boldsymbol{\Phi}(\boldsymbol{x}) + \boldsymbol{\delta}, \tag{23}$$

where

$$\boldsymbol{\delta} = (\delta_0(x), \delta_1(x), \ldots, \delta_N(x))^T,$$
$$\delta_i = \begin{cases} a+b-2x, & i \text{ even}, \\ a-b, & i \text{ odd}, \end{cases}$$

and $M = (m_{ij})_{0\leqslant i,j \leqslant N}$, is an $(N+1)\times(N+1)$ matrix whose nonzero elements can be given explicitly from relation (11) as:

$$m_{i,j} = \begin{cases} \dfrac{2}{b-a}(2j+1)(1+2H_i-2H_j), & i>j,\ (i+j) \text{ odd}, \\ 0, & \text{otherwise}. \end{cases}$$

For example, for $N=5$, we have

$$M = \frac{2}{b-a}\begin{pmatrix} 0 & 0 & 0 & 0 & 0 & 0 \\ 3 & 0 & 0 & 0 & 0 & 0 \\ 0 & 6 & 0 & 0 & 0 & 0 \\ \frac{14}{3} & 0 & \frac{25}{3} & 0 & 0 & 0 \\ 0 & \frac{19}{2} & 0 & \frac{21}{2} & 0 & 0 \\ \frac{167}{30} & 0 & \frac{77}{6} & 0 & \frac{63}{5} & 0 \end{pmatrix}.$$

Remark 1 The second derivative of the vector $\boldsymbol{\Phi}(x)$ is given by

$$\frac{d^2\boldsymbol{\Phi}(x)}{dx^2} = M^2\,\boldsymbol{\Phi}(\boldsymbol{x}) + M\,\boldsymbol{\delta} + \boldsymbol{\delta}', \tag{24}$$

where

$$\boldsymbol{\delta}' = (\delta'_0, \delta'_1, \ldots, \delta'_N)^T, \qquad \delta'_i = \begin{cases} -2, & i \text{ even}, \\ 0, & i \text{ odd}. \end{cases}$$

Solution of second-order linear two-point BVPs

In this section, both of linear and nonlinear second-order two-point BVPs are handled. For linear equations, a Petrov–Galerkin method is applied, while for nonlinear equations, the typical collocation method is applied.

Linear second-order BVPs subject to homogenous boundary conditions

Consider the linear second-order boundary value problem

$$y''(x) + f_1(x)\,y'(x) + f_2(x)\,y(x) = g(x), \quad x \in (a,b), \tag{25}$$

subject to the homogenous boundary conditions

$$y(a) = y(b) = 0. \tag{26}$$

If we approximate $y(x)$ as in (9), making use of relations (23) and (24), we have the following approximations for $y(x)$, $y'(x)$ and $y''(x)$:

$$y(x) \simeq \boldsymbol{C}^T\,\boldsymbol{\Phi}(x), \tag{27}$$

$$y'(x) \simeq \boldsymbol{C}^T M\,\boldsymbol{\Phi}(x) + \boldsymbol{C}^T\,\boldsymbol{\delta}, \tag{28}$$

$$y''(x) \simeq \boldsymbol{C}^T M^2\,\boldsymbol{\Phi}(x) + \boldsymbol{C}^T M\,\boldsymbol{\delta} + \boldsymbol{C}^T\,\boldsymbol{\delta}'. \tag{29}$$

If we substitute the relations (27), (28) and (29) into Eq. (25), then the residual $R(x)$, of this equation is given by:

$$\begin{aligned} R(x) =& \boldsymbol{C}^T M^2\boldsymbol{\Phi}(x) + \boldsymbol{C}^T M\,\boldsymbol{\delta} + \boldsymbol{C}^T\,\boldsymbol{\delta}' + f_1(x) \\ & \left(\boldsymbol{C}^T M\,\boldsymbol{\Phi}(x) + \boldsymbol{C}^T\,\boldsymbol{\delta}\right) \\ & + f_2(x)\left(\boldsymbol{C}^T\,\boldsymbol{\Phi}(x)\right) - g(x). \end{aligned} \tag{30}$$

The application of Petrov Galerkin method (see, [1]) yields the following $(N+1)$ linear equations in the unknown expansion coefficients, c_i, namely,

$$\int_a^b R(x)\,L_i^*(x)\,dx = 0, \quad i = 0, 1, \ldots, N. \tag{31}$$

Thus, Eq. (31) generates a set of $(N+1)$ linear equations which can be solved for the unknown components of the vector $\boldsymbol{C}$, and hence the approximate spectral solution $y_N(x)$ given in (9) can be obtained.

Linear second-order BVPs subject to nonhomogeneous boundary conditions

Consider the following one-dimensional second-order equation:

$$u''(x) + f_1(x)\,u'(x) + f_2(x)\,u(x) = g_1(x), \quad x \in (a,b), \tag{32}$$

subject to the nonhomogeneous boundary conditions:

$$u(a) = \alpha, \quad u(b) = \beta. \tag{33}$$

It is clear that the transformation

$$u(x) = y(x) + \frac{\alpha\,(b-x) + \beta\,(x-a)}{b-a},$$

turns the nonhomogeneous boundary conditions (33) into the homogeneous boundary conditions:

$$y(a) = y(b) = 0. \tag{34}$$

Hence it suffices to solve the following modified one-dimensional second-order equation:

$$y''(x) + f_1(x)\,y'(x) + f_2(x)\,y(x) = g(x), \quad x \in (a,b), \tag{35}$$

subject to the homogeneous boundary conditions (34), where

$$g(x) = g_1(x) - \frac{\beta-\alpha}{b-a} f_1(x) - \frac{\alpha(b-x)+\beta(x-a)}{b-a} f_2(x).$$

Solution of second-order nonlinear two-point BVPs

Consider the nonlinear differential equation

$$y''(x) = F(x, y(x), y'(x)), \tag{36}$$

subject to the homogenous conditions

$$y(a) = y(b) = 0.$$

If we follow the same procedure of "Linear second-order BVPs subject to homogenous boundary conditions", and approximate $y(x)$ as in (27), then after making use of the two relations (23) and (24), then we get the following nonlinear equation in the unknown vector $\mathbf{C}$

$$\begin{aligned} C^T\mathbf{M}^2\mathbf{\Phi}(\mathbf{x}) + \mathbf{C}^T\mathbf{M}\,\delta + \mathbf{C}^T\delta' = \mathbf{F}(\mathbf{x}, \mathbf{C}^T\mathbf{\Phi}(\mathbf{x}), \\ C^T M\,\mathbf{\Phi}(x) + C^T\delta). \end{aligned} \tag{37}$$

To find the numerical solution $y_N(x)$, we enforce (37) to be satisfied exactly at the first $(N+1)$ roots of the polynomial $L^*_{N+1}(x)$. Thus a set of $(N+1)$ nonlinear equations is generated in the expansion coefficients, c_i. With the aid of the well-known Newton's iterative method, this nonlinear system can be solved, and hence the approximate solution $y_N(x)$ can be obtained.

Remark 2 Following a similar procedure to that given in "Linear second-order BVPs subject to nonhomogeneous boundary conditions", the nonlinear second-order Eq. (36) subject to the nonhomogeneous boundary conditions given as in (33) can be tackled.

Convergence analysis

In this section, we state and prove a theorem for convergence of the proposed method.

Theorem 2 The series solutions of Eqs. (25) and (36) converge to the exact ones.

Proof Let

$$y(x) = \sum_{i=0}^{\infty} c_i\phi_i(x),$$

$$y_M(x) = \sum_{i=0}^{M} c_i\phi_i(x),$$

$$y_N(x) = \sum_{i=0}^{N} c_i\phi_i(x),$$

be the exact and approximate solutions (partial sums) to Eqs. (25) and (36) with $N \geqslant M$. Then we have

$$\begin{aligned} \left(y(x), y_N(x)\right)_{w(x)} &= \left(y(x), \sum_{i=0}^{N} c_i\,\phi_i(x)\right)_{w(x)} \\ &= \sum_{i=0}^{N} \bar{c}_i \left(y(x), \phi_i(x)\right)_{w(x)} \\ &= \sum_{i=0}^{N} \bar{c}_i\,c_i = \sum_{i=0}^{N} |c_i|^2. \end{aligned}$$

We show that $y_N(x)$ is a Cauchy sequence in the complete Hilbert space $L_0^2[a,b]$ and hence converges.

Now,

$$\left\|y_N(x) - y_M(x)\right\|^2_{w(x)} = \sum_{i=M+1}^{N} |c_i|^2.$$

From Bessel's inequality, we have $\sum_{i=0}^{\infty} |c_i|^2$ is convergent, which yields $\left\|y_N(x) - y_M(x)\right\|^2_{w(x)} \to 0$ as $M, N \to \infty$ and hence $y_N(x)$ converges to say $b(x)$. We prove that $b(x) = y(x)$,

$$\begin{aligned} \left(b(x) - y(x), \phi_i(x)\right)_{w(x)} &= \left(b(x), \phi_i(x)\right)_{w(x)} - \left(y(x), \phi_i(x)\right)_{w(x)} \\ &= \left(\lim_{N\to\infty} y_N, \phi_i(x)\right)_{w(x)} - c_i \\ &= \lim_{N\to\infty} \left(y_N, \phi_i(x)\right)_{w(x)} - c_i \\ &= 0. \end{aligned}$$

This proves $\sum_{i=0}^{\infty} c_i\phi_i(x)$ converges to $y(x)$. □

As the convergence has been proved, then consistency and stability can be easily deduced.

Numerical results and discussions

In this section, the presented algorithms in "Solution of second-order linear two point BVPs" are applied to solve regular, singular as well as singularly perturbed problems. As expected, the accuracy increases as the number of terms of the basis expansion increases.

Example 1 Consider the second-order nonlinear equation (see, [34]).

$$2y'' = (y + x + 1)^3, \quad 0<x<1, \quad y(0) = y(1) = 0. \tag{38}$$

The exact solution of (38) is

$$y(x) = \frac{2}{2-x} - x - 1.$$

Table 1 Maximum absolute error E for Example 1

N	6	10	14	18	22
E	1.371×10^{-6}	1.037×10^{-9}	5.715×10^{-13}	7.140×10^{-16}	5.312×10^{-17}

In Table 1, the maximum absolute error E is listed for various values of N, while in Table 2 a comparison between the numerical solution of problem (38) obtained by the application of CMM with the two numerical solutions obtained by using a sinc-collocation and a sinc-Galerkin methods in [34] is given.

Example 2 Consider the second-order nonlinear singular equation (see, [34]).

$$(4+x^s)(x^\sigma y')' = s x^{\sigma+s-2}(s x^s e^y - \sigma - s + 1), \quad 0<x<1,$$
$$y(0) = \ln\left(\frac{1}{4}\right), \quad y(1) = \ln\left(\frac{1}{5}\right), \qquad s = 3-\sigma, \quad \sigma \in (0,1),$$

with the exact solution

$$y(x) = -\ln(4+x^s).$$

In Table 3, the maximum absolute error E is listed for various values of σ and N, while in Table 4 a comparison between the solution of Example 2 obtained by our method (CMM) with the two numerical solutions obtained in [34] is given for the case $\sigma = \frac{1}{4}$. In addition, Fig. 1 illustrates the absolute error resulting from the application of CMM for the two cases corresponding to $N = 10, \sigma = \frac{1}{4}$ and $N = 15, \sigma = \frac{1}{4}$.

Example 3 Consider the following singularly perturbed linear second-order BVP (see, [35])

$$\epsilon y''(x) + y'(x) - y(x) = 0; \qquad 0<x<1,$$
$$y(0) = \frac{2\bar{\epsilon}\, e^{\frac{1+\bar{\epsilon}}{2\epsilon}}}{(\bar{\epsilon}+2\epsilon+1)e^{\frac{\bar{\epsilon}}{\epsilon}} + \bar{\epsilon} - 2\epsilon - 1}, \quad y(1) = 1,$$

where $\bar{\epsilon} = \sqrt{4\epsilon+1}$, with the exact solution

$$y(x) = e^{\frac{(\bar{\epsilon}+1)(1-x)}{2\epsilon}} \frac{(\bar{\epsilon}+2\epsilon+1)e^{\frac{x\bar{\epsilon}}{\epsilon}} + \bar{\epsilon} - 2\epsilon - 1}{(\bar{\epsilon}+2\epsilon+1)e^{\frac{\bar{\epsilon}}{\epsilon}} + \bar{\epsilon} - 2\epsilon - 1}.$$

In Table 5, the maximum absolute error E is listed for various values of ϵ and N, while in Table 6, we give a comparison between the solution of Example 3 obtained by our method (PGMM) with the solution obtained by the shooting method given in [35].

Example 4 Consider the following nonlinear second-order boundary value problem:

$$y''(x) - (y'(x))^2 + 16y(x) = 2 - 16x^6; \quad -1<x<1,$$
$$y(-1) = y(1) = 0, \tag{39}$$

with the exact solution $y(x) = x^2 - x^4$. Making use of (9) with $N = 2$ yields

$$y_N(x) = \boldsymbol{C}^T \boldsymbol{\Phi}(\boldsymbol{x}) = (1-x^2)\left(c_0 L_0(x) + c_1 L_1(x) + c_2 L_2(x)\right).$$

Moreover, in this case the two matrices M and M^2 take the forms

$$M = \begin{pmatrix} 0 & 0 & 0 \\ 3 & 0 & 0 \\ 0 & 6 & 0 \end{pmatrix}, \quad M^2 = \begin{pmatrix} 0 & 0 & 0 \\ 0 & 0 & 0 \\ 18 & 0 & 0 \end{pmatrix}.$$

Now, with the aid of Eq. (37), we have

$$c_0(2x^2-1) + c_1(6x^3-5x) + c_2(6x^4+5x^2-1) + \frac{1}{2}\left(2c_0 x - c_1 + 3c_1 x^2 + 6c_2 x^3 - 4c_2 x\right)^2 = 8x^6 - 1. \tag{40}$$

We enforce (40) to be satisfied exactly at the roots of $L_3(x)$, namely, $-\sqrt{\frac{3}{5}}, 0, \sqrt{\frac{3}{5}}$. This immediately yields three nonlinear algebraic equations in the three unknowns, c_0, c_1 and c_2. Solving these equations, we get

Table 2 Comparison between different solutions for Example 1

Method	CMM	Sinc-collocation [34]	Sinc-Galerkin [34]
N	22	130	130
E	5.312×10^{-17}	9.159×10^{-16}	9.992×10^{-16}

Table 3 Maximum absolute error E for Example 2

N	σ	E	σ	E	σ	E	σ	E
10		5.844×10^{-9}		3.234×10^{-7}		1.291×10^{-6}		4.698×10^{-7}
15	$\frac{1}{100}$	7.338×10^{-10}	$\frac{1}{4}$	4.927×10^{-8}	$\frac{1}{2}$	2.441×10^{-7}	$\frac{99}{100}$	1.416×10^{-7}
20		1.164×10^{-10}		4.435×10^{-9}		2.247×10^{-8}		1.229×10^{-8}

Table 4 Comparison between different solutions for Example 2, $\sigma = \frac{1}{4}$

Method	CMM	Sinc-collocation [34]	Sinc-Galerkin [34]
N	20	100	100
E	4.435×10^{-9}	1.552×10^{-6}	1.552×10^{-6}

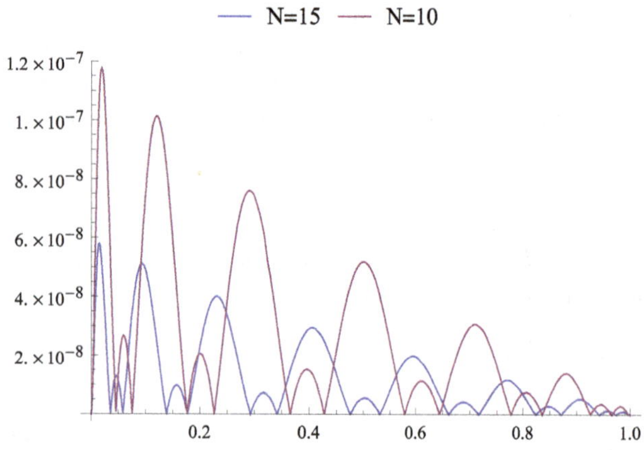

Fig. 1 The absolute error of Example 2 for $\sigma = \frac{1}{4}$

Table 5 Maximum absolute error E for Example 3

N	ϵ	E	ϵ	E
4		4.527×10^{-5}		2.942×10^{-5}
5		1.091×10^{-5}		7.974×10^{-7}
6	10^{-3}	9.202×10^{-6}	10^{-4}	9.444×10^{-7}
7		8.184×10^{-6}		8.323×10^{-7}
8		1.013×10^{-6}		7.815×10^{-8}

Table 6 Comparison between the best errors for Example 3

Method	PGMM	Shooting method [35]
Best error	7.815×10^{-8}	3.677×10^{-5}

$$c_0 = \frac{1}{3}, \quad c_1 = 0, \quad c_2 = \frac{2}{3},$$

and hence

$$y(x) = \left(\frac{1}{3},\ 0,\ \frac{2}{3}\right) \begin{pmatrix} 1 - x^2 \\ x - x^3 \\ -\frac{1}{2} + 2x^2 - \frac{3}{2}x^4 \end{pmatrix} = x^2 - x^4,$$

which is the exact solution.

Example 5 Consider the following Bratu Equation (see, [28–31]).

$$y''(x) + \lambda\, e^{y(x)} = 0, \qquad y(0) = y(1) = 0, \qquad 0 \leqslant x \leqslant 1. \tag{41}$$

Table 7 Maximum absolute error E for Example 5

N	λ	E	λ	E	λ	E
2		7.065×10^{-5}		9.420×10^{-4}		4.904×10^{-2}
4		2.274×10^{-7}		7.828×10^{-6}		1.833×10^{-3}
6		2.102×10^{-9}		1.437×10^{-7}		7.008×10^{-5}
8		1.715×10^{-11}		2.758×10^{-9}		3.761×10^{-6}
10	1	2.217×10^{-13}	2	8.082×10^{-11}	3.51	3.740×10^{-7}
12		1.984×10^{-15}		1.461×10^{-12}		2.355×10^{-8}
14		2.983×10^{-16}		5.342×10^{-14}		2.410×10^{-9}
16		3.053×10^{-16}		1.644×10^{-15}		2.975×10^{-10}
18		2.983×10^{-16}		4.024×10^{-16}		2.747×10^{-11}

With the analytical solution

$$y(x) = -2\ln\left[\frac{\cosh\left(\frac{\theta}{4}(2x-1)\right)}{\cosh\left(\frac{\theta}{4}\right)}\right], \tag{42}$$

where θ is the solution of the nonlinear equation $\theta = \sqrt{2\lambda}\cosh\theta$. The presented algorithm in Section 4.3 is applied to numerically solve Eq. (41), for the three cases corresponding to $\lambda = 1$, 2 and 3.51 which yield $\theta =$ 1.51716, 2.35755 and 4.66781, respectively. In Table 7, the maximum absolute error E is listed for various values of N, and in Table 8, we give a comparison between the best errors obtained by various methods used to solve Example 5. This table shows that our method is more accurate compared with the methods developed in [28–31]. In addition, Fig. 2 illustrates a comparison between different solutions obtained by our algorithm (CMM) in case of $\lambda = 1$ and $N = 1, 2, 3$.

Concluding remarks

In this paper, a novel matrix algorithm for obtaining numerical spectral solutions for second-order boundary value problems is presented and analyzed. The derivation of this algorithm is essentially based on choosing a set of basis functions satisfying the boundary conditions of the given boundary value problem in terms of shifted Legendre polynomials. The two spectral methods, namely, Petrov–Galerkin and collocation methods, are used for handling linear second-order and nonlinear second-order boundary value problems, respectively. One of the main advantages of the presented algorithms is their availability for application on both linear and nonlinear second-order boundary value problems including some important singular perturbed equations and also a Bratu-type equation. Another advantage of the developed algorithms is that high accurate approximate solutions are achieved by using

Table 8 Comparison between the best errors for Example 5 for $\lambda = 1$

Present error	Error in [28]	Error in [29]	Error in [30]	Error in [31]
2.98×10^{-16}	1.02×10^{-6}	8.89×10^{-6}	1.35×10^{-5}	3.01×10^{-3}

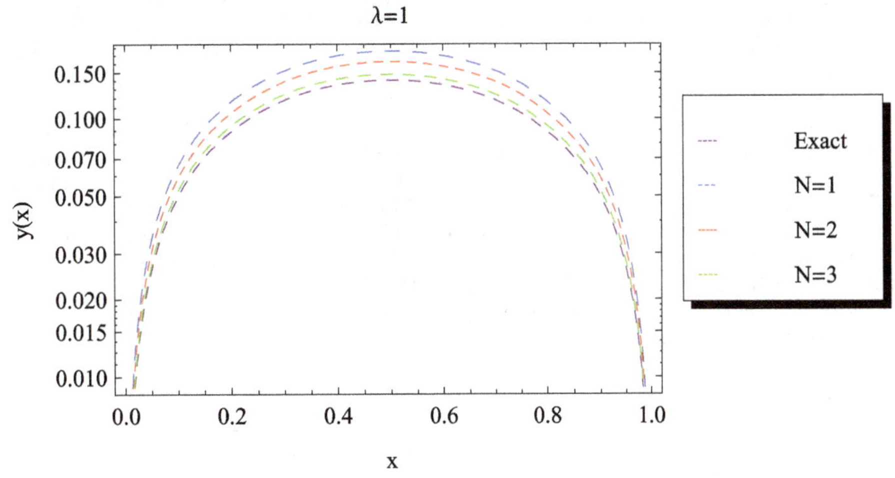

Fig. 2 Different solutions of Example 5

a few number of terms of the suggested expansion. The obtained numerical results are comparing favorably with the analytical ones. We do believe that the proposed algorithms in this article can be extended to treat other types of problems including some two-dimensional problems.

Acknowledgments The author would like to thank the referee for carefully reading the paper and for his useful comments which have improved the paper.

References

1. Canuto, C., Hussaini, M.Y., Quarteroni, A., Zang, T.A.: Spectral methods in fluid dynamics. Springer, New York (1988)
2. Fornberg, B.: A practical guide to pseudospectral methods. Cambridge University Press, Cambridge (1998)
3. Peyret, R.: Spectral methods for incompressible viscous flow. Springer, New York (2002)
4. Trefethen, L.N.: Spectral methods in MATLAB. SIAM, Philadelphia (2000)
5. Gottlieb, D., Hesthaven, J.S.: Spectral methods for hyperbolic problems. Comp. Appl. Math. **128**, 83–131 (2001)
6. Bhrawy, A.H.: An efficient Jacobi pseudospectral approximation for nonlinear complex generalized Zakharov system. Appl. Math. Comp. **247**, 30–46 (2014)
7. Guo B.-Y., Yan J.-P.: Legendre–Gauss collocation method for initial value problems of second order ordinary differential equations. Appl. Numer. Math. **59**, 1386–1408 (2009)
8. Bhrawy, A.H., Zaky, M.A.: A method based on the Jacobi tau approximation for solving multi-term timespace fractional partial differential equations. J. Comp. Phys. **281**, 876–895 (2015)
9. Doha, E.H., Abd-Elhameed, W.M.: Effcient spectral ultraspherical-dual-PetrovGalerkin algorithms for the direct solution of (2n+1)th-order linear differential equations. Math. Comput. Simulat. **79**, 3221–3242 (2009)
10. Doha, E.H., Abd-Elhameed, W.M., Youssri, Y.H.: Efficient spectral-Petrov–Galerkin methods for the integrated forms of third- and fifth-order elliptic differential equations using general parameters generalized Jacobi polynomials. Appl. Math. Comput. **218**, 7727–7740 (2012)
11. Doha, E.H., Abd-Elhameed, W.M., Youssri, Y.H.: Efficient spectral-Petrov–Galerkin methods for third- and fifth-order differential equations using general parameters generalized Jacobi polynomials. Quaest. Math. **36**, 15–38 (2013)
12. Doha, E.H., Abd-Elhameed, W.M., Bhrawy, A.H.: New spectral-Galerkin algorithms for direct solution of high even-order differential equations using symmetric generalized Jacobi polynomials. Collect. Math. **64**(3), 373–394 (2013)
13. Abd-Elhameed W. M.: Some algorithms for solving third-order boundary value problems using novel operational matrices of generalized Jacobi polynomials. Abs. Appl. Anal., vol 2015, Article ID 672703, p. 10
14. Yu, C.C., Heinrich, J.C.: Petrov–Galerkin methods for the time dependent convective transport equation. Int. J. Num. Mech. Eng. **23**, 883–901 (1986)
15. Doha, E.H., Abd-Elhameed, W.M.: Efficient spectral-Galerkin algorithms for direct solution of second-order equations using ultraspherical polynomials. SIAM J. Sci. Comput. **24**, 548–571 (2006)
16. Doha, E.H., Abd-Elhameed, W.M., Bassuony, M.A.: New algorithms for solving high even-order differential equations using third and fourth Chebyshev–Galerkin methods. J. Comput. Phys. **236**, 563–579 (2013)
17. Abd-Elhameed, W.M.: On solving linear and nonlinear sixth-order two point boundary value problems via an elegant harmonic numbers operational matrix of derivatives. CMES Comp. Model. Eng. Sci. **101**(3), 159–185 (2014)
18. Doha, E.H., Abd-Elhameed, W.M., Bhrawy, A.H.: Efficient spectral ultraspherical-Galerkin algorithms for the direct solution of 2nth-order linear differential equations. Appl. Math. Model. **33**, 1982–1996 (2009)
19. Bratu, G.: Sur les quations intgrales non linaires. Bull. Soc. Math. France **43**, 113–142 (1914)
20. Gelfand, I.M.: Some problems in the theory of quasi-linear equations. Trans. Amer. Math. Soc. Ser. **2**, 295–381 (1963)
21. Aregbesola, Y.A.S.: Numerical solution of Bratu problem using the method of weighted residual. Electron. J. South. Afr. Math. Sci. **3**, 1–7 (2003)

22. Hassan, I., Erturk, V.: Applying differential transformation method to the one-dimensional planar Bratu problem. Int. J. Contemp. Math. Sci. **2**, 1493–1504 (2007)
23. He, J.H.: Some asymptotic methods for strongly nonlinear equations. Int. J. Mod. Phys. B **20**, 1141–1199 (2006)
24. Li, S., Liao, S.: An analytic approach to solve multiple solutions of a strongly nonlinear problem. Appl. Math. Comput. **169**, 854–865 (2005)
25. Mounim, A., de Dormale, B.: From the fitting techniques to accurate schemes for the Liouville–Bratu–Gelfand problem. Numer. Meth. Part. Diff. Eq. **22**, 761–775 (2006)
26. Syam, M., Hamdan, A.: An efficient method for solving Bratu equations. Appl. Math. Comput. **176**, 704–713 (2006)
27. Wazwaz, A.M.: Adomian decomposition method for a reliable treatment of the Bratu-type equations. Appl. Math. Comput. **166**, 652–663 (2005)
28. Abbasbandy, S., Hashemi M.S., Liu C.-S.: The Lie-group shooting method for solving the Bratu equation. Appl. Math. Comput. **16**, 4238–4249 (2011)
29. Caglar, H., Caglar, N., Özer M., Valaristos A., Anagnostopoulos A.N.: B-spline method for solving Bratu's problem. Int. J. Comput. Math. **87**(8), 1885–1891 (2010)
30. Khuri, S.A.: A new approach to Bratu's problem. Appl. Math. Comput. **147**, 131–136 (2004)
31. Deeba, E., Khuri, S.A., Xie, S.: An algorithm for solving boundary value problems. J. Comput. Phys. **159**, 125–138 (2000)
32. Boyd, P.J.: One-point pseudospectral collocation for the one-dimensional Bratu equation. Appl. Math. Comput. **217**, 5553–5565 (2011)
33. Adams, R.A., Fournier, J.F.: Sobolev spaces, pure and applied mathematics series. Academic Press, New York (2003)
34. Mohsen, A., El-Gamel, M.: On the Galerkin and collocation methods for two-point boundary value problems using sinc bases. Comput. Math. Appl. **56**, 930–941 (2008)
35. Natesan, S., Ramanujam, N.: Shooting method for the solution of singularly perturbed two-point boundary-value problems having less severe boundary layer. Appl. Math. Comput. **133**, 623–641 (2002)

6

Fixed points of multivalued nonself almost contractions in metric-like spaces

Hassen Aydi[1] · Abdelbasset Felhi[2] · Slah Sahmim[2]

Abstract In this paper, we consider multivalued nonself weak contractions on convex metric-like spaces and we establish the existence of fixed point of such mappings. We provide some examples making effective our obtained result.

Keywords Hausdorff metric like · Almost contraction · Fixed point

Mathematics Subject Classification 47H10 · 54H25

Introduction and preliminaries

The study of fixed points for multivalued self mappings contractions using the Hausdorff metric was initiated by Nadler [16]. The fixed point theory for multivalued nonself mappings is developed by Assad and Kirk's [5]. They [5] proved the Banach's contraction principle for nonself multivalued mappings. For other results for multivalued nonself mappings, see [3, 9, 10, 14, 17–19]. On the other hand, Berinde [6, 7] introduced a new class of self mappings usually called weak contractions or almost contractions.

✉ Hassen Aydi
hassen.aydi@isima.rnu.tn; hmaydi@uod.edu.sa

Abdelbasset Felhi
afelhi@kfu.edu.sa

Slah Sahmim
ssahmim@kfu.edu.sa

1 Department of Mathematics, College of Education of Jubail, University of Dammam, P.O. 12020, Industrial Jubail 31961, Saudi Arabia

2 College of Sciences, Department of Mathematics, KFU University, Al-hasa, Saudi Arabia

Recently, Alghamdi et al. [3] introduced the notion of multivalued nonself almost contractions as follows.

Definition 1.1 Let (X, d) be a metric space and K a nonempty subset of X. A map $T:K \to CB(X)$ is called a multivalued almost contraction if there exist a constant $k \in (0,1)$ and some $L \geq 0$ such that

$$H(Tx,Ty) \leq kd(x,y) + Ld(y,Tx) \quad \text{for all } x,y \in K. \tag{1.1}$$

Alghamdi et al. [3] proved the following fixed point theorem for multivalued nonself almost contractions on convex metric spaces.

Theorem 1.2 [3] *Let (X, d) be a complete convex metric space and $T: X \to CB(X)$ a multivalued almost contraction with $k \in (0,1)$ and some $L \geq 0$. If $k(1+L) < 1$ and T satisfies Rothe's type condition, that is, $x \in \partial K \Rightarrow Tx \subset K$, then there exists $x \in K$ such that $x \in Tx$.*

In this paper, we extend the obtained results in [3] to the class of convex metric-like spaces. Mention that the concept of Hausdorff metric like was introduced in a very recent paper of Aydi et al. [4]. First, we need the following definitions and properties in the sequel.

Definition 1.3 Let X be a nonempty set. A function $\sigma : X \times X \to \mathbb{R}^+$ is said to be a metric like (dislocated metric) on X if for any $x, y, z \in X$, the following conditions hold:

(σ_1) $\sigma(x,y) = 0 \Longrightarrow x = y$;

(σ_2) $\sigma(x,y) = \sigma(y,x)$;

(σ_3) $\sigma(x,z) \leq \sigma(x,y) + \sigma(y,z)$.

The pair (X,σ) is then called a metric-like (dislocated metric) space.

Each metric-like σ on X generates a T_0 topology τ_σ on X which has as a base the family open σ-balls $\{B_\sigma(x,\varepsilon) :$

$x \in X, \varepsilon > 0\}$, where $B_\sigma(x,\varepsilon) = \{y \in X : |\sigma(x,y) - \sigma(x,x)| < \varepsilon\}$, for all $x \in X$ and $\varepsilon > 0$.

Observe that a sequence $\{x_n\}$ in a metric-like space (X,σ) converges to a point $x \in X$, with respect to τ_σ, if and only if $\sigma(x,x) = \lim_{n\to\infty} \sigma(x,x_n)$.

Definition 1.4 Let (X,σ) be a metric-like space.

(a) A sequence $\{x_n\}$ in X is said to be a Cauchy sequence if $\lim_{n,m\to\infty} \sigma(x_n,x_m)$ exists and is finite.

(b) (X,σ) is said to be complete if every Cauchy sequence $\{x_n\}$ in X converges with respect to τ_σ to a point $x \in X$ such that $\lim_{n\to\infty} \sigma(x,x_n) = \sigma(x,x) = \lim_{n,m\to\infty} \sigma(x_n,x_m)$.

Every metric space is a metric-like space, but the converse may not be true.

Example 1.5 Let $X = \mathbb{R}$ and $\sigma : X \times X \to \mathbb{R}^+$ defined by

$$\sigma(x,y) = |x| + |y| \quad \text{for all } x,y \in X.$$

Note that σ is a metric like, but not a metric since $\sigma(1,1) = 2 > 0$.

We need in the sequel the following trivial inequality

$$\sigma(x,x) \le 2\sigma(x,y) \quad \text{for all } x,y \in X. \tag{1.2}$$

For fixed point results for single-valued mappings in the setting of metric-like spaces, we may cite [1, 2, 11, 13, 15, 20, 21].

Very recently, Aydi et al. [4] introduced the concept of Hausdorff metric like. For instance, let $CB^\sigma(X)$ be the family of all nonempty, closed and bounded subsets of the metric-like space (X,σ), induced by the metric-like σ. Note that the boundedness is given as follows: A is a bounded subset in (X,σ) if there exist $x_0 \in X$ and $M \ge 0$ such that for all $a \in A$, we have $a \in B_\sigma(x_0,M)$, that is,

$$|\sigma(x_0,a) - \sigma(a,a)| < M.$$

The Closedness is taken in (X,τ_σ) (where τ_σ is the topology induced by σ). Let $\bar{A}$ be the closure of A with respect to the metric-like σ. Then, if $A \in CB^\sigma(X)$, then $\overline{A} = A$. For $A \subset X$ and $a \in X$, we also have

$$a \in \overline{A} \Longleftrightarrow B_\sigma(a,\varepsilon) \cap A \ne \emptyset \quad \text{for all } \varepsilon > 0$$

For $A,B \in CB^\sigma(X)$ and $x \in X$, define

$$\begin{aligned}\sigma(x,A) &= \inf\{\sigma(x,a) : a \in A\}, \ \delta_\sigma(\mathrm{A},\mathrm{B}) \\ &= \sup\{\sigma(a,B) : a \in A\}, \\ \delta_\sigma(B,A) &= \sup\{\sigma(b,A) : b \in B\} \text{and } \mathrm{H}_\sigma(\mathrm{A},\mathrm{B}) \\ &= \max\{\delta_\sigma(A,B), \delta_\sigma(B,A)\}.\end{aligned}$$

We find more details on the properties of H_σ in [4]. We also have the following useful lemmas.

Lemma 1.6 *Let $A,B \in CB^\sigma(X)$ and $a \in A$. Then, for all $\varepsilon > 0$, there exists a point $b \in B$ such that $\sigma(a,b) \le H_\sigma(A,B) + \varepsilon$.*

Lemma 1.7 [4] *Let (X,σ) be a metric-like space and A be any nonempty set in (X,σ), then*

$$\text{if } \sigma(a,A) = 0, \quad \text{then } a \in \bar{A}. \tag{1.3}$$

We give the following definition concerning the concept of convexity on metric-like spaces. One may find its analog for the metric case in [5].

Definition 1.8 A metric-like space (X,σ) is convex if for each $x,y \in X$ with $x \ne y$ there exists $z \in X$, $x \ne z \ne y$, such that

$$\sigma(x,y) = \sigma(x,z) + \sigma(z,y). \tag{1.4}$$

We also need the following concepts.

Definition 1.9 Let (X,σ) be a metric-like space and A be a set in X. We have

$$x \in \mathring{A} \Longleftrightarrow \exists\varepsilon > 0, \ B(x,\varepsilon) \subset A,$$

where $B(x,\varepsilon) = \{y \in X, \ |\sigma(x,y) - \sigma(x,x)| < \varepsilon\}$. We define the boundary of A in (X,σ) as
$\partial A = \bar{A} \backslash \mathring{A}$.

The purpose of this paper is to prove a fixed point theorem for multivalued nonself almost contractions on convex metric-like spaces. We derive many interesting corollaries on existing known results in the literature. Some examples are also presented illustrating our obtained result.

Fixed point of multivalued almost contraction

Now, we state and prove our main result.

Theorem 2.1 *Let (X,σ) be a complete metric-like space and K a nonempty closed subset of X such that if $x \in K$ and $y \notin K$, then there exists a point $z \in \partial K$ (the boundary of K) such that*

$$\sigma(x,y) = \sigma(x,z) + \sigma(z,y). \tag{2.1}$$

Suppose that $T : K \to CB^\sigma(X)$ is a multivalued almost contraction, that is,

$$H_\sigma(Tx,Ty) \le k\sigma(x,y) + L\sigma(y,Tx), \quad \text{for all } x,y \in K \tag{2.2}$$

with $k \in (0,1)$ and some $L \ge 0$ such that $(1+L)(k+2L) < 1$. If T satisfies Rothe's type condition, that is, $x \in$

$\partial K \Rightarrow Tx \subset K$, then there exists $x^\star \in K$ such that $x^\star \in Tx^\star$, that is, T has a fixed point in K.

Proof We construct a sequence $\{x_n\} \subset K$ in the following way:

Let $x_0 \in K$ and $y_1 \in Tx_0$. If $y_1 \in K$, let $x_1 = y_1$. If $y_1 \notin K$, by (2.1) there exists $x_1 \in \partial K$ such that

$$\sigma(x_0, x_1) + \sigma(x_1, y_1) = \sigma(x_0, y_1). \tag{2.3}$$

We have $x_1 \in \partial K$ and so by Definition 1.9, $x_1 \in K$. Thus, by Lemma 1.6, there exists $y_2 \in Tx_1$, such that

$$\sigma(y_1, y_2) \le H_\sigma(Tx_0, Tx_1) + k. \tag{2.4}$$

If $y_2 \in K$, let $x_2 = y_2$. If $y_2 \notin K$, by (2.1) there exists $x_2 \in \partial K$ such that

$$\sigma(x_1, x_2) + \sigma(x_2, y_2) = \sigma(x_1, y_2). \tag{2.5}$$

Therefore, $x_2 \in K$. From Lemma 1.6, there exists $y_3 \in Tx_2$, such that

$$\sigma(y_2, y_3) \le H_\sigma(Tx_1, Tx_2) + k^2. \tag{2.6}$$

Continuing in this fashion, we construct two sequences $\{x_n\}$ and $\{y_n\}$ such that

(i) $y_{n+1} \in Tx_n$;
(ii) $\sigma(y_n, y_{n+1}) \le H_\sigma(Tx_{n-1}, Tx_n) + k^n$, where
(iii) $x_n = y_n$ if $y_n \in K$;
(iv) $x_n \ne y_n$ if $y_n \notin K$ and then $x_n \in \partial K$ such that

$$\sigma(x_{n-1}, x_n) + \sigma(x_n, y_n) = \sigma(x_{n-1}, y_n). \tag{2.7}$$

Mention that in the case (iv), $x_n \in \partial K$ and by Rothe's type condition, $y_{n+1} \in Tx_n \in K$. Let

$$P_1 = \{x_i \in \{x_n\} : x_i = y_i, i = 1, 2, \ldots\} \quad \text{and}$$
$$P_2 = \{x_i \in \{x_n\} : x_i \ne y_i, i = 1, 2, \ldots\}.$$

Note that, if $x_n \in P_2$ for some n, then $x_{n+1}, x_{n-1} \in P_1$. Now, for $n \ge 2$, three cases should be considered.

Case 1 $x_n, x_{n+1} \in P_1$. Then, $y_n = x_n$ and $y_{n+1} = x_{n+1}$. Thus, using (1.2)

$$\begin{aligned}\sigma(x_n, x_{n+1}) &= \sigma(y_n, y_{n+1})\\ &\le H_\sigma(Tx_{n-1}, Tx_n) + k^n\\ &\le k\sigma(x_{n-1}, x_n) + L\sigma(x_n, Tx_{n-1}) + k^n\\ &\le k\sigma(x_{n-1}, x_n) + L\sigma(x_n, x_n) + k^n\\ &\le (k + 2L)\sigma(x_{n-1}, x_n) + k^n.\end{aligned}$$

Case 2 $x_n \in P_1$ and $x_{n+1} \in P_2$. Then, $y_n = x_n$ and $y_{n+1} \ne x_{n+1}$. In this case, we have by (iv),

$$\begin{aligned}\sigma(x_n, x_{n+1}) &\le \sigma(x_n, x_{n+1}) + \sigma(x_{n+1}, y_{n+1})\\ &= \sigma(x_n, y_{n+1}) = \sigma(y_n, y_{n+1})\\ &\le H_\sigma(Tx_{n-1}, Tx_n) + k^n\\ &\le k\sigma(x_{n-1}, x_n) + L\sigma(x_n, Tx_{n-1}) + k^n\\ &\le (k + 2L)\sigma(x_{n-1}, x_n) + k^n.\end{aligned}$$

Case 3 $x_n \in P_2$ and $x_{n+1} \in P_1$. Then, $x_n \ne y_n$, $x_{n-1} = y_{n-1}$, $x_{n+1} = y_{n+1}$ and $y_n \in Tx_{n-1}$. We have

$$\begin{aligned}\sigma(x_n, x_{n+1}) &\le \sigma(x_n, y_n) + \sigma(y_n, x_{n+1})\\ &= \sigma(x_n, y_n) + \sigma(y_n, y_{n+1})\\ &\le \sigma(x_n, y_n) + k\sigma(x_{n-1}, x_n) + L\sigma(x_n, Tx_{n-1}) + k^n.\end{aligned}$$

Since $k < 1$ and $y_n \in Tx_{n-1}$, then

$$\begin{aligned}\sigma(x_n,x_{n+1}) &< \sigma(x_n,y_n)+\sigma(x_{n-1},x_n)+L\sigma(x_n,y_n)+k^n\\ &\le \sigma(x_{n-1},y_n)+L\sigma(x_n,y_n)+k^n\\ &= \sigma(x_{n-1},y_n)+L[\sigma(x_{n-1},y_n)-\sigma(x_{n-1},x_n)]+k^n\\ &\le (1+L)\sigma(y_{n-1},y_n)+k^n\\ &\le (1+L)\left[H_\sigma(Tx_{n-2},Tx_{n-1})+k^{n-1}\right]+k^n\\ &\le (1+L)[k\sigma(x_{n-2},x_{n-1})+L\sigma(x_{n-1},Tx_{n-2})]\\ &\quad +(1+L)k^{n-1}+k^n\\ &\le (1+L)k\sigma(x_{n-2},x_{n-1})+(1+L)L\sigma(x_{n-1},x_{n-1})\\ &\quad +(1+L)k^{n-1}+k^n\\ &\le (1+L)(k+2L)\sigma(x_{n-2},x_{n-1})+(1+L)k^{n-1}+k^n.\end{aligned}$$

Since $h := (1+L)(k+2L) < 1$, then

$$\sigma(x_n, x_{n+1}) \le h\sigma(x_{n-2}, x_{n-1}) + hk^{n-2} + k^n.$$

Mention that $0 < k \le h < 1$. Thus, due to above three cases, we deduce for $n \ge 2$

$$\sigma(x_n, x_{n+1}) \le \begin{cases} h\sigma(x_{n-1}, x_n) + h^n, & \text{or}\\ h\sigma(x_{n-2}, x_{n-1}) + h^{n-1} + h^n. \end{cases} \tag{2.8}$$

Let

$$\alpha = \max\{\sigma(x_0, x_1), \sigma(x_1, x_2)\}.$$

Following [5], by induction it follows that for $n \ge 1$,

$$\sigma(x_n, x_{n+1}) \le h^{\frac{n-1}{2}}\alpha + h^{\frac{n}{2}}n. \tag{2.9}$$

Now, for $n > m$, we have

$$\begin{aligned}\sigma(x_n, x_m) &\le \sum_{i=m}^{n-1} \sigma(x_i, x_{i+1}) \le \alpha \sum_{i=m}^{n-1} h^{\frac{i-1}{2}} + \sum_{i=m}^{n-1} ih^{\frac{i}{2}}\\ &\le \alpha \sum_{i=m}^{\infty} h^{\frac{i-1}{2}} + \sum_{i=m}^{\infty} ih^{\frac{i}{2}} \to 0 \text{ as } m \to \infty.\end{aligned}$$

Since $\sum_{n=1}^{\infty} h^n$ converges, so

$$\lim_{n,m\to\infty} \sigma(x_n, x_m) = 0. \quad (2.10)$$

Hence, $\{x_n\}$ is Cauchy in (K, σ). Since K is closed and (X, σ) is complete, then (K, σ) is complete. Thus, $\{x_n\}$ converges to a point $x^\star \in K$, that is,

$$\lim_{n\to\infty} \sigma(x_n, x^\star) = \sigma(x^\star, x^\star) = \lim_{n,m\to\infty} \sigma(x_n, x_m) = 0. \quad (2.11)$$

We will show that $x^\star$ is a fixed point of T.

Observe that, by construction of $\{x_n\}$, there exists a subsequence $\{x_{n(p)}\}$ of $\{x_n\}$ each of whose terms is in the set P_1, (i.e, $x_{n(p)} = y_{n(p)}, p = 1, 2, \ldots$). Thus, by (i), $x_{n(p)} = y_{n(p)} \in Tx_{n(p)-1}$.

We have, for all $p = 1, 2, \ldots$

$$\begin{aligned}\sigma(x^\star, Tx^\star) &\le \sigma(x^\star, x_{n(p)+1}) + \sigma(x_{n(p)+1}, Tx^\star)\\ &\le \sigma(x^\star, x_{n(p)+1}) + H_\sigma(Tx_{n(p)}, Tx^\star)\\ &\le \sigma(x^\star, x_{n(p)+1}) + k\sigma(x_{n(p)}, x^\star) + L\sigma(x^\star, Tx_{n(p)}).\end{aligned}$$

Since,

$$\lim_{p\to\infty} \sigma(x^\star, x_{n(p)+1}) = \lim_{p\to\infty} \sigma(x_{n(p)}, x^\star) = \lim_{p\to\infty} \sigma(x^\star, Tx_{n(p)}) = 0,$$

then

$$\sigma(x^\star, Tx^\star) = 0.$$

Hence, by Lemma 1.7, $x^\star \in \overline{Tx^\star} = Tx^\star$. Then, $x^\star$ is a fixed point of T. □

We state the following simple corollaries as consequences of Theorem 2.1.

Corollary 2.2 *Let (X, σ) be a complete metric-like space and K a nonempty closed subset of X such that: if $x \in K$ and $y \notin K$, then there exists a point $z \in \partial K$, (the boundary of K) such that*

$$\sigma(x, y) = \sigma(x, z) + \sigma(z, y). \quad (2.12)$$

Suppose that $T : K \to CB^\sigma(X)$ is a multivalued contraction, that is,

$$H_\sigma(Tx, Ty) \le k\sigma(x, y), \quad \text{for all} \quad x, y \in X \quad (2.13)$$

with $k \in (0, 1)$. If T satisfies Rothe's type condition, that is, $x \in \partial K \Rightarrow Tx \subset K$, then there exists $x^\star \in K$ such that $x^\star \in Tx^\star$, that is, Thas a fixed point in K.

Proof It suffices to take $L = 0$ in Theorem 2.1. □

The metric case of Corollary 2.2 is

Corollary 2.3 [5] *Let (X, σ) be a complete convex metric space and K a nonempty closed subset of X. Suppose that $T : K \to CB^\sigma(X)$ is a multivalued contraction, that is,*

$$H_\sigma(Tx, Ty) \le k\sigma(x, y), \quad \text{for all} \quad x, y \in X \quad (2.14)$$

with $k \in (0, 1)$. If T satisfies Rothe's type condition, that is, $x \in \partial K \Rightarrow Tx \subset K$, then there exists $x^\star \in K$ such that $x^\star \in Tx^\star$, that is, T has a fixed point in K.

We give the following illustrated examples.

Example 2.4 Let $X = [0, \infty)$, $K = [0, 1]$ and $k = \frac{1}{6}$. Consider $T : K \to CB^\sigma(X)$ given by $Tx = \{0, \frac{x+5}{6}\}$, for all $x \in K$. Take $\sigma(x, y) = |x - y|$.

Mention that Rothe's type condition is easily verified. We show that T is a multivalued almost contraction. In fact, we have for all $x, y \in [0, 1]$

$$\begin{aligned}H_\sigma(Tx, Ty) &= \max\{\delta_\sigma(Tx, Ty), \delta_\sigma(Ty, Tx)\}\\ &= \max\left\{\max\left\{\sigma(0, Ty), \sigma\left(\frac{x+5}{6}, Ty\right)\right\},\right.\\ &\qquad \left.\max\left\{\sigma(0, Tx), \sigma\left(\frac{y+5}{6}, Tx\right)\right\}\right\}.\end{aligned}$$

Recall that

$$\begin{aligned}\sigma(0, Ty) &= \min\{\sigma(0, b) : b \in Ty\}\\ &= \min\left\{\sigma(0, 0), \sigma\left(0, \frac{y+5}{6}\right)\right\} = 0,\end{aligned}$$

and

$$\begin{aligned}\sigma\left(\frac{x+5}{6}, Ty\right) &= \min\left\{\sigma\left(\frac{x+5}{6}, b\right); b \in Ty\right\}\\ &= \min\left\{\sigma\left(\frac{x+5}{6}, 0\right), \sigma\left(\frac{x+5}{6}, \frac{y+5}{6}\right)\right\}\\ &= \min\left\{\frac{x+5}{6}, \frac{|x-y|}{6}\right\} = \frac{|x-y|}{6}.\end{aligned}$$

Therefore, $\delta_\sigma(Tx, Ty) = \frac{|x-y|}{6}$. Similarly, we find $\delta_\sigma(Ty, Tx) = \frac{|x-y|}{6}$. Then, for all $x, y \in K$,

$$H_\sigma(Tx, Ty) = \frac{|x-y|}{6}.$$

Thus, $H_\sigma(Tx, Ty) \le k\sigma(x, y) + L\sigma(y, Tx)$ for all $x, y \in K$ for all $L \ge 0$. Now, consider the case where L is chosen such that $0 < L \le \frac{3}{10}$.

Note that (2.1) is verified for $z = 1$. Moreover, the additional condition $(1 + L)(k + 2L) < 1$ is also satisfied.

Hence, T is a multivalued almost contraction that satisfies all assumptions of Theorem 2.1, and T has two fixed points; that is, $\mathrm{Fix}(T) = \{0, 1\}$.

Example 2.5 Let $X = \mathbb{R}$, $K = [0, 1]$ and $k = \frac{2}{3}$. Define $T : K \to CB^\sigma(X)$ by $Tx = \{0, \frac{x^2+2}{3}\}$, for all $x \in K$ and $\sigma(x, y) = |x - y|$. Recall that Rothe's type condition is verified. We show that T is a multivalued almost contraction. In fact, we have for all $x, y \in [0, 1]$

$$\begin{aligned}H_\sigma(Tx, Ty) &= \max\{\delta_\sigma(Tx, Ty), \delta_\sigma(Ty, Tx)\}\\ &= \max\left\{\max\left\{\sigma(0, Ty), \sigma\left(\frac{x^2+2}{3}, Ty\right)\right\},\right.\\ &\qquad \left.\max\left\{\sigma(0, Tx), \sigma\left(\frac{y^2+2}{3}, Tx\right)\right\}\right\}.\end{aligned}$$

It is easy to show that $\sigma(0, Ty) = 0$. We have

$$\begin{aligned}\sigma\left(\frac{x^2+2}{3}, Ty\right) &= \min\left\{\sigma\left(\frac{x^2+2}{3}, b\right) : b \in Ty\right\}\\ &= \min\left\{\sigma\left(\frac{x^2+2}{3}, 0\right), \sigma\left(\frac{x^2+2}{3}, \frac{y^2+2}{3}\right)\right\}\\ &= \min\left\{\frac{x^2+2}{3}, \frac{|x^2-y^2|}{3}\right\} = \frac{|x^2-y^2|}{3}.\end{aligned}$$

Therefore, $\delta_\sigma(Tx, Ty) = \frac{|x^2-y^2|}{3}$. Similarly, we find $\delta_\sigma(Ty, Tx) = \frac{|x^2-y^2|}{3}$. Then, for all $x, y \in K$,

$$H_\sigma(Tx, Ty) = \frac{|x^2-y^2|}{3}.$$

Thus,

$$\begin{aligned}H_\sigma(Tx, Ty) \leq \frac{(x+y)|x-y|}{3} &\leq \frac{2}{3}|x-y| \leq k\sigma(x, y)\\ &+ L\sigma(y, Tx)\end{aligned}$$

for all $x, y \in K$ and for all $L \geq 0$. Now, consider the case where L is chosen such that $0 \leq L \leq \frac{1}{4+\sqrt{22}}$.

Note that (2.1) is verified for $z = 0$ if $y \leq 0$ and for $z = 1$ if $y \geq 1$. Moreover, the additional condition $(1 + L)(k + 2L) < 1$ is also satisfied.

Hence, T is a multivalued almost contraction that satisfies all assumptions of Theorem 2.1, and T has two fixed points; that is, $\mathrm{Fix}(T) = \{0, 1\}$.

Example 2.6 Let $X = \mathbb{R}$, $K = [0, 1]$, $k = \frac{1}{4}$. Define $T : K \to CB^\sigma(X)$ by $Tx = \{0, \frac{1}{2+x}\}$, for all $x \in K$ and $\sigma(x, y) = |x - y|$. Recall that Rothe's type condition is verified. We show that T is a multivalued almost contraction. In fact, we have for all $x, y \in [0, 1]$

$$\begin{aligned}H_\sigma(Tx, Ty) &= \max\{\delta_\sigma(Tx, Ty), \delta_\sigma(Ty, Tx)\}\\ &= \max\left\{\max\left\{\sigma(0, Ty), \sigma\left(\frac{1}{2+x}, Ty\right)\right\},\right.\\ &\qquad \left.\max\left\{\sigma(0, Tx), \sigma\left(\frac{1}{2+y}, Tx\right)\right\}\right\}.\end{aligned}$$

It is easy to show that $\sigma(0, Ty) = 0$. We have also,

$$\begin{aligned}\sigma\left(\frac{1}{2+x}, Ty\right) &= \min\left\{\sigma\left(\frac{1}{2+x}, b\right) : b \in Ty\right\}\\ &= \min\left\{\sigma\left(\frac{1}{2+x}, 0\right), \sigma\left(\frac{1}{2+x}, \frac{1}{2+y}\right)\right\}\\ &= \min\left\{\frac{1}{2+x}, \frac{|x-y|}{(2+x)(2+y)}\right\}\\ &= \frac{|x-y|}{(2+x)(2+y)}.\end{aligned}$$

Therefore, $\delta_\sigma(Tx, Ty) = \frac{|x-y|}{(2+x)(2+y)}$. Similarly, we find $\delta_\sigma(Ty, Tx) = \frac{|x-y|}{(2+x)(2+y)}$. Then, for all $x, y \in K$,

$$H_\sigma(Tx, Ty) = \frac{|x-y|}{(2+x)(2+y)}.$$

Thus,

$$H_\sigma(Tx, Ty) \leq \frac{1}{4}|x-y| \leq k\sigma(x, y) + L\sigma(y, Tx)$$

for all $x, y \in K$ and for all $L \geq 0$. Now, consider the case where L is chosen such that $0 \leq L \leq \frac{1}{4}$.

Note that (2.1) is verified for $z = 0$ if $y \leq 0$ and for $z = 1$ if $y \geq 1$. Moreover, the additional condition $(1 + L)(k + 2L) < 1$ is also satisfied.

Then, T is a multivalued almost contraction that satisfies all assumptions of Theorem 2.1, and T has two fixed points; that is, $\mathrm{Fix}(T) = \{0, -1 + \sqrt{2}\}$.

Example 2.7 Let $X = \{0, 1, 2\}$, $K = \{0, 1\}$ and $k \in [0, 1)$. Consider $\sigma : X \times X \to [0, \infty)$ defined by

$$\sigma(0,0) = \sigma(1,1) = 0,\ \sigma(2,2) = \frac{1}{4},\ \sigma(1,0) = \sigma(0,1) = \frac{1}{3},$$

$$\sigma(2,0) = \sigma(0,2) = \frac{3}{5} \text{ and } \sigma(1,2) = \sigma(2,1) = \frac{2}{5}.$$

Then, (X, σ) is a complete metric-like space. Define $T : K \to CB^\sigma(X)$ by

$$T0 = T1 = \{0, 1\}.$$

Note that Tx is bounded for all $x \in X$ in the metric-like space (X, σ) and the Rothe's type condition is verified.

About the closedness of Tx in (X,σ), mention that $Tx = \{0,1\} \subset \overline{\{0,1\}} = \overline{Tx}$ for all $x \in K$. While, if $2 \in \overline{\{0,1\}}$, so there exists $x \in \{0,1\}$, such that

$$\sigma(2,x) < \sigma(2,2) + \varepsilon = \frac{1}{4} + \varepsilon, \quad \forall\, \varepsilon > 0,$$

which is a contradiction due to $\sigma(2,0) = \frac{3}{5}$ and $\sigma(2,1) = \frac{2}{5}$. So, $2 \notin \overline{\{0,1\}}$. Similarly, it is clear that $0 \in \overline{\{0,1\}}$ and $1 \in \overline{\{0,1\}}$, then $\overline{\{0,1\}} \subset \{0,1\}$. We conclude that $Tx = \overline{Tx}$ for all $x \in X$, that is, Tx is closed in (X,σ). We also have

$$\begin{aligned} H_\sigma(Tx,Ty) &= H_\sigma(\{0,1\},\{0,1\}) \\ &= \max\{\sigma(0,\{0,1\}),\sigma(1,\{0,1\})\} \\ &= \sigma(1,\{0,1\}) \\ &= \min\{\sigma(0,1),\sigma(1,1)\} \\ &= 0 \le k\sigma(x,y) + L\sigma(y,Tx) \end{aligned}$$

for all $x, y \in K$ and for all $L \ge 0$. Now, consider the case where L is chosen such that $(1+L)(k+2L) < 1$. Note that (2.1) is verified for $z = 0$ if $x = 0$ and $y = 2$ and for $z = 1$ if $x = 1$ and $y = 2$. Therefore, T is a multivalued almost contraction that satisfies all assumptions of Theorem 2.1, and T has two fixed points; that is, $\mathrm{Fix}(T) = \{0,1\}$.

Acknowledgments The authors are grateful to the referees for their helpful comments leading to improvement of the presentation of the work.

References

1. Aage, C.T., Salunke, J.N.: Some results of fixed point theorem in dislocated quasi metric space. Bull. Marathadawa Math. Soc. **9**, 1–5 (2008)
2. Aage, C.T., Salunke, J.N.: The results of fixed points in dislocated and dislocated quasi metric space. Appl. Math. Sci. **2**, 2941–2948 (2008)
3. Alghamdi, M.A., Berinde, V., Shahzad, N.: Fixed points of non-self almost contractions. J. Appl. Math. **2013**(621614), 6 (2013)
4. Aydi, H., Felhi, A., Karapinar, E., Sahmim, S.: Hausdorff Metric-Like, generalized Nadler's fixed point theorem on metric-like spaces and application. (in press)
5. Assad, N.A., Kirk, W.A.: Fixed point theorems for set-valued mappings of contractive type. Pac. J. Math. **43**, 553–562 (1972)
6. Berinde, V.: On the approximation of fixed points of weak contractive mappings. Carpath. J. Math. **19**(1), 7–22 (2003)
7. Berinde, V.: Approximating fixed points of weak contractions using the Picard iteration. Nonlinear Anal. Forum. **9**(1), 43–53 (2004)
8. Berinde, M., Berinde, V.: On a general class of multi-valued weakly Picard mappings. J. Math. Anal. Appl. **326**(2), 772–782 (2007)
9. Ćirić, L.B.: A remark on Rhoades fixed point theorem for non-self mappings. Int. J. Math. Math. Sci. **16**(2), 397–400 (1993)
10. Assad, N.A.: A fixed point theorem for some non-selfmappings. Tamkang J. Math. **21**(4), 387–393 (1990)
11. Amini, A.: Harandi, Metric-like spaces, partial metric spaces and fixed points. Fixed Point Theory Appl. **2012**, 204 (2012)
12. Banach, S.: Sur les opérations dans les ensembles abstraits et leur application aux équations intégrales. Fund. Math. **3**, 133–181 (1922)
13. Karapınar, E., Salimi, P.: Dislocated metric space to metric spaces with some fixed point theorems. Fixed Point Theory Appl. **2013**, 222 (2013)
14. Khojasteh, F., Rakocevic, V.: Some new common fixed point results for generalized contractive multi-valued non-selfmappings. Appl. Math. Lett. **25**(3), 287–293 (2012)
15. Kohli, M., Shrivastava, R., Sharma, M.: Some results on fixed point theorems in dislocated quasi metric space. Int. J. Theoret. Appl. Sci. **2**, 27–28 (2010)
16. Nadler, S.B.: multivalued contraction mappings. Pac. J. Math. **30**, 475–488 (1969)
17. Phiangsungnoen, S., Kumam, P.: Ulam-Hyers stability results for fixed point problems via generalized multivalued almost contraction. Proceedings of the International MultiConference of Engineers and Computer Scientists 2014, Hong Kong (2014) 12–14 March, pp. 1222–1225
18. Phiangsungnoen, S., Kumam, P.: Fuzzy fixed point theorems for multivalued fuzzy contractions in b-metric spaces. J. Nonlinear Sci. Appl. **8**, 55–63 (2015)
19. Phiangsungnoen, S., Wairojjana, N., Kumam, P.: Fixed point theorem and stability for (α,ψ,ξ)-generalized contractive multivalued mappings. Trans. Eng. Technol. 127–139 (2015)
20. Zeyada, Z.M., Hassan, G.H., Ahmad, M.A.: A generalization of fixed point theorem due to Hitzler and Seda in dislocated quasi metric space. Arab. J. Sci. Eng. **31**, 111–114 (2005)
21. Zoto, K., Houxha, E., Isufati, A.: Some new results in dislocated and dislocated quasi metric space. Appl. Math. Sci. **6**, 3519–3526 (2012)

7

Some fixed point theorems for contractive maps in N-cone metric spaces

Jerolina Fernandez · Geeta Modi · Neeraj Malviya

Abstract In present paper, we prove unique fixed point theorems for contractive maps in N-cone metric spaces. Our results extend and generalize some well-known results of (Banach, Fund Math 3:133–181 1992; Chatterjee, Rend Acad Bulgare Sci 25:727–730 1972; Kannan, Bull Calcutta Math Soc 60:71–76 1968; Rezapour and Hamlbarani, J Math Anal Appl 345:719–724 2008) in the setting of N-cone metric spaces.

Keywords N-cone metric space · Fixed point · Contractive map

Mathematics Subject Classification 54H25 · 47H10

Introduction and preliminaries

The notion of cone metric space was introduced in [7]. In this paper, Huang and Zhang replace the real numbers by ordering Banach space and define cone metric space. They also gave an example of a function which is a contraction in the category of cone metric but not contraction if considered over metric spaces and hence by proving fixed point theorem in cone metric spaces ensured that this map must have a unique fixed point.

J. Fernandez (✉) · N. Malviya
Department of Mathematics, NRI Institute of Information Science & Technology, Bhopal 462021, MP, India
e-mail: jerolinafernandez@gmail.com

N. Malviya
e-mail: maths.neeraj@gmail.com

G. Modi
Department of Mathematics, Government M V M College, Bhopal 462001, MP, India

Subsequently, Rezapour and Hamlbarani [14] omitted the assumption of normality in cone metric space. After that a series of articles in cone metric space started to appear (see, [3, 10, 15, 16]).

Recently, Aage and Salunke [1] introduced a generalized D^*-metric space and Ismat Beg et al. [4] introduced G-cone metric space. Very recently, Malviya and Fisher [13] introduced the notion of N-cone metric space and proved fixed point theorems for asymptotically regular maps and sequence. This new notion generalized the notion of generalized G-cone metric space [4] and generalized D^*-metric space [1]. In [12], the authors defined expansive maps in N-cone metric spaces and proved various fixed point theorems.

In present paper, we prove the Banach contraction theorem [2] and fixed point theorems of Kannan [11], Chatterjee [5] and Rezapour et al. [14] in N-cone metric space. The examples and application in support of our results are also given.

Throughout this paper, let E be a real Banach space and P be a subset of E. P is called a cone, if and only if

(1) P is closed, nonempty and $P \neq 0$;
(2) $ax + by \in P$, for all $x, y \in P$ and non-negative real numbers a, b;
(3) $P \cap (-P) = \{0\}$.

For a given cone $P \subseteq E$, we can define a partial ordering $\leq$ with respect to P by $x \leq y$, if and only if $y - x \in P$, $x < y$ will stand for $x \leq y$ but $x \neq y$, while $x \ll y$ will stand for $y - x \in \text{int}P$, where int$P$ denotes the interior of P.

The cone P is called normal if there is a number $N > 0$ such that for all $x, y \in E$, $0 \leq x \leq y$ implies $\|x\| \leq N\|y\|$. The least positive number satisfying the above is called the normal constant of P [7].

The cone P is called regular if every increasing sequence which is bounded from above is convergent, that is, if $\{x_n\}_{n\geq 1}$ is a sequence such that $x_1 \leq x_2 \cdots \leq y$ for some $y \in E$, then there is $x \in E$ such that $\lim_{n\to\infty} \|x_n - x\| = 0$. Equivalently, the cone P is regular if and only if every decreasing sequence which is bounded from below is convergent.

Lemma 1.1 [14] *Every regular cone is normal.*

Definition 1.1 [13] Let X be a non-empty set. An N-cone metric on X is a function $N : X^3 \to E$, that satisfies the following conditions: for all $x, y, z, a \in X$

(1) $N(x, y, z) \geq 0$;
(2) $N(x, y, z) = 0$ if and only if $x = y = z$;
(3) $N(x, y, z) \leq N(x, x, a) + N(y, y, a) + N(z, z, a)$.

Then, the function N is called an N-cone metric and the pair (X, N) is called an N-cone metric space.

Remark 1.1 [13] It is easy to see that every generalized-D^*-metric space is an N-cone metric space but in general, the converse is not true, see the following example.

Example 1.1 [13] Let $E = R^3$, $P = \{(x, y, z) \in E, x, y, z \geq 0\}$, $X = R$, and $N : X \times X \times X \to E$ is defined by

$$N(x,y,z) = \Big(\alpha(\,|\,y+z-2x\,|+|\,y-z\,|\,),\beta(\,|\,y+z-2x\,| +|\,y-z\,|\,),\gamma(\,|\,y+z-2x\,|+|\,y-z\,|\,)\Big)$$

where α, β, γ are positive constants. Then, (X, N) is an N-cone metric space but not a generalized-D^*-metric space because N is not symmetric.

Proposition 1.1 [13] *If (X, N) is an N-cone metric space, then for all $x, y, z \in X$, we have $N(x, x, y) = N(y, y, x)$.*

Definition 1.2 [13] Let (X, N) be an N-cone metric space. Let $\{x_n\}$ be a sequence in X and $x \in X$. If for every $c \in E$ with $0 \ll c$ there is $\mathtt{N}$ such that for all $n > \mathtt{N}$, $N(x_n, x_n, x) \ll c$, then $\{x_n\}$ is said to be convergent, $\{x_n\}$ converges to x and x is the limit of $\{x_n\}$. We denote this by $x_n \to x$ as $(n \to \infty)$.

Lemma 1.2 [13] *Let (X, N) be an N-cone metric space and P be a normal cone with normal constant k. Let $\{x_n\}$ be a sequence in X. If $\{x_n\}$ converges to x and $\{x_n\}$ also converges to y then $x = y$. That is the limit of $\{x_n\}$, if exists, is unique.*

Definition 1.3 [13] Let (X, N) be an N-cone metric space and $\{x_n\}$ be a sequence in X. If for any $c \in E$ with $0 \ll c$ there is $\mathtt{N}$ such that for all $m, n > \mathtt{N}$, $N(x_n, x_n, x_m) \ll c$, then $\{x_n\}$ is called a Cauchy sequence in X.

Definition 1.4 [13] Let (X, N) be an N-cone metric space. If every Cauchy sequence in X is convergent in X, then X is called a complete N-cone metric space.

Lemma 1.3 [13] *Let (X, N) be an N-cone metric space and $\{x_n\}$ be a sequence in X. If $\{x_n\}$ converges to x, then $\{x_n\}$ is a Cauchy sequence.*

Definition 1.5 [13] Let (X, N) and (X', N') be N-cone metric spaces. Then, a function $f : X \to X'$ is said to be continuous at a point $x \in X$ if and only if it is sequentially continuous at x, that is, whenever $\{x_n\}$ is convergent to x we have $\{fx_n\}$ is convergent to $f(x)$.

Lemma 1.4 [13] *Let (X, N) be an N-cone metric space and P be a normal cone with normal constant k. Let $\{x_n\}$ and $\{y_n\}$ be two sequences in X and suppose that $x_n \to x$, $y_n \to y$ as $n \to \infty$. Then $N(x_n, x_n, y_n) \to N(x, x, y)$ as $n \to \infty$.*

Remark 1.2 [13] If $x_n \to x$ in an N-cone metric space X, then every subsequence of $\{x_n\}$ converges to x in X.

Proposition 1.2 [13] *Let (X, N) be an N-cone metric space and P be a cone in a real Banach space E. If $u \leq v$, $v \ll w$ then $u \ll w$.*

Lemma 1.5 [13] *Let (X, N) be an N-cone metric space, P be an N-cone in a real Banach space E and $k_1, k_2, k_3, k_4, k > 0$. If $x_n \to x$, $y_n \to y$, $z_n \to z$ and $p_n \to p$ in X and*

$$ka \leq k_1 N(x_n, x_n, x) + k_2 N(y_n, y_n, y) + k_3 N(z_n, z_n, z) + k_4 N(p_n, p_n, p),\ \text{then}\ a = 0.$$

The following lemmas are often used.

Lemma 1.6 [10] *Let P be a cone and $\{x_n\}$ be a sequence in E. If $c \in intP$ and $0 \leq x_n \to 0$ (as $n \to \infty$), then there exists N such that for all $n > N$, we have $x_n \ll c$.*

Lemma 1.7 [10] *Let $x, y, z \in E$, if $x \leq y$ and $y \ll z$, then $x \ll z$.*

Lemma 1.8 [9] *Let P be a cone and $0 \leq u \ll c$ for each $c \in intP$, then $u = 0$.*

Lemma 1.9 [6] *Let P be a cone. If $u \in P$ and $u \leq ku$ for some $0 \leq k < 1$, then $u = 0$.*

Lemma 1.10 [10] *Let P be a cone and $a \leq b + c$ for each $c \in intP$, then $a \leq b$.*

Topology of N-cone metric space Let (X, N) be an N-cone metric space, each N-cone metric N on X generates a topology τ_N on X whose base is the family of open N-balls defined as

$$B_N(x, c) = \{y \in X : N(y, y, x) \ll c\},$$

for $c \in E$ with $0 \ll c$ and for all $x \in X$.

Definition 1.6 Let (X,N) be an N-cone metric space. A map $f: X \to X$ is said to be a contractive mapping if there exists a constant $0 \leq k < 1$ such that

$$N(fx,fx,fy) \leq kN(x,x,y) \text{ for all } x,y \in X.$$

Example 1.2 Let $E = R^3$, $P = \{(x,y,z) \in E,\ x,y, z \geq 0\}$ and $X = R$ and $N: X \times X \times X \to E$ is defined by

$$N(x,y,z) = (\alpha(|x-z|+|y-z|), \beta(|x-z|+|y-z|), \gamma(|x-z|+|y-z|)),$$

where α, β, γ are positive constants. Then (X,N) is an N-cone metric space. Define a self-map f on X as follows $fx = \frac{x}{4}$ for all x. Clearly, f is a contractive map in X.

Main results

Theorem 2.1 *Let (X,N) be a complete N-cone metric space and the mapping $T: X \to X$ satisfies the contractive condition*

$$N(Tx,Tx,Ty) \leq kN(x,x,y) \qquad (2.1)$$

for all $x,y \in X$, where $k \in [0,1)$ is a constant. Then, T has a unique fixed point in X. For each $x \in X$, the sequence of iterates $\{T^n x\}_{n\geq 1}$ converges to the fixed point.

Proof For each $x_0 \in X$ and $n \geq 1$, set $x_1 = Tx_0$ and $x_{n+1} = T^{n+1}x_0$.

Then

$$\begin{aligned} N(x_n,x_n,x_{n+1}) &= N(Tx_{n-1},Tx_{n-1},Tx_n) \\ &\leq kN(x_{n-1},x_{n-1},x_n) \\ &\leq k^2N(x_{n-2},x_{n-2},x_{n-1}) \\ &\vdots \\ &\vdots \\ &\leq k^nN(x_0,x_0,x_1) \end{aligned}$$

So for $m > n$,

$$\begin{aligned} N(x_n,x_n,x_m) \leq\ & 2N(x_n,x_n,x_{n+1}) + 2N(x_{n+1},x_{n+1},x_{n+2}) \\ &+ \cdots + 2N(x_{m-2},x_{m-2},x_{m-1}) \\ &+ N(x_{m-1},x_{m-1},x_m) \\ \leq\ & 2N(x_n,x_n,x_{n+1}) + 2N(x_{n+1},x_{n+1},x_{n+2}) \\ &+ \cdots + 2N(x_{m-2},x_{m-2},x_{m-1}) \\ &+ 2N(x_{m-1},x_{m-1},x_m) \\ \leq\ & [2k^n + 2k^{n+1} + \cdots + 2k^{m+n-2} \\ &+ 2k^{m+n-1}]N(x_0,x_0,x_1) \\ =\ & 2k^n[1+k+k^2+\cdots+k^{m+n-1}]N(x_0,x_0,x_1) \\ <\ & \frac{2k^n}{1-k}N(x_0,x_0,x_1) \end{aligned}$$

Let $0 \ll c$ be given. Choose a natural number N_1 such that $\frac{2k^n}{1-k}N(x_0,x_0,x_1) \ll c$ for all $n \geq N_1$. Thus, $N(x_n,x_n,x_m) \ll c$ for all $n > m$.

Therefore, $\{x_n\}_{n\geq 1}$ is a cauchy sequence in (X,N). Since (X,N) is a complete N-cone metric space, there exists $x^* \in X$ such that $x_n \to x^*$. Choose a natural number N_2 such that $N(x^*,x^*,x_n) \ll \frac{c}{4k}$ and $N(x^*,x^*,x_{n+1}) \ll \frac{c}{2}$ for all $n \geq N_2$.

Hence, for all $n \geq N_2$, we have

$$\begin{aligned} N(Tx^*,Tx^*,x^*) &\leq N(Tx^*,Tx^*,Tx_n) + N(Tx^*,Tx^*,Tx_n) \\ &\quad + N(x^*,x^*,Tx_n) \\ &\leq 2N(Tx^*,Tx^*,Tx_n) + N(x^*,x^*,Tx_n) \\ &\leq 2kN(x^*,x^*,x_n) + N(x^*,x^*,x_{n+1}) \\ &\ll 2k\frac{c}{4k} + \frac{c}{2} \\ &= c \end{aligned}$$

for all $n \geq N_2$. Thus, $N(Tx^*,Tx^*,x^*) \ll \frac{c}{m}$ for all $m \geq 1$. So $\frac{c}{m} - N(Tx^*,Tx^*,x^*) \in P$ for all $m \geq 1$. Since $\frac{c}{m} \to 0$ (as $m \to \infty$) and P is closed, $-N(Tx^*,Tx^*,x^*) \in P$, but $N(Tx^*,Tx^*,x^*) \in P$. Therefore, $N(Tx^*,Tx^*,x^*) = 0$ and so $Tx^* = x^*$.

To prove uniqueness, let y^* be another fixed point of T, then

$$\begin{aligned} N(x^*,x^*,y^*) &= N(Tx^*,Tx^*,Ty^*) \\ &\leq kN(x^*,x^*,y^*), \end{aligned}$$

which implies that by Lemma (1.9) $N(x^*,x^*,y^*) = 0$. Hence the fixed point of T is unique. □

Corollary 2.1 *Let (X,N) be a complete N-cone metric space. Suppose the mapping $T: X \to X$ satisfies for some positive integer n,*

$$N(T^n x, T^n x, T^n y) \leq k.N(x,x,y),$$

for all $x,y \in X$, where $k \in [0,1)$ is a constant. Then, T has a unique fixed point in X.

Proof From Theorem (2.1), T^n has a unique fixed point x^*. But $T^n(Tx^*) = T(T^n x^*) = Tx^*$. So Tx^* is also a fixed point of T^n. Hence $Tx^* = x^*$, x^* is a fixed point of T. Since the fixed point of T is also fixed point of T^n, then fixed point of T is unique. □

Theorem 2.2 *Let (X,N) be a complete N-cone metric space. Suppose the mapping $T: X \to X$ satisfies the contractive condition*

$$N(Tx,Tx,Ty) \leq k[N(Tx,Tx,x) + N(Ty,Ty,y)],$$

for all $x,y \in X$, where $k \in [0,\frac{1}{2})$ is a constant. Then, T has a unique fixed point in X. For each $x \in X$, the iterative sequence $\{T^n x\}_{n\geq 1}$ converges to the fixed point.

Proof For each $x_0 \in X$ and $n \geq 1$, set $x_1 = Tx_0$ and $x_{n+1} = T^{n+1}x_0$.

Then, we have

$$\begin{aligned}N(x_n,x_n,x_{n+1}) &= N(Tx_{n-1},Tx_{n-1},Tx_n)\\ &\leq k[N(Tx_{n-1},Tx_{n-1},x_{n-1}) + N(Tx_n,Tx_n,x_n)]\\ &= k[N(x_n,x_n,x_{n-1}) + N(x_{n+1},x_{n+1},x_n)]\\ &= k[N(x_n,x_n,x_{n-1}) + N(x_n,x_n,x_{n+1})]\\ &\quad [\text{by Proposition 1.1.}]\end{aligned}$$

So

$$\begin{aligned}N(x_n,x_n,x_{n+1}) &\leq \frac{k}{1-k}N(x_n,x_n,x_{n-1})\\ &= hN(x_n,x_n,x_{n-1}) \quad \text{where } h = \frac{k}{1-k}\\ &= hN(x_{n-1},x_{n-1},x_n) \quad [\text{by Proposition 1.1.}]\\ &\vdots\\ &\leq h^n N(x_0,x_0,x_1), \qquad (2.2)\end{aligned}$$

Now using (2.2), we can prove $\{x_n\}_{n\geq 1}$ is a Cauchy sequence as proved in Theorem (2.1).

Since (X,N) is a complete N-cone metric space, there exists $x^* \in X$ such that $x_n \to x^*$. Choose a natural number N_2 such that $N(x_{n+1},x_{n+1},x_n) \ll \frac{(1-2k)c}{4k}$ and $N(x^*,x^*,x_{n+1}) \ll \frac{(1-2k)c}{2}$, for all $n \geq N_2$.

Hence, for $n \geq N_2$, we have

$$\begin{aligned}N(Tx^*,Tx^*,x^*) &\leq N(Tx^*,Tx^*,Tx_n) + N(Tx^*,Tx^*,Tx_n)\\ &\quad + N(x^*,x^*,Tx_n)\\ &= 2N(Tx^*,Tx^*,Tx_n) + N(x^*,x^*,Tx_n)\\ &\leq 2k[N(Tx^*,Tx^*,x^*) + N(Tx_n,Tx_n,x_n)]\\ &\quad + N(x^*,x^*,Tx_n).\end{aligned}$$

Thus,

$$\begin{aligned}N(Tx^*,Tx^*,x^*) &\leq \frac{1}{1-2k}[2kN(x_{n+1},x_{n+1},x_n) + N(x^*,x^*,x_{n+1})]\\ &\ll \frac{c}{2} + \frac{c}{2} = c.\end{aligned}$$

Thus, $N(Tx^*,Tx^*,x^*) \ll \frac{c}{m}$ for all $m \geq 1$.

So $\frac{c}{m} - N(Tx^*,Tx^*,x^*) \in P$ for all $m \geq 1$. Since $\frac{c}{m} \to 0$ as $m \to \infty$ and P is closed, $-N(Tx^*,Tx^*,x^*) \in P$. But $N(Tx^*,Tx^*,x^*) \in P$. Therefore, $N(Tx^*,Tx^*,x^*) = 0$ and so $Tx^* = x^*$.

Now, if y^* is another fixed point of T, then

$$\begin{aligned}N(x^*,x^*,y^*) &= N(Tx^*,Tx^*,Ty^*)\\ &\leq k[N(Tx^*,Tx^*,x^*) + N(Ty^*,Ty^*,y^*)]\\ &= k[N(x^*,x^*,x^*) + N(y^*,y^*,y^*)]\\ &= 0 \quad [\text{by Definition 1.1. and by Lemma 1.5}]\end{aligned}$$

Hence $x^* = y^*$. Therefore, the fixed point of T is unique. □

Theorem 2.3 *Let (X,N) be a complete N-cone metric space. Suppose the mapping $T: X \to X$ satisfies the contractive condition*

$$N(Tx,Tx,Ty) \leq k[N(Tx,Tx,y) + N(x,x,Ty)],$$

for all $x,y \in X$, where $k \in [0,\frac{1}{2})$ is a constant. Then, T has a unique fixed point in X. For each $x \in X$, the iterative sequence $\{T^n x\}_{n\geq 1}$ converges to the fixed point.

Proof For each $x_0 \in X$ and $n \geq 1$, set $x_1 = Tx_0$ and $x_{n+1} = T^{n+1}x_0$.

Then, we have

$$\begin{aligned}N(x_n,x_n,x_{n+1}) &= N(Tx_{n-1},Tx_{n-1},Tx_n)\\ &\leq k[N(Tx_{n-1},Tx_{n-1},x_n) + N(x_{n-1},x_{n-1},Tx_n)]\\ &= k[N(x_n,x_n,x_n) + N(x_{n-1},x_{n-1},x_{n+1})]\\ &= kN(x_{n-1},x_{n-1},x_{n+1})\\ &\leq k[2N(x_{n-1},x_{n-1},x_n) + N(x_n,x_n,x_{n+1})]\\ &\quad [\text{by Definition 1.1}]\\ &\leq \frac{2k}{1-k}N(x_{n-1},x_{n-1},x_n)\\ &= hN(x_{n-1},x_{n-1},x_n) \quad \text{where, } h = \frac{2k}{1-k}\\ &\leq h^n N(x_0,x_0,x_1).\end{aligned} \qquad (2.3)$$

Now using (2.3), we can prove $\{x_n\}_{n\geq 1}$ is a Cauchy sequence as proved in Theorem (2.1).

Since (X,N) is a complete N-cone metric space, there exists $x^* \in X$ such that $x_n \to x^*$.

Now, we have

$$\begin{aligned}N(Tx^*,Tx^*,x^*) &\leq N(Tx^*,Tx^*,Tx_n) + N(Tx^*,Tx^*,Tx_n)\\ &\quad + N(x^*,x^*,Tx_n)\\ &\leq 2N(Tx^*,Tx^*,Tx_n) + N(x^*,x^*,Tx_n)\\ &\leq 2k[N(Tx^*,Tx^*,x_{n+1}) + N(x^*,x^*,Tx_n)]\\ &\quad + N(x^*,x^*,Tx_n)\\ &= 2kN(Tx^*,Tx^*,x_n) + (2k+1)N(x^*,x^*,x_{n+1})\\ &= 2kN(Tx^*,Tx^*,x^*) + (2k+1)N(x^*,x^*,x^*)\\ &\quad \text{as } n \to \infty\\ &\leq 2kN(Tx^*,Tx^*,x^*), \quad \left[\text{As } 0 \leq k < \frac{1}{2}\right]\end{aligned}$$

which implies that, by Lemma 1.9, $N(Tx^*,Tx^*,x^*) = 0$. Hence $Tx^* = x^*$.

Now, if y^* is another fixed point of T, then

$$\begin{aligned}N(x^*,x^*,y^*) &= N(Tx^*,Tx^*,Ty^*)\\ &\leq k[N(Tx^*,Tx^*,y^*) + N(x^*,x^*,Ty^*)]\\ &= k[N(x^*,x^*,y^*) + N(x^*,x^*,y^*)]\\ &= 2kN(x^*,x^*,y^*) \quad [\text{by Lemma (1.9)}].\end{aligned}$$

Hence, $N(x^*, x^*, y^*) = 0$ and so $x^* = y^*$. Therefore, the fixed point of T is unique. □

Theorem 2.4 *Let (X, N) be a complete N-cone metric space. Suppose the mapping $T : X \to X$ satisfies the contractive condition*

$$N(Tx, Tx, Ty) \le kN(x, x, y) + lN(x, x, Ty),$$

for all $x, y \in X$, where $k, l \in [0, 1)$ is a constant. Then, T has a fixed point in X. Also the fixed point of T is unique whenever $k + 3l < 1$.

Proof For each $x_0 \in X$ and $n \ge 1$, set $x_1 = Tx_0$ and $x_{n+1} = T^{n+1}x_0$.

$$\begin{aligned} N(x_n, x_n, x_{n+1}) &= N(Tx_{n-1}, Tx_{n-1}, Tx_n) \\ &\le kN(x_{n-1}, x_{n-1}, x_n) + lN(x_{n-1}, x_{n-1}, Tx_n) \\ &= kN(x_{n-1}, x_{n-1}, x_n) + lN(x_{n-1}, x_{n-1}, x_{n+1}) \\ &\le kN(x_{n-1}, x_{n-1}, x_n) + l[N(x_{n-1}, x_{n-1}, x_n) \\ &\quad + N(x_{n-1}, x_{n-1}, x_n) \\ &\quad + N(x_{n+1}, x_{n+1}, x_n)] \quad \text{[byDefinition1.1]} \\ &\le kN(x_{n-1}, x_{n-1}, x_n) + l[2N(x_{n-1}, x_{n-1}, x_n) \\ &\quad + N(x_{n+1}, x_{n+1}, x_n)] \end{aligned}$$

So

$$\begin{aligned} N(x_n, x_n, x_{n+1}) &\le \frac{k + 2l}{1 - l} N(x_{n-1}, x_{n-1}, x_n) \\ &\quad \text{[by Proposition(1.1)]} \qquad (2.4) \\ &\le hN(x_{n-1}, x_{n-1}, x_n) \\ &\le h^n N(x_0, x_0, x_1), \end{aligned}$$

where $h = \frac{k+2l}{1-l}$.

Now using (2.4), we can prove $\{x_n\}_{n\ge1}$ is a Cauchy sequence as proved in Theorem (2.1).

Since (X, N) is a complete N-cone metric space, there exists $x^* \in X$ such that $x_n \to x^*$. Choose a natural number N_2 such that $N(x^*, x^*, x_n) \ll \frac{c}{4k}$ and $N(x^*, x^*, x_{n+1}) \ll \frac{c}{2(2l+1)}$, for all $n \ge N_2$.

Hence, for $n \ge N_2$, we have

$$\begin{aligned} N(Tx^*, Tx^*, x^*) &\le N(Tx^*, Tx^*, Tx_n) + N(Tx^*, Tx^*, Tx_n) \\ &\quad \text{[by Definition 1.1]} \\ &\quad + N(x^*, x^*, Tx_n) \\ &\le 2N(Tx^*, Tx^*, Tx_n) + N(x^*, x^*, Tx_n) \\ &= 2kN(x^*, x^*, x_n) + 2lN(x^*, x^*, Tx_n) \\ &\quad + N(x^*, x^*, Tx_n) \\ &= 2kN(x^*, x^*, x_n) + (2l + 1)N(x^*, x^*, x_{n+1}) \\ &\ll \frac{c}{2} + \frac{c}{2} = c \end{aligned}$$

Thus, $N(Tx^*, Tx^*, x^*) \ll \frac{c}{m}$ for all $m \ge 1$.

Hence, $\frac{c}{m} - N(Tx^*, Tx^*, x^*) \in P$ for all $m \ge 1$. Since $\frac{c}{m} \to 0$ as $m \to \infty$ and P is closed, $-N(Tx^*, Tx^*, x^*) \in P$, but $N(Tx^*, Tx^*, x^*) \in P$. Therefore, $N(Tx^*, Tx^*, x^*) = 0$ and so $Tx^* = x^*$.

Now, if y^* is another fixed point of T, then

$$\begin{aligned} N(x^*, x^*, y^*) &= N(Tx^*, Tx^*, Ty^*) \\ &\le kN(x^*, x^*, y^*) + lN(x^*, x^*, Ty^*) \\ &= (k + l)N(x^*, x^*, y^*) \\ &\quad \text{[by Lemma 1.9 and since } k + 3l < 1] \end{aligned}$$

Hence, $N(x^*, x^*, y^*) = 0$ and so $x^* = y^*$. Therefore, the fixed point of T is unique. □

Example 2.1 Let $E = R^3$, $P = \{(x, y, z) \in E,\ x, y, z \ge 0\}$ and $X = R$ and $N : X \times X \times X \to E$ is defined by

$$\begin{aligned} N(x, y, z) = (&\alpha(|x - z| + |y - z|), \beta(|x - z| \\ &+ |y - z|), \gamma(|x - z| + |y - z|)), \end{aligned}$$

where α, β, γ are positive constants. Then, (X, N) is an N-cone metric space. Define a self-map T on X as follows $Tx = \frac{x}{3}$ for all x. If we take $\alpha = \frac{1}{3}$, then the contractive condition (2.1) holds trivially good and 0 is the unique fixed point of the map T.

Application

In this section, we shall apply Theorem 2.1 to the following first-order periodic boundary value problem:

$$\frac{dx}{dt} = F(t, x(t)), \quad \text{with } x(0) = \xi, \qquad (3.5)$$

where $F : [-h, h] \times [\xi - \delta, \xi + \delta]$ is a continuous function.

Example 3.1 Consider the boundary value problem (3.5) with the continuous function F and suppose $F(x, y)$ satisfies the local Lipschitz condition, i.e. if $|x| \le h$, $y_1, y_2 \in [\xi - \delta, \xi + \delta]$ it induces

$$|F(x, y_1) - F(x, y_2)| \le L|y_1 - y_2|$$

Set $M = \max_{[-h,h]\times[\xi-\delta,\xi+\delta]} |F(x, y)|$ such that $2h < \min[\frac{\delta}{M}, \frac{1}{L}]$, then there exists a unique solution of (3.5).

Proof Let $X = E = C([-h, h])$ and $P = \{u \in E : u \ge 0\}$. Put $N : X \times X \times X \to E$ as

$$\begin{aligned} N(x, x, y) &= f(t) \max_{-h \le t \le h} (|x(t) - y(t)| + |x(t) - y(t)|) \\ &= f(t) \max_{-h \le t \le h} 2|x(t) - y(t)| \end{aligned}$$

with $f : [-h, h] \to R$ such that $f(t) = e^t$.

It is clear that (X, N) is a complete N-cone metric space.

Note that (3.5) is equivalent to the integral equation

$$x(t) = \xi + \int_0^t F(\tau, x(\tau))d\tau.$$

Define a mapping $T : C([-h,h]) \to R$ by

$$Tx(t) = \xi + \int_0^t F(\tau, x(\tau))d\tau.$$

If $x(t), y(t) \in B(\xi, \delta f) \triangleq \{\psi(t) \in C([-h,h]) : N(\xi, \xi, \psi) \leq \delta f\}$, then we have

$$\begin{aligned} N(Tx, Tx, Ty) &= f(t) \max_{-h \leq t \leq h} 2\left| \int_0^t F(\tau, x(\tau))d\tau - \int_0^t F(\tau, y(\tau))d\tau \right| \\ &= 2f(t) \max_{-h \leq t \leq h} \left| \int_0^t [F(\tau, x(\tau)) - F(\tau, y(\tau))]d\tau \right| \\ &\leq 2hf(t) \max_{-h \leq \tau \leq h} |F(\tau, x(\tau)) - F(\tau, y(\tau))| \\ &= 2hf(t) \max_{-h \leq \tau \leq h} L|x(\tau) - y(\tau)| \\ &\leq 2hLf(t) \max_{-h \leq \tau \leq h} |x(\tau) - y(\tau)| \\ &= hLN(x, x, y) \end{aligned}$$

and

$$\begin{aligned} N(Tx, Tx, \xi) &= 2f(t) \max_{-h \leq t \leq h} \left| \int_0^t F(\tau, x(\tau))d\tau \right| \\ &\leq 2hf \max_{-h \leq \tau \leq h} |F(\tau, x(\tau))| \\ &\leq 2hMf \\ &\leq \delta f \end{aligned}$$

We speculate $T : B(\xi, \delta f) \to B(\xi, \delta f)$ is a contractive mapping.

Finally, we prove that $(B(\xi, \delta f), N)$ is complete. In fact, suppose $\{x_n\}$ is a Cauchy sequence in $B(\xi, \delta f)$. Then $\{x_n\}$ is also a Cauchy sequence in X. Since (X, N) is complete, there is $x \in X$ such that $x_n \to x$ as $n \to \infty$. So for each $c \in intP$, there exists N_1, whenever $n > N_1$ we obtain $N(x, x, x_n) \ll \frac{c}{2}$. Thus, it follows that

$$\begin{aligned} N(x, x, \xi) &\leq N(x, x, x_n) + N(x, x, x_n) + N(\xi, \xi, x_n) \\ &\quad \text{[by Definition1.1]} \\ &\leq 2N(x, x, x_n) + N(Tx_{n-1}, Tx_{n-1}, \xi) \\ &\quad \text{[by Proposition 1.1 and definition of T in Theorem 2.1]} \\ &= c + \delta f \end{aligned}$$

and by Lemma (1.10), $N(x, x, \xi) \leq \delta f$ which means $x \in B(\xi, \delta f)$, that is, $(B(\xi, \delta f), N)$ is complete.

Owing to the above statement, all the conditions of Theorem 2.1 are satisfied. Hence, T has a unique fixed point $x(t) \in B(\xi, \delta f)$. That is to say, there exists a unique solution of Example (3.1).

We notice that the above-mentioned application of fixed point theorem in b-cone metric space was given by [8]. □

Conclusion

In this paper, we define topology in N-cone metric space and extend various famous results such as Banach contraction theorem and Chatterjee's theorem in this newly defined space with applications in integral equations.

Acknowledgments The first author is thankful to Prof. Pankaj Kumar Jhade, Department of Mathematics, NRI Institute of Information Science & Technology, Bhopal (MP), India and Dr. Kalpana Saxena, Department of Mathematics, Government MVM College, Bhopal (MP), India for his/her suggestions and constant encouragement. The authors are also thankful to the referee for their valuable comments and suggestions which improved greatly the quality of this paper.

References

1. Aage, C.T., Salunke, J.N.: Some fixed point theorem in generalized D^*-metric spaces. Appl. Sci. **12**, 1–13 (2010)
2. Banach, S.: Sur les operations dans les ensambles abstraits et leur application aux equations integrals. Fund. Math. **3**, 133–181 (1922)
3. Babu, G.V.R., Alemayehu, G.N., Vara Prasand, K.N.V.V.: Common fixed point theorems of generalized contraction, Zamfirescu pair of maps in cone metric spaces. Albenian J. Math. **4**(1), 19–29 (2010)
4. Ismat, Beg, Mujahid, Abbas, Talat, Nazir: Generalized cone metric spaces. J. Nonlinear Sci. Appl. **3**, 21–31 (2010)
5. Chatterjee, S.K.: Fixed point theorems. Rend. Acad. Bulgare Sci. **25**, 727–730 (1972)
6. Cho, S.H., Bae, J.S.: Common fixed point theorems for mappings satisfying property (E.A.) on cone metric space. Math. Comput. Model. **53**, 945–951 (2011)
7. Huang, L.G., Zhang, X.: Cone metric spaces and fixed point theorems for contractive mappings. J. Math. Anal. Appl. **332**(2), 1468–1476 (2007)
8. Huaping, Huang, Shaoyuan, Xu: Fixed point theorems of contractive mappings in cone b-metric spaces and applications. Fixed Point Theory Appl. **10**, 220 (2012)
9. Hussain, N., Shah, M.H.: KKM mapping in cone b-metric spaces. Comput. Math. Appl. **62**, 1677–1684 (2011)
10. Jankovic, S., Kodelburg, Z., Radenovic, S.: On cone metric spaces: a survey. Nonlinear Anal. **4**(7), 2591–2601 (2011)
11. Kannan, R.: Some results on fixed points. Bull. Calcutta Math. Soc. **60**, 71–76 (1968)
12. Malviya, N., Chouhan, P.: Fixed point of expansive type mapping in N-cone metric space. Math. Theory Model. 3(7) (2013)
13. Malviya, N., Fisher, B.: N-cone metric space and fixed points of asymptotically regular maps. Accepted in Filomat J. Math. (Preprint)
14. Rezapour, Sh, Hamlbarani, R.: Some notes on the paper, "Cone metric spaces and fixed point theorems of contractive mappings". J. Math. Anal. Appl. **345**, 719–724 (2008)
15. Mujahid, A., Rhoades, B.E.: Fixed and periodic point results in cone metric spaces. Appl. Math. Lett. **22**, 511–515 (2009)
16. Stojan, Radenovic: Common fixed points under contractive conditions in cone metric spaces. Comput. Math. Appl. **58**, 1273–1278 (2009)

8

Coupled and tripled coincidence point results under (F, g)-invariant sets in G_b-metric spaces and G-α-admissible mappings

Nawab Hussain · Vahid Parvaneh · Farhan Golkarmanesh

Abstract In this paper, we prove that coupled and tripled coincidence point theorems under (F, g)-invariant sets for weakly contractive mappings defined on a G-metric space are immediate consequences of corresponding results via rectangular G-α-admissible mappings. This idea can also be applied to obtain coupled and tripled fixed point theorems in other spaces under various contractive conditions which reduces the proof considerably.

Keywords G_b-metric space · Coupled coincidence point · Common coupled fixed point · Tripled coincidence point · Common tripled fixed point · F-invariant set · Admissible mapping

Mathematics Subject Classification Primary 47H10 · Secondary 54H25

Introduction and mathematical preliminaries

The concept of generalized metric space, or a G-metric space, was introduced by Mustafa and Sims.

N. Hussain
Department of Mathematics, King Abdulaziz University, P.O. Box 80203, Jeddah 21589, Saudi Arabia
e-mail: nhusain@kau.edu.sa

V. Parvaneh (✉)
Department of Mathematics, Gilan-E-Gharb Branch, Islamic Azad University, Gilan-E-Gharb, Iran
e-mail: zam.dalahoo@gmail.com

F. Golkarmanesh
Department of Mathematics, Sanandaj Branch, Islamic Azad University, Sanandaj, Iran
e-mail: fgolkarmanesh@yahoo.com

Definition 1.1 (*G-Metric Space*, [14]) Let X be a nonempty set and $G: X \times X \times X \to \mathbb{R}^+$ be a function satisfying the following properties:

(G1) $G(x, y, z) = 0$ iff $x = y = z$;
(G2) $0 < G(x, x, y)$, for all $x, y \in X$ with $x \neq y$;
(G3) $G(x, x, y) \leq G(x, y, z)$, for all $x, y, z \in X$ with $y \neq z$;
(G4) $G(x, y, z) = G(x, z, y) = G(y, z, x) = \ldots$, (symmetry in all three variables);
(G5) $G(x, y, z) \leq G(x, a, a) + G(a, y, z)$, for all $x, y, z, a \in X$ (rectangle inequality).

Then, the function G is called a G-metric on X and the pair (X, G) is called a G-metric space.

Recently, Aghajani et al. [1] motivated by the concept of b-metric [27] introduced the concept of generalized b-metric spaces (G_b-metric spaces) and then they presented some basic properties of G_b-metric spaces.

The following is their definition of G_b-metric spaces.

Definition 1.2 [1] Let X be a nonempty set and $s \geq 1$ be a given real number. Suppose that a mapping $G: X \times X \times X \to \mathbb{R}^+$ satisfies:

(G_b1) $G(x, y, z) = 0$ if $x = y = z$,
(G_b2) $0 < G(x, x, y)$ for all $x, y \in X$ with $x \neq y$,
(G_b3) $G(x, x, y) \leq G(x, y, z)$ for all $x, y, z \in X$ with $y \neq z$,
(G_b4) $G(x, y, z) = G(p\{x, y, z\})$, where p is a permutation of x, y, z (symmetry),
(G_b5) $G(x, y, z) \leq s[G(x, a, a) + G(a, y, z)]$ for all $x, y, z, a \in X$ (rectangle inequality).

Then, G is called a generalized b-metric and the pair (X, G) is called a generalized b-metric space or a G_b-metric space.

Each G-metric space is a G_b-metric space with $s = 1$.

Example 1.3 [1] Let (X, G) be a G-metric space and $G_*(x, y, z) = G(x, y, z)^p$, where $p > 1$ is a real number. Then, G_* is a G_b-metric with $s = 2^{p-1}$.

Example 1.4 [13] Let $X = \mathbb{R}$ and $d(x, y) = |x - y|^2$. We know that (X, d) is a b-metric space with $s = 2$. Let $G(x, y, z) = d(x, y) + d(y, z) + d(z, x)$, it is easy to see that (X, G) is not a G_b-metric space. Indeed, (G_b3) is not true for $x = 0$, $y = 2$ and $z = 1$. However, $G(x, y, z) = \max\{d(x, y), d(y, z), d(z, x)\}$ is a G_b-metric on $\mathbb{R}$ with $s = 2$.

Proposition 1.5 [1] *Let X be a G_b-metric space. Then, for each $x, y, z, a \in X$, it follows that:*

1. if $G(x, y, z) = 0$ *then* $x = y = z$,
2. $G(x, y, z) \le s(G(x, x, y) + G(x, x, z))$,
3. $G(x, y, y) \le 2sG(y, x, x)$,
4. $G(x, y, z) \le s(G(x, a, z) + G(a, y, z))$.

Definition 1.6 [1] Let X be a G_b-metric space. A sequence $\{x_n\}$ in X is said to be:

1. G_b-Cauchy if, for each $\varepsilon > 0$ there exists a positive integer n_0 such that for all $m, n, l \ge n_0$, $G(x_n, x_m, x_l) < \varepsilon$;
2. G_b-convergent to a point $x \in X$ if, for each $\varepsilon > 0$ there exists a positive integer n_0 such that for all $m, n \ge n_0$, $G(x_n, x_m, x) < \varepsilon$.

Definition 1.7 [1] A G_b-metric space X is called G_b-complete, if every G_b-Cauchy sequence is G_b -convergent in X.

Proposition 1.8 *Let (X, G) and (X', G') be two G_b-metric spaces. Then, a function $f: X \to X'$ is G_b-continuous at a point $x \in X$ if and only if it is G_b-sequentially continuous at x, that is, whenever $\{x_n\}$ is G_b-convergent to x, $\{f(x_n)\}$ is G'_b-convergent to $f(x)$.*

Proposition 1.9 *Let (X, G) be a G_b-metric space. A mapping $F: X \times X \to X$ is said to be continuous if for any two G_b-convergent sequences $\{x_n\}$ and $\{y_n\}$ converging to x and y, respectively, $\{F(x_n, y_n)\}$ is G_b-convergent to $F(x, y)$.*

Existence of fixed points, coupled fixed points and tripled fixed points for contractive type mappings in partially ordered metric spaces has been considered recently by several authors (see, [2, 3, 5–8, 11, 17, 18, 20, 22, 24–26, 40–44]).

Lakshmikantham and Ćirić [11] introduced the notions of mixed g-monotone mapping and coupled coincidence point and proved some coupled coincidence point and common coupled fixed point theorems in partially ordered complete metric spaces.

Definition 1.10 [11] Let $(X, \preceq)$ be a partially ordered set and let $F: X \times X \to X$ and $g: X \to X$ be two mappings. F has the mixed g-monotone property, if F is monotone g-nondecreasing in its first argument and is monotone g-nonincreasing in its second argument, that is, for all $x_1, x_2 \in X$, $gx_1 \preceq gx_2$ implies $F(x_1, y) \preceq F(x_2, y)$ for any $y \in X$ and for all $y_1, y_2 \in X$, $gy_1 \preceq gy_2$ implies $F(x, y_1) \succeq F(x, y_2)$ for any $x \in X$.

Definition 1.11 [3, 11] An element $(x, y) \in X \times X$ is called

1. a coupled fixed point of mapping $F: X \times X \to X$ if $x = F(x, y)$ and $y = F(y, x)$.
2. a coupled coincidence point of mappings $F: X \times X \to X$ and $g: X \to X$ if $g(x) = F(x, y)$ and $g(y) = F(y, x)$.
3. a common coupled fixed point of mappings $F: X \times X \to X$ and $g: X \to X$ if $x = g(x) = F(x, y)$ and $y = g(y) = F(y, x)$.

Choudhury and maity [6] have established some coupled fixed point results for mappings with mixed monotone property in partially ordered G-metric spaces. They obtained the following results.

Theorem 1.12 ([6], Theorem 3.1) *Let $(X, \preceq)$ be a partially ordered set and G be a G-metric on X such that (X, G) is a complete G-metric space. Let $F: X \times X \to X$ be a continuous mapping having the mixed monotone property on X. Assume that there exists $k \in [0, 1)$ such that*

$$G(F(x, y), F(u, v), F(w, z)) \le \frac{k}{2}[G(x, u, w) + G(y, v, z)], \tag{1.1}$$

for all $x \preceq u \preceq w$ and $y \succeq v \succeq z$, where either $u \ne w$ or $v \ne z$.

If there exist $x_0, y_0 \in X$ such that $x_0 \preceq F(x_0, y_0)$ and $y_0 \succeq F(y_0, x_0)$, then, F has a coupled fixed point in X, that is, there exist $x, y \in X$ such that $x = F(x, y)$ and $y = F(y, x)$.

Theorem 1.13 ([6], Theorem 3.2) *If in the above theorem, in place of the continuity of F, we assume the following conditions, namely,*

(i) *if a nondecreasing sequence $\{x_n\} \to x$, then $x_n \preceq x$ for all n, and*
(ii) *if a nonincreasing sequence $\{y_n\} \to y$, then $y_n \succeq y$ for all n, then, F has a coupled fixed point.*

Definition 1.14 [4] Let $(X, \preceq)$ be a partially ordered set and G be a G-metric on X. We say that $(X, G, \preceq)$ is regular if the following conditions hold:

(i) If $\{x_n\}$ is a nondecreasing sequence with $x_n \to x$, then $x_n \preceq x$ for all $n \in \mathbb{N}$.

(ii) If $\{x_n\}$ is a nonincreasing sequence with $x_n \to x$, then $x_n \succeq x$ for all $n \in \mathbb{N}$.

Definition 1.15 [6] Let (X, G) be a generalized b-metric space. Mappings $f : X^2 \to X$ and $g : X \to X$ are called compatible if

$$\lim_{n\to\infty} G(gf(x_n, y_n), f(gx_n, gy_n), f(gx_n, gy_n)) = 0$$

and

$$\lim_{n\to\infty} G(gf(y_n, x_n), f(gy_n, gx_n), f(gy_n, gx_n)) = 0$$

hold whenever $\{x_n\}$ and $\{y_n\}$ are sequences in X such that $\lim_{n\to\infty} f(x_n, y_n) = \lim_{n\to\infty} gx_n$ and $\lim_{n\to\infty} f(y_n, x_n) = \lim_{n\to\infty} gy_n$.

On the other hand, Berinde and Borcut [24] introduced the concept of tripled fixed point and obtained some tripled fixed point theorems for contractive type mappings in partially ordered metric spaces. For a survey of tripled fixed point theorems and related topics, we refer the reader to [24–26].

Definition 1.16 ([24, 25]) Let $(X, \preceq)$ be a partially ordered set, $f : X^3 \to X$ and $g : X \to X$. An element $(x, y, z) \in X^3$ is called

1. a tripled fixed point of f if $f(x, y, z) = x$, $f(y, x, y) = y$, and $f(z, y, x) = z$.
2. a tripled coincidence point of the mappings f and g if $f(x, y, z) = gx$, $f(y, x, y) = gy$ and $f(z, y, x) = gz$.
3. a tripled common fixed point of f and g if $x = g(x) = f(x, y, z)$, $y = g(y) = f(y, x, y)$ and $z = g(z) = f(z, y, x)$.
4. We say that f has the mixed g-monotone property if $f(x, y, z)$ is g-nondecreasing in x, g-nonincreasing in y and g-nondecreasing in z, that is, if for any $x, y, z \in \mathcal{X}$,

$$x_1, x_2 \in X, gx_1 \preceq gx_2 \Rightarrow f(x_1, y, z) \preceq f(x_2, y, z),$$
$$y_1, y_2 \in X, gy_1 \preceq gy_2 \Rightarrow f(x, y_1, z) \succeq f(x, y_2, z)$$

and

$$z_1, z_2 \in X, gz_1 \preceq gz_2 \Rightarrow f(x, y, z_1) \preceq f(x, y, z_2).$$

Definition 1.17 Let (X, G) be a generalized b-metric space. Mappings $f : X^3 \to X$ and $g : X \to X$ are called compatible if

$$\lim_{n\to\infty} G(gf(x_n, y_n, z_n), f(gx_n, gy_n, gz_n), f(gx_n, gy_n, gz_n)) = 0,$$
$$\lim_{n\to\infty} G(gf(y_n, x_n, y_n), f(gy_n, gx_n, gy_n), f(gy_n, gx_n, gy_n)) = 0$$

and

$$\lim_{n\to\infty} G(gf(z_n, y_n, x_n), f(gz_n, gy_n, gx_n), f(gz_n, gy_n, gx_n)) = 0$$

hold whenever $\{x_n\}$, $\{y_n\}$ and $\{z_n\}$ are sequences in X such that $\lim_{n\to\infty} f(x_n, y_n, z_n) = \lim_{n\to\infty} gx_n$, $\lim_{n\to\infty} f(y_n, x_n, y_n) = \lim_{n\to\infty} gy_n$ and $\lim_{n\to\infty} f(z_n, y_n, x_n) = \lim_{n\to\infty} gz_n$.

let $\psi : [0, +\infty) \to [0, +\infty)$ satisfies:

(i) ψ is continuous and nondecreasing,
(ii) $\psi(t) = 0$ if and only if t = 0.

That is, ψ is an altering distance function.

Batra and Vashistha [36] introduced the concept of an (F, g)-invariant set which is a generalization of the F-invariant set introduced by Samet and Vetro [37].

Definition 1.18 [36] Let (X, d) be a metric space and let $F : X \times X \to X$ and $g : X \to X$ be given mappings. Let M be a nonempty subset of X^4. We say that M is an (F, g)-invariant subset of X^4 if and only if, for all $x, y, z, w \in X$,

(i) $(x, y, z, w) \in M$ iff $(w, z, y, x) \in M$;
(ii) $(g(x), g(y), g(z), g(w)) \in M$ implies that $(F(x, y), F(y, x), F(z, w), F(w, z)) \in M$.

Definition 1.19 [35] Let (X, G) be a G-metric space and let $F : X \times X \to X$ be a given mapping. Let M be a nonempty subset of X^6. We say that M is an F^*-invariant subset of X^6 if and only if, for all $x, y, z, u, v, w \in X$,

1. $(x, y, z, u, v, w) \in M$ iff $(w, v, u, z, y, x) \in M$;
2. $(x, y, z, u, v, w) \in M$ implies that $(F(x, y), F(y, x), F(z, u), F(u, z), F(v, w), F(w, v)) \in M$.

Definition 1.20 [35] Let (X, G) be a G-metric space and let $F : X \times X \to X$ and $g : X \to X$ are given mappings. Let M be a nonempty subset of X^6. We say that M is an (F^*, g)-invariant subset of X^6 if and only if, for all $x, y, z, u, v, w \in X$,

1. $(x, y, z, u, v, w) \in M$ iff $(w, v, u, z, y, x) \in M$;
2. $(gx, gy, gz, gu, gv, gw) \in M$ implies that $(F(x, y), F(y, x), F(z, u), F(u, z), F(v, w), F(w, v)) \in M$.

Definition 1.21 (*corrected from* [35]) Let (X, G) be a G-metric space and let M be a subset of X^6. We say that M satisfies the *transitive property* if and only if, for all $x, y, z, u, v, w, a, b \in X$,

$(x, y, z, u, z, u) \in M$ and $(z, u, v, w, a, b) \in M$ then $(x, y, v, w, a, b) \in M$.

Samet et al. [29] defined the notion of α-admissible mapping as follows.

Definition 1.22 Let T be a self-mapping on X and let $\alpha : X \times X \to [0, +\infty)$ be a function. We say that T is an α-admissible mapping if

$$x, y \in X, \quad \alpha(x, y) \geq 1 \quad \Longrightarrow \quad \alpha(Tx, Ty) \geq 1.$$

Definition 1.23 [30] Let (X, G) be a G-metric space, T be a self-mapping on X and $\alpha : X^3 \to [0, +\infty)$ be a function. We say that T is an G-α-admissible mapping if

$$x, y, z \in X, \quad \alpha(x, y, z) \geq 1 \quad \Longrightarrow \quad \alpha(Tx, Ty, Tz) \geq 1.$$

Following the recent work in [31], Hussain et al. [23] presented the following definition in the setting of G-metric spaces.

Definition 1.24 [23] Let (X, G) be a G-metric space, $f, g : X \to X$ and $\alpha : X^3 \to [0, +\infty)$. We say that f is a rectangular G-α-admissible mapping with respect to g if

(R1) $\alpha(gx, gy, gz) \geq 1$ implies $\alpha(fx, fy, fz) \geq 1$, $x, y, z \in X$,

(R2) $\left\{ \begin{array}{l} \alpha(gx, gy, gy) \geq 1 \\ \alpha(gy, gz, gw) \geq 1 \end{array} \right.$ implies $\alpha(gx, gz, gw) \geq 1$, $x, y, z, w \in X$.

Definition 1.25 Let (X, G) be a G-metric space, $F : X \times X \to X$, $g : X \to X$ and $\alpha : (X^2)^3 \to [0, +\infty)$. We say that F is a rectangular G-α-admissible mapping with respect to g if

(R1)

$$\alpha((gx, gy), (gu, gv), (ga, gb)) \geq 1 \text{ implies } \alpha((F(x, y), F(y, x)), (F(u, v), F(v, u)), (F(a, b), F(b, a))) \geq 1, \tag{1.2}$$

where $x, y, u, v, a, b \in X$,

(R2)

$$\begin{aligned} &\alpha((gx, gy), (gu, gv), (gu, gv)) \geq 1, \quad \text{and} \\ &\alpha((gu, gv), (ga, gb), (gc, gd)) \geq 1 \quad \text{implies that} \\ &\alpha((gx, gy), (ga, gb), (gc, gd)) \geq 1, \end{aligned} \tag{1.3}$$

where $x, y, u, v, a, b, c, d \in X$.

Definition 1.26 Let (X, G) be a G-metric space, $F : X \times X \times X \to X$, $g : X \to X$ and $\alpha : (X^3)^3 \to [0, +\infty)$. We say that F is a rectangular G-α-admissible mapping with respect to g if

(R1)

$$\begin{aligned} &\alpha((gx, gy, gz), (gu, gv, gw), (ga, gb, gc)) \geq 1 \text{ implies} \\ &\alpha((F(x, y, z), F(y, x, y), F(z, y, x)), (F(u, v, w), F(v, u, v), \\ &F(w, v, u)), (F(a, b, c), F(b, a, b), F(c, b, a))) \geq 1, \end{aligned} \tag{1.4}$$

where $x, y, z, u, v, w, a, b, c \in X$,

(R2)

$$\begin{aligned} &\alpha((gx, gy, gz), (gu, gv, gw), (gu, gv, gw)) \geq 1 \text{ and} \\ &\alpha((gu, gv, gw), (ga, gb, gc), (gd, ge, gf)) \geq 1 \text{ implies} \\ &\alpha((gx, gy, gz), (ga, gb, gc), (gd, ge, gf)) \geq 1, \end{aligned} \tag{1.5}$$

where $x, y, z, u, v, w, a, b, c, d, e, f \in X$.

Using the following coincidence point result, Hussain et al. obtained some interesting coupled and tripled coincidence point results which we use them in obtaining the main results here.

Theorem 1.27 [23] *Let (X, G) be a generalized b-metric space and let $f, g : X \to X$ satisfy the following condition:*

$$\begin{aligned} \alpha(gx, gy, gz)\psi(sG(fx, fy, fz)) \leq \psi(G(gx, gy, gz)) \\ - \varphi(G(gx, gy, gz)) \end{aligned} \tag{1.6}$$

for all $x, y, z \in X$, where $\psi, \varphi : [0, \infty) \to [0, \infty)$ are two altering distance mappings, $\alpha : X^3 \to [0, +\infty)$ and f is a rectangular G-α-admissible mapping w.r.t. g. Then, f and g have a coincidence point if,

(i) *$f(X) \subseteq g(X)$;*

(ii) *there exists $x_0 \in X$ such that $\alpha(gx_0, fx_0, fx_0) \geq 1$;*

(iii) *f and g are continuous and compatible and (X, G) is complete, or.*

(iii)′ *one of $f(X)$ or $g(X)$ is complete and whenever $\{x_n\}$ in X be a sequence such that $\alpha(x_n, x_{n+1}, x_{n+1}) \geq 1$ for all $n \in \mathbb{N} \cup \{0\}$ and $x_n \to x$ as $n \to +\infty$, we have $\alpha(x_n, x, x) \geq 1$ for all $n \in \mathbb{N} \cup \{0\}$*

Lemma 1.28 [23] *Let (X, G) be a generalized b-metric space (with the parameter s).*

(a) *If a mapping $\Omega_2^m : X^2 \times X^2 \times X^2 \to \mathbb{R}^+$ is given by*

$$\Omega_2^m(X, U, A) = \max\{G(x, u, a), G(y, v, b)\},$$
$$X = (x, y), \quad U = (u, v) \text{ and } A = (a, b) \in X^2,$$

then (X^2, Ω_2^m) is a generalized b-metric space (with the same parameter s). The space (X^2, Ω_2^m) is G_b-complete iff (X, G) is G_b-complete.

Let (X, G) be a generalized b-metric space, $F : X^2 \to X$ and $g : X \to X$. In the rest of this paper unless otherwise stated, for all $x, y, u, v, z, w \in X$, let

$$N_F^m(x,y,u,v,z,w) = \max\{G(F(x,y),F(u,v),F(z,w)),\\ G(F(y,x),F(v,u),F(w,z))\}$$

and

$$N_g^m(x,y,u,v,z,w) = \max\{G(gx,gu,gz),G(gy,gv,gw)\}.$$

Theorem 1.29 [23] *Let (X, G) be a generalized b-metric space with the parameter s and let $F : X^2 \to X$ and $g : X \to X$. Assume that*

$$\alpha((gx,gy),(gu,gv),(gz,gw))\\ \psi(sN_F^m(x,y,u,v,z,w)) \le \psi(N_g^m(x,y,u,v,z,w)) - \varphi(N_g^m(x,y,u,v,z,w)), \tag{1.7}$$

for all $x, y, u, v, z, w \in X$, where $\psi, \varphi : [0,\infty) \to [0,\infty)$ are altering distance functions, $\alpha : (X^2)^3 \to [0,\infty)$ and F is a rectangular G-α-admissible mapping with respect to g. Assume also that

1. $F(X^2) \subseteq g(X)$;
2. *there exist $x_0, y_0 \in X$ such that*

$$\alpha((gx_0,gy_0),(F(x_0,y_0),F(y_0,x_0)),(F(x_0,y_0),F(y_0,x_0))) \ge 1$$

and

$$\alpha((gy_0,gx_0),(F(y_0,x_0),F(x_0,y_0)),(F(y_0,x_0),F(x_0,y_0))) \ge 1.$$

Also, suppose that either

(a) *F and g are continuous, the pair (F, g) is compatible and (X, G) is G_b-complete, or*
(b) *$(g(X), G)$ is G_b-complete and assume that whenever $\{x_n\}$ and $\{y_n\}$ in X be sequences such that*

$$\alpha((x_n,y_n),(x_{n+1},y_{n+1}),(x_{n+1},y_{n+1})) \ge 1$$

and

$$\alpha((y_n,x_n),(y_{n+1},x_{n+1}),(y_{n+1},x_{n+1})) \ge 1$$

for all $n \in \mathbb{N} \cup \{0\}$ and $x_n \to x$, $y_n \to y$ as $n \to +\infty$, we have

$$\alpha((x_n,y_n),(x,y),(x,y)) \ge 1$$

and

$$\alpha((y_n,x_n),(y,x),(y,x)) \ge 1$$

for all $n \in \mathbb{N} \cup \{0\}$. Then, F and g have a coupled coincidence point in X.

Theorem 1.30 [23] *In addition to the hypotheses of Theorem 1.29, suppose that for all (x, y) and $(x^*, y^*) \in X^2$, there exists $(u, v) \in X^2$, such that $\alpha((gx,gy),(gu,gv),(gu,gv)) \ge 1$ and $\alpha((gx^*,gy^*),(gu,gv),(gu,gv)) \ge 1$. Then, F and g have a unique common coupled fixed point of the form (a, a).*

Let $\Omega_2^a : X^2 \times X^2 \times X^2 \to \mathbb{R}^+$ is given by

$$\Omega_2^a(X,U,A) = \frac{G(x,u,a)+G(y,v,b)}{2},$$
$$X=(x,y),\ U=(u,v) \text{ and } \mathrm{A}=(\mathrm{a},\mathrm{b}) \in \mathrm{X}^2,$$

then (X^2, Ω_2^a) is a generalized b-metric space (with the same parameter s).

Let (X, G) be a generalized b-metric space, $F : X^2 \to X$ and $g : X \to X$. For all $x, y, u, v, z, w \in X$, let

$$N_F^a(x,y,u,v,z,w)\\ = \frac{G(F(x,y),F(u,v),F(z,w))+G(F(y,x),F(v,u),F(w,z))}{2}$$

and

$$N_g^a(x,y,u,v,z,w) = \frac{G(gx,gu,gz)+G(gy,gv,gw)}{2}.$$

Remark 1.31 [23] The result of Theorems 1.29 and 1.30 holds, if we replace Ω_2^m, N_F^m and N_g^m by Ω_2^a, N_F^a and N_g^a, respectively.

Coupled fixed point results under (F^*, g)-invariant sets

Definition 2.1 Let (X, G) be a G_b-metric space and $M \subseteq X^6$. We say that X is M-regular if and only if the following hypothesis holds:

Whenever $\{x_n\}$ and $\{y_n\}$ in X be sequences such that

$$(x_n,y_n,x_{n+1},y_{n+1},x_{n+1},y_{n+1}) \in M$$

and

$$(y_n,x_n,y_{n+1},x_{n+1},y_{n+1},x_{n+1}) \in M$$

for all $n \in \mathbb{N} \cup \{0\}$ and $x_n \to x$, $y_n \to y$ as $n \to +\infty$, we have

$$(x_n,y_n,x,y,x,y) \in M$$

and

$$(y_n,x_n,y,x,y,x) \in M$$

for all $n \in \mathbb{N} \cup \{0\}$.

Theorem 2.2 *Let (X, G_b) be a G_b-metric space with the parameter s, $F : X^2 \to X$, $g : X \to X$ and M be a nonempty subset of X^6. Assume that*

$$\psi(sN_F^m(x,y,u,v,z,w)) \le \psi(N_g^m(x,y,u,v,z,w))\\ -\varphi(N_g^m(x,y,u,v,z,w)), \tag{2.1}$$

for all $x, y, u, v, z, w \in X$ with $(gx, gu, gz, gy, gv, gw) \in M$, where $\psi, \varphi : [0, \infty) \to [0, \infty)$ are altering distance functions. Assume also that

1. $F(X^2) \subseteq g(X)$;
2. *M is an (F^*, g)-invariant set which satisfies the transitive property;*
3. there exist $x_0, y_0 \in X$ *such that* $(gx_0, gy_0, F(x_0, y_0), F(y_0, x_0), F(x_0, y_0), F(y_0, x_0)) \in M$.

Also, suppose that either

(a) *F and g are continuous, the pair (F, g) is compatible and (X, G) is G_b-complete, or*

(b) *(X, G_b) is M-regular and $(g(X), G)$ is G_b-complete. Then, F and g have a coupled coincidence point in X.*

Proof Define $\alpha : (X^2)^3 \to [0, +\infty)$ by

$$\alpha((x, y), (u, v), (a, b)) = \begin{cases} 1, & \text{if } (x, y, u, v, a, b) \in M \\ 0, & \text{otherwise.} \end{cases}$$

First, we prove that F is a rectangular G-α-admissible mapping w.r.t. g. Hence, we assume that $\alpha((gx, gy), (gu, gv), (ga, gb)) \geq 1$. Therefore, we have $(gx, gy, gu, gv, ga, gb) \in M$. Since, M is an (F^*, g)-invariant subset of X^6, then

$$(F(x, y), F(y, x), F(u, v), F(v, u), F(a, b), F(b, a)) \in M$$

which implies that

$$\alpha((F(x, y), F(y, x)), (F(u, v), F(v, u)), (F(a, b), F(b, a))) \geq 1.$$

Now, let $\alpha((x, y), (a, b), (a, b)) \geq 1$ and $\alpha((a, b), (u, v), (u, v)) \geq 1$, then $(x, y, a, b, a, b) \in M$ and $(a, b, u, v, u, v) \in M$. Consequently, as M satisfies the transitive property, we deduce that $(x, y, a, b, u, v) \in M$, that is, $\alpha((x, y), (a, b), (u, v)) \geq 1$. Thus, F is a rectangular G-α-admissible mapping w.r.t. g.

From (2.1) and the definition of α,

$$\begin{aligned} &\alpha((gx, gy), (gu, gv), (gz, gw)) \\ &\psi(sN_F^m(x, y, u, v, z, w)) \leq \psi(N_g^m(x, y, u, v, z, w)) \\ &- \varphi(N_g^m(x, y, u, v, z, w)), \end{aligned} \tag{2.2}$$

for all $x, y, u, v, z, w \in X$. Moreover, from (2) there exist $x_0, y_0 \in \mathcal{X}$ such that

$$\alpha((gx_0, gy_0), (F(x_0, y_0), F(y_0, x_0)), (F(x_0, y_0), F(y_0, x_0))) > 1$$

and

$$\alpha((gy_0, gx_0), (F(y_0, x_0), F(x_0, y_0)), (F(y_0, x_0), F(x_0, y_0))) \geq 1.$$

Hence, all the conditions of Theorem 1.29 are satisfied and so F and g have a coupled coincidence point. □

In the following theorem, we give a sufficient condition for the uniqueness of the common coupled fixed point (see also [24]).

Theorem 2.3 *In addition to the hypotheses of Theorem 2.2, suppose that for all (x, y) and $(x^*, y^*) \in X^2$, there exists $(u, v) \in X^2$, such that $(gx, gy, gu, gv, gu, gv) \in M$ and $(gx^*, gy^*, gu, gv, gu, gv) \in M$. Then, F and g have a unique common coupled fixed point of the form (a, a).*

Remark 2.4 In Theorem 2.2, we can replace the contractive condition (2.1) by the following:

$$\begin{aligned} \psi(sN_F^a(x, y, u, v, z, w)) &\leq \psi(N_g^a(x, y, u, v, z, w)) \\ &- \varphi(N_g^a(x, y, u, v, z, w)). \end{aligned} \tag{2.3}$$

In Theorem 2.2, if we take $\psi(t) = t$ for all $t \in [0, \infty)$, we obtain the following result.

Corollary 2.5 *Let (X, G_b) be a G_b-metric space with the parameter s, $F : X^2 \to X$, $g : X \to X$ and M be a nonempty subset of X^6. Assume that*

$$\begin{aligned} &\frac{G(F(x, y), F(u, v), F(z, w)) + G(F(y, x), F(v, u), F(w, z))}{2} \\ &\leq \frac{1}{s} \frac{G(gx, gu, gz) + G(gy, gv, gw)}{2} \\ &\quad - \frac{1}{s} \varphi\left(\frac{G(gx, gu, gz) + G(gy, gv, gw)}{2}\right), \end{aligned} \tag{2.4}$$

for all $x, y, u, v, z, w \in X$ with $(gx, gu, gz, gy, gv, gw) \in M$, where $\varphi : [0, \infty) \to [0, \infty)$ is an altering distance function. Assume also that

1. $F(X^2) \subseteq g(X)$;
2. M is an *(F^*, g)-invariant set which satisfies the transitive property*;
3. there exist $x_0, y_0 \in X$ *such that* $(gx_0, gy_0, F(x_0, y_0), F(y_0, x_0), F(x_0, y_0), F(y_0, x_0)) \in M$.

Also, suppose that either

(a) *F and g are continuous, the pair (F, g) is compatible and (X, G) is G_b-complete, or*

(b) *(X, G_b) is M-regular and $(g(X), G)$ is G_b-complete. Then, F and g have a coupled coincidence point in X.*

Tripled coincidence point results via (F^*, g)-invariant sets

In this section, we prove some tripled coincidence and tripled common fixed point results.

Definition 3.1 Let (X, G) be a G-metric space and let $F : X \times X \times X \to X$ and $g : X \to X$ are given mappings. Let M be a nonempty subset of X^9. We say that M is an (f^*, g)-invariant subset of X^9 if and only if, for all $x, y, z, u, v, w, a, b, c \in X$,

1. $(x, y, z, u, v, w, a, b, c) \in M$ iff $(c, b, a, w, v, u, z, y, x) \in M$;
2. $$(gx, gy, gz, gu, gv, gw, ga, gb, gc) \in M$$
   implies that
   $$(F(x, y, z), F(y, x, y), F(z, y, x), F(u, v, w), F(v, u, v), F(w, v, u), F(a, b, c), F(b, a, b), F(c, b, a)) \in M.$$

Definition 3.2 Let (X, G) be a G-metric space and let M be a subset of X^9. We say that M satisfies the *transitive property* if and only if, for all $x, y, z, u, v, w, a, b, c, d, e, f \in X$,

$(x, y, z, u, v, w, u, v, w) \in M$ and $(u, v, w, a, b, c, d, e, f) \in M$ then $(x, y, z, a, b, c, d, e, f) \in M$.

Definition 3.3 Let (X, G) be a G_b-metric space and $M \subseteq X^9$. We say that X is M-regular if and only if the following hypothesis holds:

Whenever $\{x_n\}$, $\{y_n\}$ and $\{z_n\}$ in X be sequences such that

$$(x_n, y_n, z_n, x_{n+1}, y_{n+1}, z_{n+1}, x_{n+1}, y_{n+1}, z_{n+1}) \in M,$$
$$(y_n, x_n, y_n, y_{n+1}, x_{n+1}, y_{n+1}, y_{n+1}, x_{n+1}, y_{n+1}) \in M$$

and

$$(z_n, y_n, x_n, z_{n+1}, y_{n+1}, x_{n+1}, z_{n+1}, y_{n+1}, x_{n+1}) \in M$$

for all $n \in \mathbb{N} \cup \{0\}$ and $x_n \to x$, $y_n \to y$ and $z_n \to z$ as $n \to +\infty$, we have

$$(x_n, y_n, z_n, x, y, z, x, y, z) \in M,$$
$$(y_n, x_n, y_n, y, x, y, y, x, y) \in M$$

and

$$(z_n, y_n, x_n, z, y, x, z, y, x) \in M$$

for all $n \in \mathbb{N} \cup \{0\}$.

Lemma 3.4 [23] *Let (X, G) be a generalized b-metric space (with the parameter s).*

If a mapping $\Omega_3^m : X^3 \times X^3 \times X^3 \to \mathbb{R}^+$ is given by

$$\Omega_3^m(X, U, A) = \max\{G(x, u, a), G(y, v, b), G(z, w, c)\},$$

for all $X = (x, y, z), U = (u, v, w)$ and $A = (a, b, c) \in X^3$, then (X^3, Ω_3^m) is an generalized b-metric space (with the same parameter s). The space (X^3, Ω_3^m) is G_b-complete iff (X, G) is G_b-complete.

Let (X, G) be a generalized b-metric space, $f : X^3 \to X$ and $g : X \to X$. For all $x, y, z, u, v, w, a, b, c \in X$, let

$$\begin{aligned} M_F^m&(x, y, z, u, v, w, a, b, c) \\ &= \max\{G(F(x, y, z), F(u, v, w), F(a, b, c)), G(F(y, x, y), \\ &F(v, u, v), F(b, a, b)), G(F(z, y, x), F(w, v, u), F(c, b, a))\} \end{aligned}$$

and

$$M_g^m(x, y, z, u, v, w, a, b, c) = \max\{G(gx, gu, ga), G(gy, gv, gb), G(gz, gw, gc)\}.$$

Theorem 3.5 [23] *Let (X, G) be a generalized b -metric space with the parameter s, $F : X^3 \to X$ and $g : X \to X$. Assume that*

$$\begin{aligned} \alpha((gx, gy, gz), (gu, gv, gw), (ga, gb, gc))\psi& \\ (sM_F^m(x, y, z, u, v, w, a, b, c))& \\ \le \psi(M_g^m(x, y, z, u, v, w, a, b, c))& \\ -\varphi(M_g^m(x, y, z, u, v, w, a, b, c)),& \end{aligned} \tag{3.1}$$

for all $x, y, z, u, v, w, a, b, c \in X$ where $\psi, \varphi : [0, \infty) \to [0, \infty)$ are altering distance functions and $\alpha : (X^3)^3 \to [0, \infty)$ is a mapping such that F is a rectangular G-α-admissible mapping w.r.t. g. Assume also that

1. *$F(X^3) \subseteq g(X)$;*
2. *there exist $x_0, y_0, z_0 \in X$*

such that

$$\alpha((gx_0, gy_0, gz_0), (F(x_0, y_0, z_0), F(y_0, x_0, y_0), F(z_0, y_0, x_0)), (F(x_0, y_0, z_0), F(y_0, x_0, y_0), F(z_0, y_0, x_0))) \ge 1,$$
$$\alpha((gy_0, gx_0, gy_0), (F(y_0, x_0, y_0), F(x_0, y_0, z_0), F(y_0, x_0, y_0)), (F(y_0, x_0, y_0), F(x_0, y_0, z_0), F(y_0, x_0, y_0))) \ge 1$$

and

$$\alpha((gz_0, gy_0, gx_0), (F(z_0, y_0, x_0), F(y_0, x_0, y_0), F(x_0, y_0, z_0)), (F(z_0, y_0, x_0), F(y_0, x_0, y_0), F(x_0, y_0, z_0))) \ge 1.$$

Also, suppose that either

(a) *F and g are continuous, the pair (F, g) is compatible and (X, G) is G_b-complete, or*
(b) *$(g(X), G)$ is G_b-complete and assume that whenever $\{x_n\}, \{y_n\}$ and $\{z_n\}$ in X be sequences such that*

$$\alpha((x_n,y_n,z_n),(x_{n+1},y_{n+1},z_{n+1}),(x_{n+1},y_{n+1},z_{n+1}))\geq 1,$$
$$\alpha((y_n,x_n,y_n),(y_{n+1},x_{n+1},y_{n+1}),(y_{n+1},x_{n+1},y_{n+1}))\geq 1$$

and

$$\alpha((z_n,y_n,x_n),(z_{n+1},y_{n+1},x_{n+1}),(z_{n+1},y_{n+1},x_{n+1}))\geq 1$$

for all $n\in\mathbb{N}\cup\{0\}$ *and* $x_n\to x, y_n\to y$ *and* $z_n\to z$ *as* $n\to+\infty$, *we have*

$$\alpha((x_n,y_n,z_n),(x,y,z),(x,y,z))\geq 1,$$
$$\alpha((y_n,x_n,y_n),(y,x,y),(y,x,y))\geq 1$$

and

$$\alpha((z_n,y_n,x_n),(z,y,x),(z,y,x))\geq 1$$

for all $n\in\mathbb{N}\cup\{0\}$.

Then, F and g have a tripled coincidence point in X.

Theorem 3.6 [23] *In addition to the hypotheses of Theorem 3.5, suppose that for all* (x,y,z) *and* $(x^*,y^*,z^*)\in X^3$, *there exists* $(u,v,w)\in X^3$, *such that*

$$\alpha((gx,gy,gz),(gu,gv,gw),(gu,gv,gw))\geq 1$$

and

$$\alpha((gx^*,gy^*,gz^*),(gu,gv,gw),(gu,gv,gw))\geq 1.$$

Then, F and g have a unique common tripled fixed point of the form (a,a,a).

Let

$$\Omega_3^a(X,U,A)=\frac{G(x,u,a)+G(y,v,b)+G(z,w,c)}{3},$$
$$X=(x,y,z),U=(u,v,w),A=(a,b,c)\in\mathcal{X}^3,$$

$$M_F^a(x,y,z,u,v,w,a,b,c)$$
$$=\frac{G(F(x,y,z),F(u,v,w),F(a,b,c))+G(F(y,x,y),F(v,u,v),F(b,a,b))+G(F(z,y,x),F(w,v,u),F(c,b,a))}{3}$$

and

$$M_g^a(x,y,z,u,v,w,a,b,c)$$
$$=\frac{G(gx,gu,ga)+G(gy,gv,gb)+G(gz,gw,gc)}{3}.$$

Remark 3.7 [23] In Theorem 3.5, we can replace the contractive condition (3.1) by the following:

$$\begin{aligned}\alpha((gx,gy,gz),(gu,gv,gw),(ga,gb,gc))\psi\\(sM_F^a(x,y,z,u,v,w,a,b,c))\\\leq\psi(M_g^a(x,y,z,u,v,w,a,b,c))\\-\varphi(M_g^a(x,y,z,u,v,w,a,b,c)).\end{aligned}\tag{3.2}$$

Following tripled fixed point results in G_b-metric spaces can be obtained.

Theorem 3.8 *Let* (X,G) *be a* G_b *-metric space with the parameter* $s, F:X^3\to X, g:X\to X$ *and M be a nonempty subset of* X^9. *Assume that*

$$\begin{aligned}\psi(sM_F^m(x,y,z,u,v,w,a,b,c))\leq\psi(M_g^m(x,y,z,u,v,w,a,b,c))\\-\varphi(M_g^m(x,y,z,u,v,w,a,b,c)),\end{aligned}\tag{3.3}$$

for all $x,y,z,u,v,w,a,b,c\in X$ *with* $(gx,gu,ga,gy,gv,gb,gz,gw,gc)\in M$, *where* $\psi,\varphi:[0,\infty)\to[0,\infty)$ *are altering distance functions. Assume also that*

1. $F(X^3)\subseteq g(X)$;
2. M *is an* (F^*,g)*-invariant set which satisfies the transitive property;*
3. *there exist* $x_0,y_0,z_0\in X$ *such that*

$$(gx_0,gy_0,,gz_0,F(x_0,y_0,z_0),F(y_0,x_0,y_0),F(z_0,y_0,x_0),$$
$$F(x_0,y_0,z_0),$$
$$F(y_0,x_0,y_0),F(z_0,y_0,x_0))\in M.$$

Also, suppose that either

(a) *F and g are continuous, the pair* (F,g) *is compatible and* (X,G) *is* G_b*-complete, or*

(b) (X,G) *is M-regular and* $(g(X),G)$ *is* G_b*-complete.*

Then, F and g have a tripled coincidence point in X.

Theorem 3.9 *In addition to the hypotheses of Theorem 3.5, suppose that for all* (x,y,z) *and* $(x^*,y^*,z^*)\subset X^3$, *there exists* $(u,v,w)\in X^3$, *such that*

$$(gx,gy,gz,gu,gv,gw,gu,gv,gw)\in M$$

and

$(gx^*, gy^*, gz^*, gu, gv, gw, gu, gv, gw) \in M.$

Then, F and g have a unique common tripled fixed point of the form (a, a, a).

Remark 3.10 In Theorem 3.8, we can replace the contractive condition (3.3) by the following:

$$\psi(sM_F^a(x,y,z,u,v,w,a,b,c)) \le \psi(M_g^a(x,y,z,u,v,w,a,b,c)) - \varphi(M_g^a(x,y,z,u,v,w,a,b,c)). \tag{3.4}$$

Coupled and tripled fixed point results for ψ-contractions in G-metric spaces

Let Ψ denotes the set of all functions $\psi : [0,\infty) \to [0,\infty)$ satisfying $\psi^{-1}(0) = 0$, $\psi(t) < t$ for all $t > 0$ and $\lim_{r\to t^+} \psi(r) < t$ for all $t > 0$.

Using the following coincidence point result, we obtain some coupled and tripled coincidence point results under (F^*, g)-invariant sets.

Lemma 4.1 [23] *Let f be a rectangular G-α-admissible mapping w.r.t. g such that* $f(X) \subseteq g(X)$. *Assume that there exists* $x_0 \in X$ *such that* $\alpha(fx_0, fx_0, gx_0) \ge 1$. *Define sequence* $\{y_n\}$ *by* $y_n = gx_n = fx_{n-1}$. *Then,*

$$\alpha(y_n, y_m, y_m) \ge 1 \quad \text{for all} \quad n, m \in \mathbb{N} \quad \text{with} \quad n < m.$$

Theorem 4.2 *Let* (X, G) *be a G-metric space and let* $f, g : X \to X$ *satisfy the following condition:*

$$\alpha(gx, gy, gz)G(fx, fy, fz) \le \psi(G(gx, gy, gz)) \tag{4.1}$$

for all $x, y, z \in X$, *where* $\psi \in \Psi$, $\alpha : X^3 \to [0, +\infty)$ *and f is a rectangular G-α-admissible mapping w.r.t. g.*

Then, f and g have a coincidence point if,

(i) $f(X) \subseteq g(X)$;
(ii) *there exists* $x_0 \in X$ *such that* $\alpha(fx_0, fx_0, gx_0) \ge 1$;
(iii) *f and g are continuous, g commutes with f and* (X, G) *is complete, or.*
(iii)′ $(g(X), G)$ *is G-complete and assume that whenever* $\{x_n\}$ *in X be a sequence such that* $\alpha(x_n, x_{n+1}, x_{n+1}) \ge 1$ *for all* $n \in \mathbb{N} \cup \{0\}$ *and* $x_n \to x$ *as* $n \to +\infty$, *we have* $\alpha(x_n, x, x) \ge 1$ *for all* $n \in \mathbb{N} \cup \{0\}$

Proof Let $x_0 \in X$ be such that $\alpha(fx_0, fx_0, gx_0) \ge 1$. According to (i) one can define the sequence $\{y_n\}$ as $y_{n+1} = gx_{n+1} = fx_n$ for all $n = 0, 1, 2, \ldots$.

As $\alpha(gx_1, gx_1, gx_0) = \alpha(fx_0, fx_0, gx_0) \ge 1$ and since f is a G-α-admissible mapping with respect to g, then $\alpha(y_2, y_2, y_1) = \alpha(fx_1, fx_1, fx_0) \ge 1$. Continuing this process, we get $\alpha(y_{n+1}, y_{n+1}, y_n) \ge 1$ for all $n \in \mathbb{N} \cup \{0\}$.

If $y_n = y_{n+1}$, then x_n is a coincidence point of f and g.

Now, assume that $y_n \ne y_{n+1}$ for all n, that is,

$$G(y_n, y_{n+1}, y_{n+2}) > 0, \tag{4.2}$$

for all n. Let $G_n = G(y_n, y_{n+1}, y_{n+2})$. Then, from (4.1) we obtain that

$$\begin{aligned} G(y_{n+1}, y_{n+2}, y_{n+3}) &\le \alpha(y_n, y_{n+1}, y_{n+2})G(y_{n+1}, y_{n+2}, y_{n+3}) \\ &= \alpha(gx_n, gx_{n+1}, gx_{n+2})G(fx_n, fx_{n+1}, fx_{n+2}) \\ &\le \psi(G(y_n, y_{n+1}, y_{n+2})) < G(y_n, y_{n+1}, y_{n+2}). \end{aligned} \tag{4.3}$$

So, we have proved that $G_{n+1} \le G_n$ for each $n \in \mathbb{N}$, and so there exists $r \ge 0$ such that $\lim_{n\to\infty} G_n = r \ge 0$.

Suppose that $r > 0$. Then, from (4.3), by taking the limit as $n \to \infty$, since $\psi \in \Psi$ we have

$$r \le \lim_{n\to\infty} \psi(G_n) = \lim_{G_n\to r^+} \psi(G_n) < r,$$

a contradiction. Hence,

$$\lim_{n\to\infty} G_n = \lim_{n\to\infty} G(y_n, y_{n+1}, y_{n+2}) = 0. \tag{4.4}$$

Since $y_{n+1} \ne y_{n+2}$ for every n, so by property $(G3)$ we obtain

$$G(y_n, y_{n+1}, y_{n+1}) \le 2G(y_n, y_n, y_{n+1}) \le 2G(y_n, y_{n+1}, y_{n+2}).$$

Hence,

$$\lim_{n\to\infty} G(y_n, y_{n+1}, y_{n+1}) = \lim_{n\to\infty} G(y_n, y_n, y_{n+1}) = 0. \tag{4.5}$$

Now, we prove that $\{y_{2n}\}$ is a G-Cauchy sequence. Assume on contrary that $\{y_{2n}\}$ is not a G-Cauchy sequence. Then, there exists $\varepsilon > 0$ for which we can find subsequences $\{y_{2m_k}\}$ and $\{y_{2n_k}\}$ of $\{y_{2n}\}$ such that m_k is the smallest index for which $m_k > n_k > k$ and

$$G(y_{2n_k}, y_{2m_k}, y_{2m_k}) \ge \varepsilon. \tag{4.6}$$

This means that

$$G(y_{2n_k}, y_{2m_k-1}, y_{2m_k-1}) < \varepsilon. \tag{4.7}$$

Since f is a rectangular G-α-admissible mapping with respect to g, then from Lemma 4.1 $\alpha(y_{2n_k}, y_{2m_k-1}, y_{2m_k-1}) \ge 1$. Now, from (4.1) we have

$$\begin{aligned} G(y_{2n_k+1}, y_{2m_k}, y_{2m_k}) &\le \alpha(y_{2n_k}, y_{2m_k-1}, y_{2m_k-1})G(y_{2n_k+1}, y_{2m_k}, y_{2m_k}) \\ &= \alpha(gx_{2n_k}, gx_{2m_k-1}, gx_{2m_k-1})G(fx_{2n_k}, fx_{2m_k-1}, fx_{2m_k-1}) \\ &\le \psi(G(y_{2n_k}, y_{2m_k-1}, y_{2m_k-1})) \\ &< G(y_{2n_k}, y_{2m_k-1}, y_{2m_k-1}), \end{aligned} \tag{4.8}$$

as $2n_k \ne 2m_k - 1$.

Using $(G5)$, we obtain that

$$G(y_{2n_k}, y_{2m_k}, y_{2m_k}) \le G(y_{2n_k}, y_{2m_k-1}, y_{2m_k-1}) + G(y_{2m_k-1}, y_{2m_k}, y_{2m_k}).$$

Taking the limit as $k \to \infty$ and using (4.5) and (4.7), we obtain that

$$\lim_{k\to\infty} G(y_{2n_k}, y_{2m_k-1}, y_{2m_k-1}) \ge \varepsilon. \tag{4.9}$$

Hence, by (4.7), we have

$$\lim_{k\to\infty} G(y_{2n_k}, y_{2m_k-1}, y_{2m_k-1}) = \varepsilon. \tag{4.10}$$

Using $(G5)$, we obtain that

$$G(y_{2n_k}, y_{2m_k}, y_{2m_k}) \le G(y_{2n_k}, y_{2n_k+1}, y_{2n_k+1}) + G(y_{2n_k+1}, y_{2m_k}, y_{2m_k}).$$

Taking the upper limit as $k \to \infty$ and using (4.5) and (4.7), we obtain that

$$\lim_{k\to\infty} G(y_{2n_k+1}, y_{2m_k}, y_{2m_k}) \ge \varepsilon. \tag{4.11}$$

Using $(G5)$, we obtain that

$$\begin{aligned} G(y_{2n_k+1}, y_{2m_k}, y_{2m_k}) &\le G(y_{2n_k+1}, y_{2n_k}, y_{2n_k}) + G(y_{2n_k}, y_{2m_k}, y_{2m_k}) \\ &\le G(y_{2n_k+1}, y_{2n_k}, y_{2n_k}) + G(y_{2n_k}, y_{2m_k-1}, y_{2m_k-1}) + G(y_{2m_k-1}, y_{2m_k}, y_{2m_k}). \end{aligned}$$

Taking the upper limit as $k \to \infty$ and using (4.5) and (4.7), we obtain that

$$\lim_{k\to\infty} \sup G(y_{2n_k+1}, y_{2m_k}, y_{2m_k}) \le \varepsilon. \tag{4.12}$$

Consequently, from (4.11),

$$\lim_{k\to\infty} \sup G(y_{2n_k+1}, y_{2m_k}, y_{2m_k}) = \varepsilon. \tag{4.13}$$

Taking the upper limit as $k \to \infty$ in (4.8) and using (4.7) and (4.9), we obtain that

$$\epsilon \le \lim_{k\to\infty} G(y_{2n_k+1}, y_{2m_k}, y_{2m_k}) \le \lim_{k\to\infty} \psi(G(y_{2n_k}, y_{2m_k-1}, y_{2m_k-1})) < \epsilon,$$

a contradiction. It follows that $\{y_n\}$ is a G-Cauchy sequence in X. Suppose first that (iii) holds. Then, there exists

$$\lim_{n\to\infty} fx_n = \lim_{n\to\infty} gx_n = z \in X.$$

Further, since f and g are continuous and g commutes with f, we get

$$fz = \lim_{n\to\infty} fgx_n = \lim_{n\to\infty} gfx_n = gz.$$

It means that f and g have a coincidence point.

In the case (iii′), if we assume that $g(X)$ is G-complete, then

$$\lim_{n\to\infty} fx_n = \lim_{n\to\infty} gx_n = gu = z$$

for some $u \in X$. Also, from (iii′) we have $\alpha(gx_n, gu, gu) \ge 1$. Applying (4.1) with $x = x_n$ and $y = z = u$, we have:

$$\begin{aligned} G(fx_n, fu, fu) &\le \alpha(gx_n, gu, gu) G(fx_n, fu, fu)) \\ &\le \psi(G(gx_n, gu, gu)) < G(gx_n, gu, gu). \end{aligned} \tag{4.14}$$

It follows that $G(fx_n, fu, fu) \to 0$ when $n \to \infty$, that is, $fx_n \to fu$. Uniqueness of the limit yields that $fu = z = gu$. Hence, f and g have a coincidence point $u \in X$.

Theorem 4.3 *Let (X, G) be a G-metric space and let $F : X^2 \to X$ and $g : X \to X$. Assume that*

$$\alpha((gx, gy), (gu, gv), (gz, gw)) N_F^m(x, y, u, v, z, w) \le \psi(N_g^m(x, y, u, v, z, w)), \tag{4.15}$$

for all $x, y, u, v, z, w \in X$, where $\psi \in \Psi$, $\alpha : (X^2)^3 \to [0, \infty)$ and F is a rectangular G-α-admissible mapping with respect to g. Assume also that

1. $F(X^2) \subseteq g(X)$;
2. *there exist $x_0, y_0 \in X$ such that*

$$\alpha((F(x_0, y_0), F(y_0, x_0)), (F(x_0, y_0), F(y_0, x_0)), (gx_0, gy_0),) \ge 1$$

and

$$\alpha((F(y_0, x_0), F(x_0, y_0)), (F(y_0, x_0), F(x_0, y_0)), (gy_0, gx_0),) \ge 1.$$

Also, suppose that either

(a) *F and g are continuous, g commutes with F and (X, G) is complete, or*

(b) *$(g(X), G)$ is complete and assume that whenever $\{x_n\}$ and $\{y_n\}$ in X be sequences such that*

$$\alpha((x_n, y_n), (x_{n+1}, y_{n+1}), (x_{n+1}, y_{n+1})) \ge 1$$

and

$$\alpha((y_n, x_n), (y_{n+1}, x_{n+1}), (y_{n+1}, x_{n+1})) \ge 1$$

for all $n \in \mathbb{N} \cup \{0\}$ and $x_n \to x$, $y_n \to y$ as $n \to +\infty$, we have

$$\alpha((x_n, y_n), (x, y), (x, y)) \ge 1$$

and

$$\alpha((y_n, x_n), (y, x), (y, x)) \geq 1$$

for all $n \in \mathbb{N} \cup \{0\}$. *Then, F and g have a coupled coincidence point in X.*

Theorem 4.4 *In addition to the hypotheses of Theorem 4.3, suppose that for all* (x, y) *and* $(x^*, y^*) \in X^2$, *there exists* $(u, v) \in X^2$, *such that* $\alpha((gx, gy), (gu, gv), (gu, gv)) \geq 1$ *and* $\alpha((gx^*, gy^*), (gu, gv), (gu, gv)) \geq 1$. *Then, F and g have a unique common coupled fixed point of the form* (a, a).

Remark 4.5 The result of Theorems 4.3 and 4.4 holds, if we replace N_F^m and N_g^m by N_F^a and N_g^a, respectively.

Theorem 4.6 *Let* (X, G) *be a G-metric space and let* $F : X^2 \to X$ *and* $g : X \to X$ *and M be a nonempty subset of* X^6. *Assume that*

$$N_F^m(x, y, u, v, z, w) \leq \psi(N_g^m(x, y, u, v, z, w)), \tag{4.16}$$

for all $x, y, u, v, z, w \in X$ *with* $(gx, gu, gz, gy, gv, gw) \in M$, *where* $\psi \in \Psi$. *Assume also that*

1. $F(X^2) \subseteq g(X)$;
2. *M is an* (F^*, g)*-invariant set which satisfies the transitive property;*
3. *there exist* $x_0, y_0 \in X$ *such that* $(F(x_0, y_0), F(y_0, x_0), F(x_0, y_0), F(y_0, x_0), gx_0, gy_0) \in M$.

Also, suppose that either

(a) *F and g are continuous, g commutes with F and* (X, G) *is complete, or*

(b) (X, G) *is M-regular and* $(g(X), G)$ *is G-complete. Then, F and g have a coupled coincidence point in X.*

Theorem 4.7 *In addition to the hypotheses of Theorem 4.6, suppose that for all* (x, y) *and* $(x^*, y^*) \in X^2$, *there exists* $(u, v) \in X^2$, *such that* $(gx, gy, gu, gv, gu, gv) \in M$ *and* $(gx^*, gy^*, gu, gv, gu, gv) \in M$. *Then, F and g have a unique common coupled fixed point of the form* (a, a).

We have the following corollary which is Theorems 3.1 and 3.2 of [35], but more general in contractive condition.

Corollary 4.8 *Let* (X, G) *be a G-metric space and let* $F : X^2 \to X$ *and* $g : X \to X$ *and M be a nonempty subset of* X^6. *Assume that*

$$\frac{G(gx, gu, gz) + G(gy, gv, gw)}{2} \leq \psi\left(\frac{G(F(x, y), F(u, v), F(z, w)) + G(F(y, x), F(v, u), F(w, z))}{2}\right), \tag{4.17}$$

for all $x, y, u, v, z, w \in X$ *with* $(gx, gu, gz, gy, gv, gw) \in M$, *where* $\psi \in \Psi$. *Assume also that*

1. $F(X^2) \subseteq g(X)$;
2. *M is an* (F^*, g)*-invariant set which satisfies the transitive property;*
3. there exist $x_0, y_0 \in X$ *such that* $(F(x_0, y_0), F(y_0, x_0), F(x_0, y_0), F(y_0, x_0), gx_0, gy_0) \in M$.

Also, suppose that either

(a) *F and g are continuous, g commutes with F and* (X, G) *is complete, or*

(b) (X, G) *is M-regular and* $(g(X), G)$ *is G-complete. Then, F and g have a coupled coincidence point in X.*

Theorem 4.9 *Let* (X, G) *be a G-metric space and let* $F : X^3 \to X$ *and* $g : X \to X$. *Assume that*

$$\alpha((gx, gy, gz), (gu, gv, gw), (ga, gb, gc)) M_F^m(x, y, z, u, v, w, a, b, c) \leq \psi(M_g^m(x, y, z, u, v, w, a, b, c)), \tag{4.18}$$

for all $x, y, z, u, v, w, a, b, c \in X$ *where* $\psi \in \Psi$ *and* $\alpha : (X^3)^3 \to [0, \infty)$ *is a mapping such that F is a rectangular G-α-admissible mapping w.r.t. g. Assume also that*

1. $F(X^3) \subseteq g(X)$;
2. *there exist* $x_0, y_0, z_0 \in X$

such that

$$\alpha((F(x_0, y_0, z_0), F(y_0, x_0, y_0), F(z_0, y_0, x_0)), (F(x_0, y_0, z_0), F(y_0, x_0, y_0), F(z_0, y_0, x_0)), (gx_0, gy_0, gz_0)) \geq 1,$$
$$\alpha((F(y_0, x_0, y_0), F(x_0, y_0, z_0), F(y_0, x_0, y_0)), (F(y_0, x_0, y_0), F(x_0, y_0, z_0), F(y_0, x_0, y_0)), (gy_0, gx_0, gy_0)) \geq 1$$

and

$$\alpha((F(z_0, y_0, x_0), F(y_0, x_0, y_0), F(x_0, y_0, z_0)), (F(z_0, y_0, x_0), F(y_0, x_0, y_0), F(x_0, y_0, z_0)), (gz_0, gy_0, gx_0)) \geq 1.$$

Also, suppose that either

(a) *F and g are continuous, g commutes with F and* (X, G) *is G-complete, or*

(b) $(g(X), G)$ *is G-complete and assume that whenever* $\{x_n\}, \{y_n\}, \{z_n\}$ *in X be sequences such that*

$$\alpha((x_n, y_n, z_n), (x_{n+1}, y_{n+1}, z_{n+1}), (x_{n+1}, y_{n+1}, z_{n+1})) \geq 1,$$
$$\alpha((y_n, x_n, y_n), (y_{n+1}, x_{n+1}, y_{n+1}), (y_{n+1}, x_{n+1}, y_{n+1})) \geq 1$$

and

$$\alpha((z_n, y_n, x_n), (z_{n+1}, y_{n+1}, x_{n+1}), (z_{n+1}, y_{n+1}, x_{n+1})) \geq 1$$

for all $n \in \mathbb{N} \cup \{0\}$ *and* $x_n \to x$, $y_n \to y$ *and* $z_n \to z$ *as* $n \to +\infty$, *we have*

$$\alpha((x_n, y_n, z_n), (x, y, z), (x, y, z)) \geq 1,$$
$$\alpha((y_n, x_n, y_n), (y, x, y), (y, x, y)) \geq 1$$

and

$$\alpha((z_n, y_n, x_n), (z, y, x), (z, y, x)) \geq 1$$

for all $n \in \mathbb{N} \cup \{0\}$. *Then, F and g have a tripled coincidence point in X.*

Theorem 4.10 *In addition to the hypotheses of Theorem 4.9, suppose that for all* (x, y, z) *and* $(x^*, y^*, z^*) \in X^3$, *there exists* $(u, v, w) \in X^3$, *such that*

$$\alpha((gx, gy, gz), (gu, gv, gw), (gu, gv, gw)) \geq 1$$

and

$$\alpha((gx^*, gy^*, gz^*), (gu, gv, gw), (gu, gv, gw)) \geq 1.$$

Then, F and g have a unique common tripled fixed point of the form (a, a, a).

Remark 4.11 [23] In Theorem 4.9, we can replace the contractive condition (4.18) by the following:

$$\alpha((gx, gy, gz), (gu, gv, gw), (ga, gb, gc))M_F^a(x, y, z, u, v, w, a, b, c) \leq \psi(M_g^a(x, y, z, u, v, w, a, b, c)). \quad (4.19)$$

Following tripled fixed point results in G-metric spaces can be obtained.

Theorem 4.12 *Let* (X, G) *be a G-metric space,* $F : X^3 \to X$, $g : X \to X$ *and M be a nonempty subset of* X^9. *Assume that*

$$M_F^m(x, y, z, u, v, w, a, b, c) \leq \psi(M_g^m(x, y, z, u, v, w, a, b, c)), \quad (4.20)$$

for all $x, y, z, u, v, w, a, b, c \in \mathcal{X}$ *with* $(gx, gu, ga, gy, gv, gb, gz, gw, gc) \in M$, *where* $\psi \in \Psi$. *Assume also that*

1. $F(X^3) \subseteq g(X)$;
2. *M is an* (F^*, g)*-invariant set which satisfies the transitive property;*
3. *there exist* $x_0, y_0, z_0 \in X$ *such that*

$$(F(x_0, y_0, z_0), F(y_0, x_0, y_0), F(z_0, y_0, x_0), F(x_0, y_0, z_0), F(y_0, x_0, y_0), F(z_0, y_0, x_0), gx_0, gy_0, , gz_0) \in M.$$

Also, suppose that either

(a) *F and g are continuous, g commutes with F and* (X, G) *is G-complete, or*
(b) (X, G) *is M-regular and* $(g(X), G)$ *is G-complete.*

Then, f and g have a tripled coincidence point in X.

Theorem 4.13 *In addition to the hypotheses of Theorem 4.12, suppose that for all* (x, y, z) *and* $(x^*, y^*, z^*) \in X^3$, *there exists* $(u, v, w) \in \mathcal{X}^3$, *such that*

$$(gx, gy, gz, gu, gv, gw, gu, gv, gw) \in M$$

and

$$(gx^*, gy^*, gz^*, gu, gv, gw, gu, gv, gw) \in M.$$

Then, F and g have a unique common tripled fixed point of the form (a, a, a).

Remark 4.14 In Theorem 3.8, we can replace the contractive condition (4.20) by the following:

$$M_F^a(x, y, z, u, v, w, a, b, c) \leq \psi(M_g^a(x, y, z, u, v, w, a, b, c)). \quad (4.21)$$

Application to integral equations

As an application of the (coupled) fixed point theorems established in Sect. 4, we study the existence and uniqueness of a solution for a Fredholm nonlinear integral equation.

To compare our results to the ones in [12, 39], we shall consider the same integral equation, that is,

$$x(t) = \int_a^b (K_1(t, s) + K_2(t, s))(f(s, x(s)) + g(s, x(s)))ds + h(t), \quad (5.1)$$

where $t \in I = [a, b]$.

Consider the space $X = C([0, T], \mathbb{R})$ of continuous functions defined on $I = [a, b]$.

Assume that the functions K_1, K_2, f and g fulfill the following conditions:

Assumption 5.1 (*i*) $K_1(t, s) \geq 0$ and $K_2(t, s) \leq 0$, for all $t, s \in I$;

(*iii*)

$$\sup_{t \in I}\left[\left(\int_a^b K_1(t, s)ds\right)^p + \left(\int_a^b -K_2(t, s)ds\right)^p\right] \leq \frac{r}{2^{3p-2}}, \quad (5.2)$$

for an $0 \leq r < 1$.

(*ii*) There exists $M \subseteq X^6$ which satisfies the transitive property and

1. $(x, y, z, u, v, w) \in M$ iff $(w, v, u, z, y, x) \in M$;

2. $(x,y,z,u,v,w) \in M$ implies that

$$\Bigg(\int_a^b K_1(t,s)[f(s,x(s))+g(s,y(s))]ds + \int_a^b K_2(t,s)[f(s,y(s))+g(s,x(s))]ds,$$
$$\int_a^b K_1(t,s)[f(s,y(s))+g(s,x(s))]ds + \int_a^b K_2(t,s)[f(s,x(s))+g(s,y(s))]ds,$$
$$\int_a^b K_1(t,s)[f(s,z(s))+g(s,u(s))]ds + \int_a^b K_2(t,s)[f(s,u(s))+g(s,z(s))]ds,$$
$$\int_a^b K_1(t,s)[f(s,u(s))+g(s,z(s))]ds + \int_a^b K_2(t,s)[f(s,z(s))+g(s,u(s))]ds,$$
$$\int_a^b K_1(t,s)[f(s,v(s))+g(s,w(s))]ds + \int_a^b K_2(t,s)[f(s,w(s))+g(s,v(s))]ds,$$
$$\int_a^b K_1(t,s)[f(s,w(s))+g(s,v(s))]ds + \int_a^b K_2(t,s)[f(s,v(s))+g(s,w(s))]ds\Bigg) \in M$$

for all $x,y,z,u,v,w \in X$.

(iii) For all $x,y \in X$, the following Lipschitzian type conditions hold:

$$0 \le f(t,x(t)) - f(t,y(t)) \le |x(t)-y(t)| \tag{5.3}$$

and

$$-|x(t)-y(t)| \le g(t,x(t)) - g(t,y(t)) \le 0; \tag{5.4}$$

(iv) For all (x,y) and $(x^*,y^*) \in X^2$, there exists $(u,v) \in X^2$, such that $(x,y,u,v,u,v) \in M$ and $(x^*,y^*,u,v,u,v) \in M$.

Motivated by [12], we present the following definition.

Definition 5.2 A pair $(\alpha,\beta) \in X^2$ is called an M-coupled solution of Eq. (5.1) if, for all $s \in I$,

$$\Bigg(\alpha, \beta, \int_a^b K_1(\cdot,s)[f(s,\alpha(s))+g(s,\beta(s))]ds + \int_a^b K_2(\cdot,s)[f(s,\beta(s))+g(s,\alpha(s))]ds + h(\cdot),$$
$$\int_a^b K_1(\cdot,s)[f(s,\beta(s))+g(s,\alpha(s))]ds + \int_a^b K_2(\cdot,s)[f(s,\alpha(s))+g(s,\beta(s))]ds + h(\cdot),$$
$$\int_a^b K_1(\cdot,s)[f(s,\alpha(s))+g(s,\beta(s))]ds + \int_a^b K_2(\cdot,s)[f(s,\beta(s))+g(s,\alpha(s))]ds + h(\cdot),$$
$$\int_a^b K_1(\cdot,s)[f(s,\beta(s))+g(s,\alpha(s))]ds + \int_a^b K_2(\cdot,s)[f(s,\alpha(s))+g(s,\beta(s))]ds + h(\cdot)\Bigg) \in M.$$

Theorem 5.3 *Consider the integral equation* (5.1) *with* $K_1, K_2 \in C(I \times I, \mathbb{R})$ *and* $h \in C(I,\mathbb{R})$.

Suppose that there exists an M*-coupled solution* (α,β) *of* (5.1) *and that Assumption* 5.1 *is satisfied. Then, the integral equation* (5.1) *has a unique solution in* $C(I,\mathbb{R})$.

Proof It is well known that X is a complete b-metric space with respect to the sup metric

$$d(x,y) = \sup_{t\in I} |x(t)-y(t)|^p, \quad x,y \in C(I,\mathbb{R}).$$

Define

$$G(x,y,z) = \max\{d(x,y), d(y,z), d(z,x)\}.$$

It is easy to see that (X,G) is a complete G_b−metric space with $s = 2^{p-1}$(see, Example 1.3).

Define now the mapping $F : X \times X \to X$ by

$$F(x,y)(t) = \int_a^b K_1(t,s)[f(s,x(s))+g(s,y(s))]ds + \int_a^b K_2(t,s)[f(s,y(s))+g(s,x(s))]ds + h(t), \quad \text{for all } t \in I.$$

According to the computations done by Berinde in [39],

$$
\begin{aligned}
&F(x,y)(t)-F(u,v)(t)\\
&=\int_a^b K_1(t,s)[f(s,x(s))+g(s,y(s))]\mathrm{d}s\\
&\quad+\int_a^b K_2(t,s)[f(s,y(s))+g(s,x(s))]\mathrm{d}s\\
&\quad-\int_a^b K_1(t,s)[f(s,u(s))+g(s,v(s))]\mathrm{d}s\\
&\quad-\int_a^b K_2(t,s)[f(s,v(s))+g(s,u(s))]\mathrm{d}s\\
&=\int_a^b K_1(t,s)[f(s,x(s))-f(s,u(s))+g(s,y(s))\\
&\quad-g(s,v(s))]\mathrm{d}s+\int_a^b K_2(t,s)[f(s,y(s))-f(s,v(s))\\
&\quad+g(s,x(s))-g(s,u(s))]\mathrm{d}s\\
&=\int_a^b K_1(t,s)[(f(s,x(s))-f(s,u(s)))\\
&\quad-(g(s,v(s))-g(s,y(s)))]\mathrm{d}s\\
&\quad-\int_a^b K_2(t,s)[(f(s,y(s))-f(s,v(s)))\\
&\quad-(g(s,u(s))-g(s,x(s)))]\mathrm{d}s\\
&\leq\int_a^b K_1(t,s)[|x(s)-u(s)|+|v(s)-y(s)|]\mathrm{d}s\\
&\quad-\int_a^b K_2(t,s)[|y(s)-v(s)|+|u(s)-x(s)|]\mathrm{d}s.
\end{aligned}
\tag{5.5}
$$

Hence, by (5.5), in view of the fact that $K_2(t,s)\leq 0$, we obtain that

$$
\begin{aligned}
|F(x,y)(t)-F(u,v)(t)|\leq&\int_a^b K_1(t,s)[|x(s)-u(s)|+|v(s)-y(s)|]\mathrm{d}s\\
&-\int_a^b K_2(t,s)[|y(s)-v(s)|+|u(s)-x(s)|]\mathrm{d}s,
\end{aligned}
\tag{5.6}
$$

as all quantities in the right-hand side of (5.5) are non-negative.

Now, from (5.5) we have

$$
\begin{aligned}
&|F(x,y)(t)-F(u,v)(t)|^p\\
&\leq\begin{pmatrix}\int_a^b K_1(t,s)[|x(s)-u(s)|+|v(s)-y(s)|]\mathrm{d}s\\ -\int_a^b K_2(t,s)[|y(s)-v(s)|+|u(s)-x(s)|]\mathrm{d}s\end{pmatrix}^p\\
&\leq 2^{p-1}\begin{pmatrix}\left(\int_a^b K_1(t,s)\mathrm{d}s\right)^p(|x(s)-u(s)|+|v(s)-y(s)|)^p\\ +\left(-\int_a^b K_2(t,s)\mathrm{d}s\right)^p(|y(s)-v(s)|+|u(s)-x(s)|)^p\end{pmatrix}\\
&\leq 2^{p-1}\begin{pmatrix}\left(\int_a^b K_1(t,s)\mathrm{d}s\right)^p 2^{p-1}(|x(s)-u(s)|^p+|v(s)-y(s)|^p)\\ +\left(-\int_a^b K_2(t,s)\mathrm{d}s\right)^p 2^{p-1}(|y(s)-v(s)|^p+|u(s)-x(s)|^p)\end{pmatrix}\\
&\leq 2^{2p-2}\begin{bmatrix}\left(\left(\int_a^b K_1(t,s)ds\right)^p+\left(-\int_a^b K_2(t,s)\mathrm{d}s\right)^p\right)d(x,u)\\ +\left(\left(\int_a^b K_1(t,s)\mathrm{d}s\right)^p+\left(-\int_a^b K_2(t,s)\mathrm{d}s\right)^p\right)d(v,y)\end{bmatrix}.
\end{aligned}
$$

So, we have

$$
\begin{aligned}
|F(x,y)(t)-F(u,v)(t)|^p\leq 2^{2p-2}&\left(\left(\int_a^b K_1(t,s)\mathrm{d}s\right)^p\right.\\
&\left.+\left(-\int_a^b K_2(t,s)\mathrm{d}s\right)^p\right)[d(x,u)+d(y,v)].
\end{aligned}
\tag{5.7}
$$

Similarly, one can obtain that

$$
\begin{aligned}
|F(u,v)(t)-F(z,t)(t)|^p\leq 2^{2p-2}&\left(\left(\int_a^b K_1(t,s)\mathrm{d}s\right)^p\right.\\
&\left.+\left(-\int_a^b K_2(t,s)\mathrm{d}s\right)^p\right)[d(u,z)+d(v,t)]
\end{aligned}
\tag{5.8}
$$

and

$$
\begin{aligned}
|F(z,w)(t)-F(x,y)(t)|^p\leq 2^{2p-2}&\left(\left(\int_a^b K_1(t,s)\mathrm{d}s\right)^p\right.\\
&\left.+\left(-\int_a^b K_2(t,s)\mathrm{d}s\right)^p\right)[d(x,z)+d(y,w)].
\end{aligned}
\tag{5.9}
$$

Taking the supremum with respect to t and using (5.2) we get,

$$
\begin{aligned}
&G(F(x,y),F(u,v),F(z,w))\\
&\quad=\max\{\sup_{t\in I}|F(x,y)(t)-F(u,v)(t)|^p,\sup_{t\in I}|F(u,v)(t)\\
&\quad\quad-F(z,w)(t)|^p,\sup_{t\in I}|F(z,w)(t)-F(x,y)(t)|^p\}\\
&\quad\leq 2^{2p-2}\sup_{t\in I}\left[\left(\int_a^b K_1(t,s)\mathrm{d}s\right)^p+\left(\int_a^b -K_2(t,s)\mathrm{d}s\right)^p\right]\\
&\quad\quad\max\{d(x,u)+d(y,v),d(u,z)+d(v,w),d(x,z)+d(y,w)\}\\
&\quad\leq 2^{2p-2}\sup_{t\in I}\left[\left(\int_a^b K_1(t,s)\mathrm{d}s\right)^p+\left(\int_a^b -K_2(t,s)\mathrm{d}s\right)^p\right]\\
&\quad\quad\times[G(x,u,z)+G(y,v,w)]\\
&\quad\leq\frac{2^{2p-2}\cdot r}{2^{3p-2}}[G(x,u,z)+G(y,v,w)]\\
&\quad=\frac{r}{2^p}[G(x,u,z)+G(y,v,w)]\\
&\quad\leq\frac{r}{2^{p-1}}[\max\{G(x,u,z),G(y,v,w)\}]
\end{aligned}
$$

and analogously,

$$
G(F(y,x),F(v,u),F(w,z))\leq\frac{r}{2^{p-1}}[\max\{G(x,u,z),G(y,v,w)\}].
$$

Combination of the above inequalities is just the contractive condition (2.1) in Theorem 2.2 with $\psi(t)=t$ and $\varphi(t)=(1-r)t$, for all $t>0$.

Now, let $(\alpha,\beta)\in X^2$ be an M-coupled solution of (5.1). Thus, all hypotheses of Theorem 2.2 are satisfied.

This proves that F has a coupled fixed point (x_*,y_*) in X^2. From (iv) and by Theorem 2.3 it follows that $x_*=y_*$, that is, $x_* = F(x_*, x_*)$, and therefore $x_* \in C(I, \mathbb{R})$ is the solution of the integral equation (5.1). □

Conclusion

After that Samet and Vetro [37] introduced the concept of F-invariant set and used it to obtain coupled fixed point results in usual metric spaces, several authors have obtained different coincidence point results in various classes of generalized metric spaces (see, e.g., [32–38]). As we saw in the present paper, (also, see [22, 23]) we showed that

1. coupled and tripled fixed point results can be deduced from corresponding fixed point theorems,
2. coupled and tripled fixed point results via invariant subsets can be deduced from coupled and tripled fixed point results via the concept of α-admissible mappings.

Acknowledgments This article was funded by the Deanship of Scientific Research (DSR), King Abdulaziz University, Jeddah. Therefore, the first author acknowledge with thanks DSR, KAU for financial support.

References

1. Aghajani, A., Abbas, M., Roshan, J.R.: Common fixed point of generalized weak contractive mappings in partially ordered G_b-metric spaces. Filomat **28**(6), 1087–1101 (2014). doi:10.2298/FIL1406087A
2. Aydi, H., Postolache, M., Shatanawi, W.: Coupled fixed point results for (ψ, φ)-weakly contractive mappings in ordered G-metric spaces. Comput. Math. Appl. **63**, 298–309 (2012)
3. Bhaskar, T.G., Lakshmikantham, V.: Fixed point theorems in partially ordered metric spaces and applications. Nonlinear Anal. **65**(7), 1379–1393 (2006)
4. Cho, Y.J., Rhoades, B.E., Saadati, R., Samet, B., Shatanawi, W.: Nonlinear coupled fixed point theorems in ordered generalized metric spaces with integral type. Fixed Point Theory Appl. **2012**, 8 (2012)
5. Shah, M.H., Hussain, N.: Coupled fixed points of weakly F-contractive mappings in topological spaces. Appl. Math. Lett. **24**, 1185–1190 (2011)
6. Choudhury, B.S., Maity, P.: Coupled fixed point results in generalized metric spaces. Math. Comput. Model. **54**, 73–79 (2011)
7. Guo, D., Lakshmikantham, V.: Coupled fixed points of nonlinear operators with applications. Nonlinear Anal. **11**, 623–632 (1987)
8. Hussain, N., Latif, A., Shah, M.H.: Coupled and tripled coincidence point results without compatibility. Fixed Point Theory Appl. **2012**, 77 (2012)
9. Hussain, N., Dorić, D., Kadelburg, Z., Radenović, S.: Suzuki-type fixed point results in metric type spaces. Fixed Point Theory Appl. **2012**, 126 (2012)
10. Khan, M.S., Swaleh, M., Sessa, S.: Fixed point theorems by altering distancces between the points. Bull. Aust. Math. Soc. **30**, 1–9 (1984)
11. Lakshmikantham, V., Ćirić, L.: Coupled fixed point theorems for nonlinear contractions in partially ordered metric spaces. Nonlinear Anal. **70**(12), 4341–4349 (2009)
12. Luong, N.V., Thuan, N.X.: Coupled fixed point theorems in partially ordered G-metric spaces. Math. Comput. Model. **55**, 1601–1609 (2012)
13. Mustafa, Z., Roshan, J.R., Parvaneh, V.: Coupled coincidence point results for (ψ, φ)-weakly contractive mappings in partially ordered G_b-metric spaces. Fixed Point Theory Appl. **2013**, 206 (2013). doi:10.1186/1687-1812-2013-206
14. Mustafa, Z., Sims, B.: A new approach to generalized metric spaces. J. Nonlinear Convex Anal. **7**(2), 289–297 (2006)
15. Nieto, J.J., Rodriguez-López, R.: Existence and uniqueness of fixed point in partially ordered sets and applications to ordinary diferential equations. Order **22**(3), 223–239 (2005)
16. Nieto, J.J., Rodriguez-Lopez, R.: Contractive mapping theorems in partially ordered sets and applications to ordinary differential equations. Order **22**(3), 223–239 (2005)
17. Nieto, J.J., Rodriguez-Lopez, R.: Existence and uniqueness of fixed point in partially ordered sets and applications to ordinary differential equations. Acta Math. Sin. (Engl. Ser.) **23**(12), 2205–2212 (2007)
18. Ran, A.C.M., Reurings, M.C.B.: A fixed point theorem in partially ordered sets and some applications to matrix equations. Proc. Am. Math. Soc. **132**(5), 1435–1443 (2004)
19. Mustafa, Z., Roshan, J.R., Parvaneh, V.: Existence of tripled coincidence point in ordered G_b-metric spaces and applications to a system of integral equations. J. Inequal. Appl. **453** (accepted) (2013)
20. Mursaleen, M., Mohiuddine, S.A., Agarwal, R.P.: Coupled fixed point theorems for α-ψ-contractive type mappings in partially ordered metric spaces. Fixed Point Theory Appl. **2012**, 124 (2012)
21. Agarwal, P.: A note on 'Coupled fixed point theorems for α-ψ-contractive-type mappings in partially ordered metric spaces'. Fixed Point Theory Appl. **2013**, 216 (2013)
22. Parvaneh, V., Roshan, J.R., Radenović, S.: Existence of tripled coincidence points in ordered b-metric spaces and an application to a system of integral equations. Fixed Point Theory Appl. **2013**, 130 (2013)
23. Kutbi, M.A., Hussain, N., Roshan, J.R., Parvaneh, V.: Coupled and Tripled Coincidence Point Results with Application to Fredholm Integral Equations, Abstract and Applied Analysis, vol. 2014, p. 18 (Article ID 568718)
24. Berinde, V., Borcut, M.: Tripled fixed point theorems for contractive type mappings in partially ordered metric spaces. Nonlinear Anal. **74**, 4889–4897 (2011)
25. Borcut, M.: Tripled coincidence theorems for contractive type mappings in partially ordered metric spaces. Appl. Math. Comput. **218**, 7339–7346 (2012)
26. Borcut, M., Berinde, V.: Tripled coincidence theorems for contractive type mappings in partially ordered metric spaces. Appl. Math. Comput. **218**, 5929–5936 (2012)
27. Czerwik, S.: Contraction mappings in b-metric spaces. Acta Math. Inf. Univ. Ostrav. **1**, 5–11 (1993)
28. Jungck, G.: Compatible mappings and common fixed points. Int. J. Math. Math. Sci. **9**(4), 771–779 (1986)
29. Samet, B., Vetro, C., Vetro, P.: Fixed point theorem for $\alpha - \psi$-contractive type mappings. Nonlinear Anal. **75**, 2154–2165 (2012)
30. Alghamdi, M.A., Karapınar, E.: G-β-ψ-contractive-type mappings and related fixed point theorems. J. Inequal. Appl. **2013**, 70 (2013)
31. Hussain, N., Karapinar, E., Salimi, P., Vetro, P.: Fixed point results for G^m-Meir-Keeler contractive and G-(α, ψ)-Meir–Keeler contractive mappings. Fixed Point Theory Appl. **2013**, 34 (2013)

32. Charoensawan, P., Klanarong, C.: Coupled coincidence point theorems for φ-contractive under (f, g)-invariant set in complete metric space. Int. J. Math. Anal. **7**(33–36), 1685–1701 (2013)
33. Charoensawan, P.: Tripled fixed points theorems for ϕ-contractive mixed monotone operators on partially ordered metric spaces. Appl. Math. Sci. **6**, 5229–5239 (2012)
34. Charoensawan, P.: Tripled fixed points theorems for a ϕ-contractive mapping in a complete metric spaces without the mixed g-monotone property. Fixed Point Theory Appl. **2013** (Article ID 252) (2013)
35. Thangthong, C., Charoensawan, P.: Coupled coincidence point theorems for a φ-contractive mapping in partially ordered G-metric spaces without mixed g-monotone property. Fixed Point Theory Appl. **2014**, 128 (2014)
36. Batra, R., Vashistha, S.: Coupled coincidence point theorems for nonlinear contractions under (F, g)-invariant set in cone metric spaces. J. Nonlinear Sci. Appl. **6**, 86–96 (2013)
37. Samet, B., Vetro, C.: Coupled fixed point F-invariant set and fixed point of N-order. Ann. Funct. Anal. **1**, 46–56 (2010)
38. Sintunavarat, W., Radenović, S., Golubović, Z., Kuman, P.: Coupled fixed point theorems for F-invariant set. Appl. Math. Inf. Sci. **7**(1), 247–255 (2013)
39. Berinde, V.: Coupled fixed point theorems for contractive mixed monotone mappings in partially ordered metric spaces. Nonlinear Anal. **75**, 3218–3228 (2012)
40. Roshan, J.R., Shobkolaei, N., Sedghi, S., Parvaneh, V., Radenović, S.: Common fixed point theorem for three maps in discontinuous G_b-metric spaces. Acta Mathematica Scientia 34B(5), 1–12 (2014)
41. Radenović, S: A note on tripled coincidence and tripled common fixed point theorems in partially ordered metric spaces. Appl. Math. Comput. **2**(236), 367–372 (2014)
42. Agarwal, R.P., Kadelburg, Z., Radenović, S.: On coupled fixed point results in asymmetric G-metric spaces. J. Inequal. Appl. **2013**, 528 (2013)
43. Radenović, S.: Remarks on some recent coupled coincidence point results in symmetric G-metric spaces. J. Oper. **2013**, 8 (2013). Art. ID 290525
44. Alsulami, H.H, Roldan, A., Karapınar, E., Radenović, S.: Some inevitable remarks on "Tripled fixed point theorems for mixed monotone Kannan type contractive mappings". J. Appl. Math. **2014**, 7 (2014). Art. ID 392301

9

An order theoretic approach in fixed point theory

Yaé Ulrich Gaba

Abstract In the present article, we show the existence of a coupled fixed point for an order preserving mapping in a preordered left K-complete quasi-pseudometric space using a preorder induced by an appropriate function. We also define the concept of left-weakly related mappings on a preordered space and discuss common coupled fixed points for two and three left-weakly related mappings in the same space. Similar results are given for right-weakly related mappings, the dual notion of left-weakly related mappings.

Keywords Quasi-pseudometric space · Left K-complete · Preordered space · Weakly left-related

Mathematics Subject Classification 47H10 · 47H03

Introduction

Fixed point theory plays a major role in many applications including variational and linear inequalities, optimization and applications in the field of approximation theory and minimum norm problem. The first important result on fixed points for contractive-type mapping was the well-known Banach's contraction principle that appeared in explicit form in his thesis in 1922, where it was used to establish the existence of a solution for an integral equation. This theorem is a key result in nonlinear analysis. Another interesting result on fixed points for contractive-type mapping is due to Edelstein (1962) who actually obtained slightly more general versions. Many years later, another direction of such generalizations (see [1, 2]) has been obtained by weakening the requirements in the contractive condition and in compensation, by simultaneously enriching the metric space structure with a partial order.

In this process of generalization, the study of common fixed points of mappings satisfying certain contractive conditions has also been at the center of rigorous research activity, see [3].

Bhashkar and Lakshmikantham [4] introduced the concept of a coupled fixed point of a mapping $F : X \times X \to X$ (where X a non-empty set) and established some coupled fixed point theorems in partially ordered complete metric spaces. By doing so, they opened the way to a flourishing sub-area in the fixed point theory.

Moreover, in the last few years, there has been a growing interest in the theory of quasi-metric spaces and other related structures such as asymmetric normed linear spaces (see for instance [5]). This theory provides a convenient framework in the study of several problems in theoretical computer science and approximation theory. It is in this setting that we give our results.

The aim of this paper is to analyze the existence of common and coupled fixed points for mapping defined on a left K-complete quasi-pseudometric space (X, d). The technique of proof is different and more natural in the sense we do not use any contractive conditions. In our work, we show the existence of a coupled fixed point for an isotone mapping in a preordered left K-complete quasi-pseudometric space using a preorder induced by an appropriate function ϕ. Furthermore, common coupled fixed point for two and three mappings satisfying a certain relation that we specify later is also discussed in the same space.

Y. U. Gaba (✉)
Department of Mathematics and Applied Mathematics,
University of Cape Town, Rondebosch 7701, South Africa
e-mail: gabayae2@gmail.com

Preliminaries

In this section, we recall some elementary definitions from the asymmetric topology and the order theory, which are necessary for a good understanding of the work below.

Definition 2.1 Consider a non-empty set X and a binary relation $\preceq$ on X. Then, $\preceq$ is a preorder, or quasiorder, if it is reflexive and transitive, i.e., for all a, b and $c \in X$, we have that:

- $a \preceq a$ (reflexivity);
- if $a \preceq b$ and $b \preceq c$, then $a \preceq c$ (transitivity).

A set that is equipped with a preorder is called a preordered space (or proset).

Definition 2.2 Let $(X, \preceq_X)$ and $(Y, \preceq_Y)$ be two prosets. A map $T : X \to Y$ is said to be preorder-preserving or isotone if for any $x, y \in X$,

$$x \preceq_X y \Longrightarrow Tx \preceq_Y Ty.$$

Similarly, for any family $(Xi, \preceq_{X_i}), i = 1, 2, \ldots, n;\ (Y, \preceq_Y)$ of posets, a mapping $F : X_1 \times X_2 \times \cdots \times X_n \to Y$ is said to be preorder-preserving or isotone if for any $(x_1, x_2, \ldots, x_n)$, $(z_1, z_2, \ldots, z_n) \in X_1 \times X_2 \times \cdots \times X_n$,

$$x_i \preceq_{X_i} z_i \text{ for all } i = 1, 2, \ldots, n \Longrightarrow F(x_1, x_2, \ldots, x_n) \preceq_Y F(z_1, z_2, \ldots, z_n).$$

Dually, we have

Definition 2.3 (*Compare* [6]) Let X be a non-empty set. A function $d : X \times X \to [0, \infty)$ is called a quasi-pseudometric on X if:

(i) $d(x,x) = 0 \quad \forall\, x \in X,$
(ii) $d(x,z) \leq d(x,y) + d(y,z) \quad \forall\, x, y, z \in X.$

Moreover, if $d(x,y) = 0 = d(y,x) \Longrightarrow x = y$, then d is said to be a T_0-quasi-pseudometric. The latter condition is referred to as the T_0-condition.

Remark 2.4

- Let d be a quasi-pseudometric on X, then the map d^{-1} defined by $d^{-1}(x,y) = d(y,x)$ whenever $x, y \in X$ is also a quasi-pseudometric on X, called the conjugate of d. In the literature, d^{-1} is also denoted d^t or $\bar{d}$.
- It is easy to verify that the function d^s defined by $d^s := d \vee d^{-1}$, i.e., $d^s(x,y) = \max\{d(x,y), d(y,x)\}$ defines a metric on X whenever d is a T_0-quasi-pseudometric on X.

Let (X, d) be a quasi-pseudometric space. For $x \in X$ and $\varepsilon > 0$,

$$B_d(x, \varepsilon) = \{y \in X : d(x,y) < \varepsilon\}$$

denotes the open ε-ball at x. The collection of all such balls yields a base for the topology $\tau(d)$ induced by d on X. Hence, for any $A \subseteq X$, we shall, respectively, denote by $\mathrm{int}_{\tau(d)}A$ and $\mathrm{cl}_{\tau(d)}A$ the interior and the closure of the set A with respect to the topology $\tau(d)$.

Similarly, for $x \in X$ and $\varepsilon \geq 0$,

$$C_d(x, \varepsilon) = \{y \in X : d(x,y) \leq \varepsilon\}$$

denotes the closed ε-ball at x.

Definition 2.5 Let (X, d) be a quasi-pseudometric space. The convergence of a sequence (x_n) to x with respect to $\tau(d)$, called d-convergence or left-convergence and denoted by $x_n \xrightarrow{d} x$, is defined in the following way

$$x_n \xrightarrow{d} x \Longleftrightarrow d(x, x_n) \longrightarrow 0. \tag{1}$$

Similarly, the convergence of a sequence (x_n) to x with respect to $\tau(d^{-1})$, called d^{-1}-convergence or right-convergence and denoted by $x_n \xrightarrow{d^{-1}} x$,
is defined in the following way

$$x_n \xrightarrow{d^{-1}} x \Longleftrightarrow d(x_n, x) \longrightarrow 0. \tag{2}$$

Finally, in a quasi-pseudometric space (X, d), we shall say that a sequence (x_n) d^s-converges to x if it is both left and right convergent to x, and we denote it as $x_n \xrightarrow{d^s} x$ or $x_n - \to x$ when there is no confusion. Hence,

$$x_n \xrightarrow{d^s} x \iff x_n \xrightarrow{d} x \text{ and } x_n \xrightarrow{d^{-1}} x.$$

Definition 2.6 A sequence (x_n) in a quasi-pseudometric (X, d) is called

(a) left d-Cauchy if for every $\epsilon > 0$, there exist $x \in X$ and $n_0 \in \mathbb{N}$ such that

$$\forall\, n \geq n_0 \quad d(x, x_n) < \epsilon;$$

(b) left K-Cauchy if for every $\epsilon > 0$, there exists $n_0 \in \mathbb{N}$ such that

$$\forall\, n, k : n_0 \leq k \leq n \quad d(x_k, x_n) < \epsilon;$$

(c) d^s-Cauchy if for every $\epsilon > 0$, there exists $n_0 \in \mathbb{N}$ such that

$$\forall n, k \geq n_0 \quad d^s(x_n, x_k) < \epsilon.$$

Dually, we define in the same way, right d-Cauchy and right K-Cauchy sequences.

Remark 2.7

- d^s-Cauchy $\Longrightarrow$ left K-Cauchy $\Longrightarrow$ left d-Cauchy. The same implications hold for the corresponding right notions. None of the above implications is reversible.

- A sequence is left d-Cauchy with respect to d if and only if it is right K-Cauchy with respect to d^{-1}.
- A sequence is left K-Cauchy with respect to d if and only if it is right K-Cauchy with respect to d^{-1}.
- A sequence is d^s-Cauchy if and only if it is both left and right d-Cauchy.

Definition 2.8 (*Compare* [6]) A quasi-pseudometric space (X,d) is called

- left-K-complete provided that any left K-Cauchy sequence is d-convergent,
- left Smyth sequentially complete if any left K-Cauchy sequence is d^s-convergent.

The dual notions of right-completeness are easily derived from the above definition.

Definition 2.9 A T_0-quasi-pseudometric space (X,d) is called bicomplete provided that the metric d^s on X is complete.

Definition 2.10 Let (X,d) be a quasi-pseudometric type space. A function $T : X \to X$ is called d-sequentially continuous or left-sequentially continuous if for any d-convergent sequence (x_n) with $x_n \xrightarrow{d} x$, the sequence (Tx_n) d-converges to Tx, i.e., $Tx_n \xrightarrow{d} Tx$.

Similarly, a function $T : X \times X \to X$ is said to be d-sequentially continuous or left-sequentially continuous if for any sequences (x_n) and (y_n) such that $x_n \xrightarrow{d} x$ and $y_n \xrightarrow{d} y$, then $F(x_n, y_n) \xrightarrow{d} F(x,y)$.

Similarly, we define a d^{-1}-sequentially continuous or right-sequentially continuous and d^s-sequentially continuous functions.

Definition 2.11 (*Compare* [2]) Let X be a non-empty set. An element $(x,y) \in X \times X$ is called:

(E1) a coupled fixed point of the mappings $F : X \times X \to X$ if $F(x,y) = x$ and $F(y,x) = y$;

(E2) a coupled coincidence point of the mappings $F : X \times X \to X$ and $T : X \to X$ if $T(x,y) = Tx$ and $T(y,x) = Ty$, in this case (Tx, Ty) is called the coupled point of coincidence;

(E3) a common coupled fixed point of the mappings $F : X \times X \to X$ and $T : X \to X$ if $F(x,y) = Tx = x$ and $F(y,x) = Ty = y$.

Definition 2.12 Let X be a non-empty set. An element $(x,y) \in X \times X$ is called:

(D1) a common coupled coincidence point of the mappings $F : X \times X \to X$ and $T, R : X \to X$ if $F(x,y) = Tx = Rx$ and $F(y,x) = Ty = Ry$;

(D2) a common coupled fixed point of the mappings $F : X \times X \to X$ and $T, R : X \to X$ if $F(x,y) = Tx = Rx = x$ and $F(y,x) = Ty = Ry = y$.

First results

We start by the following lemma.

Lemma 3.1 Let (X,d) be a quasi-pseudometric space and $\phi : X \to \mathbb{R}$ a map. Define the binary relation $'' \preceq''$ on X as follows:

$$x \preceq y \iff d(x,y) \le \phi(y) - \phi(x).$$

Then, $'' \preceq''$ is a preorder on X. It will be called the preorder induced by ϕ.

Proof 3.2

1. • Reflexivity: For all $x \in X$, $d(x,x) = 0 = \phi(x) - \phi(x)$ hence $x \preceq x$, i.e., $'' \preceq''$ is reflexive.
2. • Transitivity: For $x,y,z \in X$ s.t. $x \preceq y$ and $y \preceq z$, we have

$$d(x,y) \le \phi(y) - \phi(x),$$

and

$$d(y,z) \le \phi(z) - \phi(y),$$

and since

$$\begin{aligned} d(x,z) &\le d(x,y) + d(y,z) \\ &\le \phi(y) - \phi(x) + \phi(z) - \phi(y) \\ &= \phi(z) - \phi(x). \end{aligned}$$

we have $x \preceq z$. Thus, $'' \preceq''$ is transitive, and so the relation $'' \preceq''$ is a preorder on X.

Remark 3.3 If in addition, the space (X,d) is T_0, then the relation $\precsim$ defined by

$$x \precsim y \iff d^s(x,y) \le \phi(y) - \phi(x).$$

is a partial order on X.

Now, we prove the following theorem.

Theorem 3.4 *Let (X,d) be a Hausdorff left K-complete T_0-quasi-pseudometric space, $\phi : X \to \mathbb{R}$ be a bounded from above function and $'' \preceq''$ the preorder induced by ϕ. Let $F : X \times X \to X$ be a preorder-preserving and d-sequentially continuous mapping on X such that there exist two elements $x_0, y_0 \in X$ with*

$$x_0 \preceq F(x_0, y_0) \quad \text{and} \quad y_0 \preceq F(y_0, x_0).$$

Then, F has a coupled fixed point in X.

Proof 3.5 Let $x_0, y_0 \in X$ with $x_0 \preceq F(x_0, y_0)$ and $y_0 \preceq F(y_0, x_0)$. We construct the sequences (x_n) and (y_n) in X as follows:

$$x_{n+1} = F(x_n, y_n) \text{ and } y_{n+1} = F(y_n, x_n) \quad \text{for all } n \geq 0. \tag{3}$$

We shall show that

$$x_n \preceq x_{n+1} \qquad \text{for all } n \geq 0, \tag{4}$$

and

$$y_n \preceq y_{n+1} \qquad \text{for all } n \geq 0. \tag{5}$$

For this purpose, we use the mathematical induction.

Since $x_0 \preceq F(x_0, y_0)$ and $y_0 \preceq F(y_0, x_0)$ and as $x_1 = F(x_0, y_0)$ and $y_1 = F(y_0, x_0)$, we have $x_0 \preceq x_1$ and $y_0 \preceq y_1$. Thus, (4) and (5) hold for $n = 0$.

Suppose that (4) and (10) hold for some $k > 0$. Then, since $x_k \preceq x_{k+1}$ and $y_k \preceq y_{k+1}$ and F is preorder preserving, we have

$$x_{k+1} = F(x_k, y_k) \preceq F(x_{k+1}, y_{k+1}) = x_{k+2} \tag{6}$$

and

$$y_{k+1} = F(y_k, x_k) \preceq F(y_{k+1}, x_{k+1}) = y_{k+2}. \tag{7}$$

Thus, by mathematical induction, we conclude that (4) and (5) hold for all $n \geq 0$. Therefore,

$$x_0 \preceq x_1 \preceq x_2 \preceq \cdots \preceq x_n \preceq \cdots,$$

and

$$y_0 \preceq y_1 \preceq y_2 \preceq \cdots \preceq y_n \preceq \cdots.$$

By definition of the preorder, we have

$$\phi(x_0) \leq \phi(x_1) \leq \phi(x_2) \leq \cdots \leq \phi(x_n) \leq \cdots,$$

and

$$\phi(y_0) \leq \phi(y_1) \leq \phi(y_2) \leq \cdots \leq \phi(y_n) \leq \cdots.$$

Hence, the sequences $(\phi(x_n))$ and $(\phi(y_n))$ are non-decreasing sequences of real numbers. Since ϕ is bounded from above, the sequences $(\phi(x_n))$ and $(\phi(y_n))$ are convergent and, therefore, Cauchy. This entails that for any $\varepsilon > 0$, there exists $n_0 \in \mathbb{N}$ such that for any $m > n > n_0$, we have $\phi(x_m) - \phi(x_n) < \varepsilon$ and $\phi(y_m) - \phi(y_n) < \varepsilon$. Since whenever $m > n > n_0$, $x_n \preceq x_m$ and $y_n \preceq y_m$, it follows that

$$d(x_n, x_m) \leq \phi(x_m) - \phi(x_n) < \varepsilon,$$

and

$$d(y_n, y_m) \leq \phi(y_m) - \phi(y_n) < \varepsilon.$$

We conclude that (x_n) and (y_n) are left K-Cauchy in X and since X is left K-complete, there exist $x^*, y^* \in X$ such that $x_n \xrightarrow{d} x^*$ and $y_n \xrightarrow{d} y^*$. Since F is d-sequentially continuous, we have

$$x_n \xrightarrow{d} x^* \iff x_n = F(x_{n-1}, y_{n-1}) \xrightarrow{d} x^* \iff F(x^*, y^*) = x^*,$$

and

$$y_n \xrightarrow{d} y^* \iff y_n = F(y_{n-1}, x_{n-1}) \xrightarrow{d} y^* \iff F(y^*, x^*) = y^*.$$

Thus, we have proved that $F(x^*, y^*) = x^*$ and $F(y^*, x^*) = y^*$, i.e., (x^*, y^*) is a coupled fixed point of F.

Corollary 3.6 Let (X, d) be a Hausdorff right K-complete T_0-quasi-pseudometric space, $\phi : X \to \mathbb{R}$ be a bounded from below function and "$\preceq$" the preorder induced by ϕ. Let $F : X \times X \to X$ be a preorder-preserving and d^{-1}-sequentially continuous mapping on X such that there exist two elements $x_0, y_0 \in X$ with

$$x_0 \succeq F(x_0, y_0) \quad \text{and } y_0 \succeq F(y_0, x_0).$$

Then, F has a coupled fixed point in X.

Corollary 3.7 Let (X, d) be a bicomplete T_0-quasi-pseudometric space, $\phi : X \to \mathbb{R}$ be a bounded from below function and "$\preceq$" the preorder induced by ϕ. Let $F : X \times X \to X$ be a preorder-preserving and d^s-sequentially continuous mapping on X such that there exist two elements $x_0, y_0 \in X$ with

$$x_0 \precsim F(x_0, y_0) \quad \text{and } y_0 \precsim F(y_0, x_0).$$

Then, F has a coupled fixed point in X.

Common coupled fixed point

Now, we define the concept of weakly related mappings on preordered spaces as follows:

Definition 4.1 Let $(X, \preceq)$ be a preordered space, and $F : X \times X \to X$ and $g : X \to X$ be two mappings. Then, the pair $\{F, g\}$ is said to be weakly left-related if the two following conditions are satisfied:

(C1) $F(x, y) \preceq gF(x, y)$ and $gx \preceq F(gx, gy)$ for all $(x, y) \in X \times X$,

(C2) $F(y, x) \preceq gF(y, x)$ and $gy \preceq F(gy, gx)$ for all $(x, y) \in X \times X$.

Definition 4.2 Let $(X, \preceq)$ be a preordered space, and $F : X \times X \to X$ and $g : X \to X$ be two mappings. Then, the pair $\{F, g\}$ is said to be weakly right-related if the two following conditions are satisfied:

(D1) $gF(x, y) \preceq F(x, y)$ and $F(gx, gy) \preceq gx$ for all $(x, y) \in X \times X$,

(D2) $gF(y,x) \preceq F(y,x)$ and $F(gy,gx) \preceq gy$ for all $(x,y) \in X \times X$.

We now state and prove the first common coupled fixed point existence theorem for the weakly related mappings.

Theorem 4.3 *Let (X,d) be a Hausdorff left K-complete T_0-quasi-pseudometric space, $\phi : X \to \mathbb{R}$ be a bounded from above function and " $\preceq$" the preorder induced by ϕ. Let $F : X \times X \to X$ and $G : X \to X$ be two d-sequentially continuous mapping on X such that the pair $\{F,G\}$ is weakly left-related. If there exist two elements $x_0, y_0 \in X$ with*

$$x_0 \preceq F(x_0,y_0) \quad \text{and} \quad y_0 \preceq F(y_0,x_0).$$

Then, F and G have a common coupled fixed point in X.

Proof 4.4 Let $x_0, y_0 \in X$ with $x_0 \preceq F(x_0,y_0)$ and $y_0 \preceq F(y_0,x_0)$. We construct the sequences (x_n) and (y_n) in X as follows:

$$x_{2n+1} = F(x_{2n},y_{2n}) \text{and } x_{2n+2} = Gx_{2n+1} \text{ for all } n \geq 0, \tag{8}$$

and

$$y_{2n+1} = F(y_{2n},x_{2n}) \text{ and } y_{2n+2} = Gy_{2n+1} \text{for all } n \geq 0. \tag{9}$$

We shall show that

$$x_n \preceq x_{n+1} \qquad \text{for all } n \geq 0, \tag{10}$$

and

$$y_n \preceq y_{n+1} \qquad \text{for all } n \geq 0. \tag{11}$$

Since $x_0 \preceq F(x_0,y_0)$, using (8), we have and $x_0 \preceq x_1$. Again since the pair $\{F,G\}$ is weakly left-related, we have, from (8) $x_1 = F(x_0,y_0) \preceq GF(x_0,y_0) = Gx_1 = x_2$, i.e., $x_1 \preceq x_2$. Also, since $Gx_1 \preceq F(Gx_1,Gy_1)$, and using (8), we have $x_2 = Gx_1 \preceq F(Gx_1,Gy_1) = F(x_2,y_2) = x_3$, i.e., $x_2 \preceq x_3$. Similarly, using the fact that the pair $\{F,G\}$ is weakly left-related and the relations (8), we get

$$x_0 \preceq x_1 \preceq x_2 \preceq \cdots \preceq x_n \preceq \cdots.$$

A similar reasoning, using fact that the pair $\{F,G\}$ is weakly left-related and the relations (9), leads to

$$y_0 \preceq y_1 \preceq y_2 \preceq \cdots \preceq y_n \preceq \cdots.$$

By definition of the preorder, we have

$$\phi(x_0) \leq \phi(x_1) \leq \phi(x_2) \leq \cdots \leq \phi(x_n) \leq \cdots$$

and

$$\phi(y_0) \leq \phi(y_1) \leq \phi(y_2) \leq \cdots \leq \phi(y_n) \leq \cdots.$$

Hence, the sequences $(\phi(x_n))$ and $(\phi(y_n))$ are non-decreasing sequences of real numbers. Since ϕ is bounded from above, the sequences $(\phi(x_n))$ and $(\phi(y_n))$ are convergent and, therefore, Cauchy. This entails that for any $\varepsilon > 0$, there exists $n_0 \in \mathbb{N}$ such that for any $m > n > n_0$, we have $\phi(x_m) - \phi(x_n) < \varepsilon$ and $\phi(y_m) - \phi(y_n) < \varepsilon$. Since whenever $m > n > n_0$, $x_n \preceq x_m$ and $y_n \preceq y_m$, it follows that

$$d(x_n,x_m) \leq \phi(x_m) - \phi(x_n) < \varepsilon,$$

and

$$d(y_n,y_m) \leq \phi(y_m) - \phi(y_n) < \varepsilon.$$

It follows that (x_n) and (y_n) are left K-Cauchy in X and since X is left K-complete, there exist $x^*, y^* \in X$ such that $x_n \xrightarrow{d} x^*$ and $y_n \xrightarrow{d} y^*$.

Since F and G are d-sequentially continuous, it is easy to see, using (8), that

$$x_{2n+1} \xrightarrow{d} x^* \iff x_{2n+1} = F(x_{2n},y_{2n}) \xrightarrow{d} x^* \iff F(x^*,y^*) = x^*,$$

and

$$x_{2n+2} \xrightarrow{d} x^* \iff x_{2n+2} = Gx_{2n+1} \xrightarrow{d} x^* \iff Gx^* = x^*,$$

and hence

$$Gx^* = x^* = F(x^*,y^*).$$

Similarly, since F and G are d-sequentially continuous, using (9), we easily derive that

$$Gy^* = y^* = F(y^*,x^*).$$

Hence, (x^*,y^*) is a coupled common fixed point of F and G.

Corollary 4.5 Let (X,d) be a Hausdorff right K-complete T_0-quasi-pseudometric space, $\phi : X \to \mathbb{R}$ be a bounded from below function and " $\preceq$" the preorder induced by ϕ. Let $F : X \times X \to X$ and $G : X \to X$ be two d^{-1}-sequentially continuous mapping on X such that the pair $\{F,G\}$ is weakly right-related. If there exist two elements $x_0, y_0 \in X$ with

$$F(x_0,y_0) \preceq x_0 \quad \text{and} \quad F(y_0,x_0) \preceq y_0.$$

Then, F and G have a common coupled fixed point in X.

Corollary 4.6 Let (X,d) be a bicomplete T_0-quasi-pseudometric space, $\phi : X \to \mathbb{R}$ be a bounded from above function and "$\precsim$" the preorder induced by ϕ. Let $F : X \times X \to X$ and $G : X \to X$ be two d^s-sequentially continuous mapping on X such that the pair $\{F,G\}$ satisfies the following two conditions:

($C1'$) $F(x,y) \precsim GF(x,y)$ and $Gx \precsim F(Gx,Gy)$ for all $(x,y) \in X \times X$,

($C2'$) $F(y,x) \precsim GF(y,x)$ and $Gy \precsim F(Gy,Gx)$ for all $(x,y) \in X \times X$.

If there exist two elements $x_0, y_0 \in X$ with

$x_0 \precsim F(x_0, y_0)$ and $y_0 \precsim F(y_0, x_0)$.

Then, F and G have a common coupled fixed point in X.

Theorem 4.7 *Let (X, d) be a Hausdorff left K-complete T_0-quasi-pseudometric space, $\phi : X \to \mathbb{R}$ be a bounded from above function and "$\preceq$" the preorder induced by ϕ. Let $F : X \times X \to X$ and $G, H : X \to X$ be three d-sequentially continuous mapping on X such that the pairs $\{F, G\}$ and $\{F, H\}$ are weakly left-related. Then, F, G and H have a common coupled fixed point in X.*

Proof 4.8 Let $x_0, y_0 \in X$. We construct the sequences (x_n) and (y_n) in X as follows:

$$Hx_{3n-3} = x_{3n-2},\ x_{3n-1} = F(x_{3n-2}, y_{3n-2}) \text{and} x_{3n} = Gx_{3n-1} \quad \text{for all } n \geq 1, \tag{12}$$

and

$$Hy_{3n-3} = y_{3n-2},\ y_{3n-1} = F(y_{3n-2}, x_{3n-2}) \text{ and } y_{3n} = Gy_{3n-1} \quad \text{for all } n \geq 1. \tag{13}$$

We shall show that

$$x_n \preceq x_{n+1} \qquad \text{for all } n \geq 0, \tag{14}$$

and

$$y_n \preceq y_{n+1} \qquad \text{for all } n \geq 0. \tag{15}$$

We have $x_1 = Hx_0$. Since the pair $\{F, H\}$ is weakly left-related, we have $x_1 = Hx_0 \preceq F(Hx_0, Hy_0) = F(x_1, y_1) = x_2$. Again since the pair $\{F, G\}$ is weakly left-related, we have $x_2 = F(x_1, y_1) \preceq GF(x_1, y_1) = Gx_2 = x_3$. Similarly, using the fact that the pairs $\{F, G\}$ and $\{F, H\}$ are weakly left-related and the relations (12), we get

$$x_0 \preceq x_1 \preceq x_2 \preceq x_3 \preceq \cdots \preceq x_n \preceq \cdots.$$

A similar reasoning, using fact that the pair $\{F, g\}$ is weakly left-related and the relations (13), leads to

$$y_0 \preceq y_1 \preceq y_2 \preceq y_3 \preceq \cdots \preceq y_n \preceq \cdots.$$

By definition of the preorder, we have

$$\phi(x_0) \leq \phi(x_1) \leq \phi(x_2) \leq \cdots \leq \phi(x_n) \leq \cdots$$

and

$$\phi(y_0) \leq \phi(y_1) \leq \phi(y_2) \leq \cdots \leq \phi(y_n) \leq \cdots.$$

Hence, the sequences $(\phi(x_n))$ and $(\phi(y_n))$ are non-decreasing sequences of real numbers. Since ϕ is bounded from above, the sequences $(\phi(x_n))$ and $(\phi(y_n))$ are convergent and, therefore, Cauchy. This entails that for any $\varepsilon > 0$, there exists $n_0 \in \mathbb{N}$ such that for any $m > n > n_0$, we have $\phi(x_m) - \phi(x_n) < \varepsilon$ and $\phi(y_m) - \phi(y_n) < \varepsilon$. Since whenever $m > n > n_0$, $x_n \preceq x_m$ and $y_n \preceq y_m$, it follows that

$$d(x_n, x_m) \leq \phi(x_m) - \phi(x_n) < \varepsilon,$$

and

$$d(y_n, y_m) \leq \phi(y_m) - \phi(y_n) < \varepsilon.$$

It follows that (x_n) and (y_n) are left K-Cauchy in X and since X is left K-complete, there exist $x^*, y^* \in X$ such that $x_n \xrightarrow{d} x^*$ and $y_n \xrightarrow{d} y^*$.

Since F, G and H are d-sequentially continuous, it is easy to see, using (12), that

$$x_{3n-1} \xrightarrow{d} x^* \Longleftrightarrow x_{3n-1} = F(x_{3n-2}, y_{3n-2}) \xrightarrow{d} x^* \Longleftrightarrow F(x^*, y^*) = x^*,$$

and

$$x_{3n} \xrightarrow{d} x^* \Longleftrightarrow x_{3n} = Gx_{3n-1} \xrightarrow{d} x^* \Longleftrightarrow Gx^* = x^*,$$

also

$$x_{3n-2} \xrightarrow{d} x^* \Longleftrightarrow x_{3n-2} = Hx_{3n-3} \xrightarrow{d} x^* \Longleftrightarrow Hx^* = x^*,$$

and hence

$$Hx^* = Gx^* = x^* = F(x^*, y^*).$$

Similarly, since F, G and H are d-sequentially continuous, using (13), we easily derive that

$$Hy^* = Gy^* = y^* = F(y^*, x^*).$$

Hence, (x^*, y^*) is a coupled common fixed point of F, G and H.

Corollary 4.9 Let (X, d) be a Hausdorff right K-complete T_0-quasi-pseudometric space, $\phi : X \to \mathbb{R}$ be a bounded from below function and "$\preceq$" the preorder induced by ϕ. Let $F : X \times X \to X$ and $G, H : X \to X$ be three d^{-1}-sequentially continuous mapping on X such that the pairs $\{F, G\}$ and $\{F, H\}$ are weakly right-related. Then, F, G and H have a common coupled fixed point in X.

Corollary 4.10 Let (X, d) be a bicomplete T_0-quasi-pseudometric space, $\phi : X \to \mathbb{R}$ be a bounded from above function and "$\precsim$" the preorder induced by ϕ. Let $F : X \times X \to X$ and $G, H : X \to X$ be three d^s-sequentially continuous mapping on X such that the pairs $\{F, G\}$ and $\{F, H\}$ satisfy conditions $(C1')$ and $(C2')$. Then, F, G and H have a common coupled fixed point in X.

Concluding remarks and open problem

All the results given remain true when we replace accordingly the bicomplete quasi-pseudometric space (X, d) by a left Smyth sequentially complete/left K-complete or a right Smyth sequentially complete/right K-complete space.

Moreover, the reader can convince himself that the proofs are quite straight forward; it is enough to get the right sequence. The major challenge comes when we have more than three maps. Indeed, it is not obvious to see how to construct an appropriate sequence, following the same patent as developed above. More precisely.

Let (X, d) be a Hausdorff left K-complete T_0-quasi-pseudometric space, $\phi : X \to \mathbb{R}$ be a bounded from above function and $'' \preceq''$ the preorder induced by ϕ. Let $F : X \times X \to X$ and $G_i : X \to X$, $i = 1, 2, \ldots, K$ for $K > 2$ be $K + 1$ d-sequentially continuous mapping on X such that the pairs $\{F, G_i\}$, $i = 1, 2, \ldots, K$ are weakly left-related.

(1) Can we prove that $F, G_1, \ldots, G_K$ have a common coupled fixed point in X?
(2) Alternatively, what could be a correct formulation of the statement, using the induced preorder and the weakly left-related property that guarantees a positive answer?

References

1. Agarwal, R.P., El-Gebeily, M.A., O'Regan, D.: Generalized contractions in partially ordered metric spaces. Appl. Anal. **87**(1), 1–8 (2008)
2. Lakshmikantham, V., Ciric, LjB: Coupled fixed point theorems for nonlinear contractions in partially ordered metric space. Nonlinear Anal. **70**(12), 4341–4349 (2009)
3. Mustafa, Z., Sims, B.: Fixed point theorems for contractive mapping in complete G-metric spaces. Fixed Point Theory Appl. (2009). doi:10.1155/2009/917175
4. Bhaskar, T.G., Lakshmikantham, V.: Fixed point theorems in partially ordered metric spaces and applications. Nonlinear Anal. **65**(7), 1379–1393 (2006)
5. Włodarczyk, K., Plebaniak, R.: Asymmetric structures, discontinuous contractions and iterative approximation of fixed and periodic points. Fixed Point Theory Appl. **128**, 1–18 (2013). doi:10.1186/1687-1812-2013-128
6. Gaba, Y.U.: Startpoints and (α, γ)-contractions in quasi-pseudometric spaces. J. Math. (2014). doi:10.1155/2014/709253 (Article ID 709253, 8 pages)

10

On the solutions of three-point boundary value problems using variational-fixed point iteration method

A. Kilicman[1] · **M. Wadai**[1]

Abstract Given a three-point fourth-order boundary value problems

$$y^{(iv)} + p(x)y''' + q(x)y'' + r(x)y' + s(x)y = f(x), a \leq x \leq b$$

such that

$$y(a) = y(b) = y''(b) = y''(\alpha) = 0, a \leq \alpha \leq b;$$

where $p, q, r, s, f \in C[a, b]$, we combine the application of variational iteration method and fixed point iteration process to construct an iterative scheme called variational-fixed point iteration method that approximates the solution of three-point boundary value problems. The success of the variational or weighted residual method of approximation from a practical point of view depends on the suitable selection of the basis function. The method is self correcting one and leads to fast convergence. Problems were experimented to show the effectiveness and accuracy of the proposed method.

Keywords Fixed point iteration · Variational iteration method · Lagrange's multiplier and boundary value problems.

✉ A. Kilicman
akilicman@science.upm.edu.my; akilicman@yahoo.com

[1] Department of Mathematics and Institute for Mathematical Research, University Putra Malaysia, UPM, 43400, Serdang, Selangor, Malaysia

Introduction

The numerical solution of boundary value problems is of great importance as a result of its wide application in scientific and technological research [1]. Many researchers have developed various numerical methods, especially iterative methods to approximate different types of differential equations, see [2–5]. In recent years, there has been a growing interest in the treatment of iterative approximation such as variational iteration method, fixed point iteration and so on [6], variational iteration method has been used over the years to obtain an approximate solution of some boundary value problems, see [7, 8]. On the other hand, fixed point iteration is a method of computing fixed point of iterated function, it is a well-known method of approximation whose version is the variation iteration method, see [9, 10]. Obviously, these methods have been proved by many researchers to be powerful tools in solving boundary value problems [11]. However, there are noticeable shortcomings in implementations of these methods especially the use of arbitrary function as a starting value, an in-appropriate choice of starting function may affect the rate of convergence, see [11, 12].

In this paper, the propose method is an elegant combination of variational iteration method and fixed point iteration method with the use of finite element method to determine the starting function.

Analysis of variational-fixed point iterative scheme

The variational-fixed point iteration is the combination and application of variational iteration method and fixed point iterative process endowed with finite element method.

To illustrate the basic technique of variational iteration method, we consider the following general differential equation

$$Lu + Nu = g(x) \quad (1)$$

where L is linear operator, N is nonlinear operator and g(x) is forcing term. According to variational iteration method, see [13, 14], a correctional functional of (1) can be constructed as follows

$$u_{n+1}(x) = u_n(x) + \int_0^x \lambda(Lu_n(s) + Nu_n(s) - g(s))\mathrm{d}s \quad (2)$$

where λ is the Lagrange multiplier [15, 16] which can be determined by variational theory, thus

$$\delta u_{n+1}(x) = \delta u_n(x) + \delta \int_0^x \lambda(Lu_n(s) + Nu_n(s) - g(s))\mathrm{d}s$$

$$\delta u_{n+1}(x) = \delta u_n(x) + \int_0^x \delta\lambda(Lu_n(s)\mathrm{d}s$$

its stationary conditions can be obtained using integration by parts. The second term in the right is called correction and $\tilde{u}u_n$ is considered as a restricted variation, i.e $\delta u_n = 0$. Thus, the Lagrange's multiplier [17, 19] can be identified as

$$\lambda(x) = (-1)^n(s-x)^{n-1}/(n-1)!.$$

The given solution is considered as the fixed point of the following functional [11] under the suitable choice of initial approximation [21] $u_0(x)$ at $n = 0$. We use finite element method to determine the starting function to avoid the arbitrary choice of starting function. After single iteration process we obtained u_1; repeating the process iteratively the term to be integrated become larger and cumbersome to operate and the result of each iteration step diverges from exact solution. Based on this fact we introduce the application of fixed point iterative process to overcome it. Thus we have the following theorem:

Theorem 1 *Let (E, d) be a complete metric space and T be a self map on E. Further, let $y_o \in E$ and let $y_{n+1} = f(T, y_n)$ denotes an iteration procedure which gives a sequence $\{y_n\}$. Then T is an iteration process and defined for arbitrary y_0 by*

$$y_{n+1} = f(T, y_n) = (1-\alpha_n)y_n + \alpha_n Ty_n, \quad n \geq 0 \quad (3)$$

where $\{a_n\}$ is a real sequence satisfying $\alpha_n = 1, 0 \leq \alpha_n \leq 1$ for $n \geq 0$ and $\sum_{n=0}^{\infty} \alpha_n = \infty$.

Proof Let

$$y^{(iv)} + p(x)y''' + q(x)y'' + r(x)y' + s(x)y = t(x)$$

such that

$$y(a) = y(b) = y''(b) = y''(\alpha) = 0, \quad a \leq \alpha \leq b;$$

where $p, q, r, s, f \in C[a, b]$, then the scheme

$$y_{n+1}^{(iv)} = (1-\lambda_n)y_n^{(iv)} + \lambda_n(1-\alpha)y_n^{(iv)} \quad (4)$$

obtained by harnessing Mann and Banach fixed point iteration [6, 20], to yield

$$y_{n+1}^{(iv)} = \lambda(t(x) - p(x)y_n''' - q(x)y_n'' - r(x)y_n' - s(x))y_n + (1-\lambda)y_n^{(iv)} \quad (5)$$

and converges for $0 \leq \lambda_n \leq 1$. Now let

$$y^{(iv)} = f(x, y, y', y'', y'''). \quad (6)$$

Therefore, for any $y(x)$ solution of the integral equation on $[a, b]$

$$y(x) = \int_a^b (G(x, s)f(t, y(t), y'(t), y''(t), y'''(t))\mathrm{d}t + v(x) \quad (7)$$

where $G(x, t)$ is a green function of the associated boundary value problem and $v(x)$ is a solution of $y^{(iv)} = 0$ that satisfies boundary conditions. Now if we let $T : C^1[a, b] \to C^1[a, b]$ be defined by

$$(Ty)(x) = \int_a^b (G(x, s)f(t, y(t), y'(t), y''(t), y'''(t))\mathrm{d}t + v(x) \quad (8)$$

then T is an operator such that any $y(x)$ is a solution of (6) at fixed point of T and can be referred as a fixed point operator. For convergence of (3) and (5) we let

$$y_{n+1} = (1-\alpha_n)y_n + \alpha_n Ty_n$$
$$y'_{n+1} = (1-\alpha_n)y'_n + \alpha_n Ty'_n$$
$$y''_{n+1} = (1-\alpha_n)y''_n + \alpha_n Ty''_n$$
$$y'''_{n+1} = (1-\alpha_n)y'''_n + \alpha_n Ty'''_n$$
$$y_{n+1}^{(iv)} = (1-\alpha_n)y_n^{(iv)} + \alpha_n Ty_n^{(iv)}. \quad (9)$$

From Eq. (8) it follows that

$$(Ty_n)'(x) = \int_{x_1}^{x_2} \delta/\delta x(G(x, s)f(t, y(t), y'(t), y''(t), y'''(t))\mathrm{d}t + v(x)'$$

$$(Ty_n)''(x) = \int_{x_1}^{x_2} \delta^2/\delta x^2(G(x, s)f(t, y(t), y'(t), y''(t), y'''(t))\mathrm{d}t + v(x)''$$

$$(Ty_n)'''(x) = \int_{x_1}^{x_2} \delta^3/\delta x^3(G(x, s)f(t, y(t), y'(t), y''(t), y'''(t))\mathrm{d}t + v(x)'''$$

$$(Ty_n)^{(iv)}(x) = \int_{x_1}^{x_2} \delta^4/\delta x^4 (G(x,s)f(t,y(t),y'(t),y''(t), y'''(t))dt + v(x)^{(iv)}. \quad (10)$$

Therefore Eqs. (9) and 10 become

$$y_{n+1}^{(iv)} = (1-\alpha_n)y_n^{(iv)} + \alpha_n \int_{x_1}^{x_2} \delta^4/\delta x^4 (G(x,s)f(t,y(t),y'(t), \times y''(t),y'''(t))dt + v(x)^{(iv)}. \quad (11)$$

Also, combining Eqs. (7) and (11) yields

$$y_{n+1}^{(iv)} = (1-\alpha_n)y_n^{(iv)} + \alpha_n Ty_n^{(iv)}. \quad (12)$$

Therefore scheme (3) and (5) are convergent. This scheme is use to approximate boundary value problems iteratively with the use of arbitrary initial approximation y_0 at $n = 0$. Since the variational iteration method and fixed point iteration methods are similar [10]. We let $y_0 = u_1$ to avoid the assumption of the arbitrary function y_0 where the process will be carried out iteratively until convergence is obtained or the iteration is terminated. However, the finite element methods are the Galerkin method, collocation method, Raleigh-Ritz method, etc. Galerkin method is an approximate solution of boundary value problems suggested by Galerkin [18], based on the requirement that the basis function $\phi_0, \phi_1, \phi_2, \ldots, \phi_n$ be orthogonal to the residual

$$\int \psi(x_i, a_0, a_1, \ldots, a_n)\phi_i dx = 0$$

$i = 0, 1, 2, 3, \ldots, n$. This gives rise to the following system of linear algebraic equations for the coefficients of the approximation solution

$$y_n(x) = \phi_0(x) + a_1\phi_1(x) + a_2\phi_2(x) + \cdots + a_n\phi_n(x)$$

of the boundary value problem

$$Ly = y^{(iv)} + p(x)y''' + q(x)y'' + r(x)y' + s(x)y = f(x), \quad 0 \le x \le 1$$

such that

$$y(0) = y(0) = y''(1) = y''(\alpha) = 0, \quad 0 \le \alpha \le 1.$$

Therefore,

$$a_1L(\phi_1,\phi_1) + a_2L\phi_1,\phi_2 + \cdots + a_nL\phi_n,\phi_1 = (f - L(\phi_0,\phi_1))$$
$$a_1L(\phi_1,\phi_2) + a_2L\phi_2,\phi_2 + \cdots + a_nL\phi_n,\phi_2 = (f - L(\phi_0,\phi_2))$$
$$\vdots$$
$$a_1L(\phi_1,\phi_n) + a_2L\phi_2,\phi_n + \cdots + a_nL\phi_n,\phi_n = (f - L(\phi_0,\phi_n)).$$

The weight functions are taken with the concept of inner product and orthogonality, it is obvious that the inner product of the two function in a certain domain is

$$\langle f, g\rangle = \int_a^b f(x)g(x)dx = 0$$

which is used to determine the starting function of the variational iteration method instead of an arbitrary choice. □

Numerical examples

In this section, two experiments are considered to demonstrate present methods:

Example 1 Consider

$$y^{(iv)} - y'' = -2, \quad 0 \le x \le 1 \quad (13)$$

subject to the boundary conditions

$$y(0) = y''(1/2) = y'''(0) = 0, \quad y(1) = 1.$$

The following must be observed: Galerkin method of approximation is used to determine the initial approximation of variational iteration method whose trial function

$$U(x) = U_0(x) + \sum_{i=1}^{n} C_i U_i(x)$$

is called the basis function, the approximate solution we sought. Where $U_0(x) = x, U_1 = (x - x^4), U_2 = (x - x^5)$

$$U(x) = x + C_1(x - x^4) + C_2(x - x^5). \quad (14)$$

We differentiate the Eq. (14) successively to obtain second and fourth derivatives and then substitute in Eq. (13) to get the residual

$$R(x, C_1, C_2) = -24C_1 - 120C_2x + 12C_1x^2 + 20C_2x^3 + 2. \quad (15)$$

A weight function is chosen within the bases function with the concepts of inner product and orthogonality

$$\langle U, R\rangle = \int_a^b U_i(x)D[U_0(X) + \sum_{i=1}^{n} C_iU_i(x)]dx = 0.$$

These are sets of n-order linear equations which is solved to obtain all C_i coefficients as follows:

$$\int_0^1 (x - x^4)(-24C_1 - 120C_2x + 12C_1x^2 + 20C_2x^3 + 2)dx = 0$$

we obtain

$$-37/2C_2 - 207/35C_1 = -3/5. \tag{16}$$

Also

$$\int_0^1 (x - x^5)(-24C_1 - 120C_2x + 12C_1x^2 + 20C_2x^3 + 2)\mathrm{d}x = 0$$

we get

$$-132/63C_2 - 13/2C_1 = -2/3 \tag{17}$$

solving these Eqs. (16) and (17) simultaneously we obtain $C = 308/4331, C_2 = 42/4331$. We substitute these constants in (14), hence the approximation solution we sought for.

$$U(x) = x + 308/4331(x - x^4) + 42/4331(x - x^5). \tag{18}$$

The second step is the use variational iteration method to determine the starting function of the fixed point iterative procedure; we construct the correct functional of (13) as follows:

$$t_{n+1}(x) = t_n(x) + \int_0^x \lambda(t_n^{(4)}(s) - t_n^{(2)}(s) - g(s))\mathrm{d}s$$

at $n = 0$

$$t_1(x) = t_0(x) + \int_0^x \lambda(t_0^{(4)}(s) - t_0^{(2)}(s) - g(s))\mathrm{d}s. \tag{19}$$

We let

$$t_0(x) = U(x) = x + 308/4331(x - x^4) + 42/4331(x - x^5) \tag{20}$$

we differentiate (20), i.e $t_0(x)$ successively to obtain its second and fourth derivative and substitute same in Eq. (19) to have

$$t_1(x) = x + 308/4331(x - x^4) + 42/4331(x - x^5) + \int_0^x (s - x^3)/25986(-7392x - 5040x + 3696x^2 - 840x^3 + 2)\mathrm{d}s$$

$$t_1(x) = 1/4331(4681x - 308x^4 - 42x^5) + 1/1732(635x^4/3 - 840x^5 + 616x^6 + 140x^7) - 1/51972(1270x^4 - 5040x^5 + 3696x^6 + 840x^7). \tag{21}$$

When the process is repeated for further iterations, the function to be integrated is getting larger and complex, where iterated values diverge from the analytical solution. Based on this fact, We let $y_0 = U_1(x)$, where $U_1(x)$ is the iterative function obtained after single iteration taken as an initial values for fixed point iterative technique. The scheme in (12) the fixed point iterative procedure can be used as follows

$$y_{n+1}^{(iv)}(x) = y_n''(x) - 2 \tag{22}$$

at $n = 0$

$$y_1^{(iv)}(x) = y_0''(x) - 2. \tag{23}$$

But

$$y_0(x) = t_1(x) = 1/4331(4681x - 308x^4 - 42x^5) + 1/1732(635x^4/3 - 840x^5 + 616x^6 + 140x^7) - 1/51972(1270x^4 - 5040x^5 + 3696x^6 + 840x^7). \tag{24}$$

We differentiate the Eq. (24) twice to get y_0'' and then substituted it into the given Eq. (23) to obtain

$$y_1^{(iv)}(x) = -1.29324819x^2 + 1.939505888x^3 - 3.057954283x^4 - 0.4040637266x^5 - 2. \tag{25}$$

To obtain $y_1(x)$, we integrate (25) four times successively and imposing the boundary conditions to get

$$y_1(x) = -35x^9/311832 - 11x^8/17324 + 4x^7 - x^6/360 - x^4/12 - 0.1273074324x^2 + 0.9586273016x.$$

The repeat the process at $n = 2, 3\ldots$ until the iteration is terminated or it converges with the analytical solution, after few iterations we get

$$y_5(x) = 0.9632277022x + 0.1131826518x^2 - 0.0739028367x^4 - 0.002462970407x^6 - 0.00004406199500x^8 - 4.8098132 \times 10^{-7}x^{10} - 4.175351396 \times 10^{-9}x^{12} - 1.223616824 \times 10^{-12}x^{16} - 1.145095958 \times 10^{-13}x^{17} + 3.559612579 \times 10^{-12}x^{15} - 2.294149120 \times 10^{-11}x^{14}.$$

Example 2 Consider

$$y^{(iv)} - y'' = -12x^2, \quad 0 \le x \le 1 \tag{26}$$

subject to the boundary conditions

$$y(0) = y''(0) = y'''(1/2) = 0, \quad y(1) = 13.$$

We let

$$U(x) = U_0(x) + \sum_{i=1}^{n} C_iU_i(x)$$

which is the basis function where $U_0(x) = 13x, U_1 = (x - x^4), U_2 = (x - x^5)$

$$U(x) = 13x + C_1(x - x^4) + C_2(x - x^5). \quad (27)$$

We differentiate Eq. (27) successively to obtain second and fourth derivatives and then substitute same in (26) the given differential equation to get the residual

$$R(x, C_1, C_2) = -24C_1 - 120C_2x + 12C_1x^2 + 20C_2x^3 + 12x^2. \quad (28)$$

A weight function are chosen within the basis function with the concepts of inner product and orthogonality

$$\langle U, R\rangle = \int_a^b U_i(x)D[U_0(X) + \sum_{i=1}^{n} C_iU_i(x)]dx = 0.$$

These are sets of n-order linear equations which is solved to obtain all C_i coefficients as follows:

$$\int_0^1 (x - x^4)(-24C_1 - 120C_2x + 12C_1x^2 + 20C_2x^3 + 12x^2)dx = 0$$

we obtain

$$-37/2C_2 - 207/35C_1 = -9/7. \quad (29)$$

Also

$$\int_0^1 (x - x^5)(-24C_1 - 120C_2x + 12C_1x^2 + 20C_2x^3 + 12x^2)dx = 0$$

we get

$$-132/63C_2 - 13/2C_1 = -3/2 \quad (30)$$

solving Eqs. (29) and (30) simultaneously to obtain $C_1 = -635/4331, C_2 = 504/4331$. Then we substitute these constants in (27) called the basis function and is the approximate solution we want

$$U(x) = 13x - 635/4331(x - x^4) + 504/4331(x - x^5)$$

and

$$U(x) = 56172/4331x - 635/4331x^4 - 504/4331x^5. \quad (31)$$

Next we apply variational iteration method by constructing a correctional functional of (26)

$$t_{n+1}(x) = t_n(x) + \int_0^x \lambda(t_n^{(4)}(s) - t_n^{(2)}(s) - g(s))ds$$

at $n = 0$

$$t_1(x) = t_0(x) + \int_0^x \lambda(t_0^{(4)}(s) - t_0^{(2)}(s) - g(s))ds. \quad (32)$$

We let

$$t_0(x) = U(x) = 13x - 635/4331(x - x^4) + 504/4331(x - x^5 \quad (33)$$

we differentiate (33), i.e $t_0(x)$ successively to obtain it second and fourth derivative and substitute in the given Eq. (32) to have

$$t_1(x) = 13x - 635/4331(x - x^4) + 504/4331(x - x^5) + \int_0^x (s - x^3)/25986(56172x + 5040x + 3696x^2 - 840x^3 + 12x^2)ds$$

$$t_1(x) = 1/4331(56172x + 635x^4 - 504x^5) + 1/17324(2540x^4 - 10080x^5 + 7392x^6 + 1680x^7) - 1/51972(15240x^2 - 60480x^3 + 44352x^4 + 10080x^5)$$

we simplify to get

$$t_1(x) = 12.96975294x - 0.2932348187x^2 + 1.163703533x^3 + 0.7597475208x^4 - 6.130182374x^5 + 4.267898383x^6 + 0.9699769053x^7. \quad (34)$$

As the number of iterations increases, the function to be integrated is getting larger and cumbersome making the functions diverging from the analytical solutions at each iteration steps. Based on this fact, we apply the fixed point iterative technique where we use $U_1(x) = y_0$ as a starting value after single iteration to avoid the use of arbitrary function (Fig. 1).

$$y_{n+1}^{(iv)}(x) = y_n''(x) - 12x^2$$

at $n = 0$

$$y_1^{(iv)}(x) = y_0''(x) - 12x^2. \quad (35)$$

But

$$y_0(x) = t_1(x) = 12.96975294x - 0.2932348187x^2 + 1.163703533x^3 + 0.7597475208x^4 - 6.130182374x^5 + 4.267898383x^6 + 0.9699769053x^7. \quad (36)$$

We differentiate (36) twice to get y_0'' and then substituted it into the given Eq. (35) to obtain

$$y_1^{(iv)} = 1/4331(31492x^2 \quad 60480x^3 - 55440x^4 - 17640x^5) - 12x^2. \quad (37)$$

To obtain $y_1(x)$, we integrate (37) four times successively and imposing the boundary conditions to get

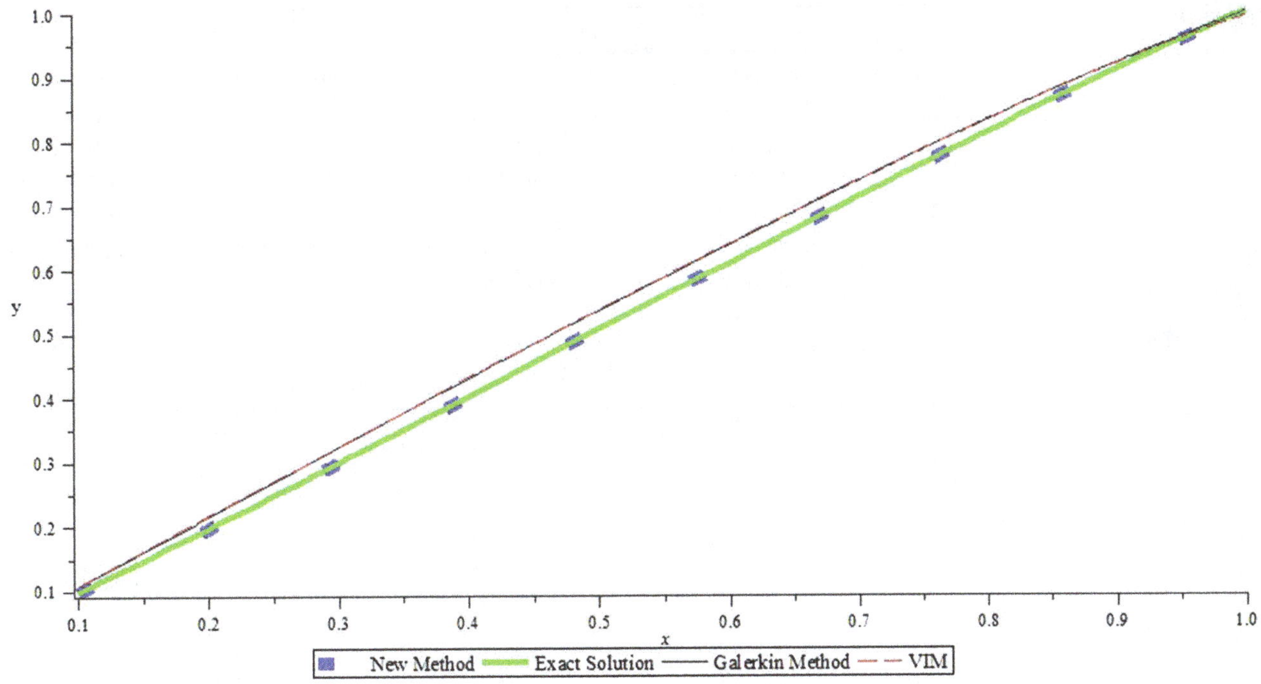

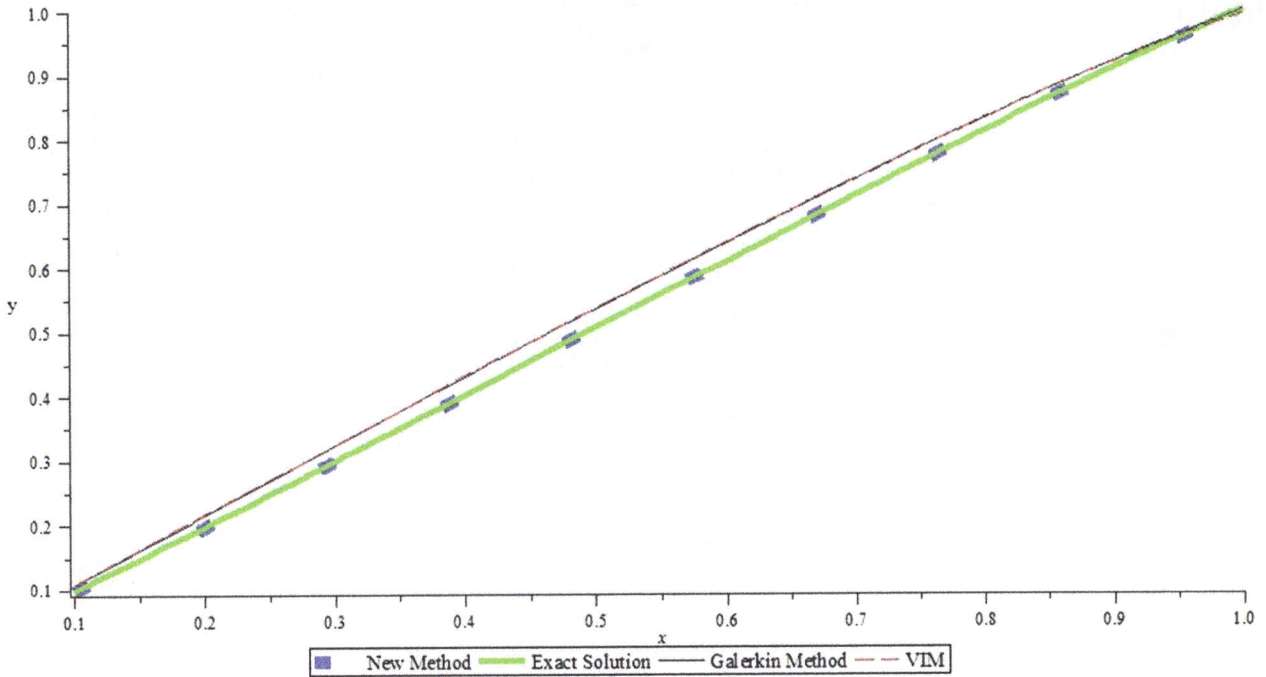

Fig. 1 Example 1

$$y_1(x) = 12.95702542x + 0.07419139152x^3 - x^6/30 + 48/4331x^7 - 93/4331x^8 - 35/25986x^9.$$

Repeat the process at $n = 2, 3\ldots$ until the iteration is terminated or convergence is achieved, after few iterations we get

$$\begin{aligned} y_8(x) = {} & 12.95527250x + 0.07483086130x^3 \\ & + 0.03742542966x^5 - 0.033333333x^6 \\ & + 0.00008908437939x^7 - 0.0005952380951x^8 \\ & + 0.000001237279890x^9 - 0.000006613756612x^{10} \\ & + 1.124828201 \times 10^{-8}x^{11} - 5.010421676 \times 10^{-8}x^{12} \\ & + 7.208657571 \times 10^{-11}x^{13} - 2752978943 \times 10^{-10}x^{14} \\ & + 3.440595345 \times 10^{-13}x^{15} - 1.147074559 \\ & \times 10^{-12}x^{16} + 1.251515555 \times 10^{-15}x^{17} \\ & - 3.748609670 \times 10^{-15}x^{18} - 9.864762290 \\ & \times 10^{-18}x^{20} + 1.093300832 \times 10^{-18}x^{21} \\ & - 2.733252078 \times 10^{-19}x^{22} - 1.890589383 \times 10^{-20}x^{23}. \end{aligned}$$

Thus we have Tables 1 and 2.

Table 1 For problem 1

x	y_e	y_g	y_v	y_n
0.1	0.097447250	0.1080740660	0.1080732927	0.0974472556
0.2	0.197054509	0.2160456616	0.2160392496	0.1970545158
0.3	0.298554395	0.3236442254	0.3236353947	0.2985544023
0.4	0.401498309	0.4304052459	0.4304300913	0.4014983126
0.5	0.505251929	0.5356586238	0.5357915078	0.5052519302
0.6	0.608988879	0.6385170353	0.6388187670	0.6089888756
0.7	0.711682441	0.7378642946	0.7382311988	0.7116824356
0.8	0.812095295	0.8323437174	0.8322116291	0.8120952881
0.9	0.908767121	0.9203464835	0.9181996281	0.9087671172
1.0	1.000000000	0.9999999996	0.9926306466	0.9999999996

Discussion

The accuracy and convergence of the method are of great significance in a numerical experiment of such type. Accuracy measures the degree of closeness of the numerical solution to the theoretical solution while convergence measures even the closer approach of successive iteration to the exact solution as the number of iteration increases. To asses the success of our method, the scheme was tested with some numerical examples whose results presented as Tables 1 and 2. These tables show the comparison with exact method, Galerkin method, variational iteration method and the approximate solution obtained by variational-fixed point iteration method. It is observed that with few iterations, the order of the error is quite encouraging, which indicates fast rate of convergence. It is clearly seen that as the iterations proceeds, the error decreases and convergence is assured (Fig. 2).

Table 2 For problem 2

x	y_e	y_g	y_v	y_n
0.1	0.093357741	0.09894632328	0.09894554621	0.0933357741
0.2	0.187337320	0.1990462157	0.1990395573	0.187337320
0.3	0.282617019	0.3002722124	0.3002644079	0.282617019
0.4	0.379679951	0.4024981796	0.4025375113	0.379679951
0.5	0.478868890	0.5054304348	0.5056211345	0.478868890
0.6	0.580307687	0.6085388686	0.6089657555	0.580307687
0.7	0.683842458	0.7109880646	0.7114628525	0.683842458
0.8	0.788979999	0.8115684206	0.8110830174	0.788979999
0.9	0.894822799	0.9086272697	0.9043752866	0.894822799
1.0	1.000000000	1.0000000000	0.9858035810	1.000000000

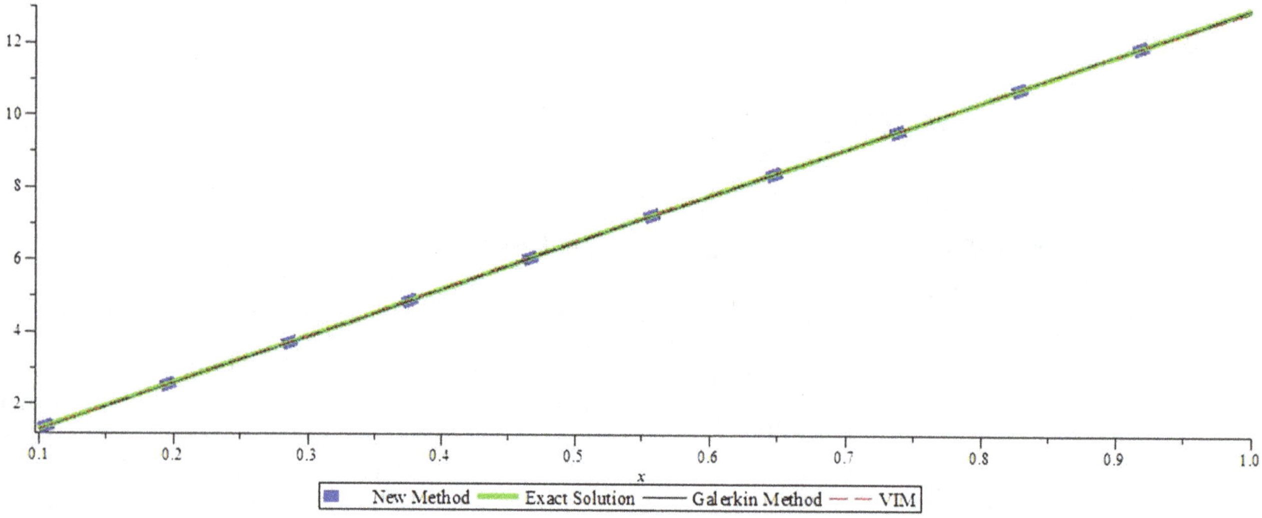

Fig. 2 Example 2

Conclusion

In this paper we have shown the performance of the variational-fixed point iterative scheme for the solution three-point boundary value problems with help of some experiments which indicates to be very powerful and an efficient technique with good convergence property when compared with the existing method. We can conclude the method have an advantage over the existing methods.

Compliance with ethical standards

Conflict of interest The authors declare that they have no competing interests.

Author's contributions Both authors jointly worked on deriving the results and approved the final manuscript.

References

1. Wazwaz, A.M.: The variational iteration method for analytic treatment for linear and nonlinear ordinary differential equations. Appl. Math. Comput. **212**, 120–134 (2009)
2. Mohyud-Din, S.T., Yildirim, A., Hossein, M.M.: Variational iteration method for initial and Boundary Value Problems Using He's Polynomials. Int. J. Differ. Equ. **2010**, Article ID426213, p 28. doi:10.1155/2010/426213 (2010)
3. Turkyilmazoglu, M.: An optimal variational Iteration method. Appl. Math. Lett. **24**, 762–765 (2011)
4. Porshokouhi, M.G., Ghambari, B.: He's variational iteration method for solving differential equation of fifth order. Gen. Math. Notes **1**(2), 153–158 (2010)
5. Kılıçman, A., Adeboye, K.R., Wadai, M.: A Variational-Fixed Point Iterative Technique for the solution of second order Linear differential equations. Malays. J. Sci. **34**(2) (2015) (in press)
6. Bildik, N., Bakir, Y., Mutlu, A.: The New Modified Ishikawa Iteration Method for the Approximate Solution of Different

Types of Differential Equations. Fixed Point Theory Appl. **52** (2013)
7. Biaza, J., Ghazuvini, H.: He's variational iteration method for fourth-order parabolic equations. Comput. Math. Appl. **54**, 790–797 (2007)
8. Ghorbani, A., Nadjafi, J.S.: An effective modifications of He's variational iteration method. Nonlinear Anal. Real World Anal. **10**, 2828–2833 (2009)
9. Altintan, D., Ugur, O.: Solution of initial and boundary value problems by the variational iteration method. J. Comput. Appl. Math. **259**, 790–797 (2014)
10. Khuri, S.A., Sayfy, A.: Variational iteration method: greens functions and fixed point iteration perspective. Appl. Math. Lett. **32**, 28–34 (2014)
11. Lu, J.: Variational iteration method for two-point boundary value problems. J. Comput. Appl. Math. **207**, 92–95 (2007)
12. He, J.H.: Variational iteration method—some recent results and new interpretation. J. Comput. Appl. Math. **207**, 3–17 (2007)
13. Noor, M.A., Mohyud-Din, S.T.: Variational itertion method for solving higher order boundary value problems. Appl. Math. Comput. **189**, 1929–1942 (2007)
14. He, J.H.: Variational iteration method: a kind of nonlinear analytical technique: some examples. Int. J. Nonlinear Mech. **34**, 699–708 (1999)
15. He, J.H.: Variational iteration method for the autonomous ordinary differential system. Appl. Math. Comput. **114**(2—-3), 115–123 (2009)
16. Inokuti, M., Sekine, H., Mura, T.: General use of the Lagrange multiplier in nonlinear mathematical physics. In: Nemat-Nasser, S. (ed.) Variational method in the mechanics of Solids, pp. 156–162. Pergamon Press, Oxford (1978)
17. Wu, G.C.: Challenge in the variational iteraation method-a new approach to identification of the lagrange multipliers. J. King Saud Univ. Sci. **25**(2), 175–178 (2013)
18. Burden, R.L., Faires, J.D.: Numerical Analysis. 120–134 (2009)
19. Shang, X., Han, D.: Application of variational iteration method for the solving nth-order integro-differential equations. J. Comput. Appl. Math. **234**, 1442–1447 (2010)
20. Herceg, D., Krejic, N.: Convergence Results for Fixed point Iteration in R. Comput. Math. Appl. **31**(2), 7–10 (1996)
21. Momani, S., Abuasad, S., Odibat, Z.: Variational iteration method for the solving nonlinear boundary value problems. Appl. Math. Comput. **183**, 1351–1358 (2006)

11

Fixed point theorems for generalized almost contractions in partial metric spaces

Ishak Altun · Kishin Sadarangani

Abstract In the present paper, we give some fixed point results for generalized Ćirić type strong almost contractions on partial metric spaces which generalizes some recent results appearing in the literature. Particularly, our result has as a particular case, mappings satisfying a general contractive condition of integral type.

Keywords Fixed point · Partial metric space · Almost contraction

Mathematics Subject Classification (2000) Primary 54H25 · Secondary 47H10

Introduction

Partial metric spaces were introduced by Matthews in [20] as a part of the study of denotational semantics of dataflow networks. These spaces are a generalization of usual metric spaces where the self distance for any point need not be equal to zero.

Let us recall that a partial metric on a set X is a function $p : X \times X \to [0,\infty)$ such that for all $x,y,z \in X$: (1) $x = y \Longleftrightarrow p(x,x) = p(x,y) = p(y,y)$ (T_0-separation axiom), (2) $p(x,x) \le p(x,y)$ (small self-distance axiom), (3) $p(x,y) = p(y,x)$ (symmetry), (4) $p(x,y) \le p(x,z) + p(z,y) - p(z,z)$ (modified triangular inequality).

I. Altun (✉)
Department of Mathematics, Faculty of Science and Arts, Kirikkale University, Yahsihan, 71540 Kirikkale, Turkey
e-mail: ishakaltun@yahoo.com

K. Sadarangani
Departamento de Matemáticas, Universidad de Las Palmas de Gran Canaria, Campus de Tafira Baja,
35017 Las Palmas de Gran Canaria, Spain
e-mail: ksadaran@dma.ulpgc.es

A partial metric space (for short PMS) is a pair (X,p) such that X is a nonempty set and p is a partial metric on X.

It is clear that, if $p(x,y) = 0$, then $x = y$. But if $x = y$, $p(x,y)$ may not be 0.

At this point it seems interesting to remark the fact that partial metric spaces play an important role in constructing models in the theory of computation (see for instance [15–17], etc).

Example 1 Let $X = [0,\infty)$ and $p(x,y) = \max\{x,y\}$ for all $x,y \in X$. Then (X,p) is a PMS.

Example 2 Let I denote the set of all intervals $[a,b]$ for some real numbers $a \le b$. Let $p : I \times I \to [0,\infty)$ be the function such that $p([a,b],[c,d]) = \max\{b,d\} - \min\{a,c\}$. Then (I,p) is a PMS.

Example 3 Let $X = \mathbb{R}$ and $p(x,y) = e^{\max\{x,y\}}$ for all $x,y \in X$. Then (X,p) is a PMS.

Other examples of partial metric spaces may be found in [16, 18, 20, 22], etc.

Each partial metric p on X generates a T_0 topology τ_p on X which has as a base the family open p-balls

$$\{B_p(x,\varepsilon) : x \in X, \varepsilon > 0\},$$

where

$$B_p(x,\varepsilon) = \{y \in X : p(x,y) < p(x,x) + \varepsilon\},$$

for all $x \in X$ and $\varepsilon > 0$.

Observe that a sequence $\{x_n\}$ in a PMS (X,p), converges to a point $x \in X$, with respect to τ_p, if and only if $p(x,x) = \lim_{n\to\infty} p(x,x_n)$.

If p is a partial metric on X, then the functions $p^s, p^w : X \times X \to \mathbb{R}^+ := [0,\infty)$, given by

$$p^s(x,y) = 2p(x,y) - p(x,x) - p(y,y)$$

and

$$p^w(x,y) = \max\{p(x,y) - p(x,x), p(x,y) - p(y,y)\}$$
$$= p(x,y) - \min\{p(x,x), p(y,y)\}$$

are ordinary metrics on X. It is easy to see that p^s and p^w are equivalent metrics on X.

According to [20], a sequence $\{x_n\}$ in a PMS (X,p) converges, with respect to τ_{p^s}, to a point $x \in X$ if and only if

$$\lim_{n,m\to\infty} p(x_n, x_m) = \lim_{n\to\infty} p(x_n, x) = p(x,x).$$

A sequence $\{x_n\}$ in a PMS (X,p) is called a Cauchy sequence if $\lim_{n,m\to\infty} p(x_n, x_m)$ exists (and is finite). (X,p) is called complete if every Cauchy sequence $\{x_n\}_{n\in\omega}$ in X converges, with respect to τ_p, to a point $x \in X$ such that $p(x,x) = \lim_{n,m\to\infty} p(x_n, x_m)$.

Finally, the following crucial facts are shown in [20]:

1. $\{x_n\}$ is a Cauchy sequence in (X,p) if and only if it is a Cauchy sequence in the metric space (X,p^s).
2. (X,p) is complete if and only if (X,p^s) is complete.

Matthews obtained, among other results, a partial metric version of the Banach fixed point theorem ([20, Theorem 5.3]) as follows.

Theorem 1 ([20]) Let (X,p) be a complete partial metric space and let $T: X \to X$ be a contraction mapping, that is, there exists $\lambda \in [0,1)$ such that

$$p(Tx, Ty) \leq \lambda p(x,y)$$

for all $x, y \in X$. Then T has a unique fixed point $z \in X$. Moreover, $p(z,z) = 0$.

Later on, Abdeljawad et al. [1], Acar et al. [2, 3], Altun et al. [6–8], Karapinar and Erhan [19], Oltra and Valero [21] and Valero [27], gave some generalizations of the result of Matthews. Also, Ćirić et al. [14], Samet et al. [25] and Shatanawi et al. [26] proved some common fixed point results in partial metric spaces. The best two generalizations of it were given by Romaguera [23, 24].

Theorem 2 Let (X,p) be a complete partial metric space and let $T: X \to X$ be a map such that

$$p(Tx, Ty) \leq \varphi(M(x,y)) \tag{1.1}$$

for all $x, y \in X$, where

$$M(x,y) = \max\left\{p(x,y), p(x,Tx), p(y,Ty), \frac{1}{2}[p(x,Ty) + p(y,Tx)]\right\}$$

and φ satisfies one of the following:

1. $\varphi: [0,\infty) \to [0,\infty)$ is [23] an upper semicontinuous from the right such that $\varphi(t) < t$ for all $t > 0$,
2. $\varphi: [0,\infty) \to [0,\infty)$ is a [24] nondecreasing function such that $\varphi^n(t) \to 0$ as $n \to \infty$ for all $t > 0$.

Then T has a unique fixed point $z \in X$. Moreover, $p(z,z) = 0$.

In [22], Romaguera defined the 0-complete PMS as follows: A sequence $\{x_n\}$ in a PMS (X,p) is called 0-Cauchy if

$$\lim_{m,n\to\infty} p(x_n, x_m) = 0$$

and (X,p) is called 0-complete if every 0-Cauchy sequence in X converges, with respect to τ_p, to a point $z \in X$ such that $p(z,z) = 0$. It is clear that every complete PMS is 0-complete, but as it was shown in [22] the converse is not true.

On the other hand Berinde [9–11] defined weak contraction (or (δ, L)-weak contraction) mappings in a metric space as follows.

Definition 1 Let (X,d) be a metric space and $T: X \to X$ be a self operator. T is said to be a weak contraction (or (δ, L)-weak contraction) if there exists a constant $\delta \in (0,1)$ and some $L \geq 0$ such that

$$d(Tx, Ty) \leq \delta d(x,y) + Ld(y, Tx) \tag{1.2}$$

for all $x, y \in X$.

Note that, by the symmetry property of the distance, the weak contraction condition implicitly includes the following dual one

$$d(Tx, Ty) \leq \delta d(x,y) + Ld(x, Ty) \tag{1.3}$$

for all $x, y \in X$. So, in order to check the weak contractiveness of a mapping T, it is necessary to check both (1.2) and (1.3).

In [9] and [11], Berinde showed that any Banach, Kannan, Chatterjea and Zamfirescu mappings are weak contraction. Using the concept of weak contraction mappings, Berinde [9] proved that if T is a (δ, L)-weak contraction self mapping of a complete metric space X, then T has a fixed point. Also, Berinde shows that any (δ, L)-weak contraction mapping is a Picard operator. Then, Berinde [12] introduced the nonlinear type weak contraction using a comparison function and proved the following fixed point theorem. A map $\varphi: \mathbb{R}^+ \to \mathbb{R}^+$, where $\mathbb{R}^+ = [0,\infty)$, is called comparison function if it satisfies:

1. φ is monotone increasing,
2. $\lim_{n\to\infty} \varphi^n(t) = 0$ for all $t \in \mathbb{R}^+$.

If φ satisfies (1) and

3. $\sum_{n=0}^{\infty} \varphi^n(t)$ converges for all $t \in \mathbb{R}^+$,

then φ is said to be (c)-comparison function.

It is clear that (c)-comparison function implies comparison function, but the converse may not be true. We can find some properties and examples of comparison and (c)-comparison functions in [11].

Definition 2 Let (X,d) be a metric space and $T:X\to X$ is a self operator. T be said to be a weak φ-contraction (or (φ,L)-weak contraction) if there exists a comparison function φ and some $L\geq 0$ such that

$$d(Tx,Ty)\leq\varphi(d(x,y))+Ld(y,Tx) \tag{1.4}$$

for all $x,y\in X$.

Similar to the case of weak contraction, in order to check the weak φ-contractiveness of a mapping T, it is necessary to check both (1.4) and

$$d(Tx,Ty)\leq\varphi(d(x,y))+Ld(x,Ty) \tag{1.5}$$

for all $x,y\in X$.

Clearly any weak contraction is a weak φ-contraction, but the converse may not be true. Also the class of weak φ-contractions includes Matkowski type nonlinear contractions.

Theorem 3 Let (X,d) be a complete metric space and $T:X\to X$ be (φ,L)-weak contraction with φ is (c)-comparison function. Then T has a fixed point.

Let (X,d) be a metric space and $T:X\to X$ be a map such that

$$d(Tx,Ty)\leq\alpha M_d(x,y)+Ld(y,Tx)$$

for all $x,y\in X$, where $\alpha\in[0,1)$, $L\geq 0$,

$$M_d(x,y) = \max\left\{d(x,y),d(x,Tx),d(y,Ty),\frac{1}{2}[d(x,Ty)+p(y,Tx)]\right\}.$$

Then T is called Ciric type strong almost contraction [13].

In light of the above information, Altun and Acar [5] introduced the concepts of weak and weak φ-contractions in the sense of Berinde on partial metric space, showed that any Banach, Kannan, Chatterjea and Zamfirescu mappings are weak contraction and proved some fixed point theorems in this interesting space.

Let (X,p) be a partial metric space. A map $T:X\to X$ is called (φ,L)-weak contraction if there exists a comparison function φ and some $L\geq 0$ such that

$$p(Tx,Ty)\leq\varphi(p(x,y))+Lp^w(y,Tx) \tag{1.6}$$

for all $x,y\in X$.

As above, because of the symmetry of the distance, the (φ,L)-weak contraction condition implicitly includes the following dual one

$$p(Tx,Ty)\leq\varphi(p(x,y))+Lp^w(x,Ty) \tag{1.7}$$

for all $x,y\in X$. Consequently, in order to check the (φ,L)-weak contractiveness of T, it is necessary to check both (1.6) and (1.7).

Theorem 4 Let (X,p) be a 0-complete partial metric space and $T:X\to X$ be (φ,L) weak contraction with a (c)-comparison function. Then T has a fixed point.

Later, Acar et al generalized Theorem 4 to Ciric type strong almost contractions and they proved the following results.

Theorem 5 Let (X,p) be a 0-complete partial metric space and $T:X\to X$ be a map such that

$$p(Tx,Ty)\leq\varphi(M(x,y))+Lp^w(y,Tx)$$

for all $x,y\in X$, where $L\geq 0$, φ is a (c)-comparison function and $M(x,y)$ as in Theorem 2.

Then T has a fixed point in X.

Theorem 6 Let (X,p) be a 0-complete partial metric space and $T:X\to X$ be a map such that

$$p(Tx,Ty)\leq\varphi(M(x,y))+Lp^w(y,Tx)$$

for all $x,y\in X$, where $L\geq 0$, $M(x,y)$ as in Theorem 2 and $\varphi:[0,\infty)\to[0,\infty)$ is an upper semicontinuous from the right function such that $\varphi(t)<t$ for all $t>0$.

Then T has a fixed point in X.

The purpose of this paper is to present a generalization of Theorem 5 which has as a particular case mappings satisfying an integral type almost contraction condition.

Main results

Let $\mathcal{F}$ be the class of functions defined by

$$\mathcal{F}=\{\phi:[0,\infty)\to[0,\infty):\phi \text{ is continuous and non decreasing}\}.$$

Some examples of functions belonging to $\mathcal{F}$ are: $\phi(t)=kt$ with $k\in(0,\infty)$, $\phi(t)=\dfrac{t}{1+t}$, $\phi(t)=\ln(1+t)$ and $\phi(t)=\arctan t$.

Our main result is the following.

Theorem 7 Let (X,p) be a 0-complete partial metric space and $T:X\to X$ be a mapping satisfying

$$\phi(p(Tx,Ty))\leq\phi(\varphi(M(x,y)))+Lp^w(y,Tx) \tag{2.1}$$

for all $x,y\in X$, where $L\geq 0$, $M(x,y)$ is defined as in Theorem 2, $\phi\in\mathcal{F}$ and φ is a (c)-comparison function.

Then T has a fixed point z in X such that $p(z,z)=0$.

Proof We take $x_0\in X$ and consider $x_n=Tx_{n-1}=T^nx_0$ for any $n\in\mathbb{N}$. If $x_n=x_{n+1}$ for some $n\in\mathbb{N}$, then x_n is a fixed point of T and the proof is finished. Suppose that $x_n \neq x_{n+1}$ for any $n \in \mathbb{N}$.

Since

$$\frac{1}{2}[p(x_{n+1},x_{n+1})+p(x_n,x_{n+2})] \leq \frac{1}{2}[p(x_n,x_{n+1})+p(x_{n+1},x_{n+2})] \leq \max\{p(x_n,x_{n+1}),p(x_{n+1},x_{n+2})\},$$

then we have

$$\begin{aligned} M(x_n,x_{n+1}) &= \max\{p(x_n,x_{n+1}),p(x_n,x_{n+1}),p(x_{n+1},x_{n+2}), \\ &\quad \frac{1}{2}[p(x_{n+1},x_{n+1})+p(x_n,x_{n+2})]\} \\ &= \max\{p(x_n,x_{n+1}),p(x_{n+1},x_{n+2})\}. \end{aligned}$$

Applying the contractive condition (2.1) we have

$$\begin{aligned} \phi(p(x_{n+1},x_{n+2})) &= \phi(p(Tx_n,Tx_{n+1})) \\ &\leq \phi(\varphi(M(x_n,x_{n+1}))) + Lp^w(x_{n+1},x_{n+1}) \\ &= \phi(\varphi(M(x_n,x_{n+1}))) \end{aligned} \tag{2.2}$$

If $M(x_n,x_{n+1}) = p(x_{n+1},x_{n+2})$ for some $n \in \mathbb{N}$, then from (2.2) we obtain,

$$\phi(p(x_{n+1},x_{n+2})) \leq \phi(\varphi(p(x_{n+1},x_{n+2})))$$

and, since ϕ is nondecreasing,

$$p(x_{n+1},x_{n+2}) \leq \varphi(p(x_{n+1},x_{n+2})) < p(x_{n+1},x_{n+2})$$

which is a contradiction. Therefore $M(x_n,x_{n+1}) = p(x_n,x_{n+1})$ for all $n \in \mathbb{N}$. From (2.2), we get

$$\phi(p(x_{n+1},x_{n+2})) \leq \phi(\varphi(p(x_n,x_{n+1})))$$

and, since ϕ is nondecreasing,

$$p(x_{n+1},x_{n+2}) \leq \varphi(p(x_n,x_{n+1})).$$

By using mathematical induction, we obtain

$$p(x_{n+1},x_{n+2}) \leq \varphi^{n+1}(p(x_0,x_1)).$$

for all $n \in \mathbb{N}$. By triangle rule, for $m > n$, we have

$$\begin{aligned} p(x_n,x_m) &\leq \sum_{k=n}^{m-1} p(x_k,x_{k+1}) - \sum_{k=n}^{m-2} p(x_{k+1},x_{k+1}) \\ &\leq \sum_{k=n}^{m-1} p(x_k,x_{k+1}) \\ &\leq \sum_{k=n}^{\infty} p(x_k,x_{k+1}) \\ &\leq \sum_{k=n}^{\infty} \varphi^k(p(x_0,x_1)). \end{aligned}$$

Since φ is a (c)-comparison function, then $\sum_{k=0}^{\infty}\varphi^k(p(x_0,x_1))$ is convergent and so $\{x_n\}$ is a 0-Cauchy sequence in X. Since X is 0-complete, $\{x_n\}$ converges, with respect to τ_p, to a point $z \in X$ such that

$$\lim_{n\to\infty} p(x_n,z) = p(z,z) = 0.$$

Now we claim that $p(z,Tz) = 0$. Suppose on contrary $p(z,Tz) > 0$. As ϕ is a (c)-comparison function, $\phi(t) < t$ for $t > 0$, As $\lim_{n\to\infty} p(x_{n+1},x_n) = 0$ and $\lim_{n\to\infty} p(x_n,z) = 0$, there exists $n_0 \in \mathbb{N}$ such that for $n > n_0$,

$$p(x_{n+1},x_n) < \frac{1}{3}p(z,Tz) \tag{2.3}$$

and there exists $n_1 \in \mathbb{N}$ such that for $n > n_1$,

$$p(x_n,z) < \frac{1}{3}p(z,Tz). \tag{2.4}$$

If we take $n > \max\{n_0,n_1\}$ then, by (2.3), (2.4) and triangular inequality, we have

$$\begin{aligned} \frac{1}{2}[p(x_n,Tz)+p(z,Tx_n)] &\leq \frac{1}{2}[p(x_n,z)+p(z,Tz) \\ &\quad - p(z,z)+p(z,Tx_n)] \\ &\leq \frac{1}{2}\left[\frac{1}{3}p(z,Tz)+p(z,Tz)+\frac{1}{3}p(z,Tz)\right] \\ &= \frac{5}{6}p(z,Tz). \end{aligned} \tag{2.5}$$

Now for $n > \max\{n_0,n_1\}$, then, by (2.3), (2.4) and (2.5), we have

$$\begin{aligned} \phi(p(x_{n+1},Tz)) &= \phi(p(Tx_n,Tz)) \\ &\leq \phi(\varphi(M(x_n,z)) + Lp^w(z,x_{n+1}) \\ &= \phi(\varphi(p(z,Tz))) + Lp^w(z,x_{n+1}). \end{aligned}$$

Letting $n \to \infty$ in the last inequality, we have $\phi(p(z,Tz)) \leq \phi(\varphi(p(z,Tz)))$ and since ϕ is nondecreasing $p(z,Tz) \leq \varphi(p(z,Tz)) < p(z,Tz)$ which is a contradiction. Therefore $p(Tz,z) = 0$ and $z = Tz$. □

We can obtain the following corollaries from our main theorem.

Corollary 1 Theorem 4.

Proof Consider $\phi = I_{[0,\infty)}$ identity mapping in Theorem 7.

Notice that if $f : [0,\infty) \to [0,\infty)$ is a Lebesgue-integrable mapping then the function defined by

$$\phi(t) = \int_0^t f(s)ds \quad \text{for} t \in [0,\infty),$$

belongs to $\mathcal{F}$. Therefore we can obtain the following corollary.

Corollary 2 Let (X,p) be a 0-complete partial metric space and $T: X \to X$ be a mapping satisfying

$$\int_0^{p(Tx,Ty)} f(s)ds \leq \int_0^{\varphi(M(x,y))} f(s)ds + Lp^w(y,Tx)$$

for all $x, y \in X$, where $L \geq 0$, $M(x,y)$ as in Theorem 2, φ is a (c)-comparison function and $f: [0,\infty) \to [0,\infty)$ is a Lebesgue-integrable mapping.

Then T has a fixed point in X.

Now we give an illustrative example.

Example 4 Let $X = A \cup B$, where $A = \{0\} \cup \{\frac{1}{n} : n \in \{1,2,\ldots\}\}$, $B = \{2,3,4,\ldots\}$ and

$$p(x,y) = \begin{cases} \max\{x,y\}, & x \neq y \\ 0, & x = y \end{cases}.$$

Then (X,p) is a partial metric space and it is also 0-complete. Define $T: X \to X$ by

$$Tx = \begin{cases} x^3, & x \in A \\ \frac{1}{x}, & x \in B \end{cases}.$$

We show that the contractive condition (2.1) of Theorem 7 is satisfied for $\phi(t) = t$, $\varphi(t) = \frac{t}{2}$ and $L = 2$.

Now consider the following cases.

Case 1. If $x = y$, then $p(Tx,Ty) = 0$ and so the result is clear. Therefore we will assume $x \neq y$ in the following cases.

Case 2. Let $x, y \in A$. Then (note that if $x = 1$ or $y = 1$ then $\inf\{|y - x^3| : x, y \in A \text{ with } x \neq y\} = \frac{1}{2}$. If $x \neq 1$ and $y \neq 1$, then $x^3 \leq \frac{1}{2}x$)

$$\begin{aligned} p(Tx,Ty) &= \max\{x^3, y^3\} \\ &\leq \frac{1}{2}\max\{x,y\} + 2|y - x^3| \\ &= \frac{1}{2}p(x,y) + 2p^w(y,Tx) \\ &= \varphi(M(x,y)) + Lp^w(y,Tx). \end{aligned}$$

Case 3. Let $x, y \in B$. Then

$$\begin{aligned} p(Tx,Ty) &= \max\left\{\frac{1}{x}, \frac{1}{y}\right\} \\ &= \frac{1}{\min\{x,y\}} \\ &\leq \frac{1}{2}\max\{x,y\} \\ &= \frac{1}{2}p(x,y) \\ &\leq \varphi(M(x,y)) + Lp^w(y,Tx). \end{aligned}$$

Case 4. Let $x \in A$ and $y \in B$. Then

$$\begin{aligned} p(Tx,Ty) &= \max\left\{x^3, \frac{1}{y}\right\} \\ &\leq \frac{1}{2}y \\ &= \frac{1}{2}p(x,y) \\ &\leq \varphi(M(x,y)) + Lp^w(y,Tx). \end{aligned}$$

Case 5. Let $x \in B$ and $y \in A$. This case is similar to Case 4.

Hence, all conditions of Theorem 7 are satisfied. Therefore T has a fixed point in X.

Note that $p(T0,T1) = 1 = M(0,1)$, then the condition of (1.1) is not satisfied, because we can not find a function φ satisfying

$$p(T0,T1) = 1 \leq \varphi(M(0,1)) = \varphi(1)$$

and the condition (1) or (2) of Theorem 2. Therefore Theorem 2 is not applicable to this example.

In the above, we show that if T is a generalized almost contraction then it has a fixed point. But in order to guarantee the uniqueness of the fixed point of T, we have to consider an additional condition, as in the following theorem.

Theorem 8 Let (X,p) be a 0-complete partial metric space and $T: X \to X$ be a map such that (2.1) holds. Suppose T also satisfies the following condition: there exists a comparison function φ_1,some $L_1 \geq 0$ and $\phi_1 \in \mathcal{F}$ with $\phi_1(t) > 0$ for $t > 0$ such that

$$\phi_1(p(Tx,Ty)) \leq \phi_1(\varphi_1(M(x,y))) + L_1 p^w(x,Tx) \qquad (2.6)$$

holds, for all $x, y \in X$. Then T has a unique fixed point in X.

Proof Suppose that, there are two fixed points z and w of T. If $p(z,w) = 0$, it is clear that $z = w$. Assume that $p(z,w) > 0$. By (2.6) with $x = z$ and $y = w$, we have

$$\begin{aligned} 0 < \phi_1(p(z,w)) &= \phi_1(p(Tz,Tw)) \\ &\leq \phi_1(\varphi_1(M(z,w))) + L_1 p^w(z,Tz) \\ &= \phi_1(\varphi_1(M(z,w))) \\ &= \phi_1(\varphi_1(p(z,w))) \end{aligned}$$

since ϕ is nondecreasing

$$0 < p(z,w) \leq \varphi_1(p(z,w)) < p(z,w)$$

which is a contradiction. Therefore T has a unique fixed point. □

Acknowledgments The authors would like to thank the referees for their helpful advice which led them to present this paper.

References

1. Abdeljawad, T., Karapinar, E., Tas, K.: Existence and uniqueness of a common fixed point on partial metric spaces. Appl. Math. Lett. **24**, 1900–1904 (2011)
2. Acar, Ö., Altun, I.: Some generalizations of Caristi type fixed point theorem on partial metric spaces. Filomat **26**(4), 833–837 (2012)
3. Acar, Ö., Altun, I., Romaguera, S.: Caristi's type mappings on complete partial metric space. Fixed Point Theory (Cluj-Napoca) **14**(1), 3–10 (2013)
4. Acar, Ö., Berinde, V., Altun, I.: Fixed point theorems for Ciric strong almost contractions in partial metric spaces. J. Fixed Point Theory Appl. **12**, 247–259 (2012)
5. Altun, I., Acar, Ö.: Fixed point theorems for weak contractions in the sense of Berinde on partial metric spaces. Topol. Appl. **159**, 2642–2648 (2012)
6. Altun, I., Erduran, A.: Fixed point theorems for monotone mappings on partial metric spaces. Fixed Point Theory Appl. pp. 10 (2011). Article ID 508730
7. Altun, I., Romaguera, S.: Characterizations of partial metric completeness in terms of weakly contractive mappings having fixed point. Appl. Anal. Discrete Math. **6**, 247–256 (2012)
8. Altun, I., Sola, F., Simsek, H.: Generalized contractions on partial metric spaces. Topol. Appl. **157**, 2778–2785 (2010)
9. Berinde, V.: Approximating fixed points of weak contractions using the Picard iteration. Nonlinear Anal. Forum **9**(1), 43–53 (2004)
10. Berinde, V.: On the approximation of fixed points of weak contractive mappings. Carpath. J. Math. **19**(1), 7–22 (2003)
11. Berinde, V.: Iterative Approximation of Fixed Points. Springer, Berlin (2007)
12. Berinde, V.: Approximating fixed points of weak φ -contractions using the picard iteration. Fixed Point Theory, pp. 131–147 (2003)
13. Berinde, V.: Some remarks on a fixed point theorem for Ćirić-type almost contractions. Carpath. J. Math. **25**(2), 157–162 (2009)
14. Ćirić, L., Samet, B., Aydi, H., Vetro, C.: Common fixed points of generalized contractions on partial metric spaces and an application. Appl. Math. Comput. **218**, 2398–2406 (2011)
15. Cobzaş, S.: Completeness in quasi-metric spaces and Ekeland variational principle. Topol. Appl. **158**, 1073–1084 (2011)
16. Escardo, M.H.: Pcf extended with real numbers. Theor. Comput. Sci. **162**, 79–115 (1996)
17. Heckmann, R.: Approximation of metric spaces by partial metric spaces. Appl. Categ. Struct. **7**, 71–83 (1999)
18. Ilic, D., Pavlovic, V., Rakocevic, V.: Some new extensions of Banach's contraction principle to partial metric space. Appl. Math. Lett. **24**, 1326–1330 (2011)
19. Karapinar, E., Erhan, I.M.: Fixed point theorems for operators on partial metric spaces. Appl. Math. Lett. **24**, 1894–1899 (2011)
20. Matthews, S.G.: Partial metric topology. In: Proceedings of 8th Summer Conference on General Topology and Applications, Annals of the New York Academy of Sciences vol. 728, pp. 183–197 (1994)
21. Oltra, S., Valero, O.: Banach's fixed point theorem for partial metric spaces. Rend. Istid. Math. Univ. Trieste **36**, 17–26 (2004)
22. Romaguera, S.: A Kirk type characterization of completeness for partial metric spaces. Fixed Point Theory Appl. pp. 6 (2010). Article ID 493298
23. Romaguera, S.: Fixed point theorems for generalized contractions on partial metric spaces. Topol. Appl. **218**, 2398–2406 (2011)
24. Romaguera, S.: Matkowski's type theorems for generalized contractions on (ordered) partial metric spaces. Appl. Gen. Topol. **12**, 213–220 (2011)
25. Samet, B., Rajović, M., Lazović, R., Stojiljković, R.: Common fixed-point results for nonlinear contractions in ordered partial metric spaces. Fixed Point Theory Appl. **2011**:71 (2011)
26. Shatanawi, W., Samet, B., Abbas, M.: Coupled fixed point theorems for mixed monotone mappings in ordered partial metric spaces. Math. Comput. Model. **55**, 680–687 (2012)
27. Valero, O.: On Banach fixed point theorems for partial metric spaces. Appl. Gen. Topol. **6**, 229–240 (2005)

12

Quadruple fixed point theorems for compatible type mappings without mixed g-monotone property in partially ordered b-metric spaces

R. A. Rashwan · S. M. Saleh

Abstract In this paper, we establish quadruple fixed point theorems for compatible type mappings in partially ordered b-metric spaces without the mixed g-monotone property under some conditions. Also, an example is given to show our results are real generalization of known results in quadruple fixed point theory.

Keywords Quadruple fixed point · Compatible and w-compatible mappings · Partially ordered b-metric spaces

Mathematics Subject Classification 54H25 · 47H10

Introduction and preliminaries

The concept of b-metric space was introduced and studied by Bakhtin [7] and later used by Czerwik [13, 14] which is a generalization of the usual metric space. After that, several papers have dealt with fixed point theory for single-valued and multi-valued operators in b-metric spaces have been obtained (see, e.g., [2, 5, 11, 12, 26, 28]).

Existence of coupled fixed point was introduced by Guo and Lakshmikantham [16]. In 2006, Gnana-Bhaskar and Lakshmikantham [10] introduced the concept of mixed monotone property in partially ordered metric space. Afterward, Lakshmikantham and Ćirić in [24] extended these results by giving the definition of the g-monotone property. Many papers have been reported on coupled fixed point theory (see, e.g., [1, 3, 4, 9, 18, 27]). In 2011, Vasile Berinde and Marin Borcut [8] extended and generalized the results of [10] and introduced the concept of a tripled fixed point and the mixed monotone property of a mapping $F : X^3 \to X$. For more details on tripled fixed point results, we refer to [6, 26]. Recently, Karapinar and Luong [19] introduced the concept of a quadruple fixed point and the mixed monotone property of a mapping $F : X^4 \to X$ and they presented some new quadruple fixed point results. For a survey of quadruple fixed point theorems and related fixed points we refer the reader to [20–22].

R. A. Rashwan (✉) · S. M. Saleh
Department of Mathematics, Faculty of Science, Assiut University, Assiut, Egypt
e-mail: rr_rashwan54@yahoo.com

S. M. Saleh
e-mail: samirasaleh2007@yahoo.com

Definition 1.1 [14]. Let X be a nonempty set and $s \geq 1$ a given real number. A function $d : X^2 \to R^+$ is called a b-metric provided that, for all $x, y, z \in X$, the following conditions hold:

(b_1) $d(x,y) = 0$ if and only if $x = y$,
(b_2) $d(x,y) = d(y,x)$,
(b_3) $d(x,z) \leq s[d(x,y) + d(y,z)]$.

The pair (X, d) is called a b-metric space with parameter s.

Remark 1.1 It is obvious that any metric space must be a b-metric space where a b-metric space is a metric space when $s = 1$. The following example show that in general a b-metric need not necessarily be a metric space (see also [29]).

Example 1.1 [2]. Let (X, d) be a metric space and $\rho(x,y) = (d(x,y))^p$, where $p > 1$ is a real number. Then ρ is a b-metric with $s = 2^{p-1}$. However, if (X, d) is a metric space, then (X, ρ) is not necessarily a metric space. For example, if $X = R$ is the set of real numbers and $d(x,y) = |x-y|$ is the usual Euclidean metric, then $\rho(x,y) = (x-y)^2$ is a b-metric on R with $s = 2$, but is not a metric on R.

Definition 1.2 [11]. Let (X,d) be a b-metric space. Then a sequence $\{x_n\}$ in X is called

(i) b-convergent if and only if there exists $x \in X$ such that $d(x_n,x) \to 0$ as $n \to \infty$. In this case, we write $\lim_{n\to\infty} x_n = x$,
(ii) b-Cauchy if and only if $d(x_n,x_m) \to 0$ as $n,m \to \infty$.

Proposition 1.1 ([11] Remark 2.1) *In a b-metric space the following assertions hold:*

(i) *A b-convergent sequence has a unique limit.*
(ii) *Each b-convergent sequence is b-Cauchy.*
(iii) *In general, a b-metric is not continuous.*

Definition 1.3 [11]. Let (X,d) and $(\bar{X},\bar{d})$ be two b-metric spaces.

(i) The space (X,d) is b-complete if every b-Cauchy sequence in X b-converges.
(ii) A function $f: X \to \bar{X}$ is b-continuous at a point $x \in X$ if it is b-sequentially continuous at x, that is, whenever $\{x_n\}$ is b-convergent to x, $\{f(x_n)\}$ is b-convergent to $f(x)$.

Definition 1.4 [11]. The b-metric space (X,d) is b-complete if every b-Cauchy sequence in X b-converges.

It should be noted that, in general a b-metric function $d(x,y)$ for $s>1$ is not jointly continuous in all its variables. The following example on a b-metric which is not continuous.

Example 1.2 [17]. Let $X = \mathbf{N} \cup \{\infty\}$ and let $d: X \times X \to R$ be defined by

$$d(m,n) = \begin{cases} 0, & \text{if } m=n, \\ \left|\frac{1}{m}-\frac{1}{n}\right|, & \text{if one of } m,n \text{ is even and the other is even or } \infty, \\ 5, & \text{if one of } m,n \text{ is odd and the other is odd (and } m \neq n) \text{ or } \infty, \\ 2, & \text{otherwise.} \end{cases}$$

Then considering all possible cases, it can be checked that for all $m,n,p \in X$, we have

$$d(m,p) \leq \frac{5}{2}(d(m,n)+d(n,p)).$$

Thus, (X,d) is a b-metric space (with $s=\frac{5}{3}$). Let $x_n = 2n$ for each $n \in \mathbf{N}$. Then

$$d(2n,\infty) = \frac{1}{2n} \to 0 \quad \text{as } n \to \infty,$$

that is, $x_n \to \infty$, but $d(x_n,1) = 2 \nrightarrow 5 = d(\infty,1)$ as $n \to \infty$.

Since, in general, a b-metric is not continuous, we need the following lemma about the b-convergent sequences in the proof of our main result.

Lemma 1.1 [2]. *Let (X,d) be a b-metric space with $s \geq 1$, and suppose that $\{x_n\}$ and $\{y_n\}$ b-converge to x,y, respectively. Then, we have*

$$\frac{1}{s^2}d(x,y) \leq \liminf_{n\to\infty} d(x_n,y_n) \leq \limsup_{n\to\infty} d(x_n,y_n) \leq s^2 d(x,y).$$

In particular, if $x=y$ *then* $\lim_{n\to\infty} d(x_n,y_n) = 0$. *Moreover, for each $z \in X$ we have*

$$\frac{1}{s}d(x,z) \leq \liminf_{n\to\infty} d(x_n,z) \leq \limsup_{n\to\infty} d(x_n,z) \leq sd(x,z).$$

Definition 1.5 Let X be a nonempty set and let $F: X^4 \to X$, $g: X \to X$. An element $(x,y,z,w) \in X^4$ is called

(i) [19] a quadruple fixed point of F if

$$F(x,y,z,w) = x, \qquad F(y,z,w,x) = y,$$
$$F(z,w,x,y) = z \quad \text{and} \quad F(w,x,y,z) = w.$$

(ii) [25] a quadruple coincidence point of F and g if

$$F(x,y,z,w) = gx, \qquad F(y,z,w,x) = gy,$$
$$F(z,w,x,y) = gz \quad \text{and} \quad F(w,x,y,z) = gw.$$

(gx,gy,gz,gw) is said to be a quadruple point of coincidence of F and g.

(iii) [25] a quadruple common fixed point of F and g if

$$F(x,y,z,w) = gx = x, \qquad F(y,z,w,x) = gy = y,$$
$$F(z,w,x,y) = gz = z \quad \text{and} \quad F(w,x,y,z) = gw = w.$$

Definition 1.6 [25]. Let $(X,\preceq)$ be a partially ordered set and let $F: X^4 \to X$, $g: X \to X$. The mapping F is said to have the mixed g-monotone property if for any $x,y,z,w \in X$,

$$x_1,x_2 \in X,\ gx_1 \leq gx_2 \ \text{ implies } \ F(x_1,y,z,w) \leq F(x_2,y,z,w),$$
$$y_1,y_2 \in X,\ gy_1 \leq gy_2 \ \text{ implies } \ F(x,y_2,z,w) \leq F(x,y_1,z,w),$$
$$z_1,z_2 \in X,\ gz_1 \leq gz_2 \ \text{ implies } \ F(x,y,z_1,w) \leq F(x,y,z_2,w),$$
$$\text{and } \ w_1,w_2 \in X,\ gw_1 \leq gw_2 \text{ implies } F(x,y,z,w_2) \leq F(x,y,z,w_1).$$

In particular, when $g = i_X$, then from [19] we say that F has the mixed monotone property that is, for any $x,y,z,w \in X$,

$$x_1,x_2 \in X,\ x_1 \leq x_2 \ \text{ implies } \ F(x_1,y,z,w) \leq F(x_2,y,z,w),$$
$$y_1,y_2 \in X,\ y_1 \leq y_2 \ \text{ implies } \ F(x,y_2,z,w) \leq F(x,y_1,z,w),$$
$$z_1,z_2 \in X,\ z_1 \leq z_2 \ \text{ implies } \ F(x,y,z_1,w) \leq F(x,y,z_2,w),$$
$$\text{and } \ w_1,w_2 \in X,\ w_1 \leq w_2 \text{ implies } F(x,y,z,w_2) \leq F(x,y,z,w_1).$$

The concept of an altering distance function was introduced by Khan et al. [23] as follows.

Definition 1.7 [23]. A function $\psi : [0, \infty) \to [0, \infty)$ is called an altering distance function if

(i) ψ is non-decreasing and continuous,
(ii) $\psi(t) = 0$ if and only if $t = 0$.

Definition 1.8 [1]. The mappings $F : X \times X \to X$ and $f : X \to X$ are called w-compatible if $f(F(x, y)) = F(fx, fy)$ whenever $f(x) = F(x, y)$ and $f(y) = F(y, x)$.

In 2012, Dorić et al. [15] established coupled fixed point results without the mixed monotone property. This property is automatically satisfied in the case of a totally ordered space. Therefore, these results can be applied in a much wider class of problems. Now, we state a property due to Dorić et al. [15].

If elements x, y of a partially ordered set $(X, \preceq)$ are comparable (i.e., $x \preceq y$ or $y \succeq x$ holds) we will write $x \asymp y$. Let $g : X \to X$ and $F : X \times X \to X$. We will consider the following condition:

$$\text{if } x, y, u, v \in X \text{ are such that } gx \asymp F(x, y) = gu \text{ then } F(x, y) \asymp F(u, v). \tag{1.1}$$

In particular, when $g = i_X$, it reduces to

$$\text{for all } x, y, v \text{ if } x \asymp F(x, y) \text{ then } F(x, y) \asymp F(F(x, y), v). \tag{1.2}$$

The aim of this paper is to extend the property due to Dorić et al. [15] to the case of mappings $g : X \to X$, $F : X^4 \to X$, and show that a mixed monotone property in quadruple fixed point results for mappings in partially ordered b-metric spaces can be replaced by another property which is often easy to check in the case of a totally ordered space. We prove the existence of quadruple coincidence and uniqueness quadruple common fixed point theorems for a compatible and w-compatible mappings satisfying generalized contraction in partially ordered b-metric spaces without the mixed g-monotone property. Also, we state an example showing that our results are effective.

Quadruple coincidence point theorems

Let $g : X \to X$ and $F : X^4 \to X$. We consider the following condition:

$$\text{if } x, y, z, w, u, v, r, t \in X \text{ are such that } gx \asymp F(x, y, z, w) = gu \text{ then } F(x, y, z, w) \asymp F(u, v, r, t). \tag{2.1}$$

In particular, when $g = i_X$, it reduces to

$$\text{for all } x, y, v, r, t \text{ if } x \asymp F(x, y, z, w) \text{ then } F(x, y, z, w) \asymp F(F(x, y, z, w), v, r, t). \tag{2.2}$$

We will show by examples the condition (2.1), ((2.2) resp.) may be satisfied when F does not have the mixed g-monotone property, (monotone property resp.).

Example 2.1 Let $X = \{a, b, c, d\}$, $\preceq = \{(a, a), (b, b), (c, c), (d, d), (a, b), (c, d)\}$,

$$g : \begin{pmatrix} a & b & c & d \\ c & d & c & d \end{pmatrix},$$

$$F : \begin{pmatrix} (a, y, z, w) & (b, y, z, w) & (c, y, z, w) & (d, y, z, w) \\ b & a & c & d \end{pmatrix},$$

for all $y, z, w \in X$. Since $ga = c \preceq gb = d$ but $F(a, y, z, w) \succeq F(b, y, z, w)$ for all $y, z, w \in X$, the mapping F does not have the mixed g-monotone property. But it has property (2.1) where

(i) For each $y, z, w \in X$, we get $gc \asymp F(c, y, z, w)$ and $F(c, y, z, w) \asymp F(c, v, r, t)$ for all $v, r, t \in X$.
(ii) For each $y, z, w \in X$, we get $gd \asymp F(d, y, z, w)$ and $F(d, y, z, w) \asymp F(d, v, r, t)$ for all $v, r, t \in X$.
(iii) The other two cases are trivial.

Example 2.2 Let $X = \{a, b, c, d\}$, $\preceq = \{(a, a), (b, b), (c, c), (d, d), (a, b), (c, d)\}$,

$$F : \begin{pmatrix} (a, y, z, w) & (b, y, z, w) & (c, y, z, w) & (d, y, z, w) \\ b & a & c & d \end{pmatrix}$$

for all $y, z, w \in X$. Since $a \preceq b$ but $F(a, y, z, w) = b \succeq a = F(b, y, z, w)$ for all $y, z, w \in X$, the mapping F does not have the mixed monotone property. But it has property (2.2) since

(i) For each $y, z, w \in X$, we get $a \asymp F(a, y, z, w)$ and $F(a, y, z, w) = b \asymp a = F(F(a, y, z, w), v, r, t)$ for all $v, r, t \in X$.
(ii) For each $y, z, w \in X$, we get $b \asymp F(b, y, z, w)$ and $F(b, y, z, w) = a \asymp b = F(F(b, y, z, w), v, r, t)$ for all $v, r, t \in X$.
(iii) The other two cases are trivial.

Definition 2.1 Let (X,d) be a b-metric space and let $g : X \to X$, $F : X^4 \to X$. The mappings g and F are said to be compatible if

$$\lim_{n\to\infty} d(gF(x_n,y_n,z_n,w_n),F(gx_n,gy_n,gz_n,gw_n)) = 0,$$
$$\lim_{n\to\infty} d(gF(y_n,z_n,w_n,x_n),F(gy_n,gz_n,gw_n,gx_n)) = 0,$$
$$\lim_{n\to\infty} d(gF(z_n,w_n,x_n,y_n),F(gz_n,gw_n,gx_n,gy_n)) = 0$$
and $$\lim_{n\to\infty} d(gF(w_n,x_n,y_n,z_n),F(gw_n,gx_n,gy_n,gz_n)) = 0,$$

hold whenever $\{x_n\}$, $\{y_n\}$, $\{z_n\}$ and $\{w_n\}$ are sequences in X such that

$$\lim_{n\to\infty} F(x_n,y_n,z_n,w_n) = \lim_{n\to\infty} gx_n,$$
$$\lim_{n\to\infty} F(y_n,z_n,w_n,x_n) = \lim_{n\to\infty} gy_n,$$
$$\lim_{n\to\infty} F(z_n,w_n,x_n,y_n) = \lim_{n\to\infty} gz_n \quad \text{and}$$
$$\lim_{n\to\infty} F(w_n,x_n,y_n,z_n) = \lim_{n\to\infty} gw_n.$$

Definition 2.2 The mappings $F : X^4 \to X$ and $g : X \to X$ are called w-compatible if $g(F(x,y,z,w)) = F(gx, gy, gz, gw)$ whenever $gx = F(x,y,z,w)$, $gy = F(y,z,w,x)$, $gz = F(z,w,x,y)$ and $gw = F(w,x,y,z)$.

Remark 2.1 In an altering distance function $\psi : [0,\infty) \to [0,\infty)$, since ψ is non-decreasing then for any $a,b,c,d \in [0,\infty)$ the following holds.

$$\psi(\max\{a,b,c,d\}) = \max\{\psi(a),\psi(b),\psi(c),\psi(d)\}.$$

The triple $(X,d,\preceq)$ is called a partially ordered b-metric space if $(X,\preceq)$ is a partially ordered set and (X,d) is a b-metric space.

Our first result is the following.

Theorem 2.1 *Let $(X,d,\preceq)$ be a partially ordered complete b-metric space with parameter $s \geq 1$. Let $F : X^4 \to X$ and $g : X \to X$ be two mappings such that the following hold:*

(i) *g and F are b-continuous,*
(ii) *$F(X^4) \subseteq g(X)$, g and F are compatible,*
(iii) *g and F satisfy property (2.1),*
(iv) *there exist $x_0,y_0,z_0,w_0 \in X$ such that $gx_0 \asymp F(x_0,y_0,z_0,w_0)$, $gy_0 \asymp F(y_0,z_0,w_0,x_0)$, $gz_0 \asymp F(z_0,w_0,x_0,y_0)$ and $gw_0 \asymp F(w_0,x_0,y_0,z_0)$,*
(v) *there exist an altering distance function ψ and $\phi : [0,\infty)^4 \to [0,\infty)$ is continuous with $\phi(t_1,t_2,t_3,t_4) = 0$ if and only if $t_1 = t_2 = t_3 = t_4 = 0$ such that*

$$\begin{aligned}&\psi(sd(F(x,y,z,w),F(u,v,r,t)))\\ &\leq \psi(\max\{d(gx,gu),d(gy,gv),d(gz,gr),d(gw,gt)\})\\ &- \phi(d(gx,gu),d(gy,gv),d(gz,gr),d(gw,gt)),\end{aligned} \tag{2.3}$$

for all $x,y,z,w,u,v,r,t \in X$ and $gx \asymp gu$, $gy \asymp gv$, $gz \asymp gr$ and $gw \asymp gt$.

Then, F and g have a quadruple coincidence point.

Proof Let $x_0,y_0,z_0,w_0 \in X$ be such that condition (iv) holds. Since $F(X^4) \subseteq g(X)$, then we can choose $x_1,y_1,z_1,w_1 \in X$ such that

$$\begin{aligned}&gx_1 = F(x_0,y_0,z_0,w_0), \quad gy_1 = F(y_0,z_0,w_0,x_0),\\ &gz_1 = F(z_0,w_0,x_0,y_0) \quad \text{and} \quad gw_1 = F(w_0,x_0,y_0,z_0).\end{aligned} \tag{2.4}$$

By continuing this process, we can construct sequences $\{x_n\}, \{y_n\}, \{z_n\}$ and $\{w_n\}$ in X such that

$$\begin{aligned}&gx_{n+1} = F(x_n,y_n,z_n,w_n),\ gy_{n+1} = F(y_n,z_n,w_n,x_n),\\ &gz_{n+1} = F(z_n,w_n,x_n,y_n),\ \text{and}\ gw_{n+1} = F(w_n,x_n,y_n,z_n)\\ &\text{for all}\ n \geq 0.\end{aligned} \tag{2.5}$$

We show that

$$\begin{aligned}&gx_n \asymp gx_{n+1}, \quad gy_n \asymp gy_{n+1}, \quad gz_n \asymp gz_{n+1}\\ &\text{and} \quad gw_n \asymp gw_{n+1} \quad \text{for } n \geq 0.\end{aligned} \tag{2.6}$$

So, we use the mathematical induction. By condition (iv) and using (2.4) we get $gx_0 \asymp gx_1$, $gy_0 \asymp gy_1$, $gz_0 \asymp gz_1$, and $gw_0 \asymp gw_1$. So (2.6) holds for $n = 0$. We assume that (2.6) holds for some $n > 0$, that is $gx_n \asymp gx_{n+1}$, $gy_n \asymp gy_{n+1}$, $gz_n \asymp gz_{n+1}$, and $gw_n \asymp gw_{n+1}$, we get

$$gx_n = F(x_{n-1},y_{n-1},z_{n-1},w_{n-1}) \asymp F(x_n,y_n,z_n,w_n) = gx_{n+1},$$

$$gy_n = F(y_{n-1},z_{n-1},w_{n-1},x_{n-1}) \asymp F(y_n,z_n,w_n,x_n) = gy_{n+1},$$

$$gz_n = F(z_{n-1},w_{n-1},x_{n-1},y_{n-1}) \asymp F(z_n,w_n,x_n,y_n) = gz_{n+1},$$

$$gw_n = F(w_{n-1},x_{n-1},y_{n-1},z_{n-1}) \asymp F(w_n,x_n,y_n,z_n) = gw_{n+1}.$$

Hence from condition (iii) we conclude that

$$F(x_n,y_n,z_n,w_n) \asymp F(x_{n+1},y_{n+1},z_{n+1},w_{n+1}),$$
$$F(y_n,z_n,w_n,x_n) \asymp F(y_{n+1},z_{n+1},w_{n+1},x_{n+1}),$$
$$F(z_n,w_n,x_n,y_n) \asymp F(z_{n+1},w_{n+1},x_{n+1},y_{n+1}),$$
$$F(w_n,x_n,y_n,z_n) \asymp F(w_{n+1},x_{n+1},y_{n+1},z_{n+1}).$$

So, $gx_{n+1} \asymp gx_{n+2}$, $gy_{n+1} \asymp gy_{n+2}$, $gz_{n+1} \asymp gz_{n+2}$ and $gw_{n+1} \asymp gw_{n+2}$. Thus (2.6) holds for all $n \in \mathbf{N}$. Suppose that for some $k \in \mathbf{N}$,

$$gx_k = gx_{k+1}, \quad gy_k = gy_{k+1},$$
$$gz_k = gz_{k+1} \text{ and } gw_k = gw_{k+1},$$

then by (2.5) we get $gx_k = F(x_k, y_k, z_k, w_k)$, $gy_k = F(y_k, z_k, w_k, x_k)$, $gz_k = F(z_k, w_k, x_k, y_k)$ and $gw_k = F(w_k, x_k, y_k, z_k)$. Hence (x_k, y_k, z_k, w_k) is a quadruple coincidence point of F and g. So, we assume that for all $n \in \mathbf{N}$ at least $gx_n \neq gx_{n+1}$ or $gy_n \neq gy_{n+1}$ or $gz_n \neq gz_{n+1}$ or $gw_n \neq gw$ $n+1$. Since $gx_n \asymp gx_{n+1}$, $gy_n \asymp gy_{n+1}$, $gz_n \asymp gz_{n+1}$ and $gw_n \asymp gw_{n+1}$ for all $n \in \mathbf{N}$, then from (2.3) and (2.5) we obtain

$$\psi(sd(gx_n, gx_{n+1}) = \psi(sd(F(x_{n-1}, y_{n-1}, z_{n-1}, w_{n-1}), F(x_n, y_n, z_n, w_n))) \leq \psi(\max\{d(gx_{n-1}, gx_n), d(gy_{n-1}, gy_n), d(gz_{n-1}, gz_n), d(gw_{n-1}, gw_n)\}) - \phi(d(gx_{n-1}, gx_n), d(gy_{n-1}, gy_n), d(gz_{n-1}, gz_n), d(gw_{n-1}, gw_n)), \quad (2.7)$$

$$\psi(sd(gy_n, gy_{n+1})) = \psi(sd(F(y_{n-1}, z_{n-1}, w_{n-1}, x_{n-1}), F(y_n, z_n, w_n, x_n))) \leq \psi(\max\{d(gy_{n-1}, gy_n), d(gz_{n-1}, gz_n), d(gw_{n-1}, gw_n), d(gx_{n-1}, gx_n)\}) - \phi(d(gy_{n-1}, gy_n), d(gz_{n-1}, gz_n), d(gw_{n-1}, gw_n), d(gx_{n-1}, gx_n)), \quad (2.8)$$

$$\psi(sd(gz_n, gz_{n+1})) = \psi(sd(F(z_{n-1}, w_{n-1}, x_{n-1}, y_{n-1}), F(z_n, w_n, x_n, y_n))) \leq \psi(\max\{d(gz_{n-1}, gz_n), d(gw_{n-1}, gw_n), d(gx_{n-1}, gx_n), d(gy_{n-1}, gy_n)\}) - \phi(d(gz_{n-1}, gz_n), d(gw_{n-1}, gw_n), d(gx_{n-1}, gx_n), d(gy_{n-1}, gy_n)), \quad (2.9)$$

and

$$\psi(sd(gw_n, gw_{n+1})) = \psi(sd(F(w_{n-1}, x_{n-1}, y_{n-1}, z_{n-1}), F(w_n, x_n, y_n, z_n))) \leq \psi(\max\{d(gw_{n-1}, gw_n), d(gx_{n-1}, gx_n), d(gy_{n-1}, gy_n), d(gz_{n-1}, gz_n)\}) - \phi(d(gw_{n-1}, gw_n), d(gx_{n-1}, gx_n), d(gy_{n-1}, gy_n), d(gz_{n-1}, gz_n)). \quad (2.10)$$

Set

$$\delta_n = \max\{d(gx_n, gx_{n+1}), d(gy_n, gy_{n+1}), d(gz_n, gz_{n+1}), d(gw_n, gw_{n+1})\}.$$

From (2.7) to (2.10) and Remark 2.1, it follows that

$$\psi(s\delta_n) = \max\{\psi(sd(gx_n, gx_{n+1})), \psi(sd(gy_n, gy_{n+1})), \psi(sd(gz_n, gz_{n+1})), \psi(sd(gw_n, gw_{n+1}))\} \leq \psi(\delta_{n-1}) - \min\{\phi(d(gx_{n-1}, gx_n), d(gy_{n-1}, gy_n), d(gz_{n-1}, gz_n), d(gw_{n-1}, gw_n)), \phi(d(gy_{n-1}, gy_n), d(gz_{n-1}, gz_n), d(gw_{n-1}, gw_n), d(gx_{n-1}, gx_n)), \phi(d(gz_{n-1}, gz_n), d(gw_{n-1}, gw_n), d(gx_{n-1}, gx_n), d(gy_{n-1}, gy_n)), \psi(d(gw_{n-1}, gw_n), d(gx_{n-1}, gx_n), d(gy_{n-1}, gy_n), d(gz_{n-1}, gz_n))\}. \quad (2.11)$$

and since ψ is non-decreasing then from (2.11) we have that

$$\psi(\delta_n) \leq \psi(s\delta_n) \leq \psi(\delta_{n-1}) - \min\{\phi(d(gx_{n-1}, gx_n), d(gy_{n-1}, gy_n), d(gz_{n-1}, gz_n), d(gw_{n-1}, gw_n)), \phi(d(gy_{n-1}, gy_n), d(gz_{n-1}, gz_n), d(gw_{n-1}, gw_n), d(gx_{n-1}, gx_n)), \phi(d(gz_{n-1}, gz_n), d(gw_{n-1}, gw_n), d(gx_{n-1}, gx_n), d(gy_{n-1}, gy_n)),; \phi(d(gw_{n-1}, gw_n), d(gx_{n-1}, gx_n), d(gy_{n-1}, gy_n), d(gz_{n-1}, gz_n))\}. \quad (2.12)$$

Hence,

$$\psi(\delta_n) \leq \psi(\delta_{n-1}) \text{ for all } n \in \mathbf{N}. \quad (2.13)$$

Since ψ is non-decreasing, we have $\delta_n \leq \delta_{n-1}$ for all n. Therefore, $\{\delta_n\}$ is a non-increasing sequence, so there is some $\delta \geq 0$ such that

$$\lim_{n\to\infty} \delta_n = \delta.$$

Letting $n \to \infty$, in (2.12), we get

$$\psi(\delta) \leq \psi(\delta) - \min\{\lim_{n\to\infty} \phi(d(gx_{n-1}, gx_n), d(gy_{n-1}, gy_n), d(gz_{n-1}, gz_n), d(gw_{n-1}, gw_n)), \lim_{n\to\infty} \phi(d(gy_{n-1}, gy_n), d(gz_{n-1}, gz_n), d(gw_{n-1}, gw_n), d(gx_{n-1}, gx_n)), \lim_{n\to\infty} \phi(d(gz_{n-1}, gz_n), d(gw_{n-1}, gw_n), d(gx_{n-1}, gx_n), d(gy_{n-1}, gy_n)), \lim_{n\to\infty} \phi(d(gw_{n-1}, gw_n), d(gx_{n-1}, gx_n), d(gy_{n-1}, gy_n), d(gz_{n-1}, gz_n))\} \leq \psi(\delta).$$

Hence,

$$\min \lim_{n\to\infty} \phi(d(gx_{n-1}, gx_n), d(gy_{n-1}, gy_n), d(gz_{n-1}, gz_n), d(gw_{n-1}, gw_n)), \lim_{n\to\infty} \phi(d(gy_{n-1}, gy_n), d(gz_{n-1}, gz_n), d(gw_{n-1}, gw_n), d(gx_{n-1}, gx_n)), \lim_{n\to\infty} \phi(d(gz_{n-1}, gz_n), d(gw_{n-1}, gw_n), d(gx_{n-1}, gx_n), d(gy_{n-1}, gy_n)), \lim_{n\to\infty} \phi(d(gw_{n-1}, gw_n), d(gx_{n-1}, gx_n), d(gy_{n-1}, gy_n), d(gz_{n-1}, gz_n))\} = 0.$$

That is

$$\lim_{n\to\infty} \phi(d(gx_{n-1}, gx_n), d(gy_{n-1}, gy_n), d(gz_{n-1}, gz_n), d(gw_{n-1}, gw_n)) = 0 \text{ or } \lim_{n\to\infty} \phi(d(gy_{n-1}, gy_n), d(gz_{n-1}, gz_n), d(gw_{n-1}, gw_n), d(gx_{n-1}, gx_n)) = 0 \text{ or } \lim_{n\to\infty} \phi(d(gz_{n-1}, gz_n), d(gw_{n-1}, gw_n), d(gx_{n-1}, gx_n), d(gy_{n-1}, gy_n)) = 0 \quad \text{or} \quad \lim_{n\to\infty} \phi(d(gw_{n-1}, gw_n), d(gx_{n-1}, gx_n), d(gy_{n-1}, gy_n), d(gz_{n-1}, gz_n)) = 0.$$

So, using the properties of ϕ, we have

$$\lim_{n\to\infty} d(gx_{n-1}, gx_n) = 0, \qquad \lim_{n\to\infty} d(gy_{n-1}, gy_n) = 0,$$
$$\lim_{n\to\infty} d(gz_{n-1}, gz_n) = 0 \quad \text{and} \quad \lim_{n\to\infty} d(gw_{n-1}, gw_n) = 0.$$

Thus, $\lim_{n\to\infty} \delta_{n-1} = 0$. Therefore $\delta = 0$. Now, we show that $\{gx_n\}, \{gy_n\}, \{gz_n\}$, and $\{gw_n\}$ are b-Cauchy sequences in (X, d), that is, we show that for every $\varepsilon > 0$, there exists $k \in \mathbf{N}$ such that for all $m, n \geq k$,

$$\max\{d(gx_m, gx_n), d(gy_m, gy_n), d(gz_m, gz_n), d(gw_m, gw_n)\} < \varepsilon.$$

Suppose the contrary, that is at least one of the sequences $\{gx_n\}, \{gy_n\}, \{gz_n\}$ and $\{gw_n\}$ is not a b-Cauchy sequence, so there exists $\varepsilon > 0$ for which we can find subsequences $\{gx_{m(k)}\}$, $\{gx_{n(k)}\}$ of $\{gx_n\}$, $\{gy_{m(k)}\}, \{gy_{n(k)}\}$ of $\{gy_n\}$, $\{gz_{m(k)}\}$, $\{gz_{n(k)}\}$ of $\{gz_n\}$ and $\{gw_{m(k)}\}$, $\{gw_{n(k)}\}$ of $\{gw_n\}$ with $n(k) > m(k) \geq k$ such that

$$\max\{d(gx_{m(k)}, gx_{n(k)}), d(gy_{m(k)}, gy_{n(k)}), d(gz_{m(k)}, gz_{n(k)}), d(gw_{m(k)}, gw_{n(k)})\} \geq \varepsilon, \tag{2.14}$$

for every integer k, let $n(k)$ be the least positive integer with $n(k) > m(k) \geq k$ satisfying (2.14) and such that

$$\max\{d(gx_{m(k)}, gx_{n(k)-1}), d(gy_{m(k)}, gy_{n(k)-1}), d(gz_{m(k)}, gz_{n(k)-1}), d(gw_{m(k)}, gw_{n(k)-1})\} < \varepsilon. \tag{2.15}$$

Using the b-triangle inequality, we get

$$d(gx_{m(k)}, gx_{n(k)}) \leq s[d(gx_{m(k)}, gx_{n(k)-1}) + d(gx_{n(k)-1}, gx_{n(k)})] d(gy_{m(k)}, gy_{n(k)}) \leq s[d(gy_{m(k)}, gy_{n(k)-1}) + d(gy_{n(k)-1}, gy_{n(k)})] d(gz_{m(k)}, gz_{n(k)}) \leq s[d(gz_{m(k)}, gz_{n(k)-1}) + d(gz_{n(k)-1}, gz_{n(k)})] \text{ and } d(gw_{m(k)}, gw_{n(k)}) \leq s[d(gw_{m(k)}, gw_{n(k)-1}) + d(gw_{n(k)-1}, gw_{n(k)})]. \tag{2.16}$$

Hence from (2.14) and (2.16), we have

$$\varepsilon \leq \max\{d(gx_{m(k)}, gx_{n(k)}), d(gy_{m(k)}, gy_{n(k)}), d(gz_{m(k)}, gz_{n(k)}), d(gw_{m(k)}, gw_{n(k)})\} \leq s[\max\{d(gx_{m(k)}, gx_{n(k)-1}), d(gy_{m(k)}, gy_{n(k)-1}), d(gz_{m(k)}, gz_{n(k)-1}), d(gw_{m(k)}, gw_{n(k)-1})\} + \max\{d(gx_{n(k)-1}, gx_{n(k)}), d(gy_{n(k)-1}, gy_{n(k)}), d(gz_{n(k)-1}, gz_{n(k)}), d(gw_{n(k)-1}, gw_{n(k)})\}] = s\max\{d(gx_{m(k)}, gx_{n(k)-1}), d(gy_{m(k)}, gy_{n(k)-1}), d(gz_{m(k)}, gz_{n(k)-1}), d(gw_{m(k)}, gw_{n(k)-1})\} + s\delta_{n(k)-1}.$$

Taking the upper and lower limits as $k \to \infty$ in the above inequality, from (2.14), (2.15) and as $\lim_{n\to\infty} \delta_{n-1} = 0$, we conclude that

$$\frac{\varepsilon}{s} \leq \limsup_{k\to\infty} \max\{d(gx_{m(k)}, gx_{n(k)}), d(gy_{m(k)}, gy_{n(k)}), d(gz_{m(k)}, gz_{n(k)}), d(gw_{m(k)}, gw_{n(k)})\} < \varepsilon, \tag{2.17}$$

$$\frac{\varepsilon}{s} \leq \limsup_{k\to\infty} \max\{d(gx_{m(k)}, gx_{n(k)-1}), d(gy_{m(k)}, gy_{n(k)-1}), d(gz_{m(k)}, gz_{n(k)-1}), d(gw_{m(k)}, gw_{n(k)-1})\} < \varepsilon, \tag{2.18}$$

and

$$\frac{\varepsilon}{s} \leq \liminf_{k\to\infty} \max\{d(gx_{m(k)}, gx_{n(k)-1}), d(gy_{m(k)}, gy_{n(k)-1}), d(gz_{m(k)}, gz_{n(k)-1}), d(gw_{m(k)}, gw_{n(k)-1})\} < \varepsilon. \tag{2.19}$$

Also, from the b-triangle inequality we obtain

$$\varepsilon \leq \max\{d(gx_{m(k)}, gx_{n(k)}), d(gy_{m(k)}, gy_{n(k)}), d(gz_{m(k)}, gz_{n(k)}), d(gw_{m(k)}, gw_{n(k)})\} \leq s[\max\{d(gx_{m(k)}, gx_{m(k)+1}), d(gy_{m(k)}, gy_{m(k)+1}), d(gz_{m(k)}, gz_{m(k)+1}), d(gw_{m(k)}, gw_{m(k)+1})\} + \max\{d(gx_{m(k)+1}, gx_{n(k)}), d(gy_{m(k)+1}, gy_{n(k)}), d(gz_{m(k)+1}, gz_{n(k)}), d(gw_{m(k)+1}, gw_{n(k)})\}] = s\delta_{m(k)} + s\max\{d(gx_{m(k)+1}, gx_{n(k)}), d(gy_{m(k)+1}, gy_{n(k)}), d(gz_{m(k)+1}, gz_{n(k)}), d(gw_{m(k)+1}, gw_{n(k)})\},$$

and

$$\max\{d(gx_{m(k)+1}, gx_{n(k)}), d(gy_{m(k)+1}, gy_{n(k)}), d(gz_{m(k)+1}, gz_{n(k)}), d(gw_{m(k)+1}, gw_{n(k)})\} \leq s[\max\{d(gx_{m(k)+1}, gx_{m(k)}) + gy_{m(k)+1}, d(gy_{m(k)}) + d(gz_{m(k)+1}, gz_{m(k)}) + d(gw_{m(k)+1}, gw_{m(k)})\} + \max\{d(gx_{m(k)}, gx_{n(k)}), d(gy_{m(k)}, gy_{n(k)}), d(gz_{m(k)}, gz_{n(k)}), d(gw_{m(k)}, gw_{n(k)})\}] = s\delta_{m(k)} + s\max\{d(gx_{m(k)}, gx_{n(k)}), d(gy_{m(k)}, gy_{n(k)}), d(gz_{m(k)}, gz_{n(k)}), d(gw_{m(k)}, gw_{n(k)})\}.$$

Taking the upper limit as $k \to \infty$ in the above two inequalities, using (2.14) and as $\lim_{n\to\infty} \delta_{n-1} = 0$, we have

$$\frac{\varepsilon}{s} \leq \limsup_{n\to\infty} \max\{d(gx_{m(k)+1}, gx_{n(k)}), d(gy_{m(k)+1}, gy_{n(k)}), d(gz_{m(k)+1}, gz_{n(k)}), d(gw_{m(k)+1}, gw_{n(k)})\} < s\varepsilon. \tag{2.20}$$

Since $gx_n \asymp gx_{n+1}$, $gy_n \asymp gy_{n+1}$, $gz_n \asymp gz_{n+1}$ and $gw_n \asymp gw_{n+1}$ for all $n \geq 0$, then $gx_{m(k)} \asymp gx_{n(k)-1}$, $gy_{m(k)} \asymp gy\ n(k) - 1$, $gz_{m(k)} \asymp gz_{n(k)-1}$ and $gw_{m(k)} \asymp gw_{n(k)-1}$.

Putting $x = x_{m(k)}$, $y = y_{m(k)}$, $z = z_{m(k)}$, $w = w_{m(k)}$, $u = x_{n(k)-1}$, $v = y_{n(k)-1}$, $r = z_{n(k)-1}$, $t = w_{n(k)-1}$, in (2.3) for all $k \geq 0$, we conclude that

$$\begin{aligned}&\psi(sd(gx_{m(k)+1}, gx_{n(k)}))\\&= \psi(sd(F(x_{m(k)}, y_{m(k)}, z_{m(k)}, w_{m(k)}), F(x_{n(k)-1}, y_{n(k)-1},\\&\quad z_{n(k)-1}, w_{n(k)-1})))\\&\leq \psi(\max\{d(gx_{m(k)}, gx_{n(k)-1}), d(gy_{m(k)}, gy_{n(k)-1}),\\&\quad d(gz_{m(k)}, gz_{n(k)-1}), d(gw_{m(k)}, gw_{n(k)-1})\})\\&\quad - \phi(d(gx_{m(k)}, gx_{n(k)-1}), d(gy_{m(k)}, gy_{n(k)-1}),\\&\quad d(gz_{m(k)}, gz_{n(k)-1}), d(gw_{m(k)}, gw_{n(k)-1})),\end{aligned} \tag{2.21}$$

$$\begin{aligned}&\psi(sd(gy_{m(k)+1}, gy_{n(k)}))\\&= \psi(sd(F(y_{m(k)}, z_{m(k)}, w_{m(k)}, x_{m(k)}), F(y_{n(k)-1},\\&\quad z_{n(k)-1}, w_{n(k)-1}, x_{n(k)-1})))\\&\leq \psi(\max\{d(gy_{m(k)}, gy_{n(k)-1}), d(gz_{m(k)}, gz_{n(k)-1}),\\&\quad d(gw_{m(k)}, gw_{n(k)-1}), d(gx_{m(k)}, gx_{n(k)-1})\})\\&\quad - \phi(d(gy_{m(k)}, gy_{n(k)-1}), d(gz_{m(k)}, gz_{n(k)-1}), d(gw_{m(k)},\\&\quad gw_{n(k)-1}), d(gx_{m(k)}, gx_{n(k)-1})),\end{aligned} \tag{2.22}$$

$$\begin{aligned}&\psi(sd(gz_{m(k)+1}, gz_{n(k)}))\\&= \psi(sd(F(z_{m(k)}, w_{m(k)}, x_{m(k)}, y_{m(k)}), F(z_{n(k)-1},\\&\quad w_{n(k)-1}, x_{n(k)-1}, y_{n(k)-1})))\\&\leq \psi(\max\{d(gz_{m(k)}, gz_{n(k)-1}), d(gw_{m(k)}, gw_{n(k)-1}),\\&\quad d(gx_{m(k)}, gx_{n(k)-1}), d(gy_{m(k)}, gy_{n(k)-1})\})\\&\quad - \phi(d(gz_{m(k)}, gz_{n(k)-1}), d(gw_{m(k)}, gw_{n(k)-1}),\\&\quad d(gx_{m(k)}, gx_{n(k)-1}), d(gy_{m(k)}, gy_{n(k)-1})),\end{aligned} \tag{2.23}$$

and

$$\begin{aligned}&\psi(sd(gw_{m(k)+1}, gw_{n(k)}))\\&= \psi(sd(F(w_{m(k)}, x_{m(k)}, y_{m(k)}, z_{m(k)}), F(w_{n(k)-1}, x_{n(k)-1},\\&\quad y_{n(k)-1}, z_{n(k)-1})))\\&\leq \psi(\max\{d(gw_{m(k)}, gw_{n(k)-1}), d(gx_{m(k)}, gx_{n(k)-1}),\\&\quad d(gy_{m(k)}, gy_{n(k)-1}), d(gz_{m(k)}, gz_{n(k)-1})\})\\&\quad - \phi(d(gw_{m(k)}, gw_{n(k)-1}), d(gx_{m(k)}, gx_{n(k)-1}), d(gy_{m(k)},\\&\quad gy_{n(k)-1}), d(gz_{m(k)}, gz_{n(k)-1})).\end{aligned} \tag{2.24}$$

From (2.21)–(2.24) and Remark 2.1, it follows that

$$\begin{aligned}&\psi(s\max\{d(gx_{m(k)+1}, gx_{n(k)}), d(gy_{m(k)+1}, gy_{n(k)}),\\&\quad d(gz_{m(k)+1}, gz_{n(k)}), d(gw_{m(k)+1}, gw_{n(k)})\})\\&\quad\leq \psi(\max\{d(gx_{m(k)}, gx_{n(k)-1}), d(gy_{m(k)}, gy_{n(k)-1}),\\&\quad d(gz_{m(k)}, gz_{n(k)-1}), d(gw_{m(k)}, gw_{n(k)-1})\})\\&\quad - \min\{\phi(d(gx_{m(k)}, gx_{n(k)-1}), d(gy_{m(k)}, gy_{n(k)-1}), d(gz_{m(k)},\\&\quad gz_{n(k)-1}), d(gw_{m(k)}, gw_{n(k)-1})),\\&\quad \phi(d(gy_{m(k)}, gy_{n(k)-1}), d(gz_{m(k)}, gz_{n(k)-1}), d(gw_{m(k)},\\&\quad gw_{n(k)-1}), d(gx_{m(k)}, gx_{n(k)-1})),\\&\quad \phi(d(gz_{m(k)}, gz_{n(k)-1}), d(gw_{m(k)}, gw_{n(k)-1}), d(gx_{m(k)},\\&\quad gx_{n(k)-1}), d(gy_{m(k)}, gy_{n(k)-1})),\\&\quad \phi(d(gw_{m(k)}, gw_{n(k)-1}), d(gx_{m(k)}, gx_{n(k)-1}), d(gy_{m(k)}, gy_{n(k)-1}),\\&\quad d(gz_{m(k)}, gz_{n(k)-1}))\}.\end{aligned}$$

Taking the upper limit as $k \to \infty$ in the above inequality and using (2.18) and (2.20), we have

$$\begin{aligned}&\psi(\varepsilon) = \psi\left(s\frac{\varepsilon}{s}\right)\\&\leq \psi(\varepsilon) - \liminf_{k\to\infty} \min\{\phi(d(gx_{m(k)}, gx_{n(k)-1}),\\&\quad d(gy_{m(k)}, gy_{n(k)-1}), d(gz_{m(k)}, gz_{n(k)-1}), d(gw_{m(k)}, gw_{n(k)-1})),\\&\phi(d(gy_{m(k)}, gy_{n(k)-1}), d(gz_{m(k)}, gz_{n(k)-1}), d(gw_{m(k)},\\&\quad gw_{n(k)-1}), d(gx_{m(k)}, gx_{n(k)-1})),\\&\phi(d(gz_{m(k)}, gz_{n(k)-1}), d(gw_{m(k)}, gw_{n(k)-1}), d(gx_{m(k)},\\&\quad gx_{n(k)-1}), d(gy_{m(k)}, gy_{n(k)-1})),\\&\phi(d(gw_{m(k)}, gw_{n(k)-1}), d(gx_{m(k)}, gx_{n(k)-1}), d(gy_{m(k)},\\&\quad gy_{n(k)-1}), d(gz_{m(k)}, gz_{n(k)-1}))\}.\end{aligned}$$

Hence,

$$\begin{aligned}&\liminf_{k\to\infty} \min\{\phi(d(gx_{m(k)}, gx_{n(k)-1}), d(gy_{m(k)}, gy_{n(k)-1}),\\&\quad d(gz_{m(k)}, gz_{n(k)-1}), d(gw_{m(k)}, gw_{n(k)-1})),\\&\phi(d(gy_{m(k)}, gy_{n(k)-1}), d(gz_{m(k)}, gz_{n(k)-1}), d(gw_{m(k)},\\&\quad gw_{n(k)-1}), d(gx_{m(k)}, gx_{n(k)-1})),\\&\phi(d(gz_{m(k)}, gz_{n(k)-1}), d(gw_{m(k)}, gw_{n(k)-1}),\\&\quad d(gx_{m(k)}, gx_{n(k)-1}), d(gy_{m(k)}, gy_{n(k)-1})),\\&\phi(d(gw_{m(k)}, gw_{n(k)-1}), d(gx_{m(k)}, gx_{n(k)-1}), d(gy_{m(k)},\\&\quad gy_{n(k)-1}), d(gz_{m(k)}, gz_{n(k)-1}))\} = 0,\end{aligned}$$

Therefore,

$$\liminf_{k\to\infty} \phi(d(gx_{m(k)}, gx_{n(k)-1}), d(gy_{m(k)}, gy_{n(k)-1}), d(gz_{m(k)}, gz_{n(k)-1}), d(gw_{m(k)}, gw_{n(k)-1}))$$
$$\liminf_{k\to\infty} \phi(d(gy_{m(k)}, gy_{n(k)-1}), d(gz_{m(k)}, gz_{n(k)-1}), d(gw_{m(k)}, gw_{n(k)-1}), d(gx_{m(k)}, gx_{n(k)-1}))$$
$$\liminf_{k\to\infty} \phi(d(gz_{m(k)}, gz_{n(k)-1}), d(gw_{m(k)}, gw_{n(k)-1}), d(gx_{m(k)}, gx_{n(k)-1}), d(gy_{m(k)}, gy_{n(k)-1}))$$
$$\liminf_{k\to\infty} \phi(d(gw_{m(k)}, gw_{n(k)-1}), d(gx_{m(k)}, gx_{n(k)-1}), d(gy_{m(k)}, gy_{n(k)-1}), d(gz_{m(k)}, gz_{n(k)-1}))$$

Using the properties of ϕ, it follows that

$$\liminf_{k\to\infty} d(gx_{m(k)}, gx_{n(k)-1}) = 0,$$
$$\liminf_{k\to\infty} d(gy_{m(k)}, gy_{n(k)-1}) = 0,$$
$$\liminf_{k\to\infty} d(gz_{m(k)}, gz_{n(k)-1}) = 0 \quad \text{and}$$
$$\liminf_{k\to\infty} d(gw_{m(k)}, gw_{n(k)-1}) = 0.$$

Hence,

$$\liminf_{k\to\infty} \max\{d(gx_{m(k)}, gx_{n(k)-1}), d(gy_{m(k)}, gy_{n(k)-1}), d(gz_{m(k)}, gz_{n(k)-1}), d(gw_{m(k)}, gw_{n(k)-1})\} = 0,$$

which is a contradiction to (2.19). Thus, $\{gx_n\}$, $\{gy_n\}$, $\{gz_n\}$, and $\{gw_n\}$ are b-Cauchy sequences in X. Now, we show that F and g have a quadruple coincidence point. Since X is b-complete and $\{gx_n\}$, $\{gy_n\}$, $\{gz_n\}$, and $\{gw_n\}$ are b-Cauchy sequences in X, there exists $x, y, z, w \in X$ such that

$$\lim_{n\to\infty} d(gx_n, x) = 0, \quad \lim_{n\to\infty} d(gy_n, y) = 0, \quad \lim_{n\to\infty} d(gz_n, z) = 0 \quad \text{and} \quad \lim_{n\to\infty} d(gw_n, w) = 0. \tag{2.25}$$

From (2.5) and (2.25), we get

$$\lim_{n\to\infty} gx_n = \lim_{n\to\infty} F(x_n, y_n, z_n, w_n) = x,$$
$$\lim_{n\to\infty} gy_n = \lim_{n\to\infty} F(y_n, z_n, w_n, x_n) = y,$$
$$\lim_{n\to\infty} gz_n = \lim_{n\to\infty} F(z_n, w_n, x_n, y_n) = z$$
$$\lim_{n\to\infty} gw_n = \lim_{n\to\infty} F(w_n, x_n, y_n, z_n) = w.$$

Hence from the compatibility of F and g, we obtain

$$\lim_{n\to\infty} d(gF(x_n, y_n, z_n, w_n), F(gx_n, gy_n, gz_n, gw_n)) = 0,$$
$$\lim_{n\to\infty} d(gF(y_n, z_n, w_n, x_n), F(gy_n, gz_n, gw_n, gx_n)) = 0,$$
$$\lim_{n\to\infty} d(gF(z_n, w_n, x_n, y_n), F(gz_n, gw_n, gx_n, gy_n)) = 0,$$
$$\text{and} \quad \lim_{n\to\infty} d(gF(w_n, x_n, y_n, z_n), F(gw_n, gx_n, gy_n, gz_n)) = 0. \tag{2.26}$$

Further, from the continuity of F and g we get

$$\lim_{n\to\infty} gF(x_n, y_n, z_n, w_n) = \lim_{n\to\infty} ggx_{n+1} = gx,$$
$$\lim_{n\to\infty} F(gx_n, gy_n, gz_n, gw_n) = F(x, y, z, w),$$
$$\lim_{n\to\infty} gF(y_n, z_n, w_n, x_n) = \lim_{n\to\infty} ggy_{n+1} = gy,$$
$$\lim_{n\to\infty} F(gy_n, gz_n, gw_n, gx_n) = F(y, z, w, x),$$
$$\lim_{n\to\infty} gF(z_n, w_n, x_n, y_n) = \lim_{n\to\infty} ggz_{n+1} = gz,$$
$$\lim_{n\to\infty} F(gz_n, gw_n, gx_n, gy_n) = F(z, w, x, y),$$
$$\lim_{n\to\infty} gF(w_n, x_n, y_n, z_n) = \lim_{n\to\infty} ggw_{n+1} = gw \quad \text{and}$$
$$\lim_{n\to\infty} F(gw_n, gx_n, gy_n, gz_n) = F(w, x, y, z).$$

Thus from (2.26) and using Lemma 1.1, we have that $gx = F(x, y, z, w)$, $gy = F(y, z, w, x)$, $gz = F(z, w, x, y)$, $gw = F(w, x, y, z)$. Hence, (x, y, z, w) is a quadruple coincidence point of F and g. □

By removing the continuity and compatibility assumptions of F and g in Theorem 2.1, we prove the following theorem.

Theorem 2.2 *Let $(X, d, \preceq)$ be a partially ordered b-metric space with parameter $s \geq 1$. Let $F : X^4 \to X$ and $g : X \to X$ be two mappings satisfying* (2.3) *for all $x, y, z, w, u, v, r, t \in X$, such that $gx \asymp gu$, $gy \asymp gv$, $gz \asymp gr$ and $gw \asymp gt$, where ψ and ϕ are the same as in Theorem* 2.1. *Suppose that*

(i) *$F(X^4) \subseteq g(X)$,*
(ii) *F and g satisfy property* (2.1),
(iii) *there exist $x_0, y_0, z_0, w_0 \in X$ such that $gx_0 \asymp F(x_0, y_0, z_0, w_0)$, $gy_0 \asymp F(y_0, z_0, w_0, x_0)$, $gz_0 \asymp F(z_0, w_0, x_0, y_0)$ and $gw_0 \asymp F(w_0, x_0, y_0, z_0)$,*
(iv) *$g(X)$ is a b-complete subspace of X,*
(v) *if $x_n \to x$ when $n \to \infty$ in X, then $x_n \asymp x$ for n sufficiently large.*

Then there exist $x,y,z,w \in X$ *such that* $gx = F(x,y,z,w)$, $gy = F(y,z,w,x)$, $gz = F(z,w,x,y)$, *and* $gw = F(w,x,y,z)$. *Moreover, if* gx_0, gy_0, gz_0 *and* gw_0 *are comparable, then* $gx = gy = gz = gw$, *and if* F *and* g *are* w-*compatible, then* F *and* g *have a quadruple coincidence point of the form* (p,p,p,p).

Proof From Theorem 2.1, we have that $\{gx_n\}, \{gy_n\}, \{gz_n\}$ and $\{gw_n\}$ are b-Cauchy sequences in X. Since $g(X)$ is a b-complete subspace of X and $\{gx_n\}, \{gy_n\}, \{gz_n\}, \{gw_n\} \subseteq g(X)$, there exist $x,y,z,w \in X$ such that

$$\lim_{n\to\infty} d(gx_n, gx) = \lim_{n\to\infty} d(gy_n, gy) = \lim_{n\to\infty} d(gz_n, gz) = \lim_{n\to\infty} d(gw_n, gw) = 0.$$

Since $gx_n \to gx$ when $n \to \infty$ in X, then from condition (v) we obtain $gx_n \asymp gx$ for n sufficiently large. Similarly, we may show that $gy_n \asymp gy$, $gz_n \asymp gz$ and $gw_n \asymp gw$ for n sufficiently large. For such n, using (2.3) we get

$$\begin{aligned}\psi(sd(F(x,y,z,w), gx_{n+1})) &= \psi(sd(F(x,y,z,w), F(x_n,y_n,z_n,w_n)))\\ &\le \psi(\max\{d(gx,gx_n), d(gy,gy_n), d(gz,gz_n), d(gw,gw_n)\})\\ &\quad - \phi(d(gx,gx_n), d(gy,gy_n), d(gz,gz_n), d(gw,gw_n)).\end{aligned}$$

From the above inequality, using Lemma 1.1, as $n \to \infty$, and using the properties of ϕ we obtain

$$\begin{aligned}&\psi\left(\frac{1}{s}d(F(x,y,z,w), gx)\right) \le \psi(s \limsup_{n\to\infty} d(F(x,y,z,w), gx_{n+1}))\\ &= \limsup_{n\to\infty} \psi(sd(F(x,y,z,w), gx_{n+1}))\\ &\le \limsup_{n\to\infty} \psi(\max\{d(gx,gx_n), d(gy,gy_n), d(gz,gz_n), d(gw,gw_n)\})\\ &\quad - \liminf_{n\to\infty} \phi(d(gx,gx_n), d(gy,gy_n), d(gz,gz_n), d(gw,gw_n))\\ &\le \psi(0) - \phi(0,0,0,0) = 0.\end{aligned}$$

Thus, $F(x,y,z,w) = gx$. Similarly, we can show that $F(y,z,w,x) = gy$, $F(z,w,x,y) = gz$ and $F(w,x,y,z) = gw$.

Now, assume that $gx_0 \asymp gy_0 \asymp gz_0 \asymp gw_0$. From (2.6), we have

$$gx_n \asymp gx_0, \quad gy_n \asymp gy_0, \quad gz_n \asymp gz_0 \quad \text{and} \quad gw_n \asymp gw_0.$$

Then

$$\begin{aligned}&gx \asymp gx_n \asymp gx_0 \asymp gy_0 \asymp gy_n \asymp gy,\\ &gy \asymp gy_n \asymp gy_0 \asymp gz_0 \asymp gz_n \asymp gz\\ \text{and}\quad &gz \asymp gz_n \asymp gz_0 \asymp gw_0 \asymp gw_n \asymp gw\end{aligned}$$

for $n$sufficiently large.

Hence $gx \asymp gy \asymp gz \asymp gw$. Therefor by (2.3) we obtain

$$\begin{aligned}&\psi(\max\{d(gx,gy), d(gy,gz), d(gz,gw), d(gw,gx)\})\\ &\le \psi(s\max\{d(gx,gy), d(gy,gz), d(gz,gw), (gw,gx)\})\\ &\le \psi(\max\{d(gx,gy), d(gy,gz), d(gz,gw),\\ &\quad d(gw,gx)\}) - \min\{\phi(d(gx,gy), d(gy,gz), d(gz,gw),\\ &\quad d(gw,gx)), \phi(d(gy,gz), d(gz,gw), d(gw,gx),\\ &\quad d(gx,gy)), \phi(d(gz,gw), d(gw,gx), d(gx,gy), d(gy,gz)),\\ &\quad \phi(d(gw,gx), d(gx,gy), d(gy,gz), d(gz,gw))\}.\end{aligned}$$

Hence,

$$\begin{aligned}&\min\{\phi(d(gx,gy), d(gy,gz), d(gz,gw), d(gw,gx)),\\ &\phi(d(gy,gz), d(gz,gw), d(gw,gx), d(gx,gy)),\\ &\phi(d(gz,gw), d(gw,gx), d(gx,gy), d(gy,gz)),\\ &\phi(d(gw,gx), d(gx,gy), d(gy,gz), d(gz,gw))\} = 0,\end{aligned}$$

which implies that

$$\begin{aligned}&\phi(d(gx,gy), d(gy,gz), d(gz,gw), d(gw,gx))\\ &= 0 \text{ or } \phi(d(gy,gz), d(gz,gw), d(gw,gx), d(gx,gy))\\ &= 0 \text{ or } \phi(d(gz,gw), d(gw,gx), d(gx,gy), d(gy,gz))\\ &= 0 \text{ or } \phi(d(gw,gx), d(gx,gy), d(gy,gz), d(gz,gw))\} = 0.\end{aligned}$$

Then from the properties of ϕ we have $d(gx,gy) = d(gy,gz) = d(gz,gw) = d(gw,gx) = 0$, that is $gx = gy = gz = gw$. Now, suppose that $gx = gy = gz = gw = p$, since F and g are w-compatible, then

$$gp = ggx = g(F(x,y,z,w)) = F(gx,gy,gz,gw) = F(p,p,p,p).$$

So, F and g have a quadruple coincidence point of the form (p,p,p,p). □

Corollary 2.1 *Replace the contractive condition* (2.3) *of Theorem* 2.1 *(or Theorem* 2.2, *respectively) by the following condition:*

there exist $\psi : [0,\infty) \to [0,\infty)$ *such that* ψ *is an altering distance function and* $\varphi : [0,\infty) \to [0,\infty)$ *is continuous with* $\varphi(t) = 0$ *if and only if* $t = 0$ *such that*

$$\begin{aligned}&\psi(sd(F(x,y,z,w), F(u,v,r,t))) \le \psi(\max\{d(gx,gu),\\ &d(gy,gv), d(gz,gr), d(gw,gt)\})\\ &- \varphi(\max\{d(gx,gu), d(gy,gv), d(gz,gr), d(gw,gt)\}),\end{aligned}$$

for all $x,y,z,w,u,v,r,t \in X$ *and* $gx \asymp gu$, $gy \asymp gv$, $gz \asymp gr$ *and* $gw \asymp gt$. *Let the other conditions of Theorem* 2.1 *(or Theorem* 2.2, *respectively) be satisfied. Then,* F *and* g *have a quadruple coincidence point.*

Proof We replace $\phi(t_1,t_2,t_3,t_4) = \varphi(\max(t_1,t_2,t_3,t_4))$ in Theorem 2.1 (or Theorem 2.2, respectively). So φ is continuous and $\varphi(t) = 0$ if and only if $t = 0$. □

Corollary 2.2 *Replace the contractive condition* (2.3) *of Theorem* 2.1 *(or Theorem* 2.2*, respectively) by the following condition:*

$$d(F(x,y,z,w),F(u,v,r,t)) \le \frac{k}{4s}\max\{d(gx,gu),\\ d(gy,gv),d(gz,gr),d(gw,gt)\},$$

for all $x,y,z,w,u,v,r,t \in X$, *and* $gx \asymp gu$, $gy \asymp gv$, $gz \asymp gr$ *and* $gw \asymp gt$, *where* $k \in [0,1)$. *Let the other conditions of Theorem* 2.1 *(or Theorem* 2.2*) be satisfied. Then, F and g have a quadruple coincidence point.*

Proof We take $\psi(t) = \frac{t}{4}$ and $\phi(t_1,t_2,t_3,t_4) = \frac{(1-k)}{4}\max(t_1,t_2,t_3,t_4)$ in Theorem 2.1 (or Theorem 2.2, respectively). □

Corollary 2.3 *Replace the contractive condition* (2.3) *of Theorem* 2.1 *(or Theorem* 2.2*, respectively) by the following condition:*

there exist $\phi : [0,\infty)^4 \to [0,\infty)$ *is continuous with* $\phi(t_1,t_2,t_3,t_4) = 0$ *if and only if* $t_1 = t_2 = t_3 = t_4 = 0$ *such that*

$$d(F(x,y,z,w),F(u,v,r,t)) \le \frac{1}{s}\max\{d(gx,gu),d(gy,gv),\\ d(gz,gr),d(gw,gt)\} - \frac{1}{s}\phi(d(gx,gu),d(gy,gv),\\ d(gz,gr),d(gw,gt)),$$

for all $x,y,z,w,u,v,r,t \in X$ *and* $gx \asymp gu$, $gy \asymp gv$,$gz \asymp gr$ *and* $gw \asymp gt$. *Let the other conditions of Theorem* 2.1 *(or Theorem* 2.2*, respectively) be satisfied. Then, F and g have a quadruple coincidence point.*

Proof Taking $\psi(t) = t$ in Theorem 2.1 (or Theorem 2.2, respectively), we have Corollary 2.3. □

Corollary 2.4 *Replace the contractive condition* (2.3) *of Theorem* 2.1 *(or Theorem* 2.2*, respectively) by the following condition:*

there exist $\psi : [0,\infty) \to [0,\infty)$ *such that* ψ *is an altering distance function, and* $\phi : [0,\infty)^4 \to [0,\infty)$ *is continuous with* $\phi(t_1,t_2,t_3,t_4) = 0$ *if and only if* $t_1 = t_2 = t_3 = t_4 = 0$ *such that*

$$\psi(sd(F(x,y,z,w),F(u,v,r,t)))\\ \le \psi\left(\frac{d(gx,gu)+d(gy,gv)+d(gz,gr)+d(gw,gt)}{4}\right)\\ - \phi(d(gx,gu),d(gy,gv),d(gz,gr),d(gw,gt)),$$

for all $x,y,z,w,u,v,r,t \in X$ *and* $gx \asymp gu$, $gy \asymp gv$, $gz \asymp gr$ *and* $gw \asymp gt$. *Let the other conditions of Theorem* 2.1 *(or Theorem* 2.2*, respectively) be satisfied. Then, F and g have a quadruple coincidence point.*

Proof Since

$$\frac{(d(gx,gu)+d(gy,gv)+d(gz,gr)+d(gw,gt)}{4}\\ \le \max\{d(gx,gu),d(gy,gv),d(gz,gr),d(gw,gt)\},$$

and since ψ is assumed to be nondecreasing, then we apply Theorem 2.1 (or Theorem 2.2 respectively). □

Corollary 2.5 *Replace the contractive condition* (2.3) *of Theorem* 2.1 *(or Theorem* 2.2 *respectively) by the following condition*

$$d(F(x,y,z,w),F(u,v,r,t)) \le \frac{k}{4s}(d(gx,gu)+d(gy,gv)\\ +d(gz,gr)+d(gw,gt)),$$

for all $x,y,z,w,u,v,r,t \in X$ *and* $gx \asymp gu$, $gy \asymp gv$,$gz \asymp gr$ *and* $gw \asymp gt$, *where* $k \in [0,1)$. *Let the other conditions of Theorem* 2.1 *(or Theorem* 2.2 *respectively) be satisfied. Then F and g have a quadruple coincidence point.*

Proof We take $\psi(t) = t$ and $\phi(t_1,t_2,t_3,t_4) = (\frac{1-k}{4})(t_1+t_2+t_3+t_4)$ in Corollary 2.4. □

Now, we obtain some quadruple coincidence point results for mappings satisfying a contractive condition of integral type. We denote by Λ the set of all functions $\alpha : [0,+\infty) \to [0,+\infty)$ verifying the following conditions:

(i) α is a positive Lebesgue integrable mapping on each compact subset of $[0,+\infty)$

(ii) for all $\varepsilon > 0$, we have $\int_0^\varepsilon \alpha(t)dt > 0$.

Let $N \in \mathbf{N}$ be a fixed positive integer. Let $\{\alpha_i\}_{1\le i\le N}$ be a family of N functions that belong to Λ. For all $t \ge 0$, we denote $(I_i)_{i=1,\ldots,N}$ as follows:

$$I_1(t) = \int_0^t \alpha_1(s)ds,$$

$$I_2(t) = \int_0^{I_1(t)} \alpha_2(s)ds = \int_0^{\int_0^t \alpha_1(s)ds} \alpha_2(s)ds,$$

$$I_3(t) = \int_0^{I_2(t)} \alpha_3(s)ds = \int_0^{\int_0^{\int_0^t \alpha_1(s)ds} \alpha_2(s)ds} \alpha_3(s)ds,$$

$$\ldots$$

$$I_N(t) = \int_0^{I_{N-1}(t)} \alpha_N(s)ds.$$

We have the following result.

Corollary 2.6 *Replace the contractive condition* (2.3) *of Theorem* 2.1 *(or Theorem* 2.2 *respectively) by the following condition:*

$$I_N(\psi(sd(F(x,y,z,w),F(u,v,r,t))))\\ \le I_N(\psi(\max\{d(gx,gu),d(gy,gv),d(gz,gr),d(gw,gt)\}))\\ - I_N(\phi(d(gx,gu),d(gy,gv),\\ d(gz,gr),d(gw,gt))). \tag{2.27}$$

Let the other conditions of Theorem 2.1 *(or Theorem* 2.2 *respectively) be satisfied. Then F and g have a quadruple coincidence point.*

Proof Consider the function $\Psi = I_N \circ \psi$ and $\Phi = I_N \circ \phi$. Then (2.27) becomes

$$\begin{aligned}&\Psi(sd(F(x,y,z,w),F(u,v,r,t)))\\ &\leq \Psi(\max\{d(gx,gu),d(gy,gv),d(gz,gr),d(gw,gt)\})\\ &\quad -\Phi(d(gx,gu),d(gy,gv),d(gz,gr),d(gw,gt)).\end{aligned}$$

It is easy to show that Ψ is an altering distance function, Φ is continuous and $\Phi(t_1,t_2,t_3,t_4)=0$ if and only if $t_1 = t_2 = t_3 = t_4 = 0$. Applying Theorem 2.1 (or Theorem 2.2 respectively) we obtain the proof. □

In the case $N = 1$, we have the following corollary.

Corollary 2.7 *Replace the contractive condition* (2.3) *of Theorems* 2.1 *(or Theorem* 2.2 *respectively) by the following: There exists* $\alpha \in \Lambda$ *such that*

$$\begin{aligned}&\int_0^{\psi(sd(F(x,y,z,w),F(u,v,r,t)))}\alpha(t)\mathrm{d}t\\ &\leq \int_0^{\psi(\max\{d(gx,gu),d(gy,gv),d(gz,gr),d(gw,gt)\})}\alpha(t)\mathrm{d}t\\ &\quad -\int_0^{\phi(d(gx,gu),d(gy,gv),d(gz,gr),d(gw,gt))}\alpha(t)\mathrm{d}t.\end{aligned}$$

Let the other conditions of Theorem 2.1 *(or Theorem* 2.2 *respectively) be satisfied. Then F and g have a quadruple coincidence point.*

Uniqueness of quadruple fixed point

In this section, we will show the uniqueness of a quadruple common fixed point.

For a product X^4 of a partially ordered set $(X,\preceq)$, we define a partial ordering in the following way. For all $(x,y,z,w),(u,v,r,t)\in X^4$,

$$(x,y,z,w)\preceq(u,v,r,t)\Leftrightarrow x\preceq u,\ y\succeq v,\ z\preceq r,\ w\succeq t. \tag{3.1}$$

We say that (x,y,z,w) and (u,v,r,t) are comparable if

$$(x,y,z,w)\preceq(u,v,r,t)\text{ or }(u,v,r,t)\succeq(x,y,z,w).$$

Also, we say that (x,y,z,w) is equal to (u,v,r,t) if and only if $x=u,\ y=v,\ z=r,\ w=t$.

Theorem 3.1 *In addition to hypotheses of Theorem* 2.1 *(or Theorem* 2.2*, respectively) assume that for all quadruple coincidence points* $(x,y,z,w),(u,v,r,t)\in X^4$*, there exists* $(a,b,c,d)\in X^4$ *such that* $(F(a,b,c,d),F(b,c,d,a),F(c,d,a,b),F(d,a,b,c))$ *is comparable to both* $(F(x,y,z,w),F(y,z,w,x),F(z,w,x,y),F(w,x,y,z))$ and$(F(u,v,r,t),F(v,r,t,u),F(r,t,u,v),F(t,u,v,r))$. *Then F and g have a unique quadruple common fixed point* (x,y,z,w) *such that* $x=gx=F(x,y,z,w)$, $y=gy=F(y,z,w,x)$, $z=gz=F(z,w,x,y)$, *and* $w=gw=F(w,x,y,z)$.

Proof Theorem 2.1 (or Theorem 2.2 respectively) implies that The set of quadruple coincidence points of F and g is not empty. Suppose that (x,y,z,w) and (u,v,r,t) are two quadruple coincidence points of F and g, that is, $F(x,y,z,w)=gx$, $F(u,v,r,t)=gu$, $F(y,z,w,x)=gy$,$F(v,r,t,u)=gv$, $F(z,w,x,y)=gz$, $F(r,t,u,v)=gr$, $F(w,x,y,z)=gw$,$F(t,u,v,r)=gt$. We show that $(gx,gy,gz,gw)=(gu,gv,gr,gt)$. By assumption, there exists $(a,b,c,d)\in X^4$ such that $(F(a,b,c,d),F(b,c,d,a),F(c,d,a,b),F(d,a,b,c))$ is comparable to both

$$(F(x,y,z,w),F(y,z,w,x),F(z,w,x,y),F(w,x,y,z))$$
$$\text{and }(F(u,v,r,t),F(v,r,t,u),F(r,t,u,v),F(t,u,v,r)).$$

Since $F(X^4)\subseteq g(X)$, we can define the sequences $\{ga_n\},\{gb_n\},\{gc_n\}$ and $\{gd_n\}$ such that $a_0=a$, $b_0=b$, $c_0=c$, $d_0=d$, and

$$\begin{aligned}ga_{n+1}&=F(a_n,b_n,c_n,d_n), & gb_{n+1}&=F(b_n,c_n,d_n,a_n),\\ gc_{n+1}&=F(c_n,d_n,a_n,b_n), & gd_{n+1}&=F(d_n,a_n,b_n,c_n),\end{aligned}$$

for all $n\geq 0$. Also, in the same way define the sequences $\{gx_n\}$, $\{gy_n\}$, $\{gz_n\}$, $\{gw_n\}$ and $\{gu_n\}$,$\{gv_n\}$, $\{gr_n\}$, $\{gt_n\}$, such that $x_0=x$, $y_0=y$,$z_0=z$, $w_0=w$, and $u_0=u$, $v_0=v$, $r_0=r$, $t_0=t$, by

$$\begin{aligned}gx_{n+1}&=F(x_n,y_n,z_n,w_n), & gy_{n+1}&=F(y_n,z_n,w_n,x_n),\\ gz_{n+1}&=F(z_n,w_n,x_n,y_n), & gw_{n+1}&=F(w_n,x_n,y_n,z_n),\end{aligned}$$

and

$$\begin{aligned}gu_{n+1}&=F(u_n,v_n,r_n,t_n), & gv_{n+1}&=F(v_n,r_n,t_n,u_n),\\ gr_{n+1}&=F(r_n,t_n,u_n,v_n), & gt_{n+1}&=F(t_n,u_n,v_n,r_n),\end{aligned}$$

for all $n\geq 0$. Since (x,y,z,w) and (u,v,r,t) are quadruple coincidence points of F and g,, then $gx_n=F(x,y,z,w)$, $gu_n=F(u,v,r,t)$, $gy_n=F(y,z,w,x)$, $gv_n=F(v,r,t,u)$, $gz_n=F(z,w,x,y)$, $gr_n=F(r,t,u,v)$, $gw_n=F(w,x,y,z)$, $gt_n=F(t,u,v,r)$, for all $n\geq 0$.

Since $(F(x,y,z,w),F(y,z,w,x),F(z,w,x,y),F(w,x,y,z))=(gx_1,gy_1,gz_1,gw_1)=(gx,gy,gz,gw)$ is comparable to $(F(a,b,c,d),F(b,c,d,a),F(c,d,a,b),F(d,a,b,c))$ $=(ga_1,gb_1,gc_1,gd_1)$, then it is easy to show $gx\asymp ga_1$, $gy\asymp gb_1$, $gz\asymp gc_1$, $gw\asymp gd_1$. In a similar way, we get that

$$gx\asymp ga_n,\quad gy\asymp gb_n,\quad gz\asymp gc_n,\quad gw\asymp gd_n\text{ for all }n. \tag{3.2}$$

From (2.3) and (3.2), we obtain

$$\begin{aligned}
&\psi(sd(gx, ga_{n+1})) \\
&= \psi(sd(F(x,y,z,w), F(a_n,b_n,c_n,d_n))) \\
&\leq \psi(\max\{d(gx,ga_n), d(gy,gb_n), d(gz,gc_n), d(gw,gd_n)\}) \\
&\quad - \phi(d(gx,ga_n), d(gy,gb_n), d(gz,gc_n), d(gw,gd_n)),
\end{aligned} \tag{3.3}$$

$$\begin{aligned}
&\psi(sd(gy, gb_{n+1})) \\
&= \psi(sd(F(y,z,w,x), F(b_n,c_n,d_n,a_n))) \\
&\leq \psi(\max\{d(gy,gb_n), d(gz,gc_n), d(gw,gd_n), d(gx,ga_n)\}) \\
&\quad - \phi(d(gy,gb_n), d(gz,gc_n), d(gw,gd_n), d(gx,ga_n)),
\end{aligned} \tag{3.4}$$

$$\begin{aligned}
&\psi(sd(gz, gc_{n+1})) \\
&= \psi(sd(F(z,w,x,y), F(c_n,d_n,a_n,b_n))) \\
&\leq \psi(\max\{d(gz,gc_n), d(gw,gd_n), d(gx,ga_n), d(gy,gb_n)\}) \\
&\quad - \phi(d(gz,gc_n), d(gw,gd_n), d(gx,ga_n), d(gy,gb_n)),
\end{aligned} \tag{3.5}$$

and

$$\begin{aligned}
&\psi(sd(gw, gd_{n+1})) \\
&= \psi(sd(F(w,x,y,z), F(d_n,a_n,b_n,c_n))) \\
&\leq \psi(\max\{d(gw,gd_n), d(gx,ga_n), d(gy,gb_n), d(gz,gc_n)\}) \\
&\quad - \phi(d(gw,gd_n), d(gx,ga_n), d(gy,gb_n), d(gz,gc_n)).
\end{aligned} \tag{3.6}$$

Set

$$\gamma_n = \max\{d(gx,ga_n), d(gy,gb_n), d(gz,gc_n), d(gw,gd_n)\}.$$

By (3.3)–(3.6) and Remark 2.1, we obtain that

$$\begin{aligned}
&\psi(\gamma_{n+1}) \leq \psi(s\gamma_{n+1}) \\
&\quad \leq \psi(\gamma_n) - \min\{\phi(d(gx,ga_n), d(gy,gb_n), d(gz,gc_n), d(gw,gd_n)), \\
&\phi(d(gy,gb_n), d(gz,gc_n), d(gw,gd_n), d(gx,ga_n)), \\
&\phi(d(gz,gc_n), d(gw,gd_n), d(gx,ga_n), d(gy,gb_n)), \\
&\phi(d(gw,gd_n), d(gx,ga_n), d(gy,gb_n), d(gz,gc_n))\} \leq \psi(\gamma_n).
\end{aligned} \tag{3.7}$$

Hence,

$$\psi(\gamma_{n+1}) \leq \psi(\gamma_n) \qquad \text{for all n} \in \mathbf{N}.$$

Since ψ is nondecreasing, then $\gamma_{n+1} \leq \gamma_n$ for all n. This implies that γ_n is a non-increasing sequence. Therefore, there exists $\gamma \geq 0$ such that

$$\lim_{n\to\infty} \gamma_n = \gamma.$$

We show that $\gamma = 0$. Letting $n \to \infty$, in (3.7), we get

$$\begin{aligned}
&\psi(\gamma) \leq \psi(\gamma) - \min\{\lim_{n\to\infty} \phi(d(gx,ga_n), d(gy,gb_n), \\
&d(gz,gc_n), d(gw,gd_n)), \\
&\lim_{n\to\infty} \phi(d(gy,gb_n), d(gz,gc_n), d(gw,gd_n), d(gx,ga_n)), \\
&\quad \lim_{n\to\infty} \phi(d(gz,gc_n), d(gw,gd_n), d(gx,ga_n), d(gy,gb_n)), \\
&\quad \lim_{n\to\infty} \phi(d(gw,gd_n), d(gx,ga_n), d(gy,gb_n), d(gz,gc_n))\} \leq \psi(\gamma).
\end{aligned}$$

Hence,

$$\begin{aligned}
&\min\{\lim_{n\to\infty} \phi(d(gx,ga_n), d(gy,gb_n), d(gz,gc_n), d(gw,gd_n)), \\
&\quad \lim_{n\to\infty} \phi(d(gy,gb_n), d(gz,gc_n), d(gw,gd_n), d(gx,ga_n)), \\
&\quad \lim_{n\to\infty} \phi(d(gz,gc_n), d(gw,gd_n), d(gx,ga_n), d(gy,gb_n)), \\
&\lim_{n\to\infty} \phi(d(gw,gd_n), d(gx,ga_n), d(gy,gb_n), d(gz,gc_n))\} = 0.
\end{aligned}$$

Therefore,

$$\begin{aligned}
&\lim_{n\to\infty} \phi(d(gx,ga_n), d(gy,gb_n), d(gz,gc_n), d(gw,gd_n)) = 0 \quad \text{or} \\
&\lim_{n\to\infty} \phi(d(gy,gb_n), d(gz,gc_n), d(gw,gd_n), d(gx,ga_n)) = 0 \quad \text{or} \\
&\lim_{n\to\infty} \phi(d(gz,gc_n), d(gw,gd_n), d(gx,ga_n), d(gy,gb_n)) = 0 \quad \text{or} \\
&\quad \lim_{n\to\infty} \phi(d(gw,gd_n), d(gx,ga_n), d(gy,gb_n), d(gz,gc_n))\} = 0.
\end{aligned}$$

Using the properties of ϕ, we have

$$\begin{aligned}
&\lim_{n\to\infty} d(gx,ga_n) = \lim_{n\to\infty} d(gy,gb_n) = \lim_{n\to\infty} d(gz,gc_n) \\
&\quad = \lim_{n\to\infty} d(gw,gd_n) = 0.
\end{aligned} \tag{3.8}$$

Thus $\lim_{n\to\infty} \gamma_n = 0$. Similarly, we can show that

$$\begin{aligned}
&\lim_{n\to\infty} d(gu,ga_n) = \lim_{n\to\infty} d(gv,gb_n)) = \lim_{n\to\infty} d(gr,gc_n) \\
&\quad = \lim_{n\to\infty} d(gt,gd_n) = 0.
\end{aligned} \tag{3.9}$$

From (3.8) and (3.9), we conclude that $(gx,gy,gz,gw) = (gu,gv,gr,gt)$, That is the quadruple coincidence point of F and g is unique.

Denote $gx = x^*$, $gy = y^*$, $gz = z^*$, $gw = w^*$ and since $gx = F(x,y,z,w)$, $gy = F(y,z,w,x)$, $gz = F(z,w,x,y)$, $gw = F(w,x,y,z)$ so we have that

$$gx^* = ggx = gF(x,y,z,w), \qquad gy^* = ggy = gF(y,z,w,x), \tag{3.10}$$

$$gz^* = ggz = gF(z,w,x,y) \text{ and } gw^* = ggw = gF(w,x,y,z). \tag{3.11}$$

By definition of the sequences $\{gx_n\}, \{gy_n\}, \{gz_n\}$ and $\{gw_n\}$, we have

$$gx_n = F(x,y,z,w) = F(x_{n-1},y_{n-1},z_{n-1},w_{n-1}),$$
$$gy_n = F(y,z,w,x) = F(y_{n-1},z_{n-1},w_{n-1},x_{n-1}),$$
$$gz_n = F(z,w,x,y) = F(z_{n-1},w_{n-1},x_{n-1},y_{n-1}),$$
$$gw_n = F(w,x,y,z) = F(w_{n-1},x_{n-1},y_{n-1},z_{n-1}).$$

Consequently,

$$\lim_{n\to\infty} gx_n = \lim_{n\to\infty} F(x_n,y_n,z_n,w_n) = F(x,y,z,w), \quad (3.12)$$

$$\lim_{n\to\infty} gy_n = \lim_{n\to\infty} F(y_n,z_n,w_n,x_n) = F(y,z,w,x), \quad (3.13)$$

$$\lim_{n\to\infty} gz_n = \lim_{n\to\infty} F(z_n,w_n,x_n,y_n) = F(z,w,x,y), \quad (3.14)$$

$$\lim_{n\to\infty} gw_n = \lim_{n\to\infty} F(w_n,x_n,y_n,z_n) = F(w,x,y,z). \quad (3.15)$$

Case 1: In Theorem 2.1, from compatibility and continuity of F and g we obtain

$$\lim_{n\to\infty} d(gF(x_n,y_n,z_n,w_n),F(gx_n,gy_n,gz_n,gw_n)) = 0,$$
$$\lim_{n\to\infty} d(gF(y_n,z_n,w_n,x_n),F(gy_n,gz_n,gw_n,gx_n)) = 0,$$
$$\lim_{n\to\infty} d(gF(z_n,w_n,x_n,y_n),F(gz_n,gw_n,gx_n,gy_n)) = 0,$$
$$\lim_{n\to\infty} d(gF(w_n,x_n,y_n,z_n),F(gw_n,gx_n,gy_n,gz_n))) = 0, \quad (3.16)$$

where

$$\lim_{n\to\infty} gF(x_n,y_n,z_n,w_n) = gF(x,y,z,w),$$
$$\lim_{n\to\infty} F(gx_n,gy_n,gz_n,gw_n) = F(gx,gy,gz,gw)),$$
$$\lim_{n\to\infty} gF(y_n,z_n,w_n,x_n) = gF(y,z,w,x),$$
$$\lim_{n\to\infty} F(gy_n,gz_n,gw_n,gx_n) = F(gy,gz,gw,gx),$$
$$\lim_{n\to\infty} gF(z_n,w_n,x_n,y_n) = gF(z,w,x,y),$$
$$\lim_{n\to\infty} F(gz_n,gw_n,gx_n,gy_n) = F(gz,gw,gx,gy),$$
$$\lim_{n\to\infty} gF(w_n,x_n,y_n,z_n) = gF(w,x,y,z),$$
$$\lim_{n\to\infty} F(gw_n,gx_n,gy_n,gz_n)) = F(gw,gx,gy,gz).$$

Thus from Lemma 1.1, we conclude that

$$gF(x,y,z,w) = F(gx,gy,gz,gw), \quad gF(y,z,w,x) = F(gy,gz,gw,gx),$$
$$gF(z,w,x,y) = F(gz,gw,gx,gy), \quad gF(w,x,y,z) = F(gw,gx,gy,gz).$$

Moreover, from (3.10) implies that

$$gx^* = F(x^*,y^*,z^*,w^*), \ gy^* = F(y^*,z^*,w^*,x^*),$$
$$gz^* = F(z^*,w^*,x^*,y^*), \ gw^* = F(w^*,x^*,y^*,z^*).$$

Case 2: In Theorem 2.2 since F and g are w-compatible, then

$$gx^* = ggx = g(F(x,y,z,w)) = F(gx,gy,gz,gw) = F(x^*,y^*,z^*,w^*)$$
$$gy^* = ggy = g(F(y,z,w,x)) = F(gy,gz,gw,gx) = F(y^*,z^*,w^*,x^*)$$
$$gz^* = ggz = g(F(z,w,x,y)) = F(gz,gw,gx,gy) = F(z^*,w^*,x^*,y^*)$$
$$gw^* = ggw = g(F(w,x,y,z)) = F(gw,gx,gy,gz) = F(w^*,x^*,y^*,z^*).$$

Thus, in the two cases we conclude that (x^*,y^*,z^*,w^*) is another quadruple coincidence point of F and g. Hence, $(gx^*,gy^*,gz^*,gw^*) = (gx,gy,gz,gw)$. Therefore

$$gx^* = gx = x^*, \quad gy^* = gy = y^*, \quad gz^* = gz = z^*,$$
$$\text{and} \quad \mathrm{gw}^* = \mathrm{gw} = \mathrm{w}^*.$$

Hence, (x^*,y^*,z^*,w^*) is a quadruple common fixed point of F and g. The uniqueness of a quadruple common fixed point follows easily from the uniqueness of a quadruple coincidence point. □

Now, we give an example to justify the hypotheses of Theorem 2.1.

Example 3.1 Let $X = [0,\infty)$ be equipped with the b-metric $d(x,y) = (x-y)^2$ for all $x,y \in X$, where $s = 2$, and suppose that $\preceq$ is the usual ordering $\leq$ on X. Obviously, $(X,d,\preceq)$ is a partially ordered complete b-metric space. Let $F: X^4 \to X$ and $g: X \to X$ be defined by

$$F(x,y,z,w) = \frac{x^2+y^2+z^2+w^2}{16} \quad \textit{and} \quad g(x) = x^2.$$

It is easy to see that g and F are compatible. Define $\psi : [0,\infty) \to [0,\infty)$ by $\psi(t) = kt$ and $\phi : [0,\infty)^4 \to [0,\infty)$ by $\phi(t_1,t_2,t_3,t_4) = \frac{k-1}{4}(t_1+t_2+t_3+t_4)$, where $1 \leq k \leq 8$. Then, ψ and ϕ have the properties mentioned in Theorem 2.1. Further, for all $x,y,z,w,u,v,r,t \in X$, we have $gx \asymp gu$, $gy \asymp gv$, $gz \asymp gr$ and $gw \asymp gt$. Hence,

$$\begin{aligned}
&\psi(sd(F(x,y,z,w),F(u,v,r,t))) \\
&= \psi\left(2\left(\frac{x^2+y^2+z^2+w^2}{16} - \frac{u^2+v^2+r^2+t^2}{16}\right)^2\right) \\
&= \frac{k}{128}((x^2-u^2)+(y^2-v^2)+(z^2-r^2)+(w^2-t^2))^2 \\
&\leq \frac{4k}{128}\left((x^2-u^2)^2+(y^2-v^2)^2+(z^2-r^2)^2+(w^2-t^2)^2\right) \\
&= \frac{k}{32}(d(gx,gu)+d(gy,gv)+d(gz,gr)+d(gw,gt)) \\
&\leq \frac{8}{32}(d(gx,gu)+d(gy,gv)+d(gz,gr)+d(gw,gt)) \\
&= \frac{1}{4}(d(gx,gu)+d(gy,gv)+d(gz,gr)+d(gw,gt)) \\
&= \frac{k}{4}(d(gx,gu)+d(gy,gv)+d(gz,gr)+d(gw,gt)) \\
&\quad - \frac{k-1}{4}(d(gx,gu)+d(gy,gv)+d(gz,gr)+d(gw,gt)) \\
&\leq k\max\{d(gx,gu),d(gy,gv),d(gz,gr),d(gw,gt)\} \\
&\quad - \phi(d(gx,gu),d(gy,gv),d(gz,gr),d(gw,gt)) \\
&= \psi(\max\{d(gx,gu),d(gy,gv),d(gz,gr),d(gw,gt)\}) \\
&\quad - \phi(d(gx,gu),d(gy,gv),d(gz,gr),d(gw,gt)).
\end{aligned}$$

So, F and g satisfy all the conditions of Theorem and $(0, 0, 0, 0)$ is a quadruple coincidence point of F and g. Moreover, by Theorem 3.1 $(0, 0, 0, 0)$ is the unique quadruple common fixed point of F and g.

Note that, in this case F does not have the g-mixed monotone property, so the results of paper [25] cannot be applied.

References

1. Abbas, M., Khan, M.A., Radenovic, S.: Common coupled fixed point theorems in cone metric spaces for w-compatible mappings. Appl. Math. Comput. **217**(1), 195–202 (2010)
2. Aghajani, A., Abbas, M., Roshan, J.R.: Common fixed point of generalized weak contractive mappings in partially ordered b-metric spaces. Math. Slovaca **64**(4), 941–960 (2014)
3. Agarwal, R.P., Sintunavarat, W., Kumam, P.: Coupled coincidence point and common coupled fixed point theorems lacking the mixed monotone property. Fixed Point Theory Appl. Article ID 22 (2013)
4. Aydi, H., Damjanović, B., Samet, B., Shatanawi, W.: Coupled fixed point theorems for nonlinear contractions in partially ordered G-metric spaces. Math. Comput. Modell. **54**(9–10), 2443–2450 (2011)
5. Aydi, H., Bota, M.-F., Karapinar, E., Moradi, S.: A common fixed point for weak φ-contractions on b-metric spaces. Fixed Point Theory **13**(2), 337–346 (2012)
6. Aydi, H., Karapinar, E., Shatanawi, W.: Tripled coincidence point results for generalized contractions in ordered generalized metric spaces. Fixed Point Theory Appl. Article ID 101 (2012)
7. Bakhtin, I.A.: The contraction mapping principle in quasimetric spaces. Funct. Anal. **30**, 26–37 (1989)
8. Berinde, V., Borcut, M.: Tripled fixed point theorems for contractive type mappings in partially ordered metric spaces. Nonlinear Anal. **74**(15), 4889–4897 (2011)
9. Berinde, V.: Generalized coupled fixed point theorems for mixed monoton mappings in partially ordered metric spaces. Nonlinear Anal. **74**, 7347–7355 (2011)
10. Bhaskar, T.G., Lakshmikantham, V.: Fixed point theorems in partially ordered metric spaces and applications. Nonlinear Anal. **65**, 1379–1393 (2006)
11. Boriceanu, M.: Fixed point theory for multivalued contraction on a set with two b-metrics. Creative Math. **17**, 326–332 (2008)
12. Bota, M., Molnar, A., Varga, C.: On Ekeland's variational principle in b-metric spaces. Fixed Point Theory **12**, 21–28 (2011)
13. Czerwik, S.: Contraction mappings in b-metric spaces. Acta Math. Inform. Univ. Ostrav. **1**, 5–11 (1993)
14. Czerwik, S.: Nonlinear set-valued contraction mappings in b-metric spaces. Atti Semin. Mat. Fis. Univ. Modena **46**(2), 263–276 (1998)
15. Dorić, D., Kadelburg, Z., Radenovic, S.: Coupled fixed point theorems for mappings without mixed monotone property. Appl. Math. Lett. **25**, 1803–1808 (2012)
16. Guo, D., Lakshmikantham, V.: Coupled fixed points of nonlinear operators with applications. Nonlinear Anal. **11**, 623–632 (1987)
17. Hussain, N., Parvaneh, V., Roshan, J.R., Kadelburg, Z.: Fixed points of cyclic (ψ, φ, L, A, B)-contractive mappings in ordered b-metric spaces with applications. Fixed Point Theory Appl. Article ID 256 (2013)
18. Karapinar, E.: Coupled fixed point theorems for nonlinear contractions in cone metric spaces. Comput. Math. Appl. **59**, 3656–3668 (2010)
19. Karapinar, E., Luong, N.V.: Quadruple fixed point theorems for nonlinear contractions. Comput. Math. Appl. **64**(6), 1839–1848 (2012)
20. Karapinar, E.: Quartet fixed point for nonlinear contraction. http://arxiv.org/abs/1106.5472
21. Karapinar, E., Berinde, V.: Quadruple fixed point theorems for nonlinear contraction in partially ordered metric spaces. Banach J. Math. Anal. **6**(1), 74–89 (2012)
22. Karapinar, E., Shatanawi, W., Mustafa, Z.: Quadruple fixed point theorems under nonlinear contractive conditions in partially ordered metric spaces. J. Appl. Math. Article ID 951912 (2012)
23. Khan, M.S., Swaleh, M., Sessa, S.: Fixed point theorems by altering distances between the points. Bull. Aust. Math. Soc. **30**, 1–9 (1984)
24. Lakshmikantham, V., Ćirić, L.B.: Coupled fixed point theorems for nonlinear contractions in partially ordered metric spaces. Nonlinear Anal **70**, 4341–4349 (2009)
25. Mustafa, Z., Aydi, H., Karapinar, E.: Mixed g-monotone property and quadruple fixed point theorems in partially ordered metric spaces, Fixed Point Theory and Appl. Article ID 71 (2012)
26. Parvaneh, V., Roshan, J.R., Radenović, S.: Existence of tripled coincidence points in ordered b-metric spaces and an application to a system of integral equations. Fixed Point Theory and Appl. Article ID 130 (2013)
27. Rashwan, R.A., Saleh, S.M.: A coupled fixed point theorem for three pairs of w-compatible mappings in G-metric spaces. Math. Sci. Lett. **3**(1), 17–20 (2014)
28. Roshan, J.R., Parvaneh, V., Sedghi, S., Shobkolaei, N., Shatanawi, W.: Common fixed points of almost generalized $(\psi, \phi)_s$-contractive mappings in ordered b-metric spaces. Fixed Point Theory Appl. Article ID 159 (2013)
29. Singh, S.L., Prasad, B.: Some coincidence theorems and stability of iterative procedures. Comput. Math. Appl. **55**, 2512–2520 (2008)

13

The harmonic index of product graphs

B. N. Onagh[1]

Abstract The harmonic index of a graph G is defined as the sum of the weights $\frac{2}{\deg_G(u)+\deg_G(v)}$ of all edges uv of G, where $\deg_G(u)$ denotes the degree of a vertex u in G. In this paper, we investigate the harmonic index of Cartesian, lexicographic, tensor, strong, corona and edge corona product of two connected graphs.

Keywords Harmonic index · Product graphs · Inverse degree

Mathematics Subject Classification 05C07 · 05C76

Introduction

Throughout this paper, all graphs are finite, simple, undirected and connected. For a graph G, $V(G)$ and $E(G)$ denote the set of all vertices and edges, respectively. We will use P_n, C_n and K_n to denote the path, the cycle and the complete graph of order n, respectively.

The Cartesian product $G_1 \Box G_2$ of graphs G_1 and G_2 is the graph with vertex set $V(G_1) \times V(G_2)$ in which (u, v) is adjacent to (u', v') if and only if (1) $u = u'$ and $vv' \in E(G_2)$, or (2) $v = v'$ and $uu' \in E(G_1)$.

The lexicographic product (or composition) $G_1[G_2]$ of graphs G_1 and G_2 is the graph with vertex set $V(G_1) \times V(G_2)$ in which (u, v) is adjacent to (u', v') if and only if (1) $uu' \in E(G_1)$, or (2) $u = u'$ and $vv' \in E(G_2)$.

The tensor (or direct) product $G_1 \times G_2$ of graphs G_1 and G_2 is the graph with vertex set $V(G_1) \times V(G_2)$ in which (u, v) is adjacent to (u', v') if and only if $uu' \in E(G_1)$ and $vv' \in E(G_2)$.

The strong (or normal) product $G_1 \boxtimes G_2$ of graphs G_1 and G_2 is the graph with vertex set $V(G_1) \times V(G_2)$ in which (u, v) is adjacent to (u', v') if and only if (1) $u = u'$ and $vv' \in E(G_2)$, or (2) $v = v'$ and $uu' \in E(G_1)$, or (3) $uu' \in E(G_1)$ and $vv' \in E(G_2)$. Obviously, $G_1 \boxtimes G_2 = (G_1 \Box G_2) \cup (G_1 \times G_2)$.

Let $V(G_1) = \{v_1, \ldots, v_{n_1}\}$. The corona product $G_1 \circ G_2$ of disjoint graphs G_1 and G_2 is obtained by taking n_1 copies of G_2 and joining each vertex of the ith copy of G_2 with the vertex $v_i \in V(G_1)$.

Let $E(G_1) = \{e_1, \ldots, e_{m_1}\}$. The edge corona product $G_1 \bullet G_2$ of disjoint graphs G_1 and G_2 is obtained by taking m_1 copies of G_2 and joining each vertex of the ith copy of G_2 with two end vertices of the edge $e_i \in E(G_1)$.

The following propositions easily follow from the definition and structure of product graphs.

Proposition 1.1 [8, 9] *Let G_1 and G_2 be two graphs of orders n_1 and n_2, respectively. Then*

(i) $\deg_{G_1 \Box G_2}(u, v) = \deg_{G_1}(u) + \deg_{G_2}(v)$,

(ii) $\deg_{G_1[G_2]}(u, v) = n_2 \deg_{G_1}(u) + \deg_{G_2}(v)$,

(iii) $\deg_{G_1 \times G_2}(u, v) = \deg_{G_1}(u) \deg_{G_2}(v)$,

(iv) $\deg_{G_1 \boxtimes G_2}(u, v) = \deg_{G_1}(u) + \deg_{G_2}(v) + \deg_{G_1}(u) \deg_{G_2}(v)$.

Proposition 1.2 [8, 9] *Let G_1 and G_2 be two disjoint graphs of orders n_1 and n_2, respectively. Then*

(i) $$\deg_{G_1 \circ G_2}(u) = \begin{cases} \deg_{G_1}(u) + n_2 & u \in V(G_1) \\ \deg_{G_2}(u) + 1 & u \in V(G_2), \end{cases}$$

✉ B. N. Onagh
bn.onagh@gu.ac.ir

[1] Department of Mathematics, Golestan University, Gorgan, Iran

(ii) $\deg_{G_1 \bullet G_2}(u) = \begin{cases} (1+n_2)\deg_{G_1}(u) & u \in V(G_1) \\ \deg_{G_2}(u)+2 & u \in V(G_2). \end{cases}$

The inverse degree and harmonic index of a graph G are two important vertex-degree-based indices related to G, were denoted by r(G) and H(G), respectively, and defined as follows:

$$r(G) = \sum_{u \in V(G)} \frac{1}{\deg_G(u)},$$

$$H(G) = \sum_{uv \in E(G)} \frac{2}{\deg_G(u)+\deg_G(v)}.$$

In recent years, the harmonic index has been extensively studied. Shwetha et al. [9] *derived expressions for the harmonic index of the join, corona product, Cartesian product, composition and symmetric difference of graphs. Recently, Onagh investigated the harmonic index of subdivision graph $S(G)$, t-subdivision graph $S_t(G)$, vertex-semitotal graph $R(G)$, edge-semitotal graph $Q(G)$, total graph $T(G)$ and F-sum of graphs, where $F \in \{S, S_t, R, Q, T\}$* [5–7]. *More results on the harmonic index can been found in* [1–3, 10–12].

In this paper, we study the harmonic index of Cartesian, lexicographic, tensor, strong, corona and edge corona product of two graphs G_1 and G_2 and present some bounds in terms of the harmonic index and inverse degree of G_1 and G_2.

Main results

In this section, we give some bounds for the harmonic index of graphs $G_1 \square G_2$, $G_1[G_2]$, $G_1 \times G_2$, $G_1 \boxtimes G_2$, $G_1 \circ G_2$ and $G_1 \bullet G_2$ in terms of $H(G_1)$, $H(G_2)$, $r(G_1)$ and $r(G_2)$. To do this, we need the following well-known inequality.

Jensen's inequality [4] Let f be a convex function on the interval I and $x_1, \ldots, x_n \in I$. Then

$$f\Big(\frac{x_1+\cdots+x_n}{n}\Big) \le \frac{f(x_1)+\cdots+f(x_n)}{n},$$

with equality if and only if $x_1 = \cdots = x_n$.

Hereafter, G_1 and G_2 are two nontrivial graphs with $|V(G_i)| = n_i$ and $|E(G_i)| = m_i$, $1 \le i \le 2$.

Theorem 2.1 *Let G_1 and G_2 be two graphs. Then*

$$H(G_1 \square G_2) \le \frac{1}{4}(n_2 H(G_1) + n_1 H(G_2) + m_2 r(G_1) + m_1 r(G_2)),$$

with equality if and only if G_1 and G_2 are k-regular graphs.

Proof By definition of the harmonic index, we have

$$\begin{aligned} H(G_1 \square G_2) &= \sum_{u \in V(G_1)} \sum_{vv' \in E(G_2)} \frac{2}{\deg_{G_1 \square G_2}(u,v)+\deg_{G_1 \square G_2}(u,v')} \\ &+ \sum_{v \in V(G_2)} \sum_{uu' \in E(G_1)} \frac{2}{\deg_{G_1 \square G_2}(u,v)+\deg_{G_1 \square G_2}(u',v)} \\ &= \sum_{u \in V(G_1)} \sum_{vv' \in E(G_2)} \frac{2}{(\deg_{G_1}(u)+\deg_{G_2}(v))+(\deg_{G_1}(u)+\deg_{G_2}(v'))} \\ &+ \sum_{v \in V(G_2)} \sum_{uu' \in E(G_1)} \frac{2}{(\deg_{G_1}(u)+\deg_{G_2}(v))+(\deg_{G_1}(u')+\deg_{G_2}(v))} \\ &= \sum_{u \in V(G_1)} \sum_{vv' \in E(G_2)} \frac{2}{2\deg_{G_1}(u)+(\deg_{G_2}(v)+\deg_{G_2}(v'))} \\ &+ \sum_{v \in V(G_2)} \sum_{uu' \in E(G_1)} \frac{2}{(\deg_{G_1}(u)+\deg_{G_1}(u'))+2\deg_{G_2}(v)} \\ &:= \sum 1 + \sum 2. \end{aligned}$$

By Jensen's inequality, for every $u \in V(G_1)$ and $vv' \in E(G_2)$, we have

$$\frac{2}{2\deg_{G_1}(u)+(\deg_{G_2}(v)+\deg_{G_2}(v'))} \le \frac{1}{4}\frac{1}{\deg_{G_1}(u)} + \frac{1}{4}\frac{2}{\deg_{G_2}(v)+\deg_{G_2}(v')}, \tag{1}$$

with equality if and only if $2\deg_{G_1}(u) = \deg_{G_2}(v)+\deg_{G_2}(v')$.

Similarly, for every $v \in V(G_2)$ and $uu' \in E(G_1)$,

$$\frac{2}{(\deg_{G_1}(u)+\deg_{G_1}(u'))+2\deg_{G_2}(v)} \le \frac{1}{4}\frac{2}{\deg_{G_1}(u)+\deg_{G_1}(u')} + \frac{1}{4}\frac{1}{\deg_{G_2}(v)}, \tag{2}$$

with equality if and only if $\deg_{G_1}(u)+\deg_{G_1}(u')=2\deg_{G_2}(v)$.

Thus,

$$\sum 1 \le \frac{1}{4}\sum_{u\in V(G_1)}\sum_{vv'\in E(G_2)}\frac{1}{\deg_{G_1}(u)}+\frac{1}{4}\sum_{u\in V(G_1)}\sum_{vv'\in E(G_2)}\frac{2}{\deg_{G_2}(v)+\deg_{G_2}(v')}$$
$$=\frac{1}{4}\sum_{u\in V(G_1)}\left(m_2\times\frac{1}{\deg_{G_1}(u)}\right)+\frac{1}{4}\sum_{u\in V(G_1)}H(G_2)$$
$$=\frac{1}{4}m_2 r(G_1)+\frac{1}{4}n_1 H(G_2),$$

and

$$\sum 2 \le \frac{1}{4}\sum_{v\in V(G_2)}\sum_{uu'\in E(G_1)}\frac{2}{\deg_{G_1}(u)+\deg_{G_1}(u')}+\frac{1}{4}\sum_{v\in V(G_2)}\sum_{uu'\in E(G_1)}\frac{1}{\deg_{G_2}(v)}$$
$$=\frac{1}{4}\sum_{v\in V(G_2)}H(G_1)+\frac{1}{4}\sum_{v\in V(G_2)}\left(m_1\times\frac{1}{\deg_{G_2}(v)}\right)$$
$$=\frac{1}{4}n_2 H(G_1)+\frac{1}{4}m_1 r(G_2).$$

So, $H(G_1\Box G_2)\le \frac{1}{4}\left(n_2H(G_1)+n_1H(G_2)+m_2r(G_1)+m_1r(G_2)\right)$.

Moreover, equality holds in the above inequality if and only if the inequalities (1) and (2) be equalities, i.e., G_1 and G_2 are k-regular. □

Theorem 2.2 *Let G_1 and G_2 be two graphs. Then*

$$H(G_1[G_2])<\frac{1}{9}n_2H(G_1)+\frac{1}{4}n_1H(G_2)+\frac{1}{4}\frac{m_2}{n_2}r(G_1)+\frac{4}{9}n_2m_1r(G_2).$$

Proof Note that

$$H(G_1[G_2])=\sum_{u\in V(G_1)}\sum_{vv'\in E(G_2)}\frac{2}{\deg_{G_1[G_2]}(u,v)+\deg_{G_1[G_2]}(u,v')}+\sum_{v\in V(G_2)}\sum_{v'\in V(G_2)}\sum_{uu'\in E(G_1)}\frac{2}{\deg_{G_1[G_2]}(u,v)+\deg_{G_1[G_2]}(u',v')}$$
$$=\sum_{u\in V(G_1)}\sum_{vv'\in E(G_2)}\frac{2}{(n_2\deg_{G_1}(u)+\deg_{G_2}(v))+(n_2\deg_{G_1}(u)+\deg_{G_2}(v'))}+\sum_{v\in V(G_2)}\sum_{v'\in V(G_2)}\sum_{uu'\in E(G_1)}\frac{2}{(n_2\deg_{G_1}(u)+\deg_{G_2}(v))+(n_2\deg_{G_1}(u')+\deg_{G_2}(v'))}$$
$$=\sum_{u\in V(G_1)}\sum_{vv'\in E(G_2)}\frac{2}{2n_2\deg_{G_1}(u)+(\deg_{G_2}(v)+\deg_{G_2}(v'))}+\sum_{v\in V(G_2)}\sum_{v'\in V(G_2)}\sum_{uu'\in E(G_1)}\frac{2}{n_2(\deg_{G_1}(u)+\deg_{G_1}(u'))+\deg_{G_2}(v)+\deg_{G_2}(v')}$$
$$:=\sum 1+\sum 2.$$

One can see that for every $u\in V(G_1)$ and $vv'\in E(G_2)$,

$$\frac{2}{2n_2\deg_{G_1}(u)+(\deg_{G_2}(v)+\deg_{G_2}(v'))}\le\frac{1}{4n_2}\frac{1}{\deg_{G_1}(u)}+\frac{1}{4}\frac{2}{\deg_{G_2}(v)+\deg_{G_2}(v')},\tag{3}$$

with equality if and only if $2n_2\deg_{G_1}(u)=\deg_{G_2}(v)+\deg_{G_2}(v')$. □

Also, for every $v\in V(G_2)$, $v'\in V(G_2)$ and $uu'\in E(G_1)$,

$$\frac{2}{n_2(\deg_{G_1}(u)+\deg_{G_1}(u'))+\deg_{G_2}(v)+\deg_{G_2}(v')} \leq \frac{1}{9n_2}\frac{2}{\deg_{G_1}(u)+\deg_{G_1}(u')}+\frac{2}{9}\frac{1}{\deg_{G_2}(v)}+\frac{2}{9}\frac{1}{\deg_{G_2}(v')}, \quad (4)$$

with equality if and only if $n_2(\deg_{G_1}(u)+\deg_{G_1}(u')) = \deg_{G_2}(v) = \deg_{G_2}(v')$.

Thus,

$$\sum 1 \leq \frac{1}{4}\frac{m_2}{n_2} r(G_1) + \frac{1}{4} n_1 H(G_2),$$
$$\sum 2 \leq \frac{1}{9} n_2 H(G_1) + \frac{4}{9} n_2 m_1 r(G_2).$$

Therefore,

$$H(G_1[G_2]) \leq \frac{1}{9} n_2 H(G_1) + \frac{1}{4} n_1 H(G_2) + \frac{1}{4}\frac{m_2}{n_2} r(G_1) + \frac{4}{9} n_2 m_1 r(G_2).$$

Now, suppose that equality holds in the above inequality. Then, the inequalities (3) and (4) must be equalities. So, G_1 and G_2 are k_1-regular and k_2-regular graphs, respectively, such that $2n_2k_1 = k_2 + k_2$ and $n_2(k_1 + k_1) = k_2$, a contradiction. □

Theorem 2.3 *Let G_1 and G_2 be two graphs. Then*

$$H(G_1 \times G_2) \geq 2H(G_1)H(G_2),$$

with equality if and only if either G_1 or G_2 is a regular graph.

Proof By definition of the harmonic index, we have

$$H(G_1 \times G_2) = 2 \sum_{uu' \in E(G_1)} \sum_{vv' \in E(G_2)} \frac{2}{\deg_{G_1}(u)\deg_{G_2}(v) + \deg_{G_1}(u')\deg_{G_2}(v')}.$$

Note that for every $uu' \in E(G_1)$ and $vv' \in E(G_2)$,

$$\frac{2}{(\deg_{G_1}(u)+\deg_{G_1}(u'))(\deg_{G_2}(v)+\deg_{G_2}(v'))} = \frac{2}{(\deg_{G_1}(u)\deg_{G_2}(v)+\deg_{G_1}(u')\deg_{G_2}(v'))+(\deg_{G_1}(u)\deg_{G_2}(v')+\deg_{G_1}(u')\deg_{G_2}(v))} \leq \frac{1}{4}\left(\frac{2}{\deg_{G_1}(u)\deg_{G_2}(v)+\deg_{G_1}(u')\deg_{G_2}(v')}+\frac{2}{\deg_{G_1}(u)\deg_{G_2}(v')+\deg_{G_1}(u')\deg_{G_2}(v)}\right),$$

with equality if and only if

$$\deg_{G_1}(u)\deg_{G_2}(v) + \deg_{G_1}(u')\deg_{G_2}(v') = \deg_{G_1}(u)\deg_{G_2}(v') + \deg_{G_1}(u')\deg_{G_2}(v)$$

, or, equivalently,

$$(\deg_{G_1}(u) - \deg_{G_1}(u'))(\deg_{G_2}(v) - \deg_{G_2}(v')) = 0$$

. On the other hand,

$$\sum_{uu' \in E(G_1)} \sum_{vv' \in E(G_2)} \frac{2}{(\deg_{G_1}(u)+\deg_{G_1}(u'))(\deg_{G_2}(v)+\deg_{G_2}(v'))} = \frac{1}{2} H(G_1)H(G_2),$$

and

$$\sum_{uu' \in E(G_1)} \sum_{vv' \in E(G_2)} \frac{2}{\deg_{G_1}(u)\deg_{G_2}(v)+\deg_{G_1}(u')\deg_{G_2}(v')} + \sum_{uu' \in E(G_1)} \sum_{vv' \in E(G_2)} \frac{2}{\deg_{G_1}(u)\deg_{G_2}(v')+\deg_{G_1}(u')\deg_{G_2}(v)} = \frac{1}{2}H(G_1 \times G_2) + \frac{1}{2}H(G_1 \times G_2) = H(G_1 \times G_2).$$

This implies that $H(G_1 \times G_2) \geq 2H(G_1)H(G_2)$. □

Moreover, equality holds in the above inequality if and only if for every $uu' \in E(G_1)$ and $vv' \in E(G_2)$,

$$(\deg_{G_1}(u) - \deg_{G_1}(u'))(\deg_{G_2}(v) - \deg_{G_2}(v')) = 0$$

, i.e., either G_1 or G_2 is regular. □

The following corollary is an immediate consequence of Theorem 2.3.

Corollary 2.4

(i) For any $n \geq 3$ and $m \geq 3$, $H(P_n \times C_m) = \frac{4}{3}m + \frac{n-3}{2}m$,

(ii) for any $n \geq 3$ and $m \geq 2$, $H(P_n \times K_m) = \frac{4}{3}m + \frac{n-3}{2}m$,

(iii) for any $n \geq 3$ and $m \geq 3$, $H(C_n \times C_m) = \frac{nm}{2}$,
(iv) for any $n \geq 3$ and $m \geq 2$, $H(C_n \times K_m) = \frac{nm}{2}$,
(v) for any $n \geq 2$ and $m \geq 2$, $H(K_n \times K_m) = \frac{nm}{2}$.

Theorem 2.5 *Let G_1 and G_2 be two graphs. Then*

$$H(G_1 \boxtimes G_2) \leq \frac{1}{9}((n_2 + 2m_2 + r(G_2))H(G_1) + (n_1 + 2m_1 + r(G_1))H(G_2) + H(G_1 \times G_2) + m_2 r(G_1) + m_1 r(G_2)),$$

with equality if and only if G_1 and G_2 are 1-regular graphs.

Proof By definition of the harmonic index, we have

$$\begin{aligned} H(G_1 \boxtimes G_2) &= \sum_{u \in V(G_1)} \sum_{vv' \in E(G_2)} \frac{2}{\deg_{G_1 \boxtimes G_2}(u,v) + \deg_{G_1 \boxtimes G_2}(u,v')} \\ &+ \sum_{v \in V(G_2)} \sum_{uu' \in E(G_1)} \frac{2}{\deg_{G_1 \boxtimes G_2}(u,v) + \deg_{G_1 \boxtimes G_2}(u',v)} \\ &+ 2 \sum_{uu' \in E(G_1)} \sum_{vv' \in E(G_2)} \frac{2}{\deg_{G_1 \boxtimes G_2}(u,v) + \deg_{G_1 \boxtimes G_2}(u',v')} \\ &:= \sum 1 + \sum 2 + \sum 3. \end{aligned}$$

Then,

$$\begin{aligned} \sum 1 &= \sum_{u \in V(G_1)} \sum_{vv' \in E(G_2)} \frac{2}{(\deg_{G_1}(u) + \deg_{G_2}(v) + \deg_{G_1}(u)\deg_{G_2}(v)) + (\deg_{G_1}(u) + \deg_{G_2}(v') + \deg_{G_1}(u)\deg_{G_2}(v'))} \\ &= \sum_{u \in V(G_1)} \sum_{vv' \in E(G_2)} \frac{2}{2\deg_{G_1}(u) + \deg_{G_1}(u)(\deg_{G_2}(v) + \deg_{G_2}(v')) + (\deg_{G_2}(v) + \deg_{G_2}(v'))}, \end{aligned}$$

$$\begin{aligned} \sum 2 &= \sum_{v \in V(G_2)} \sum_{uu' \in E(G_1)} \frac{2}{(\deg_{G_1}(u) + \deg_{G_2}(v) + \deg_{G_1}(u)\deg_{G_2}(v)) + (\deg_{G_1}(u') + \deg_{G_2}(v) + \deg_{G_1}(u')\deg_{G_2}(v))} \\ &= \sum_{v \in V(G_2)} \sum_{uu' \in E(G_1)} \frac{2}{(\deg_{G_1}(u) + \deg_{G_1}(u')) + \deg_{G_2}(v)(\deg_{G_1}(u) + \deg_{G_1}(u')) + 2\deg_{G_2}(v)}, \end{aligned}$$

$$\begin{aligned} \sum 3 &= 2 \sum_{uu' \in E(G_1)} \sum_{vv' \in E(G_2)} \frac{2}{(\deg_{G_1}(u) + \deg_{G_2}(v) + \deg_{G_1}(u)\deg_{G_2}(v)) + (\deg_{G_1}(u') + \deg_{G_2}(v') + \deg_{G_1}(u')\deg_{G_2}(v'))} \\ &= 2 \sum_{uu' \in E(G_1)} \sum_{vv' \in E(G_2)} \frac{2}{(\deg_{G_1}(u) + \deg_{G_1}(u')) + (\deg_{G_2}(v) + \deg_{G_2}(v')) + (\deg_{G_1}(u)\deg_{G_2}(v) + \deg_{G_1}(u')\deg_{G_2}(v'))}. \end{aligned}$$

By similar argument as in the proof of Theorem 2.2, one can show that

$$\sum 1 \leq \frac{1}{9} m_2 r(G_1) + \frac{1}{9} r(G_1)H(G_2) + \frac{1}{9} n_1 H(G_2),$$

with equality if and only if $2\deg_{G_1}(u) = \deg_{G_1}(u)(\deg_{G_2}(v) + \deg_{G_2}(v')) = \deg_{G_2}(v) + \deg_{G_2}(v')$, for all $u \in V(G_1)$ and $vv' \in E(G_2)$,

$$\sum 2 \leq \frac{1}{9} n_2 H(G_1) + \frac{1}{9} r(G_2)H(G_1) + \frac{1}{9} m_1 r(G_2),$$

with equality if and only if $\deg_{G_1}(u) + \deg_{G_1}(u') = \deg_{G_2}(v)(\deg_{G_1}(u) + \deg_{G_1}(u')) = 2\deg_{G_2}(v)$, for all $v \in V(G_2)$ and $uu' \in E(G_1)$, and

$$\sum 3 \leq \frac{2}{9} m_2 H(G_1) + \frac{2}{9} m_1 H(G_2) + \frac{1}{9} H(G_1 \times G_2),$$

with equality if and only if $\deg_{G_1}(u) + \deg_{G_1}(u') = \deg_{G_2}(v) + \deg_{G_2}(v') = \deg_{G_1}(u)\deg_{G_2}(v) + \deg_{G_1}(u')\deg_{G_2}(v')$, for all $uu' \in E(G_1)$ and $vv' \in E(G_2)$. □

Therefore,

$$H(G_1 \boxtimes G_2) \leq \frac{1}{9}((n_2 + 2m_2 + r(G_2))H(G_1) + (n_1 + 2m_1 + r(G_1))H(G_2) + H(G_1 \times G_2) + m_2 r(G_1) + m_1 r(G_2)).$$

It is easy to see that equality holds in the above inequality if and only if G_1 and G_2 are 1-regular graphs. □

Theorem 2.6 *Let G_1 and G_2 be two disjoint graphs. Then*

$$H(G_1 \circ G_2) < \frac{1}{4} H(G_1) + \frac{1}{4} n_1 H(G_2) + \frac{2}{9} n_2 r(G_1) + \frac{2}{9} n_1 r(G_2) + \frac{1}{4} n_1 m_2 + \frac{1}{4}\frac{m_1}{n_2} + \frac{2}{9}\frac{n_1 n_2}{n_2 + 1}.$$

Proof Note that

$$
\begin{aligned}
H(G_1 \circ G_2) = & \sum_{uv\in E(G_1)} \frac{2}{\deg_{G_1\circ G_2}(u)+\deg_{G_1\circ G_2}(v)} + n_1 \\
& \sum_{uv\in E(G_2)} \frac{2}{\deg_{G_1\circ G_2}(u)+\deg_{G_1\circ G_2}(v)} \\
& + \sum_{u\in V(G_1)} \sum_{v\in V(G_2)} \frac{2}{\deg_{G_1\circ G_2}(u)+\deg_{G_1\circ G_2}(v)} \\
& = \sum_{uv\in E(G_1)} \frac{2}{(\deg_{G_1}(u)+n_2)+(\deg_{G_1}(v)+n_2)} + n_1 \\
& \sum_{uv\in E(G_2)} \frac{2}{(\deg_{G_2}(u)+1)+(\deg_{G_2}(v)+1)} \\
& + \sum_{u\in V(G_1)} \sum_{v\in V(G_2)} \frac{2}{(\deg_{G_1}(u)+n_2)+(\deg_{G_2}(v)+1)} \\
= & \sum_{uv\in E(G_1)} \frac{2}{(\deg_{G_1}(u)+\deg_{G_1}(v))+2n_2} + n_1 \\
quad & \sum_{uv\in E(G_2)} \frac{2}{(\deg_{G_2}(u)+\deg_{G_2}(v))+2} \\
& + \sum_{u\in V(G_1)} \sum_{v\in V(G_2)} \frac{2}{\deg_{G_1}(u)+\deg_{G_2}(v)+(n_2+1)} \\
:= & \sum 1 + \sum 2 + \sum 3.
\end{aligned}
$$

By using a similar method, one can verify that

$$\sum 1 \leq \frac{1}{4}H(G_1) + \frac{1}{4}\frac{m_1}{n_2},$$

with equality if and only if $\deg_{G_1}(u)+\deg_{G_1}(v)=2n_2$, for all $uv\in E(G_1)$,

$$\sum 2 \leq \frac{1}{4}n_1 H(G_2) + \frac{1}{4}n_1 m_2,$$

with equality if and only if $\deg_{G_2}(u)+\deg_{G_2}(v)=2$, for all $uv\in E(G_2)$, and $\sum 3 < \frac{2}{9}n_2 r(G_1) + \frac{2}{9}n_1 r(G_2) + \frac{2}{9}\frac{n_1 n_2}{n_2+1}$. □

So,

$$
\begin{aligned}
H(G_1 \circ G_2) < & \frac{1}{4}H(G_1) + \frac{1}{4}n_1 H(G_2) + \frac{2}{9}n_2 r(G_1) \\
& + \frac{2}{9}n_1 r(G_2) + \frac{1}{4}n_1 m_2 + \frac{1}{4}\frac{m_1}{n_2} + \frac{2}{9}\frac{n_1 n_2}{n_2+1}.
\end{aligned}
$$

This completes the proof. □

Theorem 2.7 *Let G_1 and G_2 be two disjoint graphs. Then*

$$
\begin{aligned}
H(G_1 \bullet G_2) < & \frac{1}{n_2+1}H(G_1) + \frac{1}{4}m_1 H(G_2) \\
& + \frac{8}{9}m_1 r(G_2) + \frac{4}{9}n_2 m_1 + \frac{1}{8}m_1 m_2 + \frac{4}{9}\frac{n_1 n_2}{n_2+1}.
\end{aligned}
$$

Proof Note that

$$
\begin{aligned}
H(G_1 \bullet G_2) = & \sum_{uv\in E(G_1)} \frac{2}{\deg_{G_1\bullet G_2}(u)+\deg_{G_1\bullet G_2}(v)} \\
& + m_1 \sum_{uv\in E(G_2)} \frac{2}{\deg_{G_1\bullet G_2}(u)+\deg_{G_1\bullet G_2}(v)} \\
& + 2 \sum_{uv\in E(G_1)} \sum_{x\in V(G_2)} \left(\frac{2}{\deg_{G_1\bullet G_1}(u)+\deg_{G_1\bullet G_2}(x)} \right. \\
& \left. + \frac{2}{\deg_{G_1\bullet G_1}(v)+\deg_{G_1\bullet G_2}(x)} \right) \\
= & \sum_{uv\in E(G_1)} \frac{2}{(1+n_2)\deg_{G_1}(u)+(1+n_2)\deg_{G_1}(v)} \\
& + m_1 \sum_{uv\in E(G_2)} \frac{2}{(\deg_{G_2}(u)+2)+(\deg_{G_2}(v)+2)} \\
& + 2 \sum_{uv\in E(G_1)} \sum_{x\in V(G_2)} \left(\frac{2}{(1+n_2)\deg_{G_1}(u)+(\deg_{G_2}(x)+2)} \right. \\
& \left. + \frac{2}{(1+n_2)\deg_{G_1}(v)+(\deg_{G_2}(x)+2)} \right) \\
= & \frac{1}{1+n_2} \sum_{uv\in E(G_1)} \frac{2}{\deg_{G_1}(u)+\deg_{G_1}(v)} \\
& + m_1 \sum_{uv\in E(G_2)} \frac{2}{(\deg_{G_2}(u)+\deg_{G_2}(v))+4} \\
& + 2 \sum_{uv\in E(G_1)} \sum_{x\in V(G_2)} \left(\frac{2}{(1+n_2)\deg_{G_1}(u)+\deg_{G_2}(x)+2} \right. \\
& \left. + \frac{2}{(1+n_2)\deg_{G_1}(v)+\deg_{G_2}(x)+2} \right) \\
:= & \frac{1}{n_2+1}H(G_1) + \sum 1 + \sum 2.
\end{aligned}
$$

Similarly, one can prove that $\sum 1 \leq \frac{1}{4}m_1 H(G_2) + \frac{1}{8}m_1 m_2$, with equality if and only if $\deg_{G_1}(u)+\deg_{G_1}(v)=4$, for all $uv\in E(G_1)$. □

Also, $\sum 2 \leq \frac{4}{9}\frac{n_1 n_2}{n_2+1} + \frac{8}{9}m_1 r(G_2) + \frac{4}{9}n_2 m_1$, with equality if and only if $(1+n_2)\deg_{G_1}(u) = (1+n_2)\deg_{G_1}(v) = \deg_{G_2}(x) = 2$, for all $uv\in E(G_1)$ and $x\in V(G_2)$.

Therefore,

$$
\begin{aligned}
H(G_1 \bullet G_2) \leq & \frac{1}{n_2+1}H(G_1) + \frac{1}{4}m_1 H(G_2) \\
& + \frac{8}{9}m_1 r(G_2) + \frac{4}{9}n_2 m_1 + \frac{1}{8}m_1 m_2 + \frac{4}{9}\frac{n_1 n_2}{n_2+1}.
\end{aligned}
$$

It is easy to show that equality cannot occur in the above inequality. □

References

1. Deng, H., Balachandran, S., Ayyaswamy, S.K., Venkatakrishnan, Y.B.: On the harmonic index and the chromatic number of a graph. Discrete Appl. Math. **161**, 2740–2744 (2013)
2. Lv, J.B., Li, J.: On the harmonic index and the matching numbers of trees. Ars Combin. **116**, 407–416 (2014)
3. Lv, J.B., Li, J., Shiu, W.C.: The harmonic index of unicyclic graphs with given matching number. Kragujevac J. Math. **38**(1), 173–182 (2014)
4. Niculescu, C., Persson, L.E.: Convex functions and their applications: a contemporary approach. Springer, New York (2006)
5. Onagh, B.N.: The harmonic index of subdivision graphs. Trans. Combin. (**to appear**)
6. Onagh, B.N.: The harmonic index for R-sum of graphs (**submitted**)
7. Onagh, B.N.: The harmonic index of edge-semitotal graphs, total graphs and related sums. Kragujevac J. Math. (**to appear**)
8. Pattabiraman, K., Nagarajan, S., Chendrasekharan, M.: Zagreb indices and coindices of product graphs. J. Prime Res. Math. **10**, 80–91 (2015)
9. Shwetha, B.S., Lokesha, V., Ranjini, P.S.: On the harmonic index of graph operations. Trans. Combin. **4**(4), 5–14 (2015)
10. Zhong, L.: The harmonic index for graphs. Appl. Math. Lett. **25**, 561–566 (2012)
11. Zhong, L.: The harmonic index on unicyclic graphs. Ars Combin. **104**, 261–269 (2012)
12. Zhong, L.: On the harmonic index and the girth for graphs. Roman. J. Inf. Sci. Technol. **16**(4), 253–260 (2013)

14

Fixed point results in generalized metric spaces without Hausdorff property

Z. Kadelburg · S. Radenović

Abstract It is well known that generalized metric spaces in the sense of Branciari might not be Hausdorff and, hence, there may exist sequences in them having more than one limit. Thus, in most of the fixed point results obtained recently in such spaces, Hausdorffness was additionally assumed. We show in this article that, nevertheless, most of these results remain valid without this additional assumption.

Keywords Generalized metric space · Rectangular space · Geraghty condition · Altering distance · Cone metric space

Mathematics Subject Classification (2010) 47H10 · 54H25

Introduction

A lot of generalizations of metric spaces exist. Most of them were introduced in an attempt to extend some fixed point theorems known from the metric case. One of the fruitful generalizations of this kind was given by Branciari in 2000 [1] who replaced the triangular inequality by a more general one, which was later usually called a rectangular or a quadrilateral inequality. These new spaces became known as generalized metric spaces (g.m.s., for short) or rectangular spaces. Several authors (e.g., [2–29]) proved various (common) fixed point results is such spaces.

In some of the first papers which dealt with fixed point theorems in g.m.s., it was sometimes implicitly assumed that the respective topology is Hausdorff and/or that the generalized metric is continuous (see, e.g., [1, 5, 11, 19]). However, as shown by examples in [26, 27], a generalized metric need not be continuous, neither the respective topology need to be Hausdorff. Hence, in further articles, usually one or both of these conditions were additionally assumed (see, e.g., [4, 7–10, 12, 14, 19, 20, 25, 27]).

In this paper, we show that a number of these results are nevertheless valid without assumptions of generalized metric being continuous or the respective topology being Hausdorff. In particular, we prove (common) fixed point results that are extensions of Geraghty-type results, results using altering distance or admissible functions and results in generalized cone metric spaces. References to some further results can be found in [30].

Preliminaries and auxiliary results

The following definition was given by Branciari in [1].

Definition 1 Let X be a nonempty set, and let $d : X \times X \to [0, +\infty)$ be a mapping such that for all $x, y \in X$ and all distinct points $u, v \in X$, each distinct from x and y:

1. $d(x,y) = 0$ iff $x = y$;
2. $d(x,y) = d(y,x)$;
3. $d(x,y) \leq d(x,u) + d(u,v) + d(v,y)$ (rectangular inequality).

Then (X, d) is called a generalized metric space (g.m.s.).

Convergent and Cauchy sequences in g.m.s., completeness, as well as open balls $B_r(p)$, can be introduced in a

Z. Kadelburg (✉)
Faculty of Mathematics, University of Belgrade, Beograd, Serbia
e-mail: kadelbur@matf.bg.ac.rs

S. Radenović
Faculty of Mechanical Engineering, University of Belgrade, Kraljice Marije 16, 11120 Beograd, Serbia
e-mail: radens@beotel.net

standard way. However, the following example, presented by Sarma et al. in [27, Example 1.1] (see also [26]), shows several possible properties of generalized metrics, different than in the standard metric case.

Example 1 [27] Let $A=\{0,2\}$, $B=\{\frac{1}{n}: n\in\mathbb{N}\}$, $X=A\cup B$. Define $d: X\times X\to[0,+\infty)$ as follows:

$$d(x,y)=\begin{cases}0, & x=y\\ 1, & x\neq y \text{ and } \{x,y\}\subset A \text{ or } \{x,y\}\subset B\\ y, & x\in A,\ y\in B\\ x, & x\in B,\ y\in A.\end{cases}$$

Then (X,d) is a complete g.m.s. However, it was shown in [27, Example 1.1] that:

- there exists a convergent sequence in (X,d) which is not a Cauchy sequence;
- there exists a sequence in (X,d) converging to two distinct points;
- there is no $r>0$ such that $B_r(0)\cap B_r(2)=\emptyset$;
- $\lim_{n\to\infty}\frac{1}{n}=0$ but $\lim_{n\to\infty}d(\frac{1}{n},\frac{1}{2})\neq d(0,\frac{1}{2})$; hence d is not a continuous function.

As shown in the previous example, a sequence in a g.m.s. may have two limits. However, there is a special situation where this is not possible, and this will be useful in some proofs. The following lemma is a variant of [31, Lemma 1.10].

Lemma 1 *Let (X,d) be a g.m.s. and let $\{x_n\}$ be a Cauchy sequence in X such that $x_m\neq x_n$ whenever $m\neq n$. Then $\{x_n\}$ can converge to at most one point.*

Proof Suppose, to the contrary, that $\lim_{n\to\infty}x_n=x$, $\lim_{n\to\infty}x_n=y$ and $x\neq y$. Since x_m and x_n are distinct elements, as well as x and y, it is clear that there exists $\ell\in\mathbb{N}$ such that x and y are different from x_n for all $n>l$. For $m,n>\ell$, the rectangular inequality implies that

$$d(x,y)\leq d(x,x_m)+d(x_m,x_n)+d(x_n,y).$$

Taking the limit as $m,n\to\infty$, it follows that $d(x,y)=0$, i.e., $x=y$. Contradiction. □

The following lemma is a g.m.s. modification of a result which is well-known in metric spaces (see, e,g,. [32, Lemma 2.1]). Using it, many known proofs of fixed point results in g.m. spaces become much shorter.

Lemma 2 *Let (X,d) be a g.m.s. and let $\{y_n\}$ be a sequence in X with distinct elements ($y_n\neq y_m$ for $n\neq m$). Suppose that $d(y_n,y_{n+1})$ and $d(y_n,y_{n+2})$ tend to 0 as $n\to\infty$ and that $\{y_n\}$ is not a Cauchy sequence. Then there exist $\varepsilon>0$ and two sequences $\{m_k\}$ and $\{n_k\}$ of positive integers such that $n_k>m_k>k$ and the following four sequences tend to ε as $k\to\infty$:*

$$d(y_{m_k},y_{n_k}),\quad d(y_{m_k},y_{n_k+1}),\quad d(y_{m_k-1},y_{n_k}),\quad d(y_{m_k-1},y_{n_k+1}). \tag{2.1}$$

Proof Since $\{y_n\}$ is not a Cauchy sequence, there exist $\varepsilon>0$ and two sequences $\{m_k\}$ and $\{n_k\}$ of positive integers such that $n_k>m_k>k$, $d(y_{m_k},y_{n_k})\geq\varepsilon$ and n_k is the smallest integer satisfying this inequality, i.e., $d(y_{m_k},y_\ell)<\varepsilon$ for $m_k<\ell<n_k$.

Let us prove that the first of the sequences in (2.1) tends to ε as $k\to\infty$. Note that, by the assumption, $d(y_{m_k},y_{m_k+1})\to0$ and $d(y_{m_k},y_{m_k+2})\to0$ as $k\to\infty$. Hence, it is impossible that $n_k=m_k+1$ or $n_k=m_k+2$ (because in either of these cases it would be impossible to have $d(y_{m_k},y_{n_k})\geq\varepsilon$). Thus, we can apply the rectangular inequality to obtain

$$\varepsilon\leq d(y_{m_k},y_{n_k})\leq d(y_{m_k},y_{n_k-2})+d(y_{n_k-2},y_{n_k-1})+d(y_{n_k-1},y_{n_k})$$
$$\leq\varepsilon+d(y_{n_k-2},y_{n_k-1})+d(y_{n_k-1},y_{n_k})\to\varepsilon,$$

as $k\to\infty$, implying that $d(y_{m_k},y_{n_k})\to\varepsilon$ as $k\to\infty$.

In order to prove that the second sequence in (2.1) tends to ε as $k\to\infty$, consider the following two rectangular inequalities:

$$d(y_{m_k},y_{n_k+1})\leq d(y_{m_k},y_{n_k})+d(y_{n_k},y_{n_k-1})+d(y_{n_k-1},y_{n_k+1})$$
$$d(y_{m_k},y_{n_k})\leq d(y_{m_k},y_{n_k+1})+d(y_{n_k+1},y_{n_k-1})+d(y_{n_k-1},y_{n_k}),$$

which, together with $d(y_{m_k},y_{n_k})\to\varepsilon$ imply that $d(y_{m_k},y_{n_k+1})\to\varepsilon$ as $k\to\infty$.

The proof for the other two sequences can be done in a similar way, using the following rectangles:

$$(y_{m_k-1},y_{n_k},y_{n_k-2},y_{m_k})\quad\text{and}\quad(y_{m_k},y_{n_k},y_{m_k-1},y_{m_k-2}),$$

resp.

$$(y_{m_k-1},y_{n_k+1},y_{n_k},y_{m_k})\quad\text{and}\quad(y_{m_k},y_{n_k},y_{m_k+1},y_{n_k-1}).$$

□

Main results

Geraghty-type conditions

In the following Geraghty-type result [33], we will use the class $\mathcal{S}$ of real functions $\beta:[0,+\infty)\to[0,1)$ satisfying the condition

$$\beta(t_n)\to1 \text{ as } n\to\infty \text{ implies } t_n\to0 \text{ as } n\to\infty.$$

Note that we neither assume that the space (X,d) is Hausdorff, nor that the mapping d is continuous.

Theorem 1 *Let (X,d) be a g.m.s. and let $f,g: X\to X$ be two self maps such that $f(X)\subseteq g(X)$, one of these two subsets of X being complete. If, for some function $\beta\in\mathcal{S}$,*

$$d(fx,fy) \leq \beta(d(gx,gy))d(gx,gy) \tag{3.1}$$

holds for all $x, y \in X$, *then* f *and* g *have a unique point of coincidence* y^*. *Moreover, for each* $x_0 \in X$, *a corresponding Jungck sequence* $\{y_n\}$ *can be chosen such that* $\lim_{n\to\infty} y_n = y^*$.

If, moreover, f *and* g *are weakly compatible, then they have a unique common fixed point.*

Proof We will prove first that f and g cannot have more than one point of coincidence. Suppose to the contrary that there exist $w_1, w_2 \in X$ such that $w_1 \neq w_2$, $w_1 = fu_1 = gu_1$ and $w_2 = fu_2 = gu_2$ for some $u_1, u_2 \in X$. Then (3.1) would imply that

$$\begin{aligned} d(w_1,w_2) &= d(fu_1,fu_2) \leq \beta(d(gu_1,gu_2))d(gu_1,gu_2) \\ &= \beta(d(w_1,w_2))d(w_1,w_2) < d(w_1,w_2), \end{aligned}$$

which is impossible.

In order to prove that f and g have a coincidence point, take an arbitrary $x_0 \in X$ and, using that $f(X) \subseteq g(X)$, choose sequences $\{x_n\}$ and $\{y_n\}$ in X such that

$$y_n = fx_n = gx_{n+1}, \qquad \text{for } n = 0, 1, 2, \ldots$$

Moreover, if $y_n = y_m$ for some $n \neq m$, then we choose $x_{n+1} = x_{m+1}$ (and hence also $y_{n+1} = y_{m+1}$).

If $y_{n_0} = y_{n_0+1}$ for some $n_0 \in \mathbb{N}$, then x_{n_0+1} is a coincidence point of f and g, and y_{n_0+1} is their (unique) point of coincidence.

Suppose now that $y_n \neq y_{n+1}$ for each $n \in \mathbb{N}$. Then, using (3.1), we get that

$$\begin{aligned} d(y_n,y_{n+1}) &= d(fx_n,fx_{n+1}) \leq \beta(d(gx_n,gx_{n+1}))d(gx_n,gx_{n+1}) \\ &= \beta(d(y_{n-1},y_n))d(y_{n-1},y_n) < d(y_{n-1},y_n). \end{aligned}$$

Hence, $\{d(y_n,y_{n+1})\}$ is a strictly decreasing sequence of positive real numbers, tending to some $\delta \geq 0$. Suppose that $\delta > 0$. Then, since

$$\frac{d(y_n,y_{n+1})}{d(y_{n-1},y_n)} \leq \beta(d(y_{n-1},y_n)) < 1,$$

taking the limit as $n \to \infty$, we get that $\beta(d(y_{n-1},y_n)) \to 1$. But this implies that $d(y_{n-1},y_n) \to 0$, a contradiction. Hence,

$$d(y_{n-1},y_n) \to 0 \text{ as } n \to \infty. \tag{3.2}$$

In a similar way, one can prove that

$$d(y_{n-2},y_n) \to 0 \text{ as } n \to \infty. \tag{3.3}$$

Suppose now that $y_n = y_m$ for some $n > m$ (and hence, by the way y_n's are chosen, $y_{n+k} = y_{m+k}$ for $k \in \mathbb{N}$). Then, (3.1) implies that

$$\begin{aligned} d(y_m,y_{m+1}) &= d(y_n,y_{n+1}) \leq \beta(d(y_{n-1},y_n))d(y_{n-1},y_n) \leq \cdots \\ &\leq \beta(d(y_{n-1},y_n)) \cdots \beta(d(y_m,y_{m+1}))d(y_m,y_{m+1}) < d(y_m,y_{m+1}), \end{aligned}$$

a contradiction. Thus, in what follows, we can assume that $y_n \neq y_m$ for $n \neq m$.

In order to prove that $\{y_n\}$ is a Cauchy sequence, suppose that it is not. Then, by Lemma 2, using (3.2) and (3.3), we conclude that there exist $\varepsilon > 0$ and two sequences $\{m_k\}$ and $\{n_k\}$ of positive integers such that $n_k > m_k > k$ and the sequences (2.1) tend to ε as $k \to \infty$. Using (3.1) with $x = x_{m_k}$ and $y = x_{n_k+1}$, one obtains

$$\frac{d(y_{m_k},y_{n_k+1})}{d(y_{m_k-1},y_{n_k})} \leq \beta(d(y_{m_k-1},y_{n_k})) < 1.$$

Letting $k \to \infty$, it follows that $\beta(d(y_{m_k-1},y_{n_k})) \to 1$, implying that $d(y_{m_k-1},y_{n_k}) \to 0$, a contradiction.

Suppose, e.g., that the subspace $g(X)$ is complete (the proof when $f(X)$ is complete is similar). Then $\{y_n\}$ is a Cauchy sequence, tending to some $y^* \in g(X)$, i.e., $y^* = gz$ for some $z \in X$. In order to prove that $fz = gz$, suppose that $fz \neq gz$. Then, by Lemma 1, it follows that y_n differs from both fz and gz for n sufficiently large. Hence, we can apply the rectangular inequality to obtain

$$\begin{aligned} d(fz,gz) &\leq d(fz,fx_n) + d(fx_n,fx_{n+1}) + d(fx_{n+1},gz) \\ &\leq \beta(d(gz,gx_n))d(gz,gx_n) + d(y_n,y_{n+1}) + d(y_{n+1},gz) \\ &< d(gz,gx_n) + d(y_n,y_{n+1}) + d(y_{n+1},gz) \to 0, \end{aligned}$$

as $n \to \infty$. It follows that $fz = gz$ is a point of coincidence of f and g.

In the case when f and g are weakly compatible, a well-known result implies that f and g have a unique common fixed point. □

The following example is inspired by [9, Example 2.4].

Example 2 Let $X = \{a, b, c, \delta, e\}$ and $d : X \times X \to [0, +\infty)$ be defined by:

$$\begin{aligned} &d(x,x) = 0 \text{ for } x \in X; \\ &d(x,y) = d(y,x) \text{ for } x, y \in X; \\ &d(a,b) = 3t, \\ &d(a,c) = d(b,c) = t, \\ &d(a,\delta) = d(b,\delta) = d(c,\delta) = 2t, \\ &d(a,e) = d(c,e) = t, \; d(b,e) = d(\delta,e) = 2t, \end{aligned}$$

where $0 < t < \frac{\log 2}{2}$, i.e., $e^{-2t} > e^{-\log 2} = \frac{1}{2}$. Then it is easy to check that (X,d) is a g.m.s. which is not a metric space since

$$d(a,b) = 3t > 2t = d(a,c) + d(c,b).$$

Consider the following mappings $f, g : X \to X$.

$$f = \begin{pmatrix} a & b & c & \delta & e \\ c & c & c & a & c \end{pmatrix} \quad g = \begin{pmatrix} a & b & c & \delta & e \\ a & a & c & \delta & a \end{pmatrix}.$$

Then $f(X) = \{a, c\} \subset \{a, c, \delta\} = g(X)$. Take the function $\beta \in \mathcal{S}$ defined by $\beta(t) = e^{-t}$ for $t > 0$ and $\beta(0) \in [0, 1)$. Let us check that f, g satisfy contractive condition (3.1) of Theorem 1. Let $x, y \in X$ with $x \neq y$ and consider the following possible cases:

1° $x, y \in \{a, b, c, e\}$. Then $fx = fy = c$ and $d(fx, fy) = 0$. Hence, (3.1) trivially holds.

2° $x \in \{a, b, e\}$, $y = \delta$. Then $fx = c$, $fy = a$ and $d(fx, fy) = t$; $gx = a$, $gy = \delta$ and $d(gx, gy) = 2t$. Hence,

$$d(fx, fy) = t < e^{-2t} \cdot 2t = \beta(2t) \cdot 2t = \beta(d(gx, gy))d(gx, gy),$$

since $1 < e^{-2t} \cdot 2$.

3° $x = c$, $y = \delta$. Then $fx = c$, $fy = a$ and $d(fx, fy) = t$; $gx = c$, $gy = \delta$ and $d(gx, gy) = 2t$. Hence, the inequality (3.1) is again satisfied.

All the conditions of Theorem 1 are satisfied and f and g have a unique point of coincidence (which is c). c is also their unique common fixed point.

Taking $g = i_X$, we get the following variant of Geraghty-theorem in generalized metric spaces.

Corollary 1 *Let (X, d) be a complete g.m.s. and let $f : X \to X$ be a self map. If, for some function $\beta \in \mathcal{S}$,*

$$d(fx, fy) \leq \beta(d(x, y))d(x, y)$$

holds for all $x, y \in X$, then f has a unique fixed point z. Moreover, for each $x_0 \in X$, the corresponding Picard sequence $\{x_n\}$ converges to z.

Remark 1 Taking $\beta(t) = \lambda \in (0, 1)$ in the previous corollary, we get the Banach contraction principle in generalized metric spaces, proved without the assumption that the space is Hausdorff and/or that the function d is continuous. In a similar way, most of the results from the papers [1, 4, 7–10, 12, 14, 19, 20, 25, 27] can be proved without the assumption of Hausdorffness.

Remark 2 It is easy to see that the result of Theorem 1 remains valid if the inequality (3.1) is replaced by the following one

$$d(fx, fy) \leq \beta(M(x, y))M(x, y),$$

where $M(x, y) = \max\{d(gx, gy), d(gx, fx), d(gy, fy)\}$.

Altering distance functions

Recall (see [34]) that a mapping $\psi : [0, +\infty) \to [0, +\infty)$ is called an altering distance function if:

(i) ψ is increasing and continuous,
(ii) $\psi(t) = 0$ iff $t = 0$.

The following theorem is a g.m.s. version of the main result from [34]. Its proof completely follows the lines of proof of Theorem 1 and hence it is omitted.

Theorem 2 *Let (X, d) be a g.m.s. and let $f, g : X \to X$ be two self maps such that $f(X) \subseteq g(X)$, one of these two subsets of X being complete. If, for some altering distance function ψ and some $c \in [0, 1)$,*

$$\psi(d(fx, fy)) \leq c\psi(d(gx, gy)) \tag{3.4}$$

holds for all $x, y \in X$, then f and g have a unique point of coincidence. If, moreover, f and g are weakly compatible, then they have a unique common fixed point.

Corollary 2 *Let (X, d) be a complete g.m.s. and let $f : X \to X$ be a self map. If, for some altering distance function ψ and some $c \in [0, 1)$,*

$$\psi(d(fx, fy)) \leq c\psi(d(x, y))$$

holds for all $x, y \in X$, then f has a unique fixed point.

Remark 3 It is easy to see that the result of Theorem 2 remains valid if the inequality (3.4) is replaced by the following one

$$\psi(d(fx, fy)) \leq c\psi(M(x, y)),$$

where $M(x, y) = \max\{d(gx, gy), d(gx, fx), d(gy, fy)\}$.

Admissible functions

In what follows, we will denote by Ψ the family of nondecreasing functions $\psi : [0, +\infty) \to [0, +\infty)$ such that $\sum_{n=1}^{\infty} \psi^n(t) < +\infty$ for each $t > 0$, where ψ^n is the n-th iterate of ψ. Note that $\psi(t) < t$ for $\psi \in \Psi$ and $t > 0$. Following [35], we adopt the following terminology:

Definition 2 Let X be a nonempty set, $f : X \to X$, $\psi \in \Psi$ and $\alpha : X \times X \to [0, +\infty)$.

1. f is said to be α-admissible if

   $$x, y \in X, \quad \alpha(x, y) \geq 1 \Longrightarrow \alpha(fx, fy) \geq 1.$$

2. If (X, d) is a metric space, then f is called α-ψ-contractive if

   $$\alpha(x, y)d(fx, fy) \leq \psi(d(x, y)) \tag{3.5}$$

   for all $x, y \in X$.

Samet et al. in [35], as well as several other authors proved various fixed point theorems for α-admissible mappings. We will prove one such result in the context of generalized metric spaces (as a modification of [35, Theorem 2.1, Theorem 2.2]).

Theorem 3 *Let (X, d) be a complete g.m.s. and $f : X \to X$ be an α-ψ-contractive mapping (for some α and $\psi \in \Psi$) satisfying the following conditions:*

(1) *f is α-admissible;*
(2) *there exists $x_0 \in X$ such that $\alpha(x_0, fx_0) \geq 1$ and $\alpha(x_0, f^2x_0) \geq 1$;*
(3) *f is continuous, or*
(3') *if $\{x_n\}$ is a sequence in X such that $\alpha(x_n, x_{n+1}) \geq 1$ for all n and $x_n \to x \in X$ as $n \to \infty$, then $\alpha(x_n, x) \geq 1$ for all n.*

Then f has a fixed point.

Proof Starting from x_0 given in (2), construct the sequence $\{x_n\}$ as $x_{n+1} = fx_n$, $n \in \mathbb{N}$. If $x_n = x_{n+1}$ for some $n \in \mathbb{N} \cup \{0\}$, then $x^* = x_n$ is a fixed point of f. Assume further that $x_n \neq x_{n+1}$ for each $n \in \mathbb{N} \cup \{0\}$.

Since f is α-admissible, it follows from (2) that

$$\alpha(x_1, x_2) = \alpha(fx_0, fx_1) \geq 1 \quad \text{and} \quad \alpha(x_1, x_3) = \alpha(fx_0, fx_2) \geq 1.$$

By induction, we get

$$\alpha(x_n, x_{n+1}) \geq 1 \quad \text{and} \quad \alpha(x_n, x_{n+2}) \geq 1 \quad \text{forall} n \in \mathbb{N}.$$

Applying the inequality (3.5) and using the previous relations, we get

$$d(x_n, x_{n+1}) = d(fx_{n-1}, fx_n) \leq \alpha(x_{n-1}, x_n)$$
$$d(fx_{n-1}, fx_n) \leq \psi(d(x_{n-1}, x_n)),$$

and by induction,

$$d(x_n, x_{n+1}) \leq \psi^n(d(x_0, x_1)) \text{for} n \in \mathbb{N}. \tag{3.6}$$

In a similar way, we obtain

$$d(x_n, x_{n+2}) \leq \psi^n(d(x_0, x_2)) \text{for} n \in \mathbb{N}. \tag{3.7}$$

Using the properties of function $\psi \in \Psi$, we get that both sequences $\{d(x_n, x_{n+1})\}$ and $\{d(x_n, x_{n+2})\}$ are nonincreasing and tend to 0 as $n \to \infty$. Suppose that $x_n = x_m$ for some $m, n \in \mathbb{N}$, $m < n$. Then

$$d(x_m, x_{m+1}) = d(x_n, x_{n+1}) \leq \psi^{n-m}(d(x_m, x_{m+1})) < d(x_m, x_{m+1}),$$

a contradiction. Hence, all elements of the Picard sequence $\{x_n\}$ are distinct.

In order to prove that $\{x_n\}$ is a Cauchy sequence in (X, d), we consider the distance $d(x_n, x_{n+p})$ in the cases $p = 2m + 1$ and $p = 2m$. In the first case, using the rectangular inequality and (3.6), we get that

$$\begin{aligned} d(x_n, x_{n+2m+1}) &\leq d(x_n, x_{n+1}) + d(x_{n+1}, x_{n+2}) + d(x_{n+2}, x_{n+2m+1}) \leq \cdots \\ &\leq d(x_n, x_{n+1}) + d(x_{n+1}, x_{n+2}) + \ldots + d(x_{n+2m}, x_{n+2m+1}) \\ &\leq \psi^n(d(x_0, x_1)) + \psi^{n+1}(d(x_0, x_1)) + \ldots \\ &\quad + \psi^{n+2m}(d(x_0, x_1)). \end{aligned}$$

The last expression tends to 0 as $n \to \infty$, since the series $\sum_{n=1}^{\infty} \psi^n(d(x_0, x_1))$ converges, by the properties of function $\psi \in \Psi$.

In the case $p = 2m$, using (3.7) we obtain in a similar way that

$$\begin{aligned} d(x_n, x_{n+2m}) &\leq d(x_n, x_{n+2}) + d(x_{n+2}, x_{n+3}) + \ldots \\ &\quad + d(x_{n+2m-1}, x_{n+2m}) \\ &\leq \psi^n(d(x_0, x_2)) + \psi^{n+2}(d(x_0, x_1)) + \ldots \\ &\quad + \psi^{n+2m-1}(d(x_0, x_1)). \end{aligned}$$

Again, by the properties of function ψ, it follows that $d(x_n, x_{n+2m}) \to$ as $n \to \infty$. Hence, $\{x_n\}$ is a Cauchy sequence that converges to some x^* in the complete g.m.s. (X, d).

In the case (3) when the function f is continuous, it immediately follows that $fx^* = x^*$, since a Cauchy sequence with distinct elements in (X, d) cannot have two limits, by Lemma 1.

Assume now that condition (3′) holds. Using that x_n differs from x^* and fx^* for n sufficiently large, we get that for such n,

$$d(x^*, fx^*) \leq d(x^*, x_n) + d(x_n, x_{n+1}) + d(fx_n, fx^*).$$

The first two terms on the right-hand side tend to 0 as $n \to \infty$, and for the third one we have that

$$\begin{aligned} d(fx_n, fx^*) &\leq \alpha(x_n, x^*) d(fx_n, fx^*) \\ &\leq \psi(d(x_n, x^*)) < d(x_n, x^*) \to 0, \end{aligned}$$

as $n \to \infty$. Hence, $d(x^*, fx^*) = 0$, i.e., x^* is a fixed point of f. □

The following example is inspired by [28, Example 2] and [36, Example 2.7].

Example 3 Let $X = \{\frac{1}{n} : n \in \mathbb{N}\} \cup \{0\}$ and define $d : X \times X \to [0, +\infty)$ by

$$d(x, y) = \begin{cases} 0, & x = y, \\ \dfrac{1}{n}, & \{x, y\} = \{0, \dfrac{1}{n}\}, \\ 1, & x \neq y,\ x, y \in X \setminus \{0\}. \end{cases}$$

Then it is easy to see that (X, d) is a complete g.m.s. which is not a metric space.

Consider now $f : X \to X$ given as

$$fx = \begin{cases} \dfrac{1}{4}, & x \in X \setminus \{1\}, \\ 0, & x = 1, \end{cases}$$

and $\alpha : X \times X \to [0, +\infty)$ given as

$$\alpha(x,y) = \begin{cases} 1, & (x,y) \in \left(\left[0,\frac{1}{4}\right] \times \left[\frac{1}{4},1\right]\right) \cup \left(\left[\frac{1}{4},1\right] \times \left[0,\frac{1}{4}\right]\right), \\ 0, & \text{otherwise.} \end{cases}$$

Finally, take $\psi \in \Psi$ defined by $\psi(t) = \frac{t}{2}$, $t \in [0,+\infty)$. We will prove that:

(a) f is an α-ψ-contractive mapping;
(b) f is α-admissible;
(c) there exists $x_0 \in X$ such that $\alpha(x_0, fx_0) \geq 1$ and $\alpha(x_0, f^2x_0) \geq 1$;
(d) if $\{x_n\}$ is a sequence in X such that $\alpha(x_n, x_{n+1}) \geq 1$ for all n and $x_n \to x \in X$ as $n \to \infty$, then $\alpha(x_n, x) \geq 1$ for all n.

Proof

(a) The only nontrivial case to check is when $x \in [0,\frac{1}{4}]$ and $y = 1$ (or vice versa). Then

$$\alpha(x,y)d(fx,fy) = d\left(\frac{1}{4},0\right) = \frac{1}{4} < \frac{1}{2}\cdot 1 = \psi(d(x,y)).$$

(b) It was proved in [36, Example 2.7].
(c) For $x_0 = 0$, we have $\alpha(x_0, fx_0) = \alpha(x_0, f^2x_0) = \alpha(0,\frac{1}{4}) = 1$.
(d) Let $\{x_n\}$ be a sequence in X such that $\alpha(x_n, x_{n+1}) \geq 1$ for all n and $x_n \to x$ as $n \to \infty$. From definition of α, it follows that

$$(x_n, x_{n+1}) \in \left(\left[0,\frac{1}{4}\right] \times \left[\frac{1}{4},1\right]\right) \cup \left(\left[\frac{1}{4},1\right] \times \left[0,\frac{1}{4}\right]\right).$$

Then, the only possibility is that $x = \frac{1}{4}$. Thus we have $\alpha(x_n, x) \geq 1$ for all n. Hence, all the conditions of Theorem 3 are satisfied, and f has a fixed point (which is $x^* = \frac{1}{4}$).

□

Cone rectangular metric spaces

Cone metric spaces were defined by L. G. Huang and X. Zhang in [37]. Following this paper, a huge number of articles appeared where various fixed point results in such spaces were proved. However, later it became clear that a lot of these results can be reduced to their standard metric counterparts using various methods. These include, among others, the so-called scalarization method [38] and the method of Minkowski functional [39].

Generalized cone metric spaces and fixed point results in them were treated in [6, 21, 22, 31]. We will show that some of these results can be deduced from the respective g.m.s. results using Minkowski functionals, similarly as in [39].

Definition 3 Let $(E, \|\cdot\|)$ be a real Banach space with θ as the zero element and a solid cone P with the respective order $\preceq$. Let X be a nonempty set and $d : X \times X \to E$ satisfy the following

1. $\theta \preceq d(x,y)$ for all $x,y \in X$ and $d(x,y) = \theta$ if and only if $x = y$;
2. $d(x,y) = d(y,x)$ for all $x,y \in X$;
3. $d(x,y) \preceq d(x,u) + d(u,v) + d(v,y)$ for all $x,y \in X$ and for all distinct points u,v, both distinct from x and y.

Then d is called a cone rectangular metric on X and (X,d) is called a cone rectangular metric space (or a cone g.m.s.).

Recall also the following (see, e.g., [39]).

If V is an absolutely convex and absorbing subset of E, its Minkowski functional q_V is defined by $E \ni x \mapsto q_V(x) = \inf\{\lambda > 0 : x \in \lambda V\}$.

It is a semi-norm on E. If V is an absolutely convex neighborhood of θ in E, then q_V is continuous and

$$\{x \in E : q_V(x) < 1\} = \operatorname{int} V \subset V \subset \overline{V} = \{x \in E : q_V(x) \leq 1\}.$$

Let $e \in \operatorname{int} P$. Then $[-e,e] = (P-e) \cap (e-P) = \{z \in E : -e \preceq z \preceq e\}$ is an absolutely convex neighborhood of θ; its Minkowski functional $q_{[-e,e]}$ will be denoted by q_e.

The following theorem can be proved in a very similar way as [39, Theorem 3.2].

Theorem 4 *Let (X,d) be a cone g.m.s. over a solid cone P and let $e \in \operatorname{int} P$. Let q_e be the corresponding Minkowski functional of $[-e,e]$. Then $d_q = q_e \circ d$ is a (real-valued) rectangular metric on X. Moreover,*

1. *For a sequence $\{x_n\}$ in X, $\lim_{n\to\infty} x_n = x$ in (X,d) if and only if $\lim_{n\to\infty} x_n = x$ in (X,d_q).*
2. *$\{x_n\}$ is a d-Cauchy sequence if and only if it is a d_q-Cauchy sequence.*
3. *(X,d) is complete if and only if (X,d_q) is complete.*

As a consequence, most of the results from the papers [6, 21, 22, 31] can be proved by reducing them to respective known results in (standard) g.m.s. As a sample, we will state the result for the quasicontraction (in the sense of Ćirić [40]). It is clear that the result will remain valid for several other contractive conditions listed in the well-known Rhoades's paper [41].

Recall the following recent result proved in [28, Theorem 6]:

Theorem 5 [28] *Let (X,ρ) be a complete partial rectangular metric space and $f : X \to X$ be a quasicontraction, i.e., there exists $\lambda \in [0,1)$ such that*

$$\rho(fx,fy) \leq \lambda \max\{\rho(x,y), \rho(x,fx), \rho(y,fy), \rho(x,fy), \rho(y,fx)\}$$

for all $x, y \in X$. Then f has a unique fixed point $x^* \in X$ (and $\rho(x^*, x^*) = 0$).

Since each g.m.s. is also a partial rectangular metric space, the previous theorem remains valid in generalized metric spaces. Now, applying Theorem 4, we get the following.

Corollary 3 *Let (X, d) be a complete cone g.m.s. and let $f : X \to X$ has the property that for some $\lambda \in [0, 1)$ and for all $x, y \in X$ there exists*

$$u(x, y) \in \{d(x, y), d(x, fx), d(y, fy), d(x, fy), d(y, fx)\}$$

such that $d(fx, fy) \preceq \lambda u(x, y)$. Then f has a unique fixed point in X.

Remark 4 As a consequence, we get that several other fixed point results (as, e.g., Kannan's, Chatterjea's, Zamfirrescu's and others, listed in [41]) can be proved in g.m.s. without using the assumption of Hausdorffness.

Fixed point and common fixed point results can be proved in ordered g.m.s., as well. Since some of them were already obtained in [21, 22], we will not treat them here.

We conclude by citing some additional recent references concerned with extensions of the mentioned results to some other types of spaces [42–45].

Acknowledgments The authors are indebted to the referees of this paper who helped us to improve its presentation. The authors are thankful to the Ministry of Education, Science and Technological Development of Serbia.

Conflict of interest The authors declare that they have no competing interests.

References

1. Branciari, A.: A fixed point theorem of Banach-Caccioppoli type on a class of generalized metric spaces. Publ. Math. Debrecen **57**, 31–37 (2000)
2. Ahmad, J., Arshad, M., Vetro, C.: On a theorem of Khan in a generalized metric space. Intern. J. Anal. **2013**, 6, Art. ID 852727 (2013)
3. Arshad, M., Ahmad, J., Karapinar, E.: Some common fixed point results in rectangular metric spaces. Int. J. Math. Anal. **2013**, 7, Art. ID 307234 (2013)
4. Aydi, H., Karapinar, E., Lakzian, H.: Fixed point results on a class of generalized metric spaces. Math. Sci. **6**, 46 (2012)
5. Azam, A., Arshad, M.: Kannan fixed point theorem on generalized metric spaces. J. Nonlinear Sci. Appl. **1**(1), 45–48 (2008)
6. Azam, A., Arshad, M., Beg, I.: Banach contraction principle on cone rectangular metric spaces. Appl. Anal. Discrete Math. **3**, 236–241 (2009)
7. Bilgili, N., Karapinar, E., Turkoglu, D.: A note on common fixed points for (ψ, α, β)-weakly contractive mappings in generalized metric spaces. Fixed Point Theory Appl. **2013**, 287 (2013)
8. Cakić, N.: Coincidence and common fixed point theorems for (ψ, φ) weakly contrative mappings in generalized metric spaces. Filomat **27**(8), 1415–1423 (2013)
9. Chen, Ch.-M.: Common fixed-point theorems in complete generalized metric spaces. J. Appl. Math. **2012**, 14, Art. ID 945915 (2012)
10. Chen, Ch.-M., Sun, W.Y.: Periodic points and fixed points for the weaker (ϕ, φ)-contractive mappings in complete generalized metric spaces. J. Appl. Math. **2012**, 7, Art. ID 856974 (2012)
11. Das, P., Dey, L.K.: A fixed point theorem in generalized metric spaces. Soochow J. Math. **33**(1), 33–39 (2007)
12. Di Bari, C., Vetro, P.: Common fixed points in generalized metric spaces. Appl. Math. Comput. **218**, 7322–7325 (2012)
13. Flora, A., Bellour, A., Al-Bsoul, A.: Some results in fixed point theory concerning generalized metric spaces. Mat. Vesnik **61**(3), 203–208 (2009)
14. Isik, H., Turkoglu, D.: Common fixed points for (ψ, α, β)-weakly contractive mappings in generalized metric spaces. Fixed Point Theory Appl. **2013**, 131 (2013)
15. Kikina, L., Kikina, K.: Fixed points on two generalized metric spaces. Int. J. Math. Anal. **5**(30), 1459–1467 (2011)
16. Kikina, L., Kikina, K., Gjino, K.: A new fixed point theorem on generalized quasimetric spaces. ISRN Math. Anal. **2012**, 9, Art. ID 457846 (2012)
17. Kirk, W.A., Shahzad, N.: Generalized metrics and Caristi's theorem. Fixed Point Theory Appl. **2013**, 129 (2013)
18. Kumar, M., Kumar, P., Kumar, S.: Some common fixed point theorems in generalized metric spaces. J. Math. **2013**, 7, Art. ID 719324 (2013)
19. Lahiri, B.K., Das, P.: Fixed point of a Ljubomir Ćirić's quasi-contraction mapping in generalized metric spaces. Publ. Math. Debr. **61**(3–4), 589–594 (2002)
20. Lakzian, H., Samet, B.: Fixed point for (ψ, φ)-weakly contractive mappings in generalized metric spaces. Appl. Math. Lett. **25**(5), 902–906 (2011)
21. Malhotra, S.K., Sharma, J.B., Shukla, S.: g-weak contraction in ordered cone rectangular metric spaces. Sci. World J. **2013**, 7, Art. ID 810732 (2013)
22. Malhotra, S.K., Shukla, S., Sen, R.: Some fixed point theorems for ordered Reich type contractions in cone rectangular metric spaces. Acta Math. Univ. Comen. **82**, 165–175 (2013)
23. Mihet, D.: On Kannan fixed point principle in generalized metric spaces. J. Nonlinear Sci. Appl. **2**(2), 92–96 (2009)
24. Moradi, S., Alimohammadi, D.: New extensions of Kannan fixed-point theorem on complete metric and generalized metric spaces. Int. J. Math. Anal. **5**(47), 2313–2320 (2011)
25. Samet, B.: A fixed point theorem in a generalized metric space for mappings satisfying a contractive condition of integral type. Int. J. Math. Anal. **3**(26), 1265–1271 (2009)
26. Samet, B.: Discussion on "A fixed point theorem of Banach–Caccioppoli type on a class of generalized metric spaces" by A. Branciari. Publ. Math. Debr. **76**, 493–494 (2010)
27. Sarma, I.R., Rao, J.M., Rao, S.S.: Contractions over generalized metric spaces. J. Nonlinear Sci. Appl. **2**(3), 180–182 (2009)
28. Shukla, S.: Partial rectangular metric spaces and fixed point theorems. Sci. World J. **2014**, 7, Art. ID 756298 (2014)
29. Turinici, M.: Functional contractions in local Branciari metric spaces. arXiv:1208.4610v1 [math.GN]. 22 Aug 2012.
30. Kadelburg, Z., Radenović, S.: On generalized metric spaces: a survey. TWMS J. Pure Appl. Math. **5**(1), 3–13 (2014)
31. Jleli, M., Samet, B.: The Kannan fixed point theorem in a cone rectangular metric space. J. Nonlinear Sci. Appl. **2**, 161–167 (2009)

32. Radenović, S., Kadelburg, Z., Jandrlić, D., Jandrlić, A.: Some results on weak contraction maps. Bull. Iran. Math. Soc. **38**(3), 625–645 (2012)
33. Geraghty, M.: On contractive mappings. Proc. Am. Math. Soc. **40**, 604–608 (1973)
34. Khan, M.S., Swaleh, M., Sessa, S.: Fixed point theorems by altering distances between the points. Bull. Aust. Math. Soc. **30**(1), 1–9 (1984)
35. Samet, B., Vetro, C., Vetro, P.: Fixed point theorems for α-ψ-contractive type mappings. Nonlinear Anal. TMA **75**, 2154–2165 (2012)
36. Karapinar, E., Samet, B.: Generalized α-ψ contractive type mappings and related fixed point theorems with applications. Abstract Appl. Anal. **2012**, 17, Art. ID 793486 (2012)
37. Huang, L.G., Zhang, X.: Cone metric spaces and fixed point theorems of contractive mappings. J. Math. Anal. Appl. **332**(2), 1468–1476 (2007)
38. Du, W.-S.: A note on cone metric fixed point theory and its equivalence. Nonlinear Anal. **72**, 2259–2261 (2010)
39. Kadelburg, Z., Radenović, S., Rakočević, V.: A note on the equivalence of some metric and cone metric fixed point results. Appl. Math. Lett. **24**, 370–374 (2011)
40. Ćirić, L.B.: A generalization of Banach's contraction principle. Proc. Am. Math. Soc. **45**, 267–273 (1974)
41. Rhoades, B.E.: A comparison of various definitions of contractive mappings. Trans. Am. Math. Soc. **336**, 257–290 (1977)
42. Mustafa, Z., Roshan, J.R., Parvaneh, V., Kadelburg, Z.: Common fixed point results in ordered partial b-metric spaces. J. Inequal. Appl. **2013**, 562 (2013)
43. Hussain, N., Roshan, J.R., Parvaneh, V., Latif, A.: A unification of G-metric, partial metric and b-metric spaces. Abstract Appl. Anal. **2014**, 14, Art. ID 180698 (2014)
44. Hussain, N., Parvaneh, V., Roshan, J.R.: Fixed point results for G-α-contractive maps with application to boundary value problems. Sci. World J. **2014**, 14, Art. ID 585964 (2014)
45. Hussain, N., Roshan, J.R., Parvaneh, V., Kadelburg, Z.: Fixed points of contractive mappings in b-metric-like spaces. Sci. World J. **2014**, 15, Art. ID 471827 (2014)

15

A new class of polynomial functions equipped with a parameter

Saeid Abbasbandy[1]

Abstract In this study, a new class of polynomial functions although equipped with a parameter is introduced. This class can be employed for computational solution of linear or non-linear functional equations, including ordinary differential equations or integral equations. The extra parameter permits us to obtain more accurate results. In the present paper, a number of numerical examples show the ability of this class of polynomial functions.

Keywords Lucas polynomials · Bessel polynomials · Boubaker polynomials · Collocation method

Introduction

In numerous cases, we can find the solution of the linear or non-linear functional equations by Polynomial Expansion Scheme (PES) or matrix methods (in the sense of collocation scheme) based on famous polynomials such as Taylor, Chebyshev, Berstein, Legendre, Laguerre, Hermite, Bessel, Lucas, and Boubaker [1–3, 10–13].

In many problems, utilizing polynomial functions can assist us in finding reasonable solutions. For example, in physics, engineering, and economics, we can find many computational methods applying polynomial functions. A list of polynomial functions and their applications can be found in [2]. In recent years, a great number of new families of polynomials are introduced, including q-analogue of Hermite polynomials [6], d-orthogonal polynomials by Cheikha and Romdhane [5], unified family of generalized Apostol-Bernoulli, and Euler and Genocchi polynomials by El-Desouky and Gomaa [7].

In numerical solution of an Ordinary Differential Equation (ODE), we usually put a linear combination of polynomial functions with unknown coefficients in the main equation and then find the unknowns. This can be done using collocation or, Galerkin or Tau methods. As we know, in this scheme, we can select a set of orthogonal or orthonormal polynomial functions for the solution.

In the current research, we aim to introduce a new class of polynomial functions equipped with a parameter that can be used for computational solution of linear or non-linear functional equations, including ordinary differential equations, partial differential equations, or integral equations. The extra parameter contributes to obtaining more accurate results. Numerical examples demonstrate the ability of this class of polynomial functions to obtain better solutions.

Polynomial functions

In this section, a new class of polynomial functions is introduced. This definition is based on the Chebyshev polynomial of the second kind, $U_n(x)$. These functions have already been employed. For example, Boubaker polynomials are defined as $B_n(x) = U_n(\frac{x}{2}) + 3U_{n-2}(\frac{x}{2})$, for $n \geq 2$. Furthermore, the modified Boubaker polynomials [13] are defined as $\tilde{B}_n(x) = 6xU_{n-1}(x) - 2U_n(x)$, for $n \geq 1$.

Definition 2.1 Suppose that a is an arbitrary constant and $U_n(x)$ is the Chebyshev polynomial of the second kind. Let

✉ Saeid Abbasbandy
abbasbandy@yahoo.com

[1] Department of Mathematics, Imam Khomeini International University, Qazvin 34149-16818, Iran

$A_0(x) = 1$ for any $n \geq 1$. The new class of polynomial functions is defined by the following:

$$A_n(x) = axU_{n-1}(x) + U_n(x).$$

We can find very easily that

$$\begin{aligned}
A_1(x) &= (2+a)x,\\
A_2(x) &= -1+2(2+a)x^2,\\
A_3(x) &= -(4+a)x+4(2+a)x^3,\\
A_4(x) &= 1-4(3+a)x^2+8(2+a)x^4,\\
A_5(x) &= (6+a)x-4(8+3a)x^3+16(2+a)x^5,\\
A_6(x) &= -1+6(4+a)x^2-16(5+2a)x^4+32(2+a)x^6,\\
&\vdots\\
A_{n+1}(x) &= 2xA_n(x)-A_{n-1}(x).
\end{aligned}$$

We can see that for $n \geq 2$, $A_n(x) = (1+\frac{a}{2})U_n(x) + \frac{a}{2}U_{n-2}(x)$.

Proposition 2.1 Let $T_n(x)$ be the Chebyshev polynomial of the first kind. Then, $A_n(x) = (a+1)\frac{x}{n}T_n'(x) + T_n(x)$ for $n \geq 1$.

Proposition 2.2 Let $\lfloor . \rfloor$ designate the floor function. Then, for $n \geq 1$,

$$A_n(x) = \sum_{k=0}^{\lfloor \frac{n}{2} \rfloor} (-1)^k \binom{n-k}{k} \frac{(a+2)n - 2(a+1)k}{2(n-k)} (2x)^{n-2k}.$$

Proposition 2.3 The polynomial $A_n(x)$ for $n \geq 0$ satisfies the following equation:

$$\begin{aligned}
&(x^2-1)\left(1+a+n+a(2+a)nx^2\right)y''(x)\\
&+x\left(3(1+n)+a(3+(2+a)n(2+x^2))\right)y'(x)\\
&-n\left((1+a+n)(2+2a+n)+a(2+a)n^2x^2\right)y(x) = 0.
\end{aligned}$$

Proposition 2.4 For $n \geq 0$, the following equations are obtained:

$$\sum_{k=0}^{n} A_k(x)A_k(y) = -\frac{a}{2} + \frac{A_{n+1}(x)A_n(y) - A_n(x)A_{n+1}(y)}{2(x-y)},$$

$$\sum_{k=0}^{n} A_k^2(x) = -\frac{a}{2} + \frac{1}{2}\left(A_{n+1}'(x)A_n(x) - A_n'(x)A_{n+1}(x)\right).$$

Proposition 2.5 By considering the inner product

$$(f, g) = \int_{-1}^{1} \sqrt{1-x^2} f(x)g(x)\mathrm{d}x,$$

we have

$$(A_n(x), A_m(x)) = \begin{cases}
\frac{\pi}{2} & n=m=0,\\
\frac{1}{8}(2+a)^2\pi & n=m=1,\\
\frac{1}{4}(2+2a+a^2)\pi & n=m\geq 2,\\
\frac{a\pi}{4} & |n-m|=2, nm=0,\\
\frac{1}{8}a(2+a)\pi & |n-m|=2, nm\neq 0,\\
0 & \text{otherwise}.
\end{cases}$$

Corollary 2.1 The system $\{A_{4n}(x)\}_{n=0}^{\infty}$ is an orthogonal system of polynomial functions, with respect to the inner product defined in Proposition 2.5.

The *m*th-order non-linear differential equation

In this section, the new class of polynomial functions is considered to find the numerical solutions of some non-linear differential equations. The solutions obtained show the efficiency of this class of functions. In this class, we have an unknown parameter, and by selecting it, we can reduce the error of PES. In all of the following examples, the least square method is applied so as to find the unknown parameter. All iterative algorithms, which we needed for our initial guess, were initialized at zero.

The *m*th-order non-linear differential equation

$$\sum_{k=0}^{m}\sum_{r=0}^{n} P_{k,r}(x)y^r(x)y^{(k)}(x) = g(x), \quad 0 \leq a \leq x \leq b, \tag{1}$$

with the mixed boundary conditions

$$\sum_{k=0}^{m-1} \left(a_{j,k}y^{(k)}(a) + b_{j,k}y^{(k)}\right) = \gamma_j, \quad j = 0, 1, \ldots, m-1, \tag{2}$$

where y is an unknown function, g and $P_{k,r}$ are continuous functions, and $a_{j,k}$, $b_{j,k}$, and γ_j are real or complex constants, is considered in [13]. By PFS collocation method, we can solve this problem. In this method, we approximate the solution of (1) in the truncated series form:

$$y(x) \approx y_N(x) = \sum_{k=0}^{N} c_k A_k(x), \tag{3}$$

where c_k, $k = 0, 1, \ldots, N$ are the unknown coefficients. In this method, $N \geq m$ is selected in any positive integer number. The matrix form of the collocation method is

Table 1 Numerical results of the absolute error functions of Example (3.1), with $a = 2.59490427$

x_i	$e_4(x_i)$, [4]	$e_5(x_i)$, [9]	$e_5(x_i)$, [13]	$e_5(x_i)$
0	0	0	0	0
0.2	5.81371970e−3	1.52386520e4	3.10830422e−4	3.10830419e−4
0.4	8.16606722e−2	1.1431338878e2	8.84574370e−5	8.84574342e−5
0.6	3.73690421e−1	1.2118642056e−1	1.61217488e−4	1.61217491e−4
0.8	1.08914877e−0	6.1674343718e−1	1.07446634e−4	1.07446646e−4
1.0	2.48769984e−0	2.1293665035e−0	3.60206386e−3	3.60206385e−4

Table 2 Numerical results of the absolute error functions of Example (3.1), with $a = 13.50332999$

x_i	Boubaker functions $e_{10}(x_i)$	Present method $e_{10}(x_i)$
0	0	0
0.2	3.2921789994145e−7	3.2921717829648e−7
0.4	2.2017646683636e−7	2.2017604695002e−7
0.6	1.4863375341623e−7	1.4863314123925e−7
0.8	7.0191910173101e−8	7.0190931511505e−8
1.0	1.1546738372115e−5	1.1546784083105e−5

discussed in [13]. In this method, a non-linear system with $N+1$ equations is constructed to find $N+1$ unknowns. However, here, we have $N+2$ unknowns, i.e., c_k, $k = 0, 1, \ldots, N$ and the unknown parameter of a, of the new class of polynomial functions of A_k. By adding one collocation point to the set of points, we can construct a non-linear system with $N+2$ equations. Via the least square method, we can minimize the L_2 norm of the residual in the augmented non-linear system to obtain a. Afterwards, we can continue the matrix collocation method to find other unknowns like [13].

Example 3.1 We consider the Riccati differential equation:

$$y'(x) = y(x) - 2y^2(x), \quad 0 \le x \le 1,$$

with the initial condition $y(0) = 1$, [4, 9, 13], which is a special case of (1). The exact solution is $y(x) = 1/(2 - e^{-x})$. Table 1 indicates the absolute error $e_N(x) = |y(x) - y_N(x)|$ for different methods, including the Taylor method [4], the decomposition method [9], the PES by Bessel polynomials [13], and the present method. Table 2 demonstrates a comparison between the new class of polynomial functions in present method and Boubaker polynomials, pointing to the efficiency of the new class of polynomial functions.

Example 3.2 We consider the non-linear Abel differential equation of the second kind:

$$y(x)y'(x) + xy(x) + y^2(x) + x^2y^3(x) = xe^{-x} + x^2e^{-3x}, \quad 0 \le x \le 1,$$

with the initial condition $y(0) = 1$, [8, 13], which is special case of (1). The exact solution is $y(x) = e^{-x}$. Table 3 shows the absolute error $e_N(x)$ for different methods. The Taylor method [8] in first column, the PES by Bessel polynomials in second column, the PES by Boubaker polynomials in third column, and the present method in last column are shown. Table 4 shows a comparison with Boubaker polynomials, and so shows the efficiency of the new class of polynomial functions.

Table 3 Numerical results of the absolute error functions of Example (3.2), with $a = 4.01764012$

x_i	Taylor method $e_5(x_i)$, [8]	Bessel functions $e_5(x_i)$	Boubaker functions $e_5(x_i)$	Present method $e_5(x_i)$
0	0	0	0	0
0.2	8.64080e−8	6.8031096090e−7	6.8031096079e−7	6.8031096090e−7
0.4	5.379366e−6	2.2898841512e−7	2.2898841490e−7	2.2898841501e−7
0.6	5.9636094e−5	4.0321424399e−7	4.0321424366e−7	4.0321424377e−7
0.8	3.26297447e−4	4.4156961293e−7	4.4156961343e−7	4.4156961315e−7
1.0	1.212774501e−3	1.4263286683e−5	1.4263286683e−5	1.4263286682e−5

Table 4 Numerical results of the absolute error functions of Example (3.1), with $a = 37.74361705$

x_i	Boubaker functions $e_{10}(x_i)$	Present method $e_{10}(x_i)$
0	0	0
0.2	2.8976820942717e−14	2.8754776337792e−14
0.4	2.1649348980191e−14	2.1649348980191e−14
0.6	1.3322676295502e−14	1.3211653993039e−14
0.8	2.6090241078691e−15	2.0539125955565e−15
1.0	1.8128276657592e−12	1.8171020244040e−12

Conclusion

In this study, a new class of polynomial functions equipped with a parameter was introduced. We found that the extra parameter permits us to obtain more accurate results. By solving Riccati and Abel non-linear differential equations, the ability and efficiency of this class of polynomial functions were supported. The application of this class of polynomials in solving complicated ordinary differential equations and partial differential equations and integral equations is expected.

Acknowledgements The authors thanks the anonymous reviewers for helpful comments, which lead to definite improvement in the manuscript.

References

1. Boubaker, K.: On modified Boubaker polynomials: some differential and analytical properties of the new polynomials issued from an attempt for solving bi-varied heat equation. Trends Appl. Sci. Res. **2**(6), 540–544 (2007)
2. Boubaker, K.: The Boubaker polynomials, a new function class for solving bi-varied second order differential equations. Far East J. Appl. Math. **31**(3), 299–320 (2008)
3. Boubaker, K., Van Gorder, R.A.: Application of the BPES to Lane-Emden equations governing polytropic and isothermal gas spheres. New Astronom. **17**, 565–569 (2012)
4. Bulut, H., Evans, D.J.: On the solution of the Riccati equation by the decomposition method. Int. J. Comp. Math. **79**, 103–109 (2002)
5. Cheikha, Y.B., Romdhane, N.B.: *d*-Symmetric *d*-orthogonal polynomials of Brenke type. J. Math. Anal. Appl. **416**(2), 735–747 (2014)
6. Cigler, J., Zeng, J.: A curious *q*-analogue of Hermite polynomials. J. Combin. Theory Ser. A **118**(1), 9–26 (2011)
7. El-Desouky, B.S., Gomaa, R.S.: A new unified family of generalized Apostol-Euler, Bernoulli and Genocchi polynomials. Appl. Math. Comput. **247**, 695–702 (2014)
8. Guler, C.: A new numerical algorithm for the Abel equation of the second kind. Int. J. Comp. Math. **84**, 109–119 (2007)
9. Gulsu, M., Sezer, M.: On the solution of the Riccati equation by the Taylor matrix method. Appl. Math. Comp. **176**, 414–421 (2006)
10. Işik, O.R., Güney, Z., Sezer, M.: Berstein series solutions of pantograph equations using polynomial interpolation. J. Diff. Equ. Appl. **18**(3), 357–374 (2012)
11. Karamete, A., Sezer, M.: A Taylor collocation method for the solution of linear integro-differential equations. Int. J. Comput. Math. **79**(9), 987–1000 (2002)
12. Labiadh, H., Boubaker, K.: A Sturm-Liouville shaped characteristic differential equation as a guide to establish a quasi-polynomial expression to the Boubaker polynomials. Differ. Equ. Control Process. **2**, 117–133 (2007)
13. Yüzbaşi, Ş., Şahin, N.: On the solutions of a class of nonlinear ordinary differential equations by the Bessel polynomials. J. Numer. Math. **20**(1), 55–79 (2012)

16

Numerical solution of nonlinear two-dimensional Volterra integral equation of the second kind in the reproducing kernel space

A. Fazli[1] · T. Allahviranloo[1] · Sh. Javadi[2]

Abstract In this article, an effective method is given to solve nonlinear two-dimensional Volterra integral equations of the second kind. First, we find the solution of integral equation in terms of reproducing kernel functions in series, then by truncating the series an approximate solution obtained. In addition, the calculation of Fourier coefficients solution of the integral equation in terms of reproducing kernel functions is notable. Numerical examples are presented, and their results are compared with the analytical solution to demonstrate the validity and applicability of the method.

Keywords Two dimentional Volterra integral equation of the second kind · Reproducing kernel · Fourier coefficients

Introduction

Many problems in engineering and mechanics appear to be two-dimensional integral equations. For example, it is usually necessary to solve Fredholm integral equations in the calculation of plasma physics [1]. Mckee et al. [2] revealed that a class of nonlinear telegraph equations is equivalent to two-dimensional Volterra integral equations. Graham [3] showed that two-dimensional Ferdholm integral equation used in solving the problems of electrical engineering. Some other applications of two-dimensional integral equations can be found in [2, 4].

✉ T. Allahviranloo
allahviranloo@yahoo.com

1 Department of Mathematic, Science and Research Branch, Islamic Azad University, Tehran, Iran

2 Department of Mathematic, Kharazmi University, Tehran, Iran

In this paper, we consider the nonlinear two-dimensional Volterra integral equation of the second kinds as follows:

$$u(x,t) = f(x,t) + Fu(x,t) \quad (x,t) \in D = [a,b] \times [c,d] \tag{1}$$

where

$$Fu(x,t) = \lambda \int_c^t \int_a^x K(x,t,y,z) N(u(y,z)) \mathrm{d}y\mathrm{d}z.$$

Here, $u(x, t)$ is unknown function and $u(x,t), f(x,t) \in W_2^{(1,1)}(D), N(.)$ the continues terms in $W_2^{(1,1)}(D)$, $W_2^{(1,1)}(D)$ is a reproducing kernel space. Little numerical methods for nonlinear integral equations are written. In this context, the methods used are the block-by-block method [5], block-pulse functions [6], and rationalized Harr function [7].

In this article, we obtain presentation of exact solution nonlinear two-dimensional Volterra integral equation of the second kind in reproducing kernel space, and then, approximate solution is obtained by cutting series. Of course, coefficients of the series on presentation of exact solution obtained with a numerical calculation. The error of the approximate solution is monotone deceasing in the sense of $\|.\|_{W_2^{(1,1)}}$.

The fundamental principles of the method

Below are some definitions and theorems that we used in the next sections.

Definition 2.1 [9] The function space $W_2^1[a,b]$ is defined as follows:

$$W_2^1[a,b] = \{f(x) | f(x) \text{ is absolutely continuous}, \; f'(x) \in L^2[a,b], x \in [a,b]\}.$$

Definition 2.2 [9] The inner product and norm in the function space $W_2^1[a,b]$ are defined as follows:

For any functions $f(x), g(x) \in W_2^1[a,b]$,

$$\langle f,g\rangle_{W_2^1} = f(a)g(a) + \int_a^b f(x)g(x)\mathrm{d}x$$

and

$$\|f\|_{W_2^1} = \sqrt{\langle f,f\rangle_{W_2^1}}.$$

It is easy to prove that $W_2^1[a,b]$ is an inner space. At [9] prove that function space $W_2^1[a,b]$ is a Hilbert space and also it is a reproducing kernel space.

Suppose $R_y(x)$ is the reproducing kernel of $W_2^1[a,b]$; in this case, we have for any $f(x) \in W_2^1[a,b]$:

$$\langle f(x), R_y(x)\rangle_{W_2^1} = f(y).$$

With calculation, we obtain that

$$R_y(x) = \begin{cases} 1-a+x & \text{if} \quad x \le y \\ 1-a+y & \text{if} \quad x > y \end{cases}$$

Let set $D = [a,b] \times [c,d] \subset \mathcal{R}^2$.

Definition 2.3 [9] The binary function space is defined as

$$W_2^{(1,1)}(D) = \{f(x,y) | f(x,y) \text{ is completely continuous in } D, \frac{\partial^2 f(x,y)}{\partial x \partial y} \in L^2(D)\}.$$

Definition 2.4 [9] The inner product and norm in $W_2^{(1,1)}(D)$ is defined as follows:

$$\langle f(x,y), g(x,y)\rangle_{W_2^{(1,1)}} = \int_c^d \frac{\partial f(a,y)}{\partial y}\frac{\partial g(a,y)}{\partial y}\mathrm{d}y + \langle f(x,c), g(x,c)\rangle_{W_2^1}$$

$$+ \int\int_D \frac{\partial^2 f(x,y)}{\partial x \partial y}\frac{\partial^2 g(x,y)}{\partial x \partial y}\mathrm{d}x\mathrm{d}y$$

and

$$\|f\|_{W_2^{(1,1)}} = \sqrt{\langle f(x,y), f(x,y)\rangle_{W_2^{(1,1)}}}.$$

It is easy to prove that $W_2^{(1,1)}(D)$ is an inner space. At [9] prove that function space $W_2^{(1,1)}(D)$ is a Hilbert space and also it is a reproducing kernel space which has reproduced kernel:

$$K_{(\xi,\eta)}(x,y) = R_\xi(x)Q_\eta(y)$$

where $R_\xi(x), Q_\eta(y)$ are the reproducing kernels of $W_2^1[a,b]$ and $W_2^1[c,d]$, respectively.

Therefore, for each $f(x,y) \in W_2^{(1,1)}(D)$

$$\langle f(x,y), K_{(\xi,\eta)}(x,y)\rangle = f(\xi,\eta).$$

Thus, if $K_{(y,z)}(x,t)$ is reproducing kernel of $W_2^{(1,1)}(D)$, then $K_{(y,z)}(x,t) = R_y(x)Q_z(t)$, where $R_y(x)$ and $Q_z(t)$ are reproducing kernels in $W_2^1[a,b]$ and $W_2^1[c,d]$, respectively.

We will display the solution of Eq. (1) in reproducing kernel space $W_2^{(1,1)}(D)$ for this purpose let $\phi_i(x,t) = K_{(x_i,t_i)}(x,t)$, where $\{(x_i,t_i)\}_{i=1}^\infty$ is dense in region D. From the property of the reproducing kernel, we have

$$\langle u(x,t), \phi_i(x,t)\rangle_{W_2^{(1,1)}(D)} = \langle u(x,t), K_{(x_i,t_i)}(x,t)\rangle_{W_2^{(1,1)}(D)}$$

$$= u(x_i,t_i).$$

Theorem 2.1 *If $\{(x_i,t_i)\}_{i=1}^\infty$ is dense in the region D, then $\{\phi_i(x,t)\}_{i=1}^\infty$ is the complete function system of $W_2^{(1,1)}(D)$.*

Proof Suppose $u(x,t) \in W_2^{(1,1)}(D)$ if

$$\langle u(x,t), \phi_i(x,t)\rangle_{W_2^{(1,1)}(D)} = u(x_i,t_i) = 0 \quad (i = 1,2,\ldots).$$

Then, we have $u(x,t) = 0$ from the density of $\{(x_i,t_i)\}_{i=1}^\infty$ and continuity of $u(x, t)$. ◇

By applying Gram–Schmidt orthonormalization process for $\{\phi_i(x,t)\}_{i=1}^\infty$

$$\bar{\phi}_i(x,t) = \sum_{k=1}^{i} \alpha_{ik}\phi_k(x,t)$$

where α_{ik} are coefficients of Gram–Schmidt orthonormalization and $\{\bar{\phi}_i(x,t)\}_{i=1}^\infty$ is a normal orthogonal basis of $W_2^{(1,1)}(D)$ [9].

Now, the following theorem is obtained.

Theorem 2.2 *We assume $\{(x_i,t_i)\}_{i=1}^\infty$ be dense in region D. If Eq. (1) has a unique solution, then it is as follows:*

$$u(x,t) = \sum_{i=1}^{\infty}\sum_{k=1}^{i} \alpha_{ik}(f(x_k,t_k) + Fu(x_k,t_k))\bar{\phi}_i(x,t). \qquad (2)$$

Proof Suppose $u(x, t)$ is the solution of Eq. (1). From Theorem 2.1 and since $\{\bar{\phi}_i(x,t)\}_{i=1}^\infty$ is a normal orthogonal basis of $W_2^{(1,1)}(D)$, therefore, we can write

$$u(x,t) = \sum_{i=1}^{\infty}\langle u(x,t), \bar{\phi}_i(x,t)\rangle_{W_2^{(1,1)}}\bar{\phi}_i(x,t)$$

$$= \sum_{i=1}^{\infty}\sum_{k=1}^{i} \alpha_{ik}\langle u(x,t), \phi_k(x,t)\rangle_{W_2^{(1,1)}}\bar{\phi}_i(x,t)$$

$$= \sum_{i=1}^{\infty}\sum_{k=1}^{i}\alpha_{ik}\langle f(x,t)+Fu(x,t),\phi_k(x,t)\rangle_{W_2^{(1,1)}}\bar{\phi}_i(x,t)$$

$$= \sum_{i=1}^{\infty}\sum_{k=1}^{i}\alpha_{ik}(f(x_k,t_k)+Fu(x_k,t_k))\bar{\phi}_i(x,t).$$

The proof is complete [9]. ◇

Implementations of the method

In this section, a method will be presented to calculate the solution (2) of the Eq. (1). To this end rewrite (2) as

$$u(x,t) = \sum_{i=1}^{\infty} A_i\bar{\phi}_i(x,t)$$

where

$$A_i = \sum_{k=1}^{i}\alpha_{ik}(f(x_k,t_k)+Fu(x_k,t_k))$$

A_i is unknown, because $u(x, t)$ is an unknown function. A_i with a numerical calculation can be approximated by known B_i. For this purpose, put initial function $u_1(x,t) = f(x,t)$ and n-term approximation for $u(x, t)$ is defined as follows:

$$u_{n+1}(x,t) = \sum_{i=1}^{n} B_i\bar{\phi}_i(x,t) \tag{3}$$

where

$$B_1 = \alpha_{11}(f(x_1,t_1)+Fu_1(x_1,t_1))$$

$$u_2(x,t) = B_1\bar{\phi}_1(x,t)$$

$$B_2 = \sum_{k=1}^{2}\alpha_{nk}(f(x_k,t_k)+Fu_2(x_k,t_k))$$

$$u_3(x,t) = B_1\bar{\phi}_1(x,t)+B_2\bar{\phi}_2(x,t)$$

...

$$B_n = \sum_{k=1}^{n}\alpha_{nk}(f(x_k,t_k)+Fu(x_k,t_k)) \tag{4}$$

$$u_{n+1}(x,t) = \sum_{k=1}^{n} B_k\,\bar{\phi}_k(x,t).$$

Lemma 3.1 *If* $u_n(x,t) \xrightarrow{\|\cdot\|_{W_2^{(1,1)}}} \bar{u}(x,t)$ $(n\to\infty)$ *and* $(x_n,t_n)\to(y,z)$, *then* $u_n(x_n,t_n)\to\bar{u}(y,z)(n\to\infty)$

Proof We have

$$|\,u_n(x_n,t_n)-\bar{u}(y,z)\,| = |\,u_n(x_n,t_n)-u_n(y,z)+u_n(y,z) - \bar{u}(y,z)\,|$$

$$\le |\,u_n(x_n,t_n)-u_n(y,z)\,| + |\,u_n(y,z)-\bar{u}(y,z)\,|.$$

For absolute first on the right side of the above inequality, we have

$$|\,u_n(x_n,t_n)-u_n(y,z)\,| = |\,\langle u_n(x,t),K_{(x_n,t_n)}(x,t)-K_{(y,z)}(x,t)\rangle_{W_2^{(1,1)}}\,|$$

$$\le \|u_n(x,t)\|_{W_2^{(1,1)}}\|K_{(x_n,t_n)}(x,t)-K_{(y,z)}(x,t)\|_{W_2^{(1,1)}}.$$

From the convergence of sequence $\{u_n(x,t)\}_{n=1}^{\infty}$, we conclude that there exists a constant N, such that

$$\|u_n(x,t)\|_{W_2^{(1,1)}} \le N\|\bar{u}(x,t)\|_{W_2^{(1,1)}}$$

when $n \ge N$. At the same time, it can be proved that

$$\|K_{(x_n,t_n)}(x,t)-K_{(y,z)}(x,t)\|_{W_2^{(1,1)}} \to 0$$

when $n\to\infty$. Thus, $|\,u_n(x_n,t_n)-u_n(y,z)\,|\to 0$ when $(x_n,t_n)\to(y,z)$. From $\|u\|_c \le M\|u\|_{W_2^{(1,1)}}$ for any $(y,z)\in D$, it holds that

$$\|u_n(y,z)-\bar{u}(y,z)\|_c \to 0 \quad (n\to\infty)$$

when

$$\|u_n(y,z)-\bar{u}(y,z)\|_{W_2^{(1,1)}} \to 0 \quad (n\to\infty).$$

Therefore, here, we conclude that

$$u_n(x_n,t_n)\to\bar{u}(y,z) \quad (n\to\infty)$$

when $(x_n,t_n)\to(y,z)$ [9].◇.

Using the continuation of $N(.)$, it will be obtained that $N(u_n(x_n,t_n))\to N(\bar{u}(y,z))$ when $n\to\infty$. In addition, this shows that

$$Fu_n(x_n,t_n)\to F\bar{u}(y,z)\ (n\to\infty).$$

From the method listed above, convergence theorem will be obtained.

Theorem 3.1 *Suppose that the sequence* $\{\|u_n(x,t)\|_{W_2^{(1,1)}}\}_{n=1}^{\infty}$ *is bounded in* (3), *if* $\{(x_i,t_i)\}_{i=1}^{\infty}$ *is dense in* D, *then n-term approximate solution* $u_n(x,t)$ *converges to the exact solution* $u(x, t)$ *of Eq.* (1) *and exact solution is expressed as*

$$u(x,t) = \sum_{i=1}^{\infty} B_i\bar{\phi}_i(x,t) \tag{5}$$

where B_i *is given by* (4).

Proof The theorem is proved in three steps:

(a) At this step, we provide that the sequence $\{u_n(x,t)\}_{n=1}^{\infty}$ with formula (3) is converged. For this purpose from (4), we have

$$u_{n+1}(x,t) = u_n(x,t) + B_n\bar{\phi}_n(x,t).$$

From recent equality and using the orthogonality of $\{\bar{\phi}_i(x,t)\}_{i=1}^{\infty}$, it follows

$$\|u_{n+1}(x,t)\|^2_{W_2^{(1,1)}} = \|u_n(x,t)\|^2_{W_2^{(1,1)}} + B_n^2.$$

Thus, sequence $\{\|u_n(x,t)\|_{W_2^{(1,1)}}\}_{n=1}^{\infty}$ is monotone increasing. In other hand, sequence $\{\|u_n(x,t)\|_{W_2^{(1,1)}}\}_{n=1}^{\infty}$ is bounded; therefore, this sequence is convergent. Therefore, there exists a constant c, such that

$$\sum_{i=1}^{\infty} B_i^2 = c.$$

This illustrate that

$$B_i = \sum_{k=1}^{i} \alpha_{ik}(f(x_k,t_k) + Fu_i(x_k,t_k)) \in \ell^2 \quad (i = 1,2,\ldots).$$

Using the orthogonality of $\{u_{n+1}(x,t) - u_n(x,t)\}_{n=1}^{\infty}$ if $m > n$, then

$$\|u_m(x,t) - u_n(x,t)\|^2_{W_2^{(1,1)}} = \|u_m(x,t) - u_{m-1}(x,t)$$
$$+ u_{m-1}(x,t) - u_{m-2}(x,t) + \cdots + u_{n+1}(x,t) - u_n(x,t)\|^2_{W_2^{(1,1)}}$$

$$= \|u_m(x,t) - u_{m-1}(x,t)\|^2_{W_2^{(1,1)}} + \cdots$$
$$+ \|u_{n+1}(x,t) - u_n(x,t)\|^2_{W_2^{(1,1)}}$$

$$= \sum_{i=n+1}^{m} B_i^2 \to 0 \quad (n \to \infty).$$

Considering that $W_2^{(1,1)}(D)$ is complete, so it follows

$$u_n(x,t) \xrightarrow{\|\cdot\|_{W_2^{(1,1)}}} \bar{u}(x,t) \quad (n \to \infty)$$

Therefore

$$\bar{u}(x,t) = \sum_{i=1}^{\infty} B_i\bar{\phi}_i(x,t). \tag{6}$$

(b) We define the projection operator as follows:

$$P_n\bar{u}(x,t) = \sum_{i=1}^{n} B_i\bar{\phi}_i(x,t).$$

Thus

$$u_{n+1}(x,t) = P_n\bar{u}(x,t).$$

We claim

$$u_{n+1}(x_k,t_k) = \bar{u}(x_k,t_k), k \le n.$$

To this end, we have

$$u_{n+1}(x_k,t_k) = \langle u_{n+1}(x,t), \phi_k(x,t)\rangle_{W_2^{(1,1)}}$$
$$= \langle P_n\bar{u}(x,t), \phi_k(x,t)\rangle_{W_2^{(1,1)}}$$
$$= \langle \bar{u}(x,t), P_n\phi_k(x,t)\rangle_{W_2^{(1,1)}}$$
$$= \langle \bar{u}(x,t), \phi_k(x,t)\rangle_{W_2^{(1,1)}}$$
$$= \bar{u}(x_k,t_k).$$

Hence

$$Fu_{n+1}(x_k,t_k) = F\bar{u}(x_k,t_k), \quad k \le n.$$

(c) It is obvious that $\bar{u}(x,t)$ is the solution of Eq. (1). From (6), it follows

$$\bar{u}(x_j,t_j) = \sum_{i=1}^{\infty} B_i\langle\bar{\phi}_i(x,t), \phi_j(x,t)\rangle_{W_2^{(1,1)}}. \tag{7}$$

By multiplying both sides of (7) by α_{nj} and summing for j from 1 to n, we have

$$\sum_{j=1}^{n} \alpha_{nj}\bar{u}(x_j,t_j) = \sum_{i=1}^{\infty} B_i\langle\bar{\phi}_i(x,t), \sum_{j=1}^{n}\alpha_{nj}\phi_j(x,t)\rangle_{W_2^{(1,1)}}$$

$$= \sum_{i=1}^{\infty} B_i\langle\bar{\phi}_i(x,t), \bar{\phi}_n(x,t)\rangle_{W_2^{(1,1)}} = B_n. \tag{8}$$

From (8) and (2), if $n = 1$, then

$$\alpha_{11}\bar{u}(x_1,t_1) = B_1 = \alpha_{11}(f(x_1,t_1) + Fu_1(x_1,t_1)).$$

If $n = 2$, then

$$\alpha_{21}\bar{u}(x_1,t_1) + \alpha_{22}\bar{u}(x_2,t_2) = B_2$$

$$= \alpha_{21}(f(x_1,t_1) + Fu_2(x_1,t_1)) + \alpha_{22}(f(x_2,t_2) + Fu_2(x_2,t_2))$$

$$= \alpha_{21}(f(x_1,t_1) + F\bar{u}_2(x_1,t_1)) + \alpha_{22}(f(x_2,t_2) + Fu_2(x_2,t_2))$$
$$= \alpha_{21}\bar{u}(x_1,t_1)) + \alpha_{22}(f(x_2,t_2) + Fu_2(x_2,t_2)).$$ From the above equality, it follows

$$\bar{u}(x_2,t_2) = f(x_2,t_2) + Fu_2(x_2,t_2).$$

In the same way, we obtain that

$$\bar{u}(x_n,t_n) = f(x_n,t_n) + Fu_n(x_n,t_n).$$

Since $\{(x_i,t_i)\}_{i=1}^{\infty}$ is dense in D, therefore, for any$(y,z) \in D$, there exists a subsequence $\{(x_{n_k},t_{n_k})\}_{k=1}^{\infty}$ converging to (y, z). From Lemma 3.1 and above equality, we obtain that

$$\bar{u}(y,z) = f(y,z) + F\bar{u}(y,z).$$

Thus, $\bar{u}(x,t)$ is the solution of Eq. (1) and

$$u(x,t) = \sum_{i=1}^{\infty} B_i\bar{\phi}_i(x,t).$$

□

Theorem 3.2 *Suppose $u(x, t)$ is the solution of Eq.* (1) *and $r_n(x,t)$ is the error in the approximate solution $u_{n+1}(x,t)$, where $u_{n+1}(x,t)$ is given by* (3). *Then, $r_n(x,t)$ is monotone decreasing in the sense of $\|.\|_{W_2^{(1,1)}}$.*

Proof If the $u(x, t)$ and $u_{n+1}(x,t)$ are, respectively, functions in (1) and (6), in this case, we have

$$\|r_n(x,t)\|^2_{W_2^{(1,1)}} = \|u(x,t) - u_{n+1}(x,t)\|^2_{W_2^{(1,1)}}$$

$$= \|\sum_{i=n+1}^{\infty} B_i\bar{\phi}_i(x,t)\|^2_{W_2^{(1,1)}} = \sum_{i=n+1}^{\infty} B_i^2.$$

This illustrate that the error $r_n(x,t)$ is monotone decreasing in the sense of $\|.\|_{W_2^{(1,1)}}$ [9]. □

Numerical examples

We implement the method presented in this article for some examples. We obtain absolute error of the approximate solution values in the selected points $(x,t) = (\frac{1}{2^i}, \frac{1}{2^i})(i = 1,2,\ldots,6)$. Of course, the choice of points is completely customized. Examples of references listed have been selected, so that we can compare the results in here with results in its references.

Example 4.1 [7] Consider the following nonlinear two-dimensional Volterra integral equation:

$$u(x,t) = f(x,t) + \int_0^t \int_0^x (xy^2 + \cos(z))u^2(y,z)\mathrm{d}y\mathrm{d}z \quad 0 \le x, t \le 1$$

where

$$f(x,t) = x\sin(t)(1 - \frac{1}{9}x^2\sin^2(t)) + \frac{1}{10}x^6\left(\frac{1}{2}\sin(2t) - t\right).$$

The exact solution is $u(x,t) = x\sin(t)$. Table 1 illustrates the numerical results for this example. In addition, for points $(x_i, t_i) = (0.001i, 0.001i)$ $(i = 1,2,\ldots,1000)$, the maximum error is as follows:

$$M = \max_{i=1,2,\ldots,1000}\{|r(x_i,t_i)|\} = 0.04873005644$$

where

$$r(x,t) = u(x,t) - u_{30}(x,t).$$

Example 4.2 [7] Consider the following nonlinear two-dimensional Volterra integral equation:

$$u(x,t) = f(x,t) + \int_0^t \int_0^x (x + t - z - y)u^2(y,z)\mathrm{d}y\mathrm{d}z \quad 0 \le x,\ t \le 1$$

where

$$f(x,t) = x + t - \frac{1}{12}xt(x^3 + 4x^2t + 4xt^2 + t^3).$$

The exact solution is $u(x,t) = x + t$. Table 2 illustrates the numerical results for this example. In addition, for points $(x_i, t_i) = (0.001i, 0.001i)(i = 1,2,\ldots,1000)$, the maximum error is as follows:

$$M = \max_{i=1,2,\ldots,1000}\{|r(x_i,t_i)|\} = 0.165027043$$

where

$$r(x,t) = u(x,t) - u_{40}(x,t).$$

Example 4.3 [8] Consider the following nonlinear two-dimensional Volterra integral equation:

$$u(x,t) = f(x,t) + \int_0^t \int_0^x u^2(y,z)\mathrm{d}y\mathrm{d}z \quad (x,t) \in [0,1) \times [0,1)$$

where

$$f(x,t) = x^2 + t^2 - \frac{1}{45}xt(9x^4 + 10x^2t^2 + 9t^4).$$

The exact solution of this problem is $u(x,t) = x^2 + t^2$. Table 3 illustrates the numerical results for this example. In addition, for points $(x_i, t_i) = (0.001i,\ 0.001i)$ $(i = 1,2,\ldots,1000)$, the maximum error is as follows:

$$M = \max_{i=1,2,\ldots,1000}\{|r(x_i,t_i)|\} = 0.3099048997$$

where

$$r(x,t) = u(x,t) - u_{30}(x,t).$$

Table 1 Absolute error of approximate solution for Example 4.1

Node	Exact solution $u(x, t)$	Approximate $u_{30}(x,t)$	Absolute error in presented method	Absolute error in [7] with $\alpha = 4(m = 32)$
$(\frac{1}{2}, \frac{1}{2})$	0.23972	0.23978	0.00006	1.4×10^{-2}
$(\frac{1}{4}, \frac{1}{4})$	0.061850	0.061979	0.000129	7.9×10^{-3}
$(\frac{1}{8}, \frac{1}{8})$	0.015584	0.015654	0.000070	4.1×10^{-3}
$(\frac{1}{16}, \frac{1}{16})$	0.0039037	0.003957	0.0000533	2.2×10^{-3}
$(\frac{1}{32}, \frac{1}{32})$	0.00097641	0.001036	0.00005959	1.2×10^{-3}
$(\frac{1}{64}, \frac{1}{64})$	0.00024412	0.000990	0.00074588	9.3×10^{-9}

Table 2 Absolute error of approximate solution for Example 4.2

Node	Exact solution $u(x, t)$	Approximate $u_{30}(x,t)$	Absolute error in presented method	Absolute error in [7] with $\alpha = 4(m = 32)$
$(\frac{1}{2}, \frac{1}{2})$	1	0.99734	0.00266	3.1×10^{-2}
$(\frac{1}{4}, \frac{1}{4})$	0.50000	0.49989	0.00011	3.1×10^{-2}
$(\frac{1}{8}, \frac{1}{8})$	0.25000	0.24995	0.00005	3.1×10^{-2}
$(\frac{1}{16}, \frac{1}{16})$	0.12500	0.12490	0.000101	3.1×10^{-2}
$(\frac{1}{32}, \frac{1}{32})$	0.062500	0.062375	0.000125	3.1×10^{-2}
$(\frac{1}{64}, \frac{1}{64})$	0.031250	0.047954	0.016704	2.2×10^{-9}

Table 3 Absolute error of approximate solution for Example 4.3

Node	Exact solution $u(x, t)$	Approximate $u_{30}(x,t)$	Absolute error in presented method	Absolute error in [8] with $m = 32$
$(\frac{1}{2}, \frac{1}{2})$	0.50000	0.49736	0.00264	$1.1D - 2$
$(\frac{1}{4}, \frac{1}{4})$	0.12500	0.12524	0.00024	$6.4D - 4$
$(\frac{1}{8}, \frac{1}{8})$	0.031250	0.031400	0.000150	$2.5D - 3$
$(\frac{1}{16}, \frac{1}{16})$	0.0078125	0.007907	0.0000945	$3.5D - 3$
$(\frac{1}{32}, \frac{1}{32})$	0.0019531	0.002090	0.0001369	$3.5D - 4$
$(\frac{1}{64}, \frac{1}{64})$	0.00048828	0.002007	0.0015187	$4.3D - 5$

Conclusion and comments

This paper deals with a computational method for approximate solution of Volterra integral equations of the second kind based on the expansion of the solution as series of reproducing kernel functions. The advantage of the present method (compared with methods based on basis sets of different kinds) is not limitation on the nonlinear term. Considering that absolute errors of approximate solution in given points are small enough, so it follows that the presentation method in this article is right. The codes were written in Maple. We think that this method can be generalized to the new inner multiply that it provides reproducing kernel space.

References

1. Farengo, R., Lee, Y.C., Guzdar, P.N.: An electromagnetic integral equation: application to microtearing modes. Phys. Fluids. **26**, 3515–3523 (1983)
2. Mckee, S., Tang, T., Diogo, T.: An Euler-type method for two-dimensional Volterra integral equations of the first kind. IMA J. Numer. Anal. **20**, 423–440 (2000)
3. Graham, I.G.: Collocation method for two-dimensional weakly singular integral equations. J. Aust. Math. Soc. (Ser. B) **22**, 456–473 (1981)
4. Hanson, R.J., Phillips, J.L.: Numerical solution of two-dimensional integral equations using linear elements. SIAM J. Numer. Anal. **15**(1), 113–121 (1978)
5. Mirzaee, F., Rafei, Z.: The block by block method for the numerical solution of the nonlinear two-dimensional Volterra integral equations. J. King Saud Univ. Sci. **23**, 191–195 (2011)
6. Mirzaee, F., Hoseini, A.A.: A computational method based on hybrid of block-pulse functions and Taylor series for solving two-dimensional nonlinear integral equations. Alex. Eng. J. **53**, 185–190 (2014)
7. Babolian, E., Bazm, S., Lima, P.: Numerical solution of nonlinear two-dimensional integral equations using rationalized Harr functions. Commun. Nonlinear Sci. Numer. Simul. **16**, 1164–1175 (2011)
8. Maleknejad, K., Sohrabi, B.B.: Two-dimensional FCBS: application to nonlinear Volterra integral equations. In: Proceedings of the World Congress on Engineering (WCE), vol II, 1–3 July 2009, London
9. Cui, M., Lin, Y.: Nonlinear Numerical Analysis in the Reproducing Kernel Space. Nova Science Publishers, New York (2008)

17

An observation on α-type F-contractions and some ordered-theoretic fixed point results

Mohammad Imdad[1] · Rqeeb Gubran[1] · Mohammad Arif[1] · Dhananjay Gopal[2]

Abstract We observe that all the results involving α-type F-contractions are not correct in their present forms. In this article, we prove some fixed point results for extended F-weak contraction mappings in metric and ordered-metric spaces. Our observations and the usability of our results are substantiated by using suitable examples. As an application, we prove an existence and uniqueness result for the solution of a first-order ordinary differential equation satisfying periodic boundary conditions in the presence of either its lower or upper solution.

Keywords Fixed point · F-contraction · αF-weak contraction · Extended F-weak contraction

Mathematics subject classification: 54H25 · 47H10

✉ Rqeeb Gubran
rqeeeb@gmail.com

Mohammad Imdad
mhimdad@yahoo.co.in

Mohammad Arif
mohdarif154c@gmail.com

Dhananjay Gopal
gopaldhananjay@yahoo.in

1 Department of Mathematics, Aligarh Muslim University, Aligarh 202002, India

2 Department of Applied Mathematics and Humanities, S.V. National Institute of Technology, Surat, Gujarat 395007, India

Introduction and preliminaries

In 2012, Wardowski [1] generalized Banach contraction principle in a novel way by introducing a new type of contraction called F-contraction:

Definition 1.1 [1] A self-mapping f on a metric space (X, d) is said to be *F-contraction* if there exists $\tau > 0$ such that

$$d(fx,fy) > 0 \Rightarrow \tau + F(d(fx,fy)) \leq F(d(x,y)), \tag{1.1}$$

for all $x, y \in X$, where $F : \mathbb{R}_+ \to \mathbb{R}$ is a mapping satisfying the following conditions:

F1: F is strictly increasing,

F2: for every sequence $\{s_n\}$ of positive real numbers,

$$\lim_{n\to\infty} s_n = 0 \Leftrightarrow \lim_{n\to\infty} F(s_n) = -\infty,$$

F3: there exists $k \in (0,1)$ such that $\lim_{s\to 0^+} s^k F(s) = 0$.

Let us denote by $\mathcal{F}$, the family of all functions F satisfying conditions F1–F3. Some well-known members of $\mathcal{F}$ are $F(s) = \ln s$, $F(s) = s + \ln s$, $F(s) = \frac{-1}{\sqrt{s}}$ and $F(s) = \ln(s^2 + s)$. Moreover, Wardowski [1] proved that every F-contraction mapping on a complete metric space possesses a unique fixed point. Further, on varying the elements of $\mathcal{F}$ suitably, a variety of known contractions in the literature can be deduced.

Example 1.1 [1] Consider $F \in \mathcal{F}$ given by $F(s) = \ln s$. Then each self-mapping f on X satisfying inequality (1.1) is an F-contraction such that

$$d(fx, fy) \leq e^{-\tau} d(x,y),$$

where $x, y \in X$ and $x \neq y$. Observe that this inequality holds trivially if $x = y$.

Using Ćirić-type generalized contraction in Definition 1.1, Wardowski and Van Dung [2] (also independently Mnak et al. [3]) introduced the notion of F-weak contraction and utilize the same to generalize the main result of [1] as well as several other results of the existence literature.

Definition 1.2 [2, 3] Let (X, d), τ and F be as in Definition 1.1. A self-mapping f on X is said to be *F-weak contraction* if

$$\tau + F(d(fx, fy)) \leq F(M_f(x, y)), \tag{1.2}$$

for all $x, y \in X$ whenever $d(fx, fy) > 0$ where

$$M_f(x, y) = \max\left\{d(x, y), d(x, fx), d(y, fy), \frac{d(x, fy) + d(y, fx)}{2}\right\}.$$

Usually, the following abbreviation, also, is utilized in the literature:

$$m_f(x, y) = \max\left\{d(x, y), \frac{d(x, fx) + d(y, fy)}{2}, \frac{d(x, fy) + d(y, fx)}{2}\right\}.$$

Theorem 1.1 [2, 3] *Let (X, d) a complete metric space and $f : X \to X$ be an F-weak contraction for some $F \in \mathcal{F}$. Then f has a unique fixed point $x \in X$ and for every $x_0 \in X$, the Picard sequence $\{f^n x_0\}$ converges to x provided either*

(a) *F is continuous or*
(b) *f is continuous.*

In 2016, Gopal et al. [4] introduced the concept of α-type F-contraction (for simplicity we write αF-contraction) as follows:

Definition 1.3 [4] Let (X, d), τ and F be as in Definition 1.1. A mapping $f : X \to X$ is said to be *an αF-weak contraction* if there exists $\alpha : X \times X \to \{-\infty\} \cup (0, +\infty)$ such that

$$\tau + \alpha(x, y)F(d(fx, fy)) \leq F(M_f(x, y)), \tag{1.3}$$

for all $x, y \in X$ whenever $d(fx, fy) > 0$.

Employing Definition 1.3, Gopal *et* al. [4] proved the following result:

Theorem 1.2 [4] *Let (X, d) be a complete metric space and $f : X \to X$ an αF-weak contraction satisfying the following conditions:*

(a) *there exists $x_0 \in X$ such that $\alpha(x_0, fx_0) \geq 1$,*
(b) *f is α-admissible,* i.e., *$\alpha(fx, fy) \geq 1$ whenever $\alpha(x, y) \geq 1$,*
(c) *f is continuous (or F is continuous and if a sequence $\{x_n\} \in X$ such that $\alpha(x_n, x_{n+1}) \geq 1$ for all $n \in \mathbb{N}$ and $x_n \to x$ as $n \to \infty$, then $\alpha(x_n, x) \geq 1$).*

Then f has a unique fixed point $x \in X$ and for every such $x_0 \in X$, the Picard sequence $\{f^n x_0\}$ converges to x.

In recent years, the idea of F-contraction has attracted the attention of several researchers and by now there exists a considerable literature on and around this concept (see [5–18] and references therein).

Definition 1.4 A metric space (X, d) together with a partially order "$\preceq$" on it is called *ordered metric space* and denoted by $(X, d, \preceq)$. Further, for arbitrary elements x, y of X and a self-mapping f on X we say that

(i) x, y are *comparable* if either $x \preceq y$ or $y \preceq x$.
(ii) f is *increasing* if $fx \preceq fy$ whenever $x \preceq y$.
(iii) (X, d) is *f-orbitally complete* if every Cauchy sequence $\{f^n x\}$ converges in X.
(iv) X is *regular* if for every increasing sequence $\{x_n\}$ in X with $x_n \to x$, we have $x_n \preceq x$ for all $n \in \mathbb{N}$.

Though Turinici [19, 20] initiated some order-theoretic results in 1986, yet it is often referred to be indicated in 2004 wherein Ran and Reurings [21] presented a more natural result which was well-followed by Nieto and Rodríguez-López [22, 23]. For the work of this kind, one can be referred to [24–31].

Remark 1.1 In the setting of ordered metric spaces, the conditions (1.1–1.3) are required to hold merely for all comparable pairs of elements $x, y \in X$.

Abbas *et* al. [32] utilized the idea of F-contraction to obtain order-theoretic common fixed point results. Very recently, Durmaz et al. [33] proved the following result which can be obtained by setting $g = I : X \to X$ in Theorem 2 of [32]:

Theorem 1.3 [33] *Let $(X, \preceq, d)$ be a complete ordered metric space and $f : X \to X$ an F-contraction for some $F \in \mathcal{F}$. If the following conditions hold:*

(a) *there exists $x_0 \in X$ such that $x_0 \preceq fx_0$,*
(b) *f is increasing,*
(c) *either f is continuous (or F is continuous and X is regular),*

then f has a fixed point.

Further, the authors in [33] gave the following condition to ensure the uniqueness of the fixed point in Theorem 1.3:

B: Every pair of elements of X has a lower bound and upper bound.

Remark 1.2 Very recently, Vetro [34] enlarged the class $\mathcal{F}$ (and denote the same $\mathbb{F}$) by withdrawing the condition F3 and replacing the constant τ by a function $\sigma : \mathbb{R}_+ \to \mathbb{R}_+$ with $\liminf_{t \to s^+} \sigma(t) > 0$ for all $s \geq 0$. Obviously, $\mathcal{F} \subseteq \mathbb{F}$

and $F(s) = -1/s$ is a member of $\mathbb{F}$ which is not in $\mathcal{F}$. We denote with $\mathbb{S}$ the family of all functions σ.

The aim of this article is to point out that all the existing results regarding α-type F-contraction are not correct in their existing forms. We also generalize Theorem 1.3 utilizing Ćirić-type contraction in two directions wherein $\sigma \in \mathbb{S}$ is utilized rather than the constant τ. In doing so, we obtain a slightly sharpened form of Theorem 1.1. We support our results by suitable examples and an application.

An observation on α-type F-contractions

We begin our observation with [4] wherein authors enlarged the co-domain of α to include $-\infty$ and at the same time assumed that the expression $-\infty\,.\,0$ has the value $-\infty$ which is quite unnatural. Inspired by this substitution, we are able to furnish the following counterexamples:

Example 2.1 Let $X = \{0, \frac{1}{4}, \frac{1}{2}, 1\}$ equipped with usual metric d. Then (X, d) is a complete metric space. Define $\alpha : X \times X \to \{-\infty\} \cup (0, \infty)$ by

$$\alpha(x,y) = \begin{cases} -\infty, & \text{for } x, y \in \{0,1\}, x \neq y; \\ 2 - \dfrac{\ln 3}{\ln 4}, & \text{for } x, y \in \{\frac{1}{4}, \frac{1}{2}\}, x \neq y; \\ 1, & \text{otherwise.} \end{cases}$$

Let f be a self-mapping on X defined as $f0 = 1, f\frac{1}{4} = \frac{1}{2}, f\frac{1}{2} = \frac{1}{4}$, and $f1 = 0$. Then f is continuous as well as α-admissible. By a routine calculation, one can verify that f satisfies the contraction condition (1.3) for $F(s) = \ln s$ and $\tau = \ln\frac{4}{3}$. Especially, for $x = 0$ and $y = 1$, we have

$$-\infty = \ln(4/3) + (-\infty)\ln(1) \leq \ln\big(\max\{1,1,1,0\}\big) = 0.$$

Observe that, f is fixed point free which disproves Theorem 1.2.

Even if we restrict the co-domain of α to $(0, \infty)$ in Definition 1.3 with a view to recover Theorem 1.2, still the theorem continues to be erroneous. The following example exhibits this fact:

Example 2.2 Consider $X = [1, \infty)$ equipped with the discrete metric D, that is,

$$D(x,y) = \begin{cases} 0, & \text{for } x = y; \\ 1, & \text{otherwise.} \end{cases}$$

Take $fx = ax$, for all $x \in X$ where $a \in (1, \infty)$. Then with $\alpha(x, y) = 2$, for all $x, y \in X$ and $F(s) = -\frac{1}{\sqrt{s}}$, f satisfies all the requirements of Theorem 1.2 (for $\tau < 1$) but f is a fixed point free.

Indeed, in all the proofs of the results on αF-contractions, e.g. in [4, line 4, page 962] and also in [12, equation (2.4)], the authors assumed that $F(s) \leq \alpha(x, y)F(s)$, for $\alpha(x, y) \geq 1$ which is not true in general (as F may have negative values).

Main results

In order to generalize Theorem 1.3, the following definitions are required:

Definition 3.1 Let (X, d) be a metric space and $\sigma \in \mathbb{S}$. A mapping $f : X \to X$ is said to be *an extended F-weak contraction* if for all $x, y \in X$, we have

$$\sigma(d(x,y)) + F(d(fx, fy)) \leq F(M_f(x,y)) \tag{3.1}$$

whenever $d(fx, fy) > 0$, where $F \in \mathbb{F}$.

Definition 3.2 An ordered metric space $(X, d, \preceq)$ is said to be *$\preceq$-regular* if for every increasing sequence $\{x_n\}$ in X with $x_n \to x$, there exists a subsequence $\{x_{n_k}\}$ of $\{x_n\}$ and a positive integer k_0 such that $x_{n_k} \preceq x$ for all $k \geq k_0$.

First, we prove the following result:

Theorem 3.1 *Let $(X, \preceq, d)$ be an ordered metric space and $f : X \to X$ an extended F-weak contraction for some function $F \in \mathbb{F}$. If (X, d) is f-orbitally complete such that the following conditions hold:*

(a) *there exists $x_0 \in X$ such that $x_0 \preceq fx_0$,*
(b) *f is increasing,*
(c) *F is continuous and X is $\preceq$-regular*

Then f has a fixed point $x \in X$. Moreover, for every $x_0 \in X$ satisfies (a), the sequence $\{f^n x_0\}$ converges to x.

Proof Let $x_0 \in X$ be such that $x_0 \preceq fx_0$. Define a sequence $\{x_n\}$ in X by $x_{n+1} =: fx_n$ for all $n \in \mathbb{N}_0 =: \mathbb{N} \cup \{0\}$. If $x_n = x_{n+1}$ for some $n \in \mathbb{N}_0$, then we are done. Otherwise, we assume $d(x_n, x_{n+1}) > 0$ for all $n \in \mathbb{N}$. As $x_0 \preceq fx_0$ and f is increasing, we have

$$x_0 \preceq x_1 \preceq x_2 \preceq \cdots \preceq x_n \preceq x_{n+1} \preceq \cdots.$$

Now, on setting $x = x_{n-1}$ and $y = x_n$ in (3.1), we have

$$\sigma(d(x_{n-1}, x_n)) + F(d(x_n, x_{n+1})) \leq F(M_f(x_{n-1}, x_n))$$
$$= F\big(\max\{d(x_{n-1}, x_n), d(x_n, x_{n+1})\}\big).$$

If $d(x_{n-1}, x_n) \leq d(x_n, x_{n+1})$ for some $n \in \mathbb{N}$, then

$$F(d(x_n, x_{n+1})) \leq F(d(x_n, x_{n+1})) - \sigma(d(x_{n-1}, x_n)),$$

a contradiction as $\sigma(d(x_{n-1}, x_n)) > 0$. Therefore,

$$F(d(x_n, x_{n+1})) \leq F(d(x_{n-1}, x_n) - \sigma(d(x_{n-1}, x_n)),$$

which, in turn, yields

$$F(d(x_n,x_{n+1})) \le F(d(x_0,x_1)) - n\sigma(d(x_0,x_1)), \tag{3.2}$$

for all $n \in \mathbb{N}$. On letting $n \to \infty$ in (3.2), we get $\lim_{n\to\infty} F(d(x_n,x_{n+1})) = -\infty$. Therefore, (due to F2)

$$\lim_{n\to\infty} d(x_n,x_{n+1}) = 0. \tag{3.3}$$

We assert that $\{x_n\}$ is a Cauchy sequence . Let us assume that $\{x_n\}$ is not so. Then there exists $\epsilon > 0$ and two subsequences $\{x_{n_k}\}$ and $\{x_{m_k}\}$ of $\{x_n\}$ such that

$$n_k > m_k \ge k, d(x_{n_k},x_{m_k}) \ge \epsilon \text{and} d(x_{n_k-1},x_{m_k}) < \epsilon \text{forall} k \in \mathbb{N}.$$

Now, we have

$$\epsilon \le d(x_{n_k},x_{m_k}) \le d(x_{n_k},x_{n_k-1}) + d(x_{n_k-1},x_{m_k}) \le d(x_{n_k},x_{n_k-1}) + \epsilon$$

so that

$$\lim_{k\to\infty} d(x_{n_k},x_{m_k}) = \epsilon.$$

Again, we have

$$\epsilon \le d(x_{n_k},x_{n_k+1}) + d(x_{n_k+1},x_{m_k+1}) + d(x_{m_k+1},x_{m_k})$$

so that (on letting $k \to \infty$)

$$\epsilon \le \liminf_{k\to\infty} d(x_{n_k+1},x_{m_k+1}).$$

Similarly, we can deduce that

$$\epsilon \le \liminf_{k\to\infty} d(x_{n_k+1},x_{m_k}) \text{and} \epsilon \le \liminf_{k\to\infty} d(x_{m_k+1},x_{n_k}).$$

It follows that there exists $l \in \mathbb{N}$ with $d(x_{n_k+1},x_{m_k+1}) > 0$, $d(x_{n_k+1},x_{m_k}) > 0$ and $d(x_{m_k+1},x_{n_k}) > 0$ for all $k \ge l$. Then for all $k \ge l$, we have (on setting $x = x_{n_k}$ and $y = x_{m_k}$ in (3.1))

$$\sigma(d(x_{n_k},x_{m_k})) + F(d(x_{n_k+1},x_{m_k+1})) \le F(M_f(x_{n_k},x_{n_k})), \tag{3.4}$$

where

$$M_f(x_{n_k},x_{m_k}) = \max\left\{d(x_{n_k},x_{m_k}), d(x_{n_k},x_{n_k+1}), d(x_{m_k},x_{m_k+1}), \frac{d(x_{n_k},x_{m_k+1}) + d(x_{m_k},x_{n_k+1})}{2}\right\}$$

$$\le \max\left\{d(x_{n_k},x_{m_k}), d(x_{n_k},x_{n_k+1}), d(x_{m_k},x_{m_k+1}), \frac{d(x_{n_k},x_{m_k}) + d(x_{m_k},x_{m_k+1}) + d(x_{m_k},x_{n_k}) + d(x_{n_k},x_{n_k+1})}{2}\right\}.$$

Letting $k \to \infty$ in presiding inequality and in view of the definition of σ and the continuity of F, we get

$$F(\epsilon) < \liminf_{k\to\infty} \sigma(d(x_{m_k},x_{n_k})) + F(\epsilon) \le F(\epsilon),$$

a contradiction so that $\{x_n\}$ is a Cauchy sequence and having a limit $x \in X$. Next, we show that x is a fixed point. Suppose that $x_n = fx$ for infinitely many $n \in \mathbb{N}$, then there exists a subsequence of $\{x_n\}$ which converges to fx and the uniqueness of the limit finish the proof. Henceforth, we assume that $fx_n \ne fx$ for all $n \in \mathbb{N}_0$. On using the $\preceq$-regularity of X, there exists a subsequence $\{x_{n_k}\}$ of $\{x_n\}$ and a positive integer k_0 such that $x_{n_k} \preceq x$ for all $n_k \ge k_0$. Now, for $n_k \ge k_0$, we can set $x = x_{n_k}$ and $y = x$ in (3.1) so that

$$\begin{aligned}\sigma(d(x_{n_k},x)) + F(d(x_{n_k+1},fx)) &\le F(M(x_n,x)) \\ &\le F\Big(\max\Big\{d(x_{n_k},x), d(x_{n_k},x_{n_k+1}), d(x,fx), \\ &\quad \frac{1}{2}[d(x_{n_k},x) + d(x,fx) + d(x,x_{n_k+1})]\Big\}\Big).\end{aligned} \tag{3.5}$$

Let it be on the contrary that $d(x,fx) > 0$. Making $n \to \infty$ in (3.4), one gets

$$\gamma + F(d(x,fx)) \le F(d(x,fx)),$$

where $0 < \gamma = \liminf_{d(x_n,x)\to 0^+} \sigma(d(x_n,x))$, a contradiction so that $d(x,fx) = 0$ which concludes the proof. □

The following result is yet another version of Theorem 3.1 :

Theorem 3.2 *Theorem* 3.1 *remains true if the condition (c) is replaced by the continuity of f whenever $F \in \mathcal{F}$.*

Proof The proof is identical to the proof of Theorem 3.1 up to (3.3), i.e.,

$$\lim_{n\to\infty} d(x_n,x_{n+1}) = 0.$$

Due to (F3), there exists $k \in (0,1)$ such that

$$\lim_{n\to\infty} (d(x_n,x_{n+1}))^k F(d(x_n,x_{n+1})) = 0. \tag{3.6}$$

Now, from (3.2), we have

$$\begin{aligned}d(x_n,x_{n+1})^k[F(d(x_n,x_{n+1})) - F(d(x_0,x_1))] \\ \le -n\sigma(d(x_0,x_1))d(x_n,x_{n+1})^k \le 0.\end{aligned} \tag{3.7}$$

On using (3.3), (3.5) and letting $n \to \infty$ in (3.6), we get

$$\lim_{n\to\infty} n\sigma(d(x_0,x_1))d(x_n,x_{n+1})^k = 0.$$

Hence, there exists $m \in \mathbb{N}_0$ such that $nd(x_n,x_{n+1})^k \le 1$ for all $n \ge m$, so that

$$d(x_n,x_{n+1}) \le \frac{1}{n^{\frac{1}{k}}} \text{foralln} \ge \text{m} \tag{3.8}$$

We assert that $\{x_n\}$ is a Cauchy sequence. Consider $s,t \in \mathbb{N}_0$ with $s > t \ge m$. Using the triangle inequality and (3.7), we have

$$d(x_t, x_s) \le \sum_{i=t}^{s-1} d(x_i, x_{i+1}) \le \sum_{i=t}^{\infty} d(x_i, x_{i+1}) \le \sum_{i=t}^{\infty} \frac{1}{i^{\frac{1}{k}}}.$$

As $\sum_{i=1}^{\infty} \frac{1}{i^{\frac{1}{k}}}$ is convergent, letting $s, t \to \infty$ gives rise to

$$\lim_{s,t\to\infty} d(x_s, x_t) = 0$$

so that the assertion is established. Since X is f-orbitally complete, there exists $x \in X$ such that $\lim_{n\to\infty} x_n = x$. The continuity of f implies

$$x = \lim_{n\to\infty} x_{n+1} = f(\lim_{n\to\infty} x_n) = fx.$$

This concludes the proof. □

Remark 3.1 Theorem 3.5 carries some advantage over Theorem 3.6 as $\mathbb{F}$ remains a relatively larger class as compared to $\mathcal{F}$, and at the same time most of the utilized functions in $\mathcal{F}$ are already continuous.

Corollary 3.1 *Theorem* 1.3 *follows from Theorems* 3.1 *and* 3.2.

The following example exhibits that Theorem 3.1 is a proper generalization of Theorem 1.3:

Example 3.1 Let $X = A \cup B \cup C$ where $A = [0, 1], B = (1, \frac{3}{2}]$ and $C = (\frac{3}{2}, 2]$. Then, $(X, d, \preceq)$ is an ordered metric space wherein d is the usual metric and the partial order '$\preceq$' on X is defined by

$$x \preceq y \Leftrightarrow \text{either} x = y \text{or} \{x \le y : (x \in A \text{and} y \in B) \text{or} (x \in B \text{and} y \in C)\}.$$

Consider $F \in \mathbb{F}$ given by $F(s) = \frac{-1}{\sqrt{s}}$, for $s > 0$ and $\sigma(t) = \frac{1}{4}$, for all $t \in \mathbb{R}_+$. Define a self-mapping f on X by

$$f(x) = \begin{cases} 1, & \text{for} x \in A; \\ \frac{3}{2}, & \text{for} x \in B; \\ 2, & \text{for} x \in C. \end{cases}$$

Now, in order to verify inequality (3.1), we distinguish the following two cases:

Case 1: $x \in A$ and $y \in B$. Here, we have

$$F\Big(\inf_{x\in A, y\in B} M_f(x, y)\Big) = F\Big(\inf_{x\in A, y\in B} \Big\{\max\Big\{y - x, 1 - x, \frac{3}{2} - y, \frac{y - x}{2} + \frac{1}{4}\Big\}\Big\}\Big) = F\Big(\inf_{y\in B} \Big\{\max\Big\{y - 1, \frac{3}{2} - y, y - \frac{1}{4}\Big\}\Big\}\Big) = F\Big(\frac{3}{4}\Big) = -\frac{2}{\sqrt{3}}.$$

Since $\sigma(d(x, y)) + F(d(fx, fy)) = \frac{1}{4} + F(\frac{1}{2}) = \frac{1}{4} - \sqrt{2}$, f verifies (3.1).

Case 2: $x \in B$ and $y \in C$. Here, we have

$$F\Big(\inf_{x\in B, y\in C} M_f(x, y)\Big) = F\Big(\inf_{x\in B, y\in C} \Big\{\max\Big\{y - x, x - \frac{3}{2}, 2 - y, \frac{y - x}{2} + \frac{1}{4}\Big\}\Big\}\Big) = F\Big(\inf_{y\in C} \Big\{\max\Big\{y - \frac{3}{2}, 2 - y, y - \frac{1}{2}\Big\}\Big\}\Big) = F(1) = -1.$$

Since $\sigma(d(x, y)) + F(d(fx, fy)) = \frac{1}{4} + F(\frac{1}{2}) = \frac{1}{4} - \sqrt{2}$, f verifies (3.1) in this case too. Therefore, in all, f is an F-contraction ensuring the existence of some fixed point of f.

Observe that for $x = \frac{3}{2}$ and $y = 2$, the right-hand side of (1.1) gets us $F(1/2) = -\sqrt{2}$. As $\frac{1}{4} + F(d(f(3/2), f2)) = \frac{1}{4} - \sqrt{2}$, the inequality (1.1) does not hold so that Theorem 1.3 is not applicable in the context of present example.

Now we prove the following uniqueness result corresponding to Theorems 3.1 and 3.2:

Theorem 3.3 *If in addition to the hypotheses of Theorem* 3.1 (*or Theorem* 3.2), *the following condition is satisfied, then f has a unique fixed point*:

B: $Fix(f) := \{x \in X, fx = x\}$ *is a totally ordered set.*

Proof We prove the conclusion for Theorem 3.1 (for Theorem 3.2, the proof is similar). If $F \in \mathcal{F}$ the proof is similar with $\sigma(d(x, y)) \equiv \tau$. Let x, y be two elements of *Fix(f)* such that $d(x, y) > 0$. Then,

$$\sigma(d(x, y)) + F(d(x, y)) \le \sigma(d(x, y)) + F(d(fx, fy)) \le F\Big(\max\Big\{d(x, y), d(x, fx), d(y, fy), \frac{d(x, fy) + d(y, fx)}{2}\Big\}\Big) = F(d(x, y)),$$

a contradiction so that $d(x, y) = 0$. □

In the following uniqueness result, we weaken the condition (B) at the cost of a relatively more stronger contraction condition.

Theorem 3.4 *If in addition to the hypotheses of Theorem* 3.1, *the condition* (B) *is satisfied, then f has a unique fixed point provided* $M_f(x, y)$ *in the contraction condition* (3.1) *is replaced by* $m_f(x, y)$.

Proof Let x, y be two elements of *Fix(f)*. Then there exists $z \in X$ such that z is comparable to both x and y. For $x \prec\succ z$, we may assume that $z \preceq x$ (similar arguments for $y \prec\succ z$). Since f is increasing, we deduce that

$f^n z \preceq x, f^n z \preceq y$.

Let $\xi_n =: d(x, f^n z)$. We assert that $\lim_{n\to\infty} \xi_n = 0$. For substitution $x = x, y = f^n z$ in the contraction condition, we have

$$\begin{aligned} F(\xi_{n+1}) &\leq \sigma(\xi_n) + F(\xi_{n+1}) \\ &\leq F(m_f(x, f^n z)) \\ &= F\left(\max\left\{\xi_n, \frac{0 + d(f^n z, f^{n+1} z)}{2}, \frac{\xi_{n+1} + \xi_n}{2}\right\}\right) \\ &= F\left(\max\left\{\xi_n, \frac{\xi_{n+1} + \xi_n}{2}\right\}\right). \end{aligned} \tag{3.9}$$

Now, if $\xi_n < \xi_{n+1}$, then (3.9) becomes

$$F(\xi_{n+1}) \leq F\left(\frac{\xi_{n+1} + \xi_n}{2}\right),$$

and since F is strictly increasing, we have $\xi_{n+1} \leq \xi_n$ which is a contradiction. Therefore, $\xi_n \geq \xi_{n+1}$ so that ξ_n is a decreasing sequence of nonnegative reals such that $\lim_{n\to\infty} \xi_n = r \geq 0$. If $r > 0$, then on letting n tends to infinity in (3.8), we get $F(r) < F(r)$ which is not possible. Thus, in all situations, $\lim_{n\to\infty} d(x, f^n z) = 0$. Similarly, we can prove that $\lim_{n\to\infty} d(y, f^n z) = 0$. Since $d(x, y) \leq d(x, f^n z) + d(f^n z, y) \to 0$ as $n \to \infty$, the uniqueness of the fixed point is established. This concludes the proof. □

Remark 3.2 As 1 and 2 are not comparable elements, in the context of Example 3.1, the fixed point is not unique supporting our uniqueness results.

The following result is immediate. Observe that by widening the class of functions $\mathcal{F}$ in Definition 1.2, one can derive the following result which remains a metric-version of Theorem 3.1:

Theorem 3.5 *Let (X, d) be a metric space and $f : X \to X$ an extended F-weak contraction for some function $F \in \mathbb{F}$. If (X, d) is f-orbitally complete and the following condition holds:*

(a) *F is continuous.*

Then f has a unique fixed point $x \in X$. Moreover, for every $x_0 \in X$, the Picard sequence $\{f^n x_0\}$ converges to x.

Proof The proof of existence part is very similar to that one of Theorem 3.1 and the uniqueness follows from Theorem 3.3 . Only we mention here that the extra conditions therein ensure the comparability between the element in which we apply to inequality (3.1). □

Remark 3.3 With a view to check the validity of Theorem 3.5 in the context of Example 3.1 (without any partial order on X), observe that for $x = 1$ and $y = 2$, (3.1) gives rise

$$-\frac{1}{2} = \frac{1}{2} + F(d(f1, f2)) \geq F(1) = -1$$

so that the inequality (3.1) is not satisfied. This demonstrates the utility of proving an ordered-version of Theorem 3.5.

The following is yet another version of Theorem 3.5 which remains a slightly sharpened form of Theorem 1.1 (proved for continuous mapping f).

Theorem 3.6 *Let (X, d) be a metric space and $f : X \to X$ an extended F-weak contraction for some function $F \in \mathcal{F}$. If (X, d) is f-orbitally complete and*

(a) *f is continuous,*

then f has a unique fixed point $x \in X$. Moreover, for every $x_0 \in X$, the Picard sequence $\{f^n x_0\}$ converges to x.

The proof is omitted as it is very similar to that of [18, Theorem 2.4] and [3, Theorem 2.2] where the completeness of the whole space is utilized rather than the completeness of the orbit of f.

Corollary 3.2 *Theorem* 1.1 *follows from Theorems* 3.5 *and* 3.6.

Corollary 3.3 *Let (X, d) be a complete metric space and $f : X \to X$. Assume there exists $F \in \mathbb{F}$ and $\sigma \in \mathbb{S}$ such that f is F-contraction of Hardy-Rogers, i.e.,*

$$\sigma(d(x,y)) + F(d(fx,fy)) \leq F\big(a_1 d(x,y) + a_2 d(x,fx) + a_3 d(y,fy) + a_4 d(x,fy) + a_5 d(y,fx)\big),$$

for all $x, y \in X$ whenever $d(fx, fy) > 0$, where $a_i \in [0, \infty) \forall i$, $a_1 + a_2 + a_3 + 2a_4 = 1$, $a_3 \neq 1$ and $a_1 + a_3 + a_5 \leq 1$. Then f has a unique fixed point $x \in X$.

Proof For all $x, y \in X$, we have

$$\begin{aligned} &a_1 d(x,y) + a_2 d(x,fx) + a_3 d(y,fy) + a_4 d(x,fy) + a_5 d(y,fx) \\ &\leq (a_1 + a_2 + a_3 + 2a_4) \max\left\{d(x,y), d(x,fx), d(y,fy), \frac{d(x,fy) + d(y,fx)}{2}\right\} \\ &= \max\left\{d(x,y), d(x,fx), d(y,fy), \frac{d(x,fy) + d(y,fx)}{2}\right\}. \end{aligned}$$

□

Applications

Inspired by [22], we establish the existence and uniqueness solution for the following first-order periodic boundary value problem with respect to its lower or upper solution:

$$\begin{cases} u'(s) = f(s, u(s)), & s \in I = [0, S] \\ u(0) = u(S), \end{cases} \tag{4.1}$$

where $S > 0$ and $f : I \times \mathbb{R} \to \mathbb{R}$ is a continuous function.

Let $\mathcal{C}(I)$ denote the space of all continuous functions defined on I. We recall the following two definitions:

Definition 4.1 [22] A function $\gamma \in \mathcal{C}^1(I)$ is called a lower solution of (4.1), if

$$\begin{cases} \gamma'(s) \leq f(s, \gamma(s)), & s \in I \\ \gamma(0) \leq \gamma(S). \end{cases}$$

Definition 4.2 [22] A function $\gamma \in \mathcal{C}^1(I)$ is called an upper solution of (4.1), if

$$\begin{cases} \gamma'(s) \geq f(s, \gamma(s)), & s \in I \\ \gamma(0) \geq \gamma(S). \end{cases}$$

Now, we prove the following result on the existence and uniqueness of solution of the problem described by (4.1) in the presence of a lower solution (or an upper solution).

Theorem 4.1 *In respect of the problem* (4.1), *suppose that the following conditions hold*:

(i) *there exists $\tau > 0$ such that for all $x, y \in \mathbb{R}$ with $x \leq y$*

$$0 \leq f(s, y) + e^{-\tau} y - [f(s, x) + e^{-\tau} x] \leq e^{-\tau}(y - x). \tag{4.2}$$

(ii) *there exists a function $\omega : \mathbb{R}^2 \to \mathbb{R}$ such that for all $s \in I$ and for all $a, b \in \mathbb{R}$ with $\omega(a, b) \geq 0$,*

$$\omega\Big(\int_0^S G(s,t)[f(t, u(t)) + e^{-\tau} u(t)]\mathrm{d}t, \gamma(s)\Big) \geq 0,$$

where $\gamma \in \mathcal{C}^1(I)$ is a lower solution of (4.1).

(iii) *for all $s \in I$ and all $x, y \in \mathcal{C}^1(I)$, $\omega(x(s), y(s)) \geq 0$* implies

$$\omega\Big(\int_0^S G(s,t)[f(t, x(t)) + e^{-\tau} x(t)]\mathrm{d}t, \int_0^S G(s,t)[f(t, y(t)) + e^{-\tau} y(t)]\mathrm{d}t\Big) \geq 0,$$

(iv) *if $x_n \to x \in \mathcal{C}^1(I)$ and $\omega(x_{n+1}, x_n) \geq 0$, then $\omega(x_n, x) \geq 0$ for all $n \in \mathbb{N}$. Then the existence of a lower solution of problem* (4.1) *ensures the existence and uniqueness of a solution of problem* (4.1).

Proof The problem described by (4.1) can be rewritten as

$$\begin{cases} u'(s) + e^{-\tau} u(s) = f(s, u(s)) + e^{-\tau} u(s) & \forall s \in I \\ u(0) = u(S) \end{cases},$$

which is equivalent to the integral equation

$$u(s) = \int_0^S G(s,t)[f(t, u(t)) + e^{-\tau} u(t)]\mathrm{d}t, \tag{4.3}$$

where Green function $G(s, t)$ is given by

$$G(s,t) = \begin{cases} \dfrac{e^{e^{-\tau}(S+t-s)}}{e^{e^{-\tau}S} - 1} & 0 \leq t < s \leq S, \\ \dfrac{e^{e^{-\tau}(t-s)}}{e^{e^{-\tau}S} - 1} & 0 \leq s < t \leq S. \end{cases}$$

Define a function $\mathcal{X} : \mathcal{C}(I) \to \mathcal{C}(I)$ by

$$(\mathcal{X}u)(s) = \int_0^S G(s,t)[f(t, u(t)) + e^{-\tau} u(t)]\mathrm{d}t \quad \forall s \in I. \tag{4.4}$$

Clearly, if $u \in \mathcal{C}(I)$ is a fixed point of $\mathcal{X}$, then $u \in \mathcal{C}^1(I)$ is a solution of (4.3) and hence of (4.1). Now, define a metric d on $\mathcal{C}(I)$ by

$$d(u, v) = \sup_{s \in I} |u(s) - v(s)| \quad \forall u, v \in \mathcal{C}(I). \tag{4.5}$$

On $\mathcal{C}(I)$, define a partial order $\preceq$ given by

$$u, v \in \mathcal{C}(I); u \preceq v \iff u(s) \leq v(s) \quad \forall s \in I. \tag{4.6}$$

Clearly, $(\mathcal{C}(I), d, \preceq)$ is a complete ordered metric space. We check that all other conditions of Theorem 3.4:

First, let $\gamma \in \mathcal{C}^1(I)$ be a lower solution of (4.1); we have

$$\gamma'(s) + e^{-\tau}\gamma(s) \leq f(s, \gamma(s)) + e^{-\tau}\gamma(s) \quad \forall s \in I.$$

Multiplying both the sides by $e^{e^{-\tau}s}$, we get

$$(\gamma(s) e^{e^{-\tau}s})' \leq [f(s, \gamma(s)) + e^{-\tau}\gamma(s)] e^{e^{-\tau}s} \quad \forall s \in I,$$

which implies that

$$\gamma(s) e^{e^{-\tau}s} \leq \gamma(0) + \int_0^s [f(t, \gamma(t)) + e^{-\tau}\gamma(t)] e^{e^{-\tau}t}\mathrm{d}t \quad \forall s \in I. \tag{4.7}$$

As $\gamma(0) \leq \gamma(S)$, we have

$$\gamma(0)e^{e^{-\tau S}} \leq \gamma(S)e^{e^{-\tau S}} \leq \gamma(0) + \int_0^S [f(t,\gamma(t)) + e^{-\tau}\gamma(t)]e^{e^{-\tau t}}\mathrm{d}t$$

so that

$$\gamma(0) \leq \int_0^S \frac{e^{e^{-\tau t}}}{e^{e^{-\tau S}} - 1}[f(t,\gamma(t)) + e^{-\tau}\gamma(t)]\mathrm{d}t. \tag{4.8}$$

On using (4.7) and (4.8), we obtain

$$\begin{aligned}\gamma(s)e^{e^{-\tau s}} \leq & \int_0^S \frac{e^{e^{-\tau t}}}{e^{e^{-\tau S}} - 1}[f(t,\gamma(t)) + e^{-\tau}\gamma(t)]\mathrm{d}t \\ & + \int_0^s e^{e^{-\tau t}}[f(t,\gamma(t)) + e^{-\tau}\gamma(t)]\mathrm{d}t \\ \leq & \int_0^s \frac{e^{e^{-\tau(S+t)}}}{e^{e^{-\tau S}} - 1}[f(t,\gamma(t)) + e^{-\tau}\gamma(t)]\mathrm{d}t \\ & + \int_s^S \frac{e^{e^{-\tau t}}}{e^{e^{-\tau S}} - 1}[f(t,\gamma(t)) + e^{-\tau}\gamma(t)]\mathrm{d}t\end{aligned}$$

so that

$$\begin{aligned}\gamma(s) \leq & \int_0^s \frac{e^{e^{-\tau(S+t-s)}}}{e^{e^{-\tau S}} - 1}[f(t,\gamma(t)) + e^{-\tau}\gamma(t)]\mathrm{d}t \\ & + \int_s^S \frac{e^{e^{-\tau(t-s)}}}{e^{e^{-\tau S}} - 1}[f(t,\gamma(t)) + e^{-\tau}\gamma(t)]\mathrm{d}t \\ = & \int_0^S G(s,t)[f(t,\gamma(t)) + e^{-\tau}\gamma(t)]\mathrm{d}t \\ = & (\mathcal{X}\gamma)(s)\end{aligned}$$

for all $s \in I$, which implies that $\gamma \preceq \mathcal{X}(\gamma)$.

Second, take $u, v \in \mathcal{C}(I)$ such that $u \preceq v$; then by (4.2), we have

$$f(s,u(s)) + e^{-\tau}u(s) \leq f(s,v(s)) + e^{-\tau}v(s) \quad \forall s \in I. \tag{4.9}$$

On using (4.4), (4.9) and the fact that $G(s,t) > 0$ for $(s,t) \in I \times I$, we get

$$\begin{aligned}(\mathcal{X}u)(s) = & \int_0^S G(s,t)[f(t,u(t)) + e^{-\tau}u(t)]\mathrm{d}t \\ \leq & \int_0^S G(s,t)[f(t,v(t)) + e^{-\tau}v(t)]\mathrm{d}t \\ = & (\mathcal{X}v)(s) \quad \forall s \in I,\end{aligned}$$

which, owing to (4.6), implies that $\mathcal{X}(u) \preceq \mathcal{X}(v)$ so that $\mathcal{X}$ is increasing.

Finally, take an increasing sequence $\{u_n\} \subset \mathcal{C}(I)$ such that $u_n \to u \in \mathcal{C}(I)$; then for each $s \in I$, $\{u_n(s)\}$ is a sequence in $\mathbb{R}$ converging to $u(s)$. Hence, for all $n \in \mathbb{N}$ and for all $s \in I$, we have $u_n(s) \leq u(s)$ for all $n \in \mathbb{N}_0$ so that $\mathcal{C}(I)$ is $\preceq$-regular.

Now we show that $\mathcal{X}$ is F-contraction for some $F \in \mathbb{F}$. Take $u, v \in \mathcal{C}(I)$ such that $u \preceq v$, using (4.2), (4.4) and (4.5), we have

$$\begin{aligned}d(\mathcal{X}u, \mathcal{X}v) = & \sup_{s\in I}|(\mathcal{X}u)(s) - (\mathcal{X}v)(s)| = \sup_{s\in I}((\mathcal{X}v)(s) - (\mathcal{X}u)(s)) \\ \leq & \sup_{s\in I}\int_0^S G(s,t)[f(t,v(t)) + e^{-\tau}v(t) - f(t,u(t)) - e^{-\tau}u(t)]\mathrm{d}t \\ \leq & \sup_{s\in I}\int_0^S G(s,t)e^{-\tau}(v(t) - u(t))\mathrm{d}t \\ = & e^{-\tau}d(u,v)\sup_{s\in I}\int_0^S G(s,t)\mathrm{d}t \\ = & e^{-\tau}d(u,v)\sup_{s\in I}\frac{1}{e^{e^{-\tau S}} - 1}\left(\left[\frac{1}{e^{-\tau}}e^{e^{-\tau(S+t-s)}}\right]_0^s + \left[\frac{1}{e^{-\tau}}e^{e^{-\tau(t-s)}}\right]_s^S\right) \\ = & e^{-\tau}d(u,v)\frac{1}{(e^{e^{-\tau S}} - 1)}(e^{e^{-\tau S}} - 1) \\ = & e^{-\tau}d(u,v) \\ \leq & e^{-\tau}\max\left\{d(u,v), \frac{d(u,\mathcal{X}u) + d(v,\mathcal{X}v)}{2}, \frac{d(u,\mathcal{X}v) + d(v,\mathcal{X}u)}{2}\right\},\end{aligned}$$

for all $u, v \in X$ with $u \preceq v$. Hence, $\mathcal{X}$ is F-weak contraction for τ chosen as in (i) and $F(s) = \ln s$. Thus, all the conditions of Theorem 3.1 are satisfied ensuring the existence of some fixed point of $\mathcal{X}$. Observe that, for arbitrary $u, v \in \mathcal{C}(I)$, $w := \max\{u, v\} \in \mathcal{C}(I)$ is comparable to both u and v. Therefore, by Theorem, 3.4, $\mathcal{X}$ has a unique fixed point which means that problem (4.1) has a unique solution. □

Theorem 4.2 *Theorem* 4.1 *remains true if we replace the existence of the lower solution of* (4.1) *by the existence of an upper solution*

Acknowledgements All the authors are very grateful to all the anonymous referees for their constructive comments and suggestions which significantly have improved the content and the presentation of this paper.

Authors contributions All authors contributed equally and significantly in writing this article. All authors read and approved the final manuscript.

Compliance with ethical standards

Conflict of interest The authors declare that they have no competing interests.

References

1. Wardowski, D.: Fixed points of a new type of contractive mappings in complete metric spaces. Fixed Point Theory Appl. **94**, 1–6 (2012)
2. Wardowski, D., Van Dung, N.: Fixed points of F-weak contractions on complete metric spaces. Demonstr. Math. **47**(1), 146–155 (2014)

3. Mınak, G., Helvacı, A., Altun, I.: Ćirić type generalized f-contractions on completemetric spaces and fixed point results. Filomat **28**(6), 1143–1151 (2014)
4. Gopal, D., Abbas, M., Patel, D.K., Vetro, C.: Fixed points of α-type F-contractive mappings with an application to nonlinear fractional differential equation. Acta Math. Sci. **36**(3), 957–970 (2016)
5. Acar, Ö., Durmaz, G., Minak, G.: Generalized multivalued F-contractions on complete metric spaces. Bull. Iran. Math. Soc. **40**(6), 1469–1478 (2014)
6. Altun, I., Minak, G., Dag, H.: Multivalued F-contractions on complete metric space. J. Nonlinear Convex Anal. **16**(4), 659–666 (2015)
7. Beg, I., Butt, A.R.: Common fixed point and coincidence point of generalized contractions in ordered metric spaces. Fixed Point Theory Appl. **2012**(229), 1–12 (2012)
8. Hussain, N., Salimi, P.: Suzuki–Wardowski type fixed point theorems for α-GF-contractions. Taiwan. J. Math. **18**(6), 1879 (2014)
9. Karapınar, E., Kutbi, M.A., Piri, H., Regan, D.O.: Fixed points of conditionally F-contractions in complete metric-like spaces. Fixed Point Theory Appl. **126**, 1 (2015)
10. Klim, D., Wardowski, D.: Fixed points of dynamic processes of set-valued f-contractions and application to functional equations. Fixed Point Theory Appl. **2015**(1), 22 (2015)
11. Latif, A., Abbas, M., Hussain, A.: Coincidence best proximity point of Fg-weak contractive mappings in partially ordered metric spaces. J. Nonlinear Sci. Appl. **9**(5), 2448–2457 (2016)
12. Padcharoen, A., Gopal, D., Chaipunya, P., Kumam, P.: Fixed point and periodic point results for α-type F-contractions in modular metric spaces. Fixed Point Theory Appl. **2016**(39), 1–12 (2016)
13. Parvaneh, V., Hussain, N., Kadelburg, Z.: Generalized Wardowski type fixed point theorems via α-admissible FG-contractions in b-metric spaces. Acta Math. Sci. **36**(5), 1445–1456 (2016)
14. Piri, H., Kumam, P.: Some fixed point theorems concerning F-contraction in complete metric spaces. Fixed Point Theory Appl. **2014**(210), 1–11 (2014)
15. Piri, H., Kumam, P.: Wardowski type fixed point theorems in complete metric spaces. Fixed Point Theory Appl. **2016**(45), 1–12 (2016)
16. Shukla, S., Radenović, S.: Some common fixed point theorems for F-contraction type mappings in 0-complete partial metric spaces. J. Math. **2013**(878730), 1–7 (2013)
17. Shukla, S., Radenović, S., Kadelburg, Z.: Some fixed point theorems for ordered F-generalized contractions in 0-f-orbitally complete partial metric spaces. Theory Appl. Math. Comput. Sci. **4**(1), 87–98 (2014)
18. Van Dung, N., Le Hang, V.T.: A fixed point theorem for generalized F-contractions on complete metric spaces. Vietnam J. Math. **4**(43), 743–753 (2015)
19. Turinici, M.: Abstract comparison principles and multivariable gronwall-bellman inequalities. J. Math. Anal. Appl. **117**(1), 100–127 (1986)
20. Turinici, M.: Fixed points for monotone iteratively local contractions. Demonstr. Math. **19**(1), 171–180 (1986)
21. Ran, A.C., Reurings, M.C.: "A fixed point theorem in partially ordered sets and some applications to matrix equations". In: Proceedings of the American Mathematical Society, pp. 1435–1443 (2004)
22. Nieto, J.J., Rodríguez-López, R.: Contractive mapping theorems in partially ordered sets and applications to ordinary differential equations. Order **22**(3), 223–239 (2005)
23. Nieto, J.J., Rodríguez-López, R.: Existence and uniqueness of fixed point in partially ordered sets and applications to ordinary differential equations. Acta Math. Sin. Engl. Ser. **23**(12), 2205–2212 (2007)
24. Alam, A., Khan, A.R., Imdad, M.: Some coincidence theorems for generalized nonlinear contractions in ordered metric spaces with applications. Fixed Point Theory Appl. **2014**(216), 1–30 (2014)
25. Ćirić, L.B., Abbas, M., Saadati, R., Hussain, N.: Common fixed points of almost generalized contractive mappings in ordered metric spaces. Appl. Math. Comput. **217**(12), 5784–5789 (2011)
26. Gubran, R., Imdad, M.: Results on coincidence and common fixed points for (ψ, φ) g-generalized weakly contractive mappings in ordered metric spaces. Mathematics **4**(4), 68 (2016)
27. Imdad, M., Gubran, R., Ahmadullah, M.: "Using an implicit function to prove common fixed point theorems". arXiv:1605.05743 (2016) (preprint)
28. Imdad, M., Gubran, R.: Ordered-theoretic fixed point results for monotone generalized Boyd-Wong and Matkowski type contractions. J. Adv. Math. Stud. **10**(1), 49–61 (2017)
29. Kutbi, M.A., Alam, A., Imdad, M.: Sharpening some core theorems of Nieto and Rodríguez-López with application to boundary value problem. Fixed Point Theory and Applications **2015**(198), 1–15 (2015)
30. Nashine, H.K., Altun, I.: A common fixed point theorem on ordered metric spaces. Bull. Iran. Math. Soc. **38**(4), 925–934 (2012)
31. O'Regan, D., Petruşel, A.: Fixed point theorems for generalized contractions in ordered metric spaces. J. Math. Anal. Appl. **341**(2), 1241–1252 (2008)
32. Abbas, M., Ali, B., Romaguera, S.: Fixed and periodic points of generalized contractions in metric spaces. Fixed Point Theory Appl. **2013**(243), 1–11 (2013)
33. Durmaz, G., Mınak, G., Altun, I.: Fixed points of ordered F-contractions. Hacet. J. Math. Stat. **45**(1), 15–21 (2016)
34. Vetro, F.: F-contractions of hardy-rogers type and application to multistage decision processes. Nonlinear Anal. Model. Control **21**(4), 531–546 (2016)
35. Hussain, N., Salimi, P., Vetro, P.: Fixed points for α-ψ-suzuki contractions with applications to integral equations. Carpathian J. Math. **30**(2), 197–207 (2014)

18

On mixed g-monotone and w-compatible mappings in ordered cone b-metric spaces

Zaid Mohammed Fadail[1] · **Abd Ghafur Bin Ahmad**[1] · **Stojan Radenović**[2] · **Miloje Rajović**[3]

Abstract In this paper, we proved common coupled fixed point results for mixed g-monotone and w-compatible mappings in ordered cone b-metric spaces. Our results extend and generalize several well-known comparable results in literature.

Keywords Cone b-metric spaces · Coupled coincidence points · Common coupled fixed points · Mixed g-monotone · w-compatible mappings · Partially ordered set

Introduction

Recently, there have been so many exciting developments in the field of existence of fixed point in partially ordered metric spaces. This trend was started by Ran and Reurings in [1], where they extended the Banach contraction principle in partially ordered sets with some applications to matrix equations. The obtained result in [1] was further extended and refined by many authors (see, for example [2, 3] and the references cited in therein).

The coupled fixed point theorem is an interesting and decisive concept in fixed point theory. In 2006, Bhaskar and Lakshmikantham [4] introduced the notions of the coupled fixed point and mixed monotone property of a given mapping $F : X \times X \longrightarrow X$, where X is a nonempty ordered set. In addition, they proved some coupled fixed point theorems for mappings which satisfy the mixed monotone property and considered some applications in the existence and uniqueness of a solution for a periodic boundary value problem. They also established the classical coupled fixed point theorems and obtained some of their applications.

Inspired by above notions, Lakshmikantham and Ćirić in [2] introduced the concepts of the coupled coincidence point, the common coupled fixed point and g-mixed monotone property for mappings $F : X \times X \longrightarrow X$ and $g : X \longrightarrow X$. In subsequent papers several authors proved various coupled and common coupled fixed point theorems (e.g., [5–14]).

On the other hand, Huang and Zhang [15] reintroduced the notion of cone metric spaces and established fixed point theorems for mappings in such spaces. Next in 2011, Hussain and Shah [16] (see also [17]) generalized the notion of cone metric spaces and introduced cone b-metric spaces. Afterwards, many authors obtained many fixed point, common fixed point and common coupled fixed point theorems in cone b-metric spaces. For some works in cone b-metric spaces, we may refer the reader to [18–29].

In 2012, Nashine et al. [3] established coupled coincidence point results for mixed g-monotone mappings under

✉ Zaid Mohammed Fadail
zaid_fatail@yahoo.com

✉ Stojan Radenović
radens@beotel.net

Abd Ghafur Bin Ahmad
ghafur@ukm.edu.my

Miloje Rajović
rajovic.m@maskv.edu.rs

1 School of Mathematical Sciences, Faculty of Science and Technology, Universiti Kebangsaan Malaysia (UKM), 43600 Bangi, Selangor Darul Ehsan, Malaysia

2 Faculty of Mathematics and Information Technology, Teacher Education, Dong Thap University, Cao Lanh, Dong Thap, Viet Nam

3 Faculty of Mechanical Engineering, Dositejeva19, Kraljevo 36000, Serbia

general contractive conditions in partially ordered cone metric spaces over solid cones. For more results on ordered metric spaces, cone and ordered cone metric spaces, the reader can be traced back to [30–39].

Theorem 1.1 [3] *Let $(X, d, \sqsubseteq)$ be an ordered cone metric space over a solid cone P. Let $F : X^2 \longrightarrow X$ and $g : X \longrightarrow X$ be mappings such that F has the mixed g-monotone property on X and there exist two elements $x_0, y_0 \in X$ with $gx_0 \sqsubseteq F(x_0, y_0)$ and $gy_0 \sqsupseteq F(y_0, x_0)$. Suppose further that F, g satisfy*

$$\begin{aligned} d\big(F(x,y),F(u,v)\big) \preceq\, & a_1 d(gx,gu) + a_2 d\big(F(x,y),gx\big) \\ & + a_3 d(gy,gv) \\ & + a_5 d\big(F(x,y),gu\big) \\ & + a_6 d\big(F(u,v),gx\big), \end{aligned}$$

for all $(x,y),(u,v) \in X \times X$ with ($gu \sqsubseteq gx$ and $gv \sqsupseteq gy$) or ($gx \sqsubseteq gu$ and $gy \sqsupseteq gv$), where $a_i \geq 0$, for i =1,2,...,6 and $\sum_{i=1}^{6} a_i < 1$. Further suppose

1. *$F(X \times X) \subseteq g(X)$;*
2. *g(X) is a complete subspace of X.*

Also, suppose that X has the following properties :

(i) *if a non-decreasing sequence $\{x_n\}$* in X is such that $x_n \longrightarrow x$,then $x_n \sqsubseteq x$ for all $n \in N$,
(ii) if a non-increasing sequence $\{y_n\}$ in X is such that $y_n \longrightarrow y$,then $y_n \sqsupseteq y$ for all $n \in N$.

Then there exist x and y such that $F(x,y) = gx$ and $F(y,x) = gy$, that is, F and g have a coupled coincidence point $(x,y) \in X \times X$.

Theorem 1.2 [3] *In addition to the hypotheses of Theorem 1.1, suppose that for every $(x,y),(y^*,x^*) \in X \times X$ there exists $(u,v) \in X \times X$ such that (F(u, v), F(v, u)) is comparable both to (F(x, y), F(y, x)) and $(F(x^*,y^*),F(y^*,x^*))$. Then F and g have a unique coupled common fixed point, that is, there exists a unique $(u,v) \in X \times X$ such that*

$$u = gu = F(u,v) \quad \text{and} \quad v = gv = F(v,u),$$

provided F and g are w^*-compatible.

In this paper, an essay is made to establish common coupled fixed point results for mixed g-monotone and w-compatible mappings satisfying more general contractive conditions in ordered cone b-metric spaces over a cone that is only solid.

Preliminaries

Let E be a real Banach space and θ denote to the zero element in E. A cone P is a subset of E such that:

1. P is nonempty closed set and $P \neq \{\theta\}$,
2. If a, b are nonnegative real numbers and $x, y \in P$ then $ax + by \in P$,
3. $x \in P$ and $-x \in P$ imply $x = \theta$.

For any cone $P \subset E$, the partial ordering $\preceq$ with respect to P is defined by $x \preceq y$ if and only if $y - x \in P$. The notation of $\prec$ stand for $x \preceq y$ but $x \neq y$. Also, we used $x \ll y$ to indicate that $y - x \in \text{int}P$, where int$P$ denotes the interior of P. A cone P is called normal if there exists a number K such that

$$\theta \preceq x \preceq y \Longrightarrow \|x\| \leq K\|y\|,$$

for all $x, y \in E$. The least positive number K satisfying the above condition is called the normal constant of P. Throughout this paper, we do not impose the normality condition for the cones, but the only assumption is that the cone P is solid, that is int$P \neq \emptyset$.

Definition 2.1 ([16]) Let X be a nonempty set and E be a real Banach space equipped with the partial ordering $\preceq$ with respect to the cone P. A vector-valued function $d : X \times X - \rightarrow E$ is said to be a cone b-metric function on X with the constant $s \geq 1$ if the following conditions are satisfied:

1. $\theta \preceq d(x,y)$ for all $x, y \in X$ and $d(x,y) = \theta$ if and only if $x = y$,
2. $d(x,y) = d(y,x)$ for all $x, y \in X$,
3. $d(x,y) \preceq s(d(x,y) + d(y,z))$ for all $x, y, z \in X$.

Then pairs (X, d) is called a cone b-metric space (or a cone metric type space), we will use the first mentioned term.

Definition 2.2 ([16]) Let (X, d) be a cone b-metric space, $\{x_n\}$ be a sequence in X and $x \in X$.

1. For all $c \in E$ with $\theta \ll c$, if there exists a positive integer N such that $d(x_n, x) \ll c$ for all $n > N$, then x_n is said to be convergent and x is the limit of $\{x_n\}$. We denote this by $x_n \longrightarrow x$.
2. For all $c \in E$ with $\theta \ll c$, if there exists a positive integer N such that $d(x_n, x_m) \ll c$ for all $n, m > N$, then $\{x_n\}$ is called a Cauchy sequence in X.
3. A cone b-metric space (X, d) is called complete if every Cauchy sequence in X is convergent.

The following lemma is helpful to prove our results.

Lemma 2.3 ([40])

1. If *E be a real Banach space with a cone P and $a \preceq \lambda a$ where $a \in P$ and $0 \leq \lambda < 1$, then $a = \theta$.*
2. If *$c \in \text{int}P$, $\theta \preceq a_n$ and $a_n \longrightarrow \theta$, then there exists a positive integer N such that $a_n \ll c$ for all $n \geq N$.*
3. If *$a \preceq b$ and $b \ll c$, then $a \ll c$.*
4. If *$\theta \preceq u \ll c$ for each $\theta \ll c$, then $u = \theta$.*

Recall the following definitions.

Definition 2.4 ([4]) Let $(X, \sqsubseteq)$ be a partially ordered set and let $F : X^2 \longrightarrow X$ be a mapping. The mapping F is said to have mixed monotone property if F is monotone non-decreasing in its first argument and monotone non-increasing in its second argument, that is, for any $x, y \in X$,

$$x_1, x_2 \in X, \quad x_1 \sqsubseteq x_2 \Longrightarrow F(x_1, y) \sqsubseteq F(x_2, y)$$

and

$$y_1, y_2 \in X, \quad y_1 \sqsubseteq y_2 \Longrightarrow F(x, y_1) \sqsupseteq F(x, y_2).$$

Definition 2.5 ([2]) Let $(X, \sqsubseteq)$ be a partially ordered set and let $F : X^2 \longrightarrow X$ and $g : X \longrightarrow X$ be two mappings. The mapping F is said to have mixed g-monotone property if F is monotone g-non-decreasing in its first argument and monotone g-non-increasing in its second argument, that is, for any $x, y \in X$,

$$x_1, x_2 \in X, \quad gx_1 \sqsubseteq gx_2 \Longrightarrow F(x_1, y) \sqsubseteq F(x_2, y)$$

and

$$y_1, y_2 \in X, \quad gy_1 \sqsubseteq gy_2 \Longrightarrow F(x, y_1) \sqsupseteq F(x, y_2).$$

Definition 2.6 ([4]) An element $(x, y) \in X^2$ is said to be a coupled fixed point of the mapping $F : X^2 \longrightarrow X$ if $F(x, y) = x$ and $F(y, x) = y$.

Definition 2.7 ([2]) An element $(x, y) \in X^2$ is called

1. a coupled coincidence point of mappings $F : X^2 \longrightarrow X$ and $g : X \longrightarrow X$ if $gx = F(x, y)$ and $gy = F(y, x)$, and (gx, gy) is called coupled point of coincidence.
2. a common coupled fixed point of mappings $F : X^2 \longrightarrow X$ and $g : X \longrightarrow X$ if $x = gx = F(x, y)$ and $y = gy = F(y, x)$.

Definition 2.8 ([5]) The mappings $F : X^2 \longrightarrow X$ and $g : X \longrightarrow X$ are called:

1. w-compatible if $g(F(x, y)) = F(gx, gy)$ whenever $gx = F(x, y)$ and $gy = F(y, x)$.
2. w^*-compatible if $g(F(x, x)) = F(gx, gx)$ whenever $gx = F(x, x)$.

Coupled coincidence point and common coupled fixed point results

In this section, we prove some coupled coincidence point and common coupled fixed point results in ordered cone b-metric spaces.

Theorem 3.1 *Let $(X, \sqsubseteq)$ (be a partially ordered set and X, d be a cone b-metric space with the coefficient) $s \geq 1$ relative to a solid cone P. Let $F : X^2 \longrightarrow X$ and $g : X \longrightarrow X$ be two mappings such that F has the mixed g-monotone property on X and suppose that there exist nonnegative constants $a_i \in [0, 1), i = 1, 2, \ldots, 10$ with $(s+1)(a_1 + a_2 + a_3 + a_4) + s(s+1)$ $(a_5 + a_6 + a_7 + a_8) + 2s(a_9 + a_{10}) < 2$ and $\sum_{i=1}^{10} a_i < 1$ such that the following contractive condition holds*

$$\begin{aligned} d(F(x,y), F(u,v)) \preceq & \left[a_1 d(gx, F(x,y)) + a_2 d(gy, F(y,x))\right] \\ & + \left[a_3 d(gu, F(u,v)) + a_4 d(gv, F(v,u))\right] \\ & + \left[a_5 d(gx, F(u,v)) + a_6 d(gy, F(v,u))\right] \\ & + \left[a_7 d(gu, F(x,y)) + a_8 d(gv, F(y,x))\right] \\ & + \left[a_9 d(gx, gu) + a_{10} d(gy, gv)\right], \end{aligned}$$

for all $(x, y), (u, v) \in X^2$ with ($gu \sqsubseteq gx$ and $gv \sqsupseteq gy$) or ($gx \sqsubseteq gu$ and $gy \sqsupseteq gv$). Assume that F and g satisfy the following conditions:

1. *$F(X^2) \subseteq g(X)$,*
2. *$g(X)$ is a complete subspace of X.*

Also, suppose that X has the following properties:

(i) if a non-decreasing sequence $\{x_n\}$ in X is such that $x_n \longrightarrow x$, then $x_n \sqsubseteq x$ for all $n \in \mathbb{N}$,

(ii) if a non-increasing sequence $\{y_n\}$ in X is such that $y_n \longrightarrow y$, then $y_n \sqsupseteq y$ for all $n \in \mathbb{N}$.

If there exist $x_0, y_0 \in X$ such that $gx_0 \sqsubseteq F(x_0, y_0)$ and $F(y_0, x_0) \sqsubseteq gy_0$, then F and g have a coupled coincidence point $(x^*, y^*) \in X^2$.

Proof Let $x_0, y_0 \in X$ such that $gx_0 \sqsubseteq F(x_0, y_0)$ and $F(y_0, x_0) \sqsubseteq gy_0$. Since $F(X^2) \subseteq g(X)$ we can Choose $x_1, y_1 \in X$ such that $gx_1 = F(x_0, y_0), gy_1 = F(y_0, x_0)$. Again Since $F(X^2) \subseteq g(X)$ we can Choose $x_2, y_2 \in X$ such that $gx_2 = F(x_1, y_1), gy_2 = F(y_1, x_1)$. Since F has the mixed g-monotone property, we have $gx_0 \sqsubseteq gx_1 \sqsubseteq gx_2$ and $gy_2 \sqsubseteq gy_1 \sqsubseteq gy_0$. Continuing this process, we can construct two sequences $\{x_n\}, \{y_n\}$ in X such that

$$gx_n = F(x_{n-1}, y_{n-1}) \sqsubseteq gx_{n+1} = F(x_n, y_n)$$

and

$$gy_{n+1} = F(y_n, x_n) \sqsubseteq gy_n = F(y_{n-1}, x_{n-1}).$$

Then we have:

$$\begin{aligned} & d(gx_n, gx_{n+1}) \\ & = d(F(x_{n-1}, y_{n-1}), F(x_n, y_n)) \\ & \preceq \left[a_1 d(gx_{n-1}, F(x_{n-1}, y_{n-1}) + a_2 d(gy_{n-1}, F(y_{n-1}, x_{n-1}))\right] \\ & \quad + \left[a_3 d(gx_n, F(x_n, y_n)) + a_4 d(gy_n, F(y_n, x_n))\right] \\ & \quad + \left[a_5 d(gx_{n-1}, F(x_n, y_n)) + a_6 d(gy_{n-1}, F(y_n, x_n))\right] \\ & \quad + \left[a_7 d(gx_n, F(x_{n-1}, y_{n-1})) + a_8 d(gy_n, F(y_{n-1}, x_{n-1}))\right] \\ & \quad + \left[a_9 d(gx_{n-1}, gx_n) + a_{10} d(gy_{n-1}, gy_n)\right]. \end{aligned}$$

So that,

$$\begin{aligned} d(gx_n, gx_{n+1}) &= d(F(x_{n-1}, y_{n-1}), F(x_n, y_n)) \\ &\preceq [a_1 d(gx_{n-1}, gx_n) + a_2 d(gy_{n-1}, gy_n)] \\ &\quad + [a_3 d(gx_n, gx_{n+1}) + a_4 d(gy_n, gy_{n+1})] \\ &\quad + [a_5 d(gx_{n-1}, gx_{n+1}) + a_6 d(gy_{n-1}, gy_{n+1})] \\ &\quad + [a_7 d(gx_n, gx_n) + a_8 d(gy_n, gy_n)] \\ &\quad + [a_9 d(gx_{n-1}, gx_n) + a_{10} d(gy_{n-1}, gy_n)] \\ &\preceq [a_1 d(gx_{n-1}, gx_n) + a_2 d(gy_{n-1}, gy_n)] \\ &\quad + [a_3 d(gx_n, gx_{n+1}) + a_4 d(gy_n, gy_{n+1})] \\ &\quad + [sa_5(d(gx_{n-1}, gx_n) + d(gx_n, gx_{n+1})) \\ &\quad + sa_6(d(gy_{n-1}, gy_n) + d(gy_n, gy_{n+1}))] \\ &\quad + [a_9 d(gx_{n-1}, gx_n) + a_{10} d(gy_{n-1}, gy_n)]. \end{aligned}$$

Hence

$$\begin{aligned} d(gx_n, gx_{n+1}) &\preceq [(a_1 + a_5 s + a_9) d(gx_{n-1}, gx_n) \\ &\quad + (a_2 + a_6 s + a_{10}) d(gy_{n-1}, gy_n)] \\ &\quad + [(a_3 + a_5 s) d(gx_n, gx_{n+1}) \\ &\quad + (a_4 + a_6 s) d(gy_n, gy_{n+1})]. \end{aligned} \tag{3.1}$$

Similarly, we can prove that

$$\begin{aligned} d(gy_n, gy_{n+1}) &\preceq [(a_1 + sa_5 + a_9) d(gy_{n-1}, gy_n) \\ &\quad + (a_2 + sa_6 + a_{10}) d(gx_{n-1}, gx_n)] \\ &\quad + [(a_3 + sa_5) d(gy_n, gy_{n+1}) \\ &\quad + (a_4 + sa_6) d(gx_n, gx_{n+1})]. \end{aligned} \tag{3.2}$$

Put,

$$d_n = d(gx_n, gx_{n+1}) + d(gy_n, gy_{n+1}).$$

Adding inequalities (3.1) and (3.2), one can assert that

$$\begin{aligned} d_n &\preceq (a_1 + a_2 + sa_5 + sa_6 + a_9 + a_{10}) d_{n-1} \\ &\quad + (a_3 + a_4 + sa_5 + sa_6) d_n. \end{aligned} \tag{3.3}$$

On the other hand, we have

$$\begin{aligned} &d(gx_{n+1}, gx_n) = d(F(x_n, y_n), F(x_{n-1}, y_{n-1})) \\ &\preceq [a_1 d(gx_n, F(x_n, y_n)) + a_2 d(gy_n, F(y_n, x_n))] \\ &\quad + [a_3 d(gx_{n-1}, F(x_{n-1}, y_{n-1}) + a_4 d(gy_{n-1}, F(y_{n-1}, x_{n-1}))] \\ &\quad + [a_5 d(gx_n, F(x_{n-1}, y_{n-1})) + a_6 d(gy_n, F(y_{n-1}, x_{n-1}))] \\ &\quad + [a_7 d(gx_{n-1}, F(x_n, y_n)) + a_8 d(gy_{n-1}, F(y_n, x_n))] \\ &\quad + [a_9 d(gx_n, gx_{n-1}) + a_{10} d(gy_n, gy_{n-1})]. \end{aligned}$$

So that,

$$\begin{aligned} d(gx_{n+1}, gx_n) &= d(F(x_n, y_n), F(x_{n-1}, y_{n-1})) \\ &\preceq [a_1 d(gx_n, gx_{n+1}) + a_2 d(gy_n, gy_{n+1})] \\ &\quad + [a_3 d(gx_{n-1}, gx_n) + a_4 d(gy_{n-1}, gy_n)] \\ &\quad + [a_5 d(gx_n, gx_n) + a_6 d(gy_n, gy_n)] \\ &\quad + [a_7 d(gx_{n-1}, gx_{n+1}) + a_8 d(gy_{n-1}, gy_{n+1})] \\ &\quad + [a_9 d(gx_n, gx_{n-1}) + a_{10} d(gy_n, gy_{n-1})] \\ &\preceq [a_1 d(gx_n, gx_{n+1}) + a_2 d(gy_n, gy_{n+1})] \\ &\quad + [a_3 d(gx_{n-1}, gx_n) + a_4 d(gy_{n-1}, gy_n)] \\ &\quad + [sa_7(d(gx_{n-1}, gx_n) + d(gx_n, gx_{n+1})) \\ &\quad + sa_8(d(gy_{n-1}, gy_n) + d(gy_n, gy_{n+1}))] \\ &\quad + [a_9 d(gx_{n-1}, gx_n) + a_{10} d(gy_{n-1}, gy_n)]. \end{aligned}$$

Hence

$$\begin{aligned} d(gx_{n+1}, gx_n) &\preceq [(a_3 + sa_7 + a_9) d(gx_{n-1}, gx_n) \\ &\quad + (a_4 + sa_8 + a_{10}) d(gy_{n-1}, gy_n)] \\ &\quad + [(a_1 + sa_7) d(gx_n, gx_{n+1}) \\ &\quad + (a_2 + sa_8) d(gy_n, gy_{n+1})]. \end{aligned} \tag{3.4}$$

Similarly

$$\begin{aligned} d(gy_{n+1}, gy_n) &\preceq [(a_3 + sa_7 + a_9) d(gy_{n-1}, gy_n) \\ &\quad + (a_4 + sa_8 + a_{10}) d(gx_{n-1}, gx_n)] \\ &\quad + [(a_1 + sa_7) d(gy_n, gy_{n+1}) \\ &\quad + (a_2 + sa_8) d(gx_n, gx_{n+1})]. \end{aligned} \tag{3.5}$$

Adding inequalities (3.4) and (3.5), one can assert that

$$\begin{aligned} d_n &\preceq (a_3 + a_4 + sa_7 + sa_8 + a_9 + a_{10}) d_{n-1} \\ &\quad + (a_1 + a_2 + sa_7 + sa_8) d_n. \end{aligned} \tag{3.6}$$

Finally, from (3.3) and (3.6), we have

$$\begin{aligned} 2d_n &\preceq (a_1 + a_2 + a_3 + a_4 + sa_5 + sa_6 + sa_7 + sa_8 \\ &\quad + 2(a_9 + a_{10})) d_{n-1} + (a_1 + a_2 + a_3 + a_4 + sa_5 \\ &\quad + sa_6 + sa_7 + sa_8) d_n, \end{aligned}$$

that is

$$d_n \preceq h d_{n-1},$$

where $h = \frac{(a_1+a_2+a_3+a_4+sa_5+sa_6+sa_7+sa_8+2(a_9+a_{10}))}{2-(a_1+a_2+a_3+a_4+sa_5+sa_6+sa_7+sa_8)} < \frac{1}{s}$.

Note that, $h = \frac{(a_1+a_2+a_3+a_4+sa_5+sa_6+sa_7+sa_8+2(a_9+a_{10}))}{2-(a_1+a_2+a_3+a_4+sa_5+sa_6+sa_7+sa_8)} < \frac{1}{s}$ equivalently $(s+1)(a_1 + a_2 + a_3 + a_4) + s(s+1)\ (a_5 + a_6 + a_7 + a_8) + 2s(a_9 + a_{10}) < 2$.

Consequently, we have

$$\begin{aligned} d_n &\preceq h d_{n-1} \\ &\preceq h^2 d_{n-2} \\ &\preceq h^3 d_{n-3} \\ &\ \ \cdot \\ &\ \ \cdot \\ &\ \ \cdot \\ &\preceq h^n d_0. \end{aligned} \tag{3.7}$$

Let $m > n \geq 1$. It follows that

$$\begin{aligned} d(gx_n, gx_m) &\preceq sd(gx_n, gx_{n+1}) + s^2 d(gx_{n+1}, gx_{n+2}) \\ &\quad + \cdots + s^{m-n} d(gx_{m-1}, gx_m), \end{aligned}$$

and

$$d(gy_n, gy_m) \preceq sd(gy_n, gy_{n+1}) + s^2 d(gy_{n+1}, gy_{n+2}) + \cdots + s^{m-n} d(gy_{m-1}, gy_m).$$

Now, (3.7) and $sh<1$ imply that

$$\begin{aligned} d(gx_n, gx_m) + d(gy_n, gy_m) &\preceq sd_n + s^2 d_{n+1} + \cdots + s^{m-n} d_{m-1} \\ &\preceq sh^n d_0 + s^2 h^{n+1} d_0 + \cdots + s^{m-n} h^{m-1} d_0 \\ &= sh^n (1 + sh + (sh)^2 + \cdots + (sh)^{m-n-1}) d_0 \\ &\preceq \frac{sh^n}{1-sh} d_0 \quad \to \theta \quad \text{as} \quad n \to \infty. \end{aligned} \tag{3.8}$$

According to Lemma 2.3 (2), and for any $c \in E$ with $c \gg \theta$, there exists $N_0 \in \mathbb{N}$ such that for any $n > N_0$, $\frac{h^n}{1-h} d_0 \ll c$. Furthermore, from (3.8) and for any $m > n > N_0$, Lemma 2.3 (3) shows that

$$d(gx_n, gx_m) + d(gy_n, gy_m) \ll c,$$

which implies that

$$d(gx_n, gx_m) \ll c,$$

and

$$d(gy_n, gy_m) \ll c.$$

Hence, by Definition 2.2 (2), $\{gx_n\}$ and $\{gy_n\}$ are Cauchy sequences in $g(X)$. Since $g(X)$ is complete, there exists x^* and $y^* \in X$ such that $gx_n \longrightarrow gx^*$ and $gy_n \longrightarrow gy^*$ as $n \longrightarrow \infty$. Since $\{gx_n\}$is nondecreasing and $\{gy_n\}$ is non-increasing, using the properties (i), (ii) of X, we have

$$gx_n \sqsubseteq gx^* \quad and \quad gy^* \sqsubseteq gy_n.$$

Now, we can apply the contractive condition

$$\begin{aligned} d(F(x^*,y^*), gx^*) &\preceq s(d(F(x^*,y^*), gx_{n+1}) + d(gx_{n+1}, gx^*)) \\ &= s(d(F(x^*,y^*), F(x_n,y_n)) + d(gx_{n+1}, gx^*)) \\ &\preceq s\big[a_1 d(gx^*, F(x^*,y^*)) + a_2 d(gy^*, F(y^*,x^*))\big] \\ &\quad + s\big[a_3 d(gx_n, F(x_n,y_n)) + a_4 d(gy_n, F(y_n,x_n))\big] \\ &\quad + s\big[a_5 d(gx^*, F(x_n,y_n)) + a_6 d(gy^*, F(y_n,x_n))\big] \\ &\quad + s\big[a_7 d(gx_n, F(x^*,y^*)) + a_8 d(gy_n, F(y^*,x^*))\big] \\ &\quad + s\big[a_9 d(gx^*, gx_n) + a_{10} d(gy^*, gy_n)\big] + sd(gx_{n+1}, gx^*) \\ &\preceq s\big[a_1 d(F(x^*,y^*), gx^*) + a_2 d(F(y^*,x^*), gy^*)\big] \\ &\quad + s\big[sa_3 d(gx_n, gx^*) + sa_3 d(gx^*, gx_{n+1}) \\ &\quad + sa_4 d(gy_n, gy^*) + sa_4 d(gy^*, gy_{n+1})\big] \\ &\quad + s\big[a_5 d(gx^*, gx_{n+1}) + a_6 d(gy^*, gy_{n+1})\big] \\ &\quad + s\big[sa_7 d(gx_n, gx^*) + sa_7 d(gx^*, F(x^*,y^*)) \\ &\quad + sa_8 d(gy_n, gy^*) + sa_8 d(gy^*, F(y^*,x^*))\big] \\ &\quad + s\big[a_9 d(gx^*, gx_n) + a_{10} d(gy^*, gy_n)\big] + sd(gx_{n+1}, gx^*) \\ &= s\big[a_1 d(F(x^*,y^*), gx^*) + a_2 d(F(y^*,x^*), gy^*)\big] \\ &\quad + s\big[sa_3 d(gx_n, gx^*) + sa_3 d(gx_{n+1}, gx^*) \\ &\quad + sa_4 d(gy_n, gy^*) + sa_4 d(gy_{n+1}, gy^*)\big] \\ &\quad + s\big[a_5 d(gx_{n+1}, gx^*) + a_6 d(gy_{n+1}, gy^*)\big] \\ &\quad + s\big[sa_7 d(gx_n, gx^*) + sa_7 d(F(x^*,y^*), gx^*) \\ &\quad + sa_8 d(gy_n, gy^*) + sa_8 d(F(y^*,x^*), gy^*)\big] \\ &\quad + s\big[a_9 d(gx_n, gx^*) + a_{10} d(gy_n, gy^*)\big] + sd(gx_{n+1}, gx^*). \end{aligned}$$

Hence,

$$\begin{aligned} d(F(x^*,y^*), gx^*) \preceq\ & (sa_1 + s^2 a_7) d(F(x^*,y^*), gx^*) \\ &+ (a_2 + s^2 a_8) d(F(y^*,x^*), gy^*) \\ &+ (s^2 a_3 + s^2 a_7 + sa_9) d(gx_n, gx^*) \\ &+ (s^2 a_3 + sa_5 + s) d(gx_{n+1}, gx^*) \\ &+ (s^2 a_4 + s^2 a_8 + sa_{10}) d(gy_n, gy^*) \\ &+ (s^2 a_4 + sa_6) d(gy_{n+1}, gy^*). \end{aligned}$$

Similarly

$$\begin{aligned} d(F(y^*,x^*), gy^*) \preceq\ & (sa_1 + s^2 a_7) d(F(y^*,x^*), gy^*) \\ &+ (a_2 + s^2 a_8) d(F(x^*,y^*), gx^*) \\ &+ (s^2 a_3 + s^2 a_7 + sa_9) d(gy_n, gy^*) \\ &+ (s^2 a_3 + sa_5 + s) d(gy_{n+1}, gy^*) \\ &+ (s^2 a_4 + s^2 a_8 + sa_{10}) d(gx_n, gx^*) \\ &+ (s^2 a_4 + sa_6) d(gx_{n+1}, gx^*). \end{aligned}$$

Put

$$\tau = d(F(x^*,y^*), gx^*) + d(F(y^*,x^*), gy^*).$$

Adding above inequalities, we get

$$\begin{aligned} \tau \preceq\ & (sa_1 + s^2 a_7 + a_2 + s^2 a_8)\tau \\ &+ (s^2 a_3 + s^2 a_7 + sa_9 + s^2 a_4 + s^2 a_8 + sa_{10}) d(gx_n, gx^*) \\ &+ (s^2 a_3 + s^2 a_7 + sa_9 + s^2 a_4 + s^2 a_8 + sa_{10}) d(gy_n, gy^*) \\ &+ (s^2 a_3 + sa_5 + s + s^2 a_4 + sa_6) d(gx_{n+1}, gx^*) \\ &+ (s^2 a_3 + sa_5 + s + s^2 a_4 + sa_6) d(gy_{n+1}, gy^*). \end{aligned}$$

Then,

$$\begin{aligned} \tau \preceq\ & \frac{A_2}{1-A_1} d(gx_n, gx^*) + \frac{A_2}{1-A_1} d(gy_n, gy^*) \\ &+ \frac{A_3}{1-A_1} d(gx_{n+1}, gx^*) + \frac{A_3}{1-A_1} d(gy_{n+1}, gy^*), \end{aligned}$$

where $A_1 = sa_1 + s^2 a_7 + a_2 + s^2 a_8$, $A_2 = s^2 a_3 + s^2 a_7 + sa_9 + s^2 a_4 + s^2 a_8 + sa_{10}$ and $A_3 = s^2 a_3 + sa_5 + s + s^2 a_4 + sa_6$. Since $gx_n \longrightarrow gx^*$ and $gy_n \longrightarrow gy^*$ as $n \longrightarrow \infty$, then by Definition 2.2 (1) and for $c \gg \theta$ there exists $N_0 \in \mathbb{N}$ such that for all $n > N_0$, $d(gx_n, gx^*) \ll c\frac{1-A_1}{4A_2}$, $d(gy_n, gy^*) \ll c\frac{1-A_1}{4A_2}$, $d(gx_{n+1}, gx^*) \ll c\frac{1-A_1}{4A_3}$ and $d(gy_{n+1}, gy^*) \ll c\frac{1-A_1}{4A_3}$. Hence,

$$\begin{aligned}\tau &\preceq \frac{A_2}{1-A_1}d(gx_n, gx^*) + \frac{A_2}{1-A_1}d(gy_n, gy^*)\\ &+ \frac{A_3}{1-A_1}d(gx_{n+1}, gx^*) + \frac{A_3}{1-A_1}d(gy_{n+1}, gy^*)\\ &\ll c\frac{1-A_1}{4A_2}\frac{A_2}{1-A_1} + c\frac{1-A_1}{4A_2}\frac{A_2}{1-A_1} + c\frac{1-A_1}{4A_3}\frac{A_3}{1-A_1}\\ &+ c\frac{1-A_1}{4A_3}\frac{A_3}{1-A_1} = c.\end{aligned}$$

Now, according to Lemma 2.3 (4) it follows that $\tau = \theta$, that is, $d(F(x^*, y^*), gx^*) + d(F(y^*, x^*), gy^*) = \theta$, which implies that $d(F(x^*, y^*), gx^*) = \theta$ and $d(F(y^*, x^*), gy^*) = \theta$. Hence, $gx^* = F(x^*, y^*)$ and $gy^* = F(y^*, x^*)$. Therefore (x^*, y^*) is a coupled coincidence point of F and g. □

From Theorem 3.1, we have the following corollaries.

Corollary 3.2 Let $(X, \sqsubseteq)$ *(be a partially ordered set and X, d be a cone b-metric space with the coefficient)* $s \geq 1$ *relative to a solid cone P. Let* $F : X^2 \longrightarrow X$ *and* $g : X \longrightarrow X$ *be two mappings and suppose that there exist nonnegative constants* $k, l \in (0, 1]$ *with* $k + l < \frac{1}{s}$ *such that the following contractive condition holds for all* $x, y, u, v \in X$:

$$d\big(F(x, y), F(u, v)\big) \preceq kd(gx, gu) + ld(gy, gv),$$

for all $(x, y), (u, v) \in X^2$ *with* ($gu \sqsubseteq gx$ *and* $gv \sqsupseteq gy$) *or* ($gx \sqsubseteq gu$ *and* $gy \sqsupseteq gv$). *Assume that F and g satisfy the following conditions*:

1. $F(X^2) \subseteq g(X)$,
2. *g(X) is a complete subspace of X.*

Also, suppose that X has the following properties:

(i) if a non-decreasing sequence $\{x_n\}$ in X is such that $x_n \longrightarrow x$, then $x_n \sqsubseteq x$ for all $n \in \mathbb{N}$,
(ii) if a non-increasing sequence $\{y_n\}$ in X is such that $y_n \longrightarrow y$, then $y_n \sqsupseteq y$ for all $n \in \mathbb{N}$.

If there exist $x_0, y_0 \in X$ such that $gx_0 \sqsubseteq F(x_0, y_0)$ and $F(y_0, x_0) \sqsubseteq gy_0$, then F and g have a coupled coincidence point $(x^*, y^*) \in X^2$.

Corollary 3.3 *Let* $(X, \sqsubseteq)$ *be a partially ordered set and (X, d) be a cone b-metric space with the coefficient* $s \geq 1$ *relative to a solid cone P. Let* $F : X^2 \longrightarrow X$ *and* $g : X \longrightarrow X$ *be two mappings and suppose that there exist nonnegative constants* $k, l \in (0, 1]$ *with* $k + l < \frac{2}{s+1}$ *such that the following contractive condition holds for all* $x, y, u, v \in X$:

$$d\big(F(x, y), F(u, v)\big) \preceq kd(gx, F(x, y)) + ld(gu, F(u, v)),$$

for all $(x, y), (u, v) \in X^2$ *with* ($gu \sqsubseteq gx$ *and* $gv \sqsupseteq gy$) *or* ($gx \sqsubseteq gu$ *and* $gy \sqsupseteq gv$). *Assume that F and g satisfy the following conditions*:

1. $F(X^2) \subseteq g(X)$,
2. *g(X) is a complete subspace of X.*

Also, suppose that X has the following properties:

(i) if a non-decreasing sequence $\{x_n\}$ in X is such that $x_n \longrightarrow x$, then $x_n \sqsubseteq x$ for all $n \in \mathbb{N}$,
(ii) if a non-increasing sequence $\{y_n\}$ in X is such that $y_n \longrightarrow y$, then $y_n \sqsupseteq y$ for all $n \in \mathbb{N}$.

If there exist $x_0, y_0 \in X$ such that $gx_0 \sqsubseteq F(x_0, y_0)$ and $F(y_0, x_0) \sqsubseteq gy_0$, then F and g have a coupled coincidence point $(x^*, y^*) \in X^2$.

Corollary 3.4 *Let (X, d) be a cone b-metric space with the coefficient* $s \geq 1$ *relative to a solid cone P. Let* $F : X^2 \longrightarrow X$ *and* $g : X \longrightarrow X$ *be two mappings and suppose that there exist nonnegative constants* $k, l \in (0, 1]$ *with* $k + l < \frac{2}{s(s+1)}$ *such that the following contractive condition holds for all* $x, y, u, v \in X$:

$$d\big(F(x, y), F(u, v)\big) \preceq kd\big(gx, F(u, v)\big) + ld\big(gu, F(x, y)\big),$$

for all $(x, y), (u, v) \in X^2$ *with* ($gu \sqsubseteq gx$ *and* $gv \sqsupseteq gy$) *or* ($gx \sqsubseteq gu$ *and* $gy \sqsupseteq gv$). *Assume that F and g satisfy the following conditions*:

1. $F(X^2) \subseteq g(X)$,
2. *g(X) is a complete subspace of X.*

Also, suppose that X has the following properties:

(i) if a non-decreasing sequence $\{x_n\}$ in X is such that $x_n \longrightarrow x$, then $x_n \sqsubseteq x$ for all $n \in \mathbb{N}$,
(ii) if a non-increasing sequence $\{y_n\}$ in X is such that $y_n \longrightarrow y$, then $y_n \sqsupseteq y$ for all $n \in \mathbb{N}$.

If there exist $x_0, y_0 \in X$ such that $gx_0 \sqsubseteq F(x_0, y_0)$ and $F(y_0, x_0) \sqsubseteq gy_0$, then F and g have a coupled coincidence point $(x^*, y^*) \in X^2$.

Now we prove the existence and uniqueness of a common coupled fixed point. Note that, if $(X, \sqsubseteq)$ is a partially ordered set, then we endow the product space $X \times X$ with the following partial order: for $(x, y), (u, v) \in X \times X$, $(u, v) \sqsubseteq (x, y) \Longleftrightarrow x \sqsupseteq u, y \sqsubseteq v$.

Theorem 3.5 *In addition to the hypotheses of Theorem 3.1, suppose that for every* $(x, y), (x^*, y^*) \in X \times X$ *there exists* $(u, v) \in X \times X$ *such that (F(u, v), F(v, u)) is comparable both to (F(x, y), F(y, x)) and* $(F(x^*, y^*), F(y^*, x^*))$. *Assume that,* $s(a_1 + a_2 + a_3 + a_4) + (a_5 + a_6 + a_7 + a_8 + a_9 + a_{10}) < 1$. *If F and g are w-compatible, then F and g have a unique common coupled fixed point. Moreover, a* *common coupled fixed point of F and g is of the form (u, u) for some $u \in X$.*

Proof From Theorem 3.1, F and g have a coupled coincidence point. Suppose (x, y) and (x^*, y^*) are coupled coincidence points of F and g, that is $gx = F(x,y), gy = F(y,x)$ and $gx^* = F(x^*,y^*), gy^* = F(y^*,x^*)$. First, we will show that

$$gx = gx^* \quad and \quad gy = gy^*.$$

By assumption, there exists $(u,v) \in X \times X$ such that $(F(u, v), F(v, u))$ is comparable both to $(F(x, y), F(y, x))$ and $(F(x^*,y^*), F(y^*,x^*))$. Put $u_0 = u, v_0 = v$ and choose $u_1, v_1 \in X$ such that $gu_1 = F(u_0, v_0)$ and $gv_1 = F(v_0, u_0)$. Continuing this process, we can construct two sequences $\{gu_n\}$ and $\{gv_n\}$ such that

$$gu_{n+1} = F(u_n, v_n) \quad and \quad gv_{n+1} = F(v_n, u_n).$$

Also, set $x_0 = x, y = y_0, x_0^* = x^*, y_0^* = y^*$. Define the sequences $\{gx_n\}, \{gy_n\}$ and $\{gx_n^*\}, \{gy_n^*\}$. Since (x, y) and (x^*, y^*) are coupled coincidence points of F and g, we have as $n \longrightarrow \infty$:

$$gx_n \longrightarrow F(x,y), \quad gy_n \longrightarrow F(y,x),$$

and

$$gx_n^* \longrightarrow F(x^*,y^*), \quad gy_n^* \longrightarrow F(y^*,x^*).$$

Since

$$\big(F(x,y), F(y,x)\big) = (gx, gy)$$

and

$$\big(F(u,v), F(v,u)\big) = (gu_1, gv_1)$$

are comparable, then $gx \sqsubseteq gu_1$ and $gy \sqsupseteq gv_1$. Similarly, we can show that (gx, gy) and (gu_n, gv_n) are comparable for all $n \geq 1$, that is, $gx \sqsubseteq gu_n$ and $gy \sqsupseteq gv_n$. Now, we can apply the contractive condition:

$$\begin{aligned}
d(gu_{n+1}, gx) &= d\big(F(u_n, v_n), F(x,y)\big) \\
&\preceq \big[a_1 d(gu_n, F(u_n, v_n)) + a_2 d(gv_n, F(v_n, u_n))\big] \\
&\quad + \big[a_3 d(gx, F(x,y) + a_4 d(gy, F(y,x))\big] \\
&\quad + \big[a_5 d(gu_n, F(x,y)) + a_6 d(gv_n, F(y,x))\big] \\
&\quad + \big[a_7 d(gx, F(u_n, v_n)) + a_8 d(gy, F(v_n, u_n))\big] \\
&\quad + \big[a_9 d(gu_n, gx) + a_{10} d(gv_n, gy)\big] \\
&= \big[a_1 d(gu_n, gu_{n+1}) + a_2 d(gv_n, gv_{n+1})\big] \\
&\quad + \big[a_3 d(gx, gx) + a_4 d(gy, gy)\big] \\
&\quad + \big[a_5 d(gu_n, gx) + a_6 d(gv_n, gy)\big] \\
&\quad + \big[a_7 d(gx, gu_{n+1}) + a_8 d(gy, gv_{n+1})\big] \\
&\quad + \big[a_9 d(gu_n, gx) + a_{10} d(gv_n, gy)\big] \\
&\preceq \big[s a_1 d(gu_n, gx) + s a_1 d(gx, gu_{n+1}) \\
&\quad + s a_2 d(gv_n, gy) + s a_2 d(gy, gv_{n+1})\big] \\
&\quad + \big[a_5 d(gu_n, gx) + a_6 d(gv_n, gy)\big] \\
&\quad + \big[a_7 d(gx, gu_{n+1}) + a_8 d(gy, gv_{n+1})\big] \\
&\quad + \big[a_9 d(gu_n, gx) + a_{10} d(gv_n, gy)\big].
\end{aligned}$$

Hence

$$\begin{aligned}
d(gu_{n+1}, gx) \preceq &\big[(s a_1 + a_5 + a_9) d(gu_n, gx) \\
&+ (s a_1 + a_7) d(gu_{n+1}, gx)\big] \\
&+ \big[(s a_2 + a_6 + a_{10}) d(gv_n, gy) \\
&+ (s a_2 + a_8) d(gv_{n+1}, gy)\big].
\end{aligned} \tag{3.9}$$

By similar way, we have

$$\begin{aligned}
d(gv_{n+1}, gy) \preceq &\big[(s a_1 + a_5 + a_9) d(gv_n, gy) \\
&+ (s a_1 + a_7) d(gv_{n+1}, gy)\big] \\
&+ \big[(s a_2 + a_6 + a_{10}) d(gu_n, gx) \\
&+ (s a_2 + a_8) d(gu_{n+1}, gx)\big].
\end{aligned} \tag{3.10}$$

Put, $\tau_n = d(gu_{n+1}, gx) + d(gv_{n+1}, gy)$. Adding above inequalities, we get

$$\begin{aligned}
\tau_n \preceq &\big[(s a_1 + a_5 + a_9 + s a_2 + a_6 + a_{10}) \tau_{n-1}\big] \\
&+ \big[(s a_1 + a_7 + s a_2 + a_8) \tau_n\big].
\end{aligned} \tag{3.11}$$

On the other hand starting by gx, we have:

$$\begin{aligned}
d(gx, gu_{n+1}) &= d\big(F(x,y), F(u_n, v_n)\big) \\
&\preceq \big[a_1 d(gx, F(x,y) + a_2 d(gy, F(y,x))\big] \\
&\quad + \big[a_3 d(gu_n, F(u_n, v_n)) + a_4 d(gv_n, F(v_n, u_n))\big] \\
&\quad + \big[a_5 d(gx, F(u_n, v_n)) + a_6 d(gy, F(v_n, u_n))\big] \\
&\quad + \big[a_7 d(gu_n, F(x,y)) + a_8 d(gv_n, F(y,x))\big] \\
&\quad + \big[a_9 d(gx, gu_n) + a_{10} d(gy, gv_n)\big] \\
&= \big[a_1 d(gx, gx) + a_2 d(gy, gy)\big] \\
&\quad + \big[a_3 d(gu_n, gu_{n+1}) + a_4 d(gv_n, gv_{n+1})\big] \\
&\quad + \big[a_5 d(gx, gu_{n+1}) + a_6 d(gy, gv_{n+1})\big] \\
&\quad + \big[a_7 d(gu_n, gx) + a_8 d(gv_n, gy)\big] \\
&\quad + \big[a_9 d(gx, gu_n) + a_{10} d(gy, gv_n)\big] \\
&\preceq \big[s a_3 d(gu_n, gx) + s a_3 d(gx, gu_{n+1}) \\
&\quad + s a_4 d(gv_n, gy) + s a_4 d(gy, gv_{n+1})\big] \\
&\quad + \big[a_5 d(gx, gu_{n+1}) + a_6 d(gy, gv_{n+1})\big] \\
&\quad + \big[a_7 d(gu_n, gx) + a_8 d(gv_n, gy)\big] \\
&\quad + \big[a_9 d(gx, gu_n) + a_{10} d(gy, gv_n)\big]
\end{aligned}$$

Hence

$$\begin{aligned}
d(gx, gu_{n+1}) \preceq &\big[(s a_3 + a_7 + a_9) d(gu_n, gx) \\
&+ (s a_3 + a_5) d(gu_{n+1}, gx)\big] + \big[(s a_4 + a_8 + a_{10}) d(gv_n, gy) \\
&+ (s a_4 + a_6) d(gv_{n+1}, gy)\big].
\end{aligned} \tag{3.12}$$

By similar way, we have

$$\begin{aligned} d(gy, gv_{n+1}) \preceq & \big[(sa_3 + a_7 + a_9)d(gv_n, gy) \\ & + (sa_3 + a_5)d(gv_{n+1}, gy)\big] \\ & + \big[(sa_4 + a_8 + a_{10})d(gu_n, gx) \\ & + (sa_4 + a_6)d(gu_{n+1}, gx)\big]. \end{aligned} \quad (3.13)$$

Adding above two inequalities, we get

$$\begin{aligned} \tau_n \preceq & \big[(sa_3 + a_7 + a_9 + sa_4 + a_8 + a_{10})\tau_{n-1}\big] \\ & + \big[(sa_3 + a_5 + sa_4 + a_6)\tau_n\big]. \end{aligned} \quad (3.14)$$

Adding inequalities (3.11) and (3.14), we have

$$\begin{aligned} 2\tau_n \preceq & \big[(sa_1 + sa_2 + sa_3 + sa_4) + (a_5 + a_6 + a_7 + a_8) \\ & + 2(a_9 + a_{10})\big]\tau_{n-1} + \big[(sa_1 + sa_2 + sa_3 + sa_4) \\ & + (a_5 + a_6 + a_7 + a_8)\big]\tau_n, \end{aligned}$$

which implies that

$$\tau_n \preceq k\tau_{n-1},$$

where $k = \frac{(sa_1+sa_2+sa_3+sa_4)+(a_5+a_6+a_7+a_8)+2(a_9+a_{10})}{2-\left[(sa_1+sa_2+sa_3+sa_4)+(a_5+a_6+a_7+a_8)\right]} < 1.$

Note that, $\frac{(sa_1+sa_2+sa_3+sa_4)+(a_5+a_6+a_7+a_8)+2(a_9+a_{10})}{2-\left[(sa_1+sa_2+sa_3+sa_4)+(a_5+a_6+a_7+a_8)\right]} < 1$ equivalently $s(a_1 + a_2 + a_3 + a_4) + (a_5 + a_6 + a_7 + a_8 + a_9 + a_{10}) < 1$.

Now, we have

$$\begin{aligned} \tau_n & \preceq k\tau_{n-1} \\ & \preceq k^2\tau_{n-2} \\ & \vdots \\ & \preceq k^n\tau_0 \quad \longrightarrow \theta \quad \text{as} \quad n \longrightarrow \infty. \end{aligned} \quad (3.15)$$

According to Lemma 2.3 (2), and for any $c \in E$ with $c \gg \theta$, there exists $N_0 \in \mathbb{N}$ such that for any $n > N_0$, $k^n \ll c$. Furthermore, from (3.15) and for any $n > N_0$, Lemma 2.3 (3) shows that

$$d(gu_{n+1}, gx) + d(gv_{n+1}, gy) \ll c,$$

which implies that

$$d(gu_{n+1}, gx) \ll c,$$

and

$$d(gv_{n+1}, gy) \ll c.$$

Hence, by Definition 2.2 (1), $gu_n \longrightarrow gx$ and $gv_n \longrightarrow gy$. By the same way, we can prove that $gu_n \longrightarrow gx^*$ and $gv_n \longrightarrow gy^*$. The uniqueness of the limit implies that $gx = gx^*$ and $gy = gy^*$. That is, the unique coupled point of coincidence of F and g is (gx, gy).

Clearly that if (gx, gy) is a coupled point of coincidence of F and g, then (gy, gx) is also a coupled points of coincidence of F and g. Then $gx = gy$ and therefore (gx, gx) is the unique coupled point of coincidence of F and g.

Now, let $u = gx = F(x,y)$. Since F and g are w-compatible, then we have

$$gu = g(gx) = gF(x,y) = F(gx, gy) = F(gx, gx) = F(u,u).$$

Then (gu, gu) is a coupled point of coincidence and also we have (u, u) is a coupled point of coincidence. The uniqueness of the coupled point of coincidence implies that $gu = u$. Therefore $u = gu = F(u,u)$. Hence (u, u) is the unique common coupled fixed point of F and g. This completes the proof. □

From Theorem 3.5, we have the following corollaries.

Corollary 3.6 *In addition to the hypotheses of Corollary 3.2, suppose that for every* $(x,y), (x^*,y^*) \in X \times X$ *there exists* $(u,v) \in X \times X$ *such that* $(F(u, v), F(v, u))$ *is comparable both to* $(F(x, y), F(y, x))$ *and* $(F(x^*,y^*), F(y^*,x^*))$. *If F and g are w-compatible, then F and g have a unique common coupled fixed point. Moreover, a common coupled fixed point of F and g is of the form (u, u) for some* $u \in X$.

Corollary 3.7 *In addition to the hypotheses of Corollary 3.3. Suppose that for every* $(x,y), (x^*,y^*) \in X \times X$ *there exists* $(u,v) \in X \times X$ *such that* $(F(u, v), F(v, u))$ *is comparable both to* $(F(x, y), F(y, x))$ *and* $(F(x^*,y^*), F(y^*,x^*))$. *Assume that,* $k + l < \frac{1}{s}$. *If F and g are w-compatible, then F and g have a unique common coupled fixed point. Moreover, a common coupled fixed point of F and g is of the form (u, u) for some* $u \in X$.

Corollary 3.8 *In addition to the hypotheses of Corollary 3.4. Suppose that for every* $(x,y), (x^*,y^*) \in X \times X$ *there exists* $(u,v) \in X \times X$ *such that* $(F(u, v), F(v, u))$ *is comparable both to* $(F(x, y), F(y, x))$ *and* $(F(x^*,y^*), F(y^*,x^*))$. *If F and g are w-compatible, then F and g have a unique common coupled fixed point. Moreover, a common coupled fixed point of F and g is of the form (u, u) for some* $u \in X$.

Remark 3.9 Theorems 3.1 and 3.5 extend and generalize Theorems 3.1 and 3.2 of Nashine et al. [3] to cone *b*-metric spaces, respectively.

Now, we present one example to illustrate our results.

Example 3.10 Let $X = \mathbb{R}$ be ordered by the following relation:

$$x \sqsubseteq y \iff x = y \text{ or } (x, y \in [0,1] \text{ and } x \le y).$$

Let $E = C^1_{\mathbb{R}}[0,1]$ with $\|u\| = \|u\|_\infty + \|u'\|_\infty, u \in E$ and suppose that, $P = \{u \in E : u(t) \ge 0 \text{on} [0,1]\}$. It is well known that this cone is solid, but it is not normal. Define a cone b-metric $d : X \times X \to E$ by $d(x,y)(t) = |x - y|^2 e^t$. Then (X, d) is a complete cone *b*-metric space with the coefficient $s=2$. Let $F: X\times X\to X$ $F(x,y)=\frac{x-y}{60}$ and define $g: X\to X$ by:

$$g(x)=\begin{cases}\frac{1}{2}x, & \text{if } x<0,\\ x, & \text{if } x\in[0,1],\\ \frac{1}{2}x+\frac{1}{3}, & \text{if } x>1.\end{cases}$$

We will check that conditions of theorems 3.1 and 3.5 are fulfilled for all $x,y,u,v\in X$ with ($gu\sqsubseteq gx$ and $gv\sqsupseteq gy$) or ($gx\sqsubseteq gu$ and $gy\sqsupseteq gv$). The following cases are possible.

case 1: $x,y,u,v\in[0,1]$. We have

$$\begin{aligned}d\big(F(x,y),F(u,v)\big)(t)&=\left|\frac{x-y}{60}-\frac{u-v}{60}\right|^2 e^t\\ &\preceq 2\left(\frac{1}{60}|x-u|^2e^t+\frac{1}{60}|y-v|^2e^t\right)\\ &=\frac{1}{30}|x-u|^2e^t+\frac{1}{30}|y-v|^2e^t\\ &=a_9d(gx,gu)(t)+a_{10}d(gy,gv)(t),\end{aligned}$$

where $a_9=\frac{1}{30}=a_{10}$ and $a_i=0, i=1,2,\ldots,8$.

case 2: $x,y\in[0,1]$ and u, v not in [0, 1]. Then gy, gv not in [0, 1] and since they must be comparable, $gy=gv$ and $y=v$. Then we have:

$$\begin{aligned}d\big(F(x,y),F(u,v)\big)(t)&=\left|\frac{x-u}{60}\right|^2 e^t\\ &\preceq a_9d(gx,gu)(t)+a_{10}d(gy,gv)(t),\end{aligned}$$

where $a_9=\frac{1}{30}=a_{10}$ and $a_i=0, i=1,2,\ldots,8$.

case 3: $u,v\in[0,1]$ and x, y not in [0, 1]. This case will be similar to case 2.

case 4: If x, y, u, v not in [0,1] then the only possibility for gx and gu, as well as gy and gv to be comparable is that $x=u$ and $y=v$. In this case conditions of Theorem 3.1 are trivially satisfied.

Note that, $2s(a_9+a_{10})=4(\frac{1}{30}+\frac{1}{30})<2$, $F(X\times X)\subseteq g(X)$ and $g(X)$ is a complete subspace of X. Also, F has the mixed g-monotone property. Hence, the conditions of Theorem 3.1 are satisfied, that is, F and g have a coupled coincidence point (0, 0). Also, F and g are w-compatible at (0, 0) and $a_9+a_{10}<1$. Hence, Theorem 3.5 shows that, (0, 0) is the unique common coupled fixed point of F and g.

Finally, we have the following result (immediate consequence of Theorems 3.1 and 3.5).

Theorem 3.11 *Let $(X,\sqsubseteq)$ be a partially ordered set and (X, d) be a complete cone b-metric space with the coefficient $s\geq 1$ relative to a solid cone P. Let $F: X^2\longrightarrow X$ be a mapping having the mixed monotone property on X and suppose that there exist nonnegative constants $a_i\in[0,1), i=1,2,\ldots,10$ with $\sum_{i=1}^{10}a_i<1$ such that the following contractive condition holds*

$$\begin{aligned}d(F(x,y),F(u,v))\preceq&\ [a_1d(x,F(x,y))+a_2d(y,F(y,x))]\\ &+[a_3d(u,F(u,v))+a_4d(v,F(v,u))]\\ &+[a_5d(x,F(u,v))+a_6d(y,F(v,u))]\\ &+[a_7d(u,F(x,y))+a_8d(v,F(y,x))]\\ &+[a_9d(x,u)+a_{10}d(y,v)],\end{aligned}$$

for all $(x,y),(u,v)\in X^2$ with ($u\sqsubseteq x$ and $v\sqsupseteq y$) or ($x\sqsubseteq u$ and $y\sqsupseteq v$) such that:

(A) $(s+1)(a_1+a_2+a_3+a_4)+s(s+1)(a_5+a_6+a_7+a_8)+2s(a_9+a_{10})<2$

(B) $s(a_1+a_2+a_3+a_4)+(a_5+a_6+a_7+a_8+a_9+a_{10})<1$.

Suppose that X has the following properties:

(i) if a non-decreasing sequence $\{x_n\}$ in X is such that $x_n\longrightarrow x$, then $x_n\sqsubseteq x$ for all $n\in\mathbb{N}$,

(ii) if a non-increasing sequence $\{y_n\}$ in X is such that $y_n\longrightarrow y$, then $y_n\sqsupseteq y$ for all $n\in\mathbb{N}$.

If there exist $x_0,y_0\in X$ such that $x_0\sqsubseteq F(x_0,y_0)$ and $F(y_0,x_0)\sqsubseteq y_0$, then F has a coupled fixed point $(x^*,y^*)\in X^2$. Moreover, the coupled fixed point is unique and of the form (x^*,x^*) for some $x^*\in X$.

The following corollaries can be obtained from Theorem 3.11.

Corollary 3.12 *Let $(X,\sqsubseteq)$ be a partially ordered set and (X, d) be a complete cone b-metric space with the coefficient $s\geq 1$ relative to a solid cone P. Let $F: X^2\longrightarrow X$ be a mapping having the mixed monotone property on X and suppose that there exist nonnegative constants $k,l\in(0,1]$ with $k+l<\frac{1}{s}$ such that the following contractive condition holds*

$$d\big(F(x,y),F(u,v)\big)\preceq kd(x,u)+ld(y,v),$$

for all $(x,y),(u,v)\in X^2$ with ($u\sqsubseteq x$ and $v\sqsupseteq y$) or ($x\sqsubseteq u$ and $y\sqsupseteq v$). Suppose that X has the following properties:

(i) if a non-decreasing sequence $\{x_n\}$ in X is such that $x_n\longrightarrow x$, then $x_n\sqsubseteq x$ for all $n\in\mathbb{N}$,

(ii) if a non-increasing sequence $\{y_n\}$ in X is such that $y_n\longrightarrow y$, then $y_n\sqsupseteq y$ for all $n\in\mathbb{N}$.

If there exist $x_0,y_0\in X$ such that $x_0\sqsubseteq F(x_0,y_0)$ and $F(y_0,x_0)\sqsubseteq y_0$, then F has a coupled fixed point $(x^*,y^*)\in X^2$. Moreover, the coupled fixed point is unique and of the form (x^*,x^*) for some $x^*\in X$.

Corollary 3.13 *Let $(X, \sqsubseteq)$ be a partially ordered set and (X, d) be a complete cone b-metric space with the coefficient $s \geq 1$ relative to a solid cone P. Let $F : X^2 \longrightarrow X$ be a mapping having the mixed monotone property on X and suppose that there exist nonnegative constants $k, l \in (0,1]$ with $k + l < \frac{1}{s}$ such that the following contractive condition holds*

$$d(F(x,y), F(u,v)) \preceq kd(x, F(x,y)) + ld(u, F(u,v)),$$

for all $(x,y), (u,v) \in X^2$ with ($u \sqsubseteq x$ and $v \sqsupseteq y$) or ($x \sqsubseteq u$ and $y \sqsupseteq v$). Suppose that X has the following properties:

(i) if a non-decreasing sequence $\{x_n\}$ in X is such that $x_n \longrightarrow x$, then $x_n \sqsubseteq x$ for all $n \in \mathbb{N}$,
(ii) if a non-increasing sequence $\{y_n\}$ in X is such that $y_n \longrightarrow y$, then $y_n \sqsupseteq y$ for all $n \in \mathbb{N}$.

If there exist $x_0, y_0 \in X$ such that $x_0 \sqsubseteq F(x_0, y_0)$ and $F(y_0, x_0) \sqsubseteq y_0$, then F has a coupled fixed point $(x^*, y^*) \in X^2$. Moreover, the coupled fixed point is unique and of the form (x^*, x^*) for some $x^* \in X$.

Note that, in above corollary, we ignore condition $k + l < \frac{2}{s+1}$ because $\frac{1}{s} \leq \frac{2}{s+1}$.

Corollary 3.14 *Let $(X, \sqsubseteq)$ be a partially ordered set and (X, d) be a complete cone b-metric space with the coefficient $s \geq 1$ relative to a solid cone P. Let $F : X^2 \longrightarrow X$ be a mapping having the mixed monotone property on X and suppose that there exist nonnegative constants $k, l \in (0,1]$ with $k + l < \frac{2}{s(s+1)}$ such that the following contractive condition holds*

$$d(F(x,y), F(u,v)) \preceq kd(x, F(x,y)) + ld(u, F(u,v)),$$

for all $(x,y), (u,v) \in X^2$ with ($u \sqsubseteq x$ and $v \sqsupseteq y$) or ($x \sqsubseteq u$ and $y \sqsupseteq v$). Suppose that X has the following properties:

(i) if a non-decreasing sequence $\{x_n\}$ in X is such that $x_n \longrightarrow x$, then $x_n \sqsubseteq x$ for all $n \in \mathbb{N}$,
(ii) if a non-increasing sequence $\{y_n\}$ in X is such that $y_n \longrightarrow y$, then $y_n \sqsupseteq y$ for all $n \in \mathbb{N}$.

If there exist $x_0, y_0 \in X$ such that $x_0 \sqsubseteq F(x_0, y_0)$ and $F(y_0, x_0) \sqsubseteq y_0$, then F has a coupled fixed point $(x^*, y^*) \in X^2$. Moreover, the coupled fixed point is unique and of the form (x^*, x^*) for some $x^* \in X$.

Example 3.15 Let $X = \mathbb{R}$ with usual order and let a cone b-metric d defined as in Example 3.10,

$$d(x,y)(t) = |x - y|^2 \cdot e^t.$$

Then (X, d) is a cone b-metric space with the coefficient $s = 2$. Now, let $F : X \times X \to X$ as

$$F(x,y) = \frac{x - 2y}{8}.$$

We shall check that this example satisfies all conditions of Corollary 3.12.

Indeed, we have

$$\begin{aligned} & d(F(x,y), F(u,v))(t) \\ &= \left| \frac{x-u}{8} - \frac{2(y-v)}{8} \right|^2 \cdot e^t \\ &\leq \frac{1}{64}\left(|x-u|^2 + 4|y-v|^2 + 4|x-u||y-v| \right) \cdot e^t \\ &\leq \frac{1}{64}\left(|x-u|^2 + 4|y-v|^2 + 2\left(|x-u|^2 + |y-v|^2\right) \right) \cdot e^t \\ &= \frac{3}{64}|x-u|^2 \cdot e^t + \frac{6}{64}|y-v|^2 \cdot e^t, \end{aligned}$$

for all $x, y, u, v \in X$ with $x \leq u$ and $y \leq v$ or $x \geq u$ and $y \leq v$ and for all $t \in [0,1]$. Taking $k = \frac{3}{64}, l = \frac{6}{64}$ we get

$$d(F(x,y), F(u,v)) \leq kd(x,u) + ld(y,v).$$

Since $k + l = \frac{9}{64} < \frac{1}{2} = \frac{1}{s}$, we have that this example of ordered cone b-metric space over (only) solid cone supports Corollary 3.12. Here, $(0,0)$ is (even) unique coupled fixed point.

Remark 3.16 It is worth to notice that using already some known methods (for example see [24, 25]) contractive condition from Theorem 3.1. implies the following contractive condition in ordered cone b-metric space $(X \times X, D, \sqsubseteq)$:

$$\begin{aligned} D(T_F Y, T_F V) \leq & A_1 D(T_g Y, T_F Y) + A_2 D(T_g V, T_F V) \\ & + A_3 D(T_g Y, T_F V) + A_4 D(T_g V, T_F Y) + A_5 D(T_g Y, T_g V), \end{aligned}$$

where

$$A_1 = a_1 + a_2, A_2 = a_3 + a_4, A_3 = a_5 + a_6, A_4 = a_7 + a_8, A_5 = a_9 + a_{10},$$

with

$$(s+1)(A_1 + A_2) + s(s+1)(A_3 + A_4) + 2sA_2 < 2,$$

and

$$D((x,y),(u,v)) = d(x,u) + d(y,v).$$

It is not hard to check that $(X \times X, D)$ is a new cone b-metric space with the same coefficient s as (X, d). Also, $(X \times X, D)$ is a ordered cone b-metric space. It is clear that the approach with new ordered cone b-metric space is much shorter, but both ordered cases are new.

Acknowledgments The authors would like to acknowledge the financial support received from Universiti Kebangsaan Malaysia under the research Grant no. UKM-DIP-2013-001 & FRGS/2/2013/ST06/UKM/02/2. The authors thank the referee for his/her careful reading of the manuscript and useful suggestions.

Authors contributions All authors contributed equally and significantly in writing this paper. All authors read and approved the final manuscript.

Compliance with ethical standards

Conflict of interest The authors declare that they have no competing interests.

References

1. Ran, A.C.M., Reurings, M.C.B.: A fixed point theorem in partially ordered sets and some applications to matrix equations. Proc. Am. Math. Soc. **132**(5), 1435–1443 (2004)
2. Lakshmikantham, V., Ćirić, L.: Coupled fixed point theorems for nonlinear contractions in partially ordered metric spaces. Nonlinear Anal. **70**(12), 4341–4349 (2009)
3. Nashine, H.K., Kadelburg, Z., Radenović, S.: Coupled common fixed point theorems for w^*-compatible mappings in ordered cone metric spaces. Appl. Math. Comput. **218**, 5422–5432 (2012)
4. Bhaskar, T.G., Lakshmikantham, V.: Fixed point theorems in partially ordered metric spaces and applications. Nonlinear Anal. **65**(7), 1379–1393 (2006)
5. Abbas, M., Khan, M.A., Radenović, S.: Common coupled fixed point theorems in cone metric spaces for w-compatible mappings. Appl. Math. Comput. **217**, 195–202 (2010)
6. Nashine, H.K., Abbas, M.: Common fixed point of mappings satisfying implicit contractive conditions in TVS-valued ordered cone metric spaces. J. Nonlinear Sci. Appl. **6**, 205–215 (2013)
7. Nashine, H.K., Aydi, H.: Common fixed point theorems for four mappings involving generalized contractions in cone metric spaces. J. Nonlinear Anal. Appl. **2012**, 1–12 (2012)
8. Nashine, H.K., Aydi, H.: Common fixed points for generalized (ψ,φ)-weak contractions in ordered cone metric spaces. Appl. Gen. Topol. **13**(2), 151–166 (2012)
9. Samet, B., Vetro, C.: Coupled fixed point, f-invariant set and fixed point of N-order. Ann. Funct. Anal. **1**(2), 46–56 (2010)
10. Shatanawi, W., Nashine, H.N.: A generalization of Banach's contraction principle for nonlinear contraction in a partial metric space. J. Nonlinear Sci. Appl. **5**, 37–43 (2012)
11. Samet, B., Karapinar, E., Aydi, H., Rajic, C.: Discussion on some coupled fixed point theorems. Fixed Point Theory Appl. **2013**, 50 (2013)
12. Sabetghadam, F., Masiha, H.P., Sanatpour, A.H.: Some coupled fixed point theorems in cone metric space. Fixed Point Theory Appl. **2009**(Article ID 125426) (2009). doi:10.1155/2009/125426
13. Aydi, H., Karapinar, E., Shatanawi, W.: Coupled fixed point results for (ϕ,ψ)-weakly contractive condition in ordered partial metric spaces. Comput. Math. Appl. **62**, 4449–4460 (2011)
14. Aydi, H., Samet, B., Vetro, C.: Coupled fixed point results in cone metric spaces for w-compatible mappings. Fixed Point Theory Appl. **2011**, 27 (2011)
15. Huang, L.G., Zhang, X.: Cone metric spaces and fixed point theorems of contractive mappings. J. Math. Anal. Appl. **332**(2), 1468–1476 (2007)
16. Hussain, N., Shah, M.H.: KKM mappings in cone b-metric spaces. Comput. Math. Appl. **62**(4), 1677–1684 (2011)
17. Radenović, S., Kadelburg, Z.: Quasi-contractions on symmetric and cone symmetric spaces. Banach J. Math. Anal. **5**(1), 38–50 (2011)
18. Cvetković, A.S., Stanić, M.P., Dimitrijević, S., Simić, S.: Common fixed point theorems for four mappings on cone metric type space. Fixed Point Theory Appl. **2011**, Article ID 589725 (2011)
19. Stanić, M.P., Cvetković, A.S., Simić, S., Dimitrijević, S.: Common fixed point under contractive condition of Cirics type on cone metric type spaces. Fixed Point Theory Appl. **2012**, 35 (2012)
20. Shah, M.H., Simić, S., Hussain, N., Sretenović, A., Radenović, S.: Common fixed points for occasionally weakly compatible pairs on cone metric type spaces. J. Comput. Anal. Appl. **14**(2), 290–297 (2012)
21. Huang, H., Xu, S.: Fixed point theorems of contractive mappings in cone b-metric spaces and applications. Fixed Point Theory Appl. **2012**, 220 (2012)
22. Shi, L., Xu, S.: Common fixed point theorems for two weakly compatible self-mappings in cone b-metric spaces. Fixed Point Theory Appl. **2013**, 120 (2013)
23. Fadail, Z.M., Ahmad, A.G.B.: Coupled coincidence point and common coupled fixed point results in cone b-metric spaces. Fixed Point Theory Appl. **2013**, 177 (2013)
24. Du, W., Karapnar, E.: A note on cone b-metric and its related results: generalizations or equivalence? Fixed Point Theory Appl. **2013**, 210 (2013)
25. Popović, B., Radenović, S., Shukla, S.: Fixed point results to tvs-cone b-metric spaces. Gulf J. Math. **1**, 51–64 (2013)
26. Azam, A., Mehmood, N., Ahmad, J., Radenović, S.: Multivalued fixed point theorems in cone b-metric spaces. J. Inequal. Appl. **2013**, 582 (2013)
27. Rad, G.S., Aydi, H., Kumam, P., Rahimi, H.: Common tripled fixed point results in cone metric type spaces. Rend. Circ. Mat. Palermo **63**, 287–300 (2014)
28. Kumam, P., Rahimi, H., Soleimani Rad, G.: The existence of fixed and periodic point theorems in cone metric type spaces. J. Nonlinear Sci. Appl. **7**, 255–263 (2014)
29. Rahimi, H., Soleimani Rad, G., Kumam, P.: Coupled common fixed point theorems under weak con tractions in cone metric type spaces. Thai. J. Math. **12**(1), 1–14 (2014)
30. Agarwal, R.P., El-Gebeily, M.A., O'Regan, D.: Generalized contractions in partially ordered metric spaces. Appl. Anal. **87**, 1–8 (2008)
31. O'Regan, D., Petrusel, A.: Fixed point theorems for generalized contractions in ordered metric spaces. J. Math. Anal. Appl. **341**(2), 1241–1252 (2008)
32. Altun, I., Damjanović, B., Djorić, D.: Fixed point and common fixed point theorems on ordered cone metric spaces. Appl. Math. Lett. **23**, 310–316 (2009)
33. Kadelburg, Z., Pavlović, M., Radenović, S.: Common fixed point theorems for ordered contractions and quasicontractions in ordered cone metric spaces. Comput. Math. Appl. **59**, 3148–3159 (2010)
34. Aydi, H., Nashine, H.K., Samet, B., Yazidi, H.: Coincidence and common fixed point results in partially ordered cone metric spaces and applications to integral equations. Nonlinear Anal. **74**, 6814–6825 (2011)
35. Radenović, S.: Some coupled coincidence points results of monotone mappings in partially ordered metric spaces. Int. J. Anal. Appl. **5**(2), 174–184 (2014)
36. Radenović, S.: Bhaskar-Lakshmikantham type-results for monotone mappings in partially ordered metric spaces. Int. J. Nonlinear Anal. Appl. **5**(2), 37–49 (2014)
37. Radenović, S.: Remarks on some coupled fixed point results in partial metric spaces. Nonlinear Funct. Anal. Appl. **18**(1), 39–50 (2013)
38. Fadail, Z.M., Ahmad, A.G.B.: Fixed point theorem of N-order in cone metric spaces. Far East J. Math. Sci. **79**(1), 49–76 (2013)
39. Fadail, Z.M., Ahmad, A.G.B.: Common fixed point theorem of N-order for W_N-compatible mappings in cone metric spaces. Far East J. Math. Sci. **96**(2), 133 166 (2015)
40. Jungck, G., Radenović, S., Radojević, S., Rakočević, V.: Common fixed point theorems for weakly compatible pairs on cone metric spaces. Fixed Point Theory Appl. **2009**, 1–13 (2009)

19

Simulation of Stochastic differential equation of geometric Brownian motion by quasi-Monte Carlo method and its application in prediction of total index of stock market and value at risk

Kianoush Fathi Vajargah[1] · **Maryam Shoghi**[1]

Abstract In the prediction of total stock index, we are faced with some parameters as they are uncertain in future and they can undergo changes, and this uncertainty has a few risks, and for a true analysis, the calculations should be performed under risk conditions. One of the evaluation methods under risk and uncertainty conditions is using geometric Brownian motion random differential equation and simulation by Monte Carlo and quasi-Monte Carlo methods as applied in this study. In Monte Carlo method, pseudo-random sequences are used to generate pseudo-random numbers, but in quasi-Monte Carlo method, quasi-random sequences are used with better uniformity and more rapid convergence compared with pseudo-random sequences. The predictions of total stock index and value at risk by this method are better and more exact than Monte Carlo method. This study at first evaluates random differential equation of geometric Brownian motion and its simulation by quasi-Monte Carlo method, and then its application in the predictions of total stock market index and value at risk can be evaluated.

Keywords Geometry Brownian motion · Quasi-Monte Carlo simulation · Sobol quasi-random sequence · Value at risk

✉ Kianoush Fathi Vajargah
k_fathi@iau-tnb.ac.ir

1 Department of Statistics, Islamic Azad University, North Branch Tehran, Tehran, Iran

Introduction

Like other fields of management knowledge and its application, risk management applies knowledge, principles, and specific rules to estimate predictions and achieve predefined goals. This field aims to help in ensuring the continuous protection of the people, and in safeguarding their assets and activities against the adverse events, as historically they are associated with risk.

In Iran, for 14 years, this issue has been paid serious attention. To fulfill the 20-year vision of Islamic Republic of Iran and to fulfill the development goals, the management and planning organization of the country emphasizes the effective implementation of plans by means of evaluations of economic and technical methods, and project management system by means of new management principles such as value engineering and risk management.

One of the famous methods to measure, predict, and manage risk is value at risk (VAR) that has been receiving great attention in recent years of managers of capital markets of various countries. VAR is a statistical criterion presenting quantitatively the maximum probable portfolio loss incurred during a definite period [5].

The calculation of VAR is done by two methods: parametric and nonparametric. In the parametric method, the main hypothesis is the normality of distribution of return on asset, and it is not an ideal hypothesis for application under real and practical conditions. As assets do not follow normal distribution under these conditions, this study applies nonparametric and Monte Carlo and quasi-Monte Carlo simulation techniques. In this method, by simulation of samples through computer and MATLAB software and random differential equation of geometric Brownian motion, the prediction is performed. The study aimed to evaluate the performance of Monte Carlo and quasi-Monte Carlo simulation methods to calculate total index and VAR of stock market to quantify the maximum probable loss with the lowest percentage error for the investment in stock market by considering the volatilities in the market.

This issue is of great importance for stock managers.

In recent years, by introducing VAR as a tool to calculate risk and Monte Carlo and quasi-Monte Carlo simulation techniques, various researches have been conducted, and some of them are described hereunder.

Giles et al. [2] in their study showed that quasi-Monte Carlo method in financial calculations is better than Monte Carlo method as the former can achieve similar precision with lower calculation cost, and their study refers to three important components of Sobol sequence and network points: dimension effect reduction by principal component analysis (PCA); random selection to present nonbiased estimators with a definite confidence interval; and the application of Brownian motion and Brownian bridge in their calculations.

Huang [3] applied Monte Carlo simulation method and Brownian motion to calculate VAR and obtained optimal VAR by adjustment coefficient, and his study results showed that optimal VAR was efficient and estimated the maximum expected loss with high confidence interval. In this paper, we begin with considering the review of the literature followed by the theoretical basics of study. Next, a comparison is made between quasi-random sequences and pseudo-random sequences, then definitions of the terms that are used regarding stock market are given, and finally conclusions are drawn.

Brownian motion

Let (P, Ω, A) be a probability space. In a random process, Wt: $\Omega \to R$, $\{W(t), \circ \leq t \leq T\}$ is called a Brownian motion, if the following conditions are satisfied:

1. For all $\Omega \in \omega$, $t \to W_t(\omega)$ is a continuous function on $[0, T]$ interval.
2. Each group of $\{W(t_\circ), W(t_1) - W(t_\circ), W(t_2) - W(t_1), \ldots, W(t_k) - W(t_{k-1})\}$intervals with $0 \leq t_0 < t_1 < \cdots < t_k \leq T$ is independent.
3. For each $\circ \leq s < t \leq T$, we have

$$W(t) - W(s) \rightarrow N(\circ, t - s) \tag{1}$$

Based on above properties, we can say for each $\circ < t \leq T$, we have $W(t) \rightsquigarrow N(\circ, t)$.

d-dimensional Brownian motion

Let (P, Ω, A) be a probability space and $d \in N$. In random process, Wt: $\Omega \to R$, $\{W(t), \circ \leq t \leq T\}$ is called *d*-dimensional Brownian motion, if the following conditions are satisfied:

1. For all, $\omega\ \Omega \in$, $t \to W_t(\omega)$ is a continuous function on $[\circ, T]$ interval.
2. Each group of $\{W(t_\circ), W(t_1) - W(t_\circ), W(t_2) - W(t_1), \ldots, W(t_k) - W(t_{k-1})\}$ intervals, by assuming $0 \leq t_0 < t_1 < \cdots < t_k \leq T$, is independent.
3. For each $\circ \leq s < t \leq T$, we have

$$W(t) - W(s) \rightarrow N(\circ, (t - s)I_d) \tag{2}$$

Brownian motion is also called a Standard Brownian Motion, if the following equation is established definitely: $W_0 = 0$.

As long as Brownian motion can have negative values, its direct use is doubtful for modeling the stock prices. Thus, we introduce a nonnegative type of Brownian motion called Geometric Brownian Motion. Geometric Brownian motion is always positive as the exponential function has positive values. Geometric Brownian Motion is defined as $S(t) = S_0\ e^{X(t)}$, where $X(t) = \sigma W(t) + \mu t$ is a Brownian motion with deviation, $S(0) = S_0 > 0$

Taking logarithm of the above equation, we have

$$\begin{aligned} X(t) &= \ln(s(t)/s_0) = \ln(s(t)) - \ln(s_0) \to \ln(s(t)) \\ &= \ln(s_0) + X(t) \end{aligned}$$

Thus, ln $(s(t))$ has normal distribution with mean ln $(s_0) + \mu t$ and variance $\sigma^2 t$, and for each t value, $s(t)$ has normal log distribution [10].

If we let $\bar{r} = \mu + \frac{\sigma^2}{2}, E(s(t)) = e^{\bar{r}t} s_0$ and $\bar{r} \gg r$ where r is the share growth rate under risk-free conditions as investment in banks, and $\bar{r}$ is the share growth rate under risk conditions as investment in stock market, then $\bar{r}$ is the share growth rate under risky conditions as investment in stock market. The share growth rate under risky conditions should be much more than that under risk-free conditions to motivate the investors to investment.

Theorem 2.1 *At constant time t, geometric Brownian motion has normal log distribution with mean ln $(s_0) + \mu t$ and variance $\sigma^2 t$.*

Proof

$$\begin{aligned} F(x) &= P(X \leq x) = P(s_0 \mathrm{esp}(\mu t + \sigma W(t)) \leq x) \\ &= P(\mu t + \sigma W(t) \leq \ln(x/s_0)) \\ &= P(W(t) \leq (\ln(x/s_0) - \mu t)/\sigma) \\ &= P\big(W(t)/\sqrt{t} \leq (\ln(x/s_0) - \mu t)/(\sigma\sqrt{t})\big) \\ &= \int_{-\infty}^{(\ln(x/s_0) - \mu t)/(\sigma\sqrt{t})} \frac{1}{\sqrt{2\pi}} \exp\left(-\frac{y^2}{2}\right) \mathrm{d}y \end{aligned}$$

If we calculate the derivative of the above equation with respect to X, then we have

$$f(x) = \frac{1}{\sqrt{t}x\sigma\sqrt{2\pi}} e^{\left(\left(\frac{-1}{2}\right)\left((\ln(x) - \ln(s_0) - \mu t)/(\sigma\sqrt{t})\right)^2\right)}$$

For $0 = t_0 < t_1 < \cdots < t_n = t$, ratios $1 \leq i \leq n, Li = \frac{s(t_i)}{s(t_{i-1})}$, random independent variables of log are normal, indicating the percent changes of the share value as it is not independent of real changes. $s(t_i) - s(t_{i-1})$.

For example:

$$L1 = \frac{s(t_1)}{s(t_0)} = e^{x(t_1)} \qquad L2 = \frac{s(t_2)}{s(t_1)} = e^{x(t_2)} - e^{x(t_1)}$$

are independent and normal log, and $x(t_1)$ and $x(t_2) - x(t_1)$ are independent with normal distribution. Now, we can rewrite geometric Brownian motion as $S(t) = S_0\ L1\ L2 \ldots Ln$ as an n-variate normal log. For example, assume that we sample share value at the end of each day. We can use $t_i = \mathrm{I}$, and thus, we have $Li = \frac{s(i)}{s(i-1)}$ which shows the percent of change during a day. Now, we can use the above formula in this regard. When for each i, $t_i - t_{i-1} = 1$, Li is distributed uniformly, ln (Li) has normal distribution with mean μ and variance σ^2.

Geometric Brownian motion not only solves negation problem, but based on basic economic principles, it is also considered as a good model for stock value. As stock price reacts rapidly to new information, geometric Brownian motion model as Markov process is used for it.

If a geometric Brownian motion is defined with differential equation $\mathrm{d}S = rS\,\mathrm{d}t + \sigma S\,\mathrm{d}W, S(0) = s_0$, then geometric Brownian motion is equal to:

$$S(t) = s_0 \exp\left(\left(r - \frac{1}{2}\sigma^2\right)t + \sigma W(t)\right)$$

As geometric Brownian motion has normal log distribution with parameters $\ln(s_0) + rt - \frac{1}{2}\sigma^2 t$ and $\sigma^2 t$, the mean and variance of geometric Brownian motion are given by

$$\begin{aligned} E(S(t)) &= s_0 \exp(rt) \\ \mathrm{Var}(S(t)) &= s_0^2 e^{2rt}\left(e^{\sigma^2 t} - 1\right) \end{aligned} \tag{3}$$

If the main issue is geometric Brownian motion $S(t) = s_0 \exp(\mu t + \sigma Wt)$, then its random differential equation formula is as follows:

$$\mathrm{d}S = \left(\mu + \frac{1}{2}\sigma^2\right)S(t)\mathrm{d}t + \sigma S(t)\,\mathrm{d}W, \qquad S(0) = s_0$$

As geometric Brownian motion has normal log distribution with parameters ln (s_0) + μt and $\sigma^2 t$, the mean of geometric Brownian motion is equal to $s_0 \exp\left(\mu t + \frac{1}{2}\sigma^2 t\right)$, and its variance is as per the following formula:

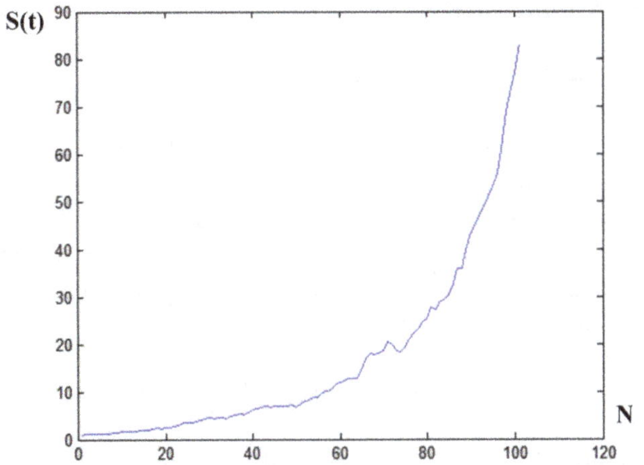

Fig. 1 Simulation of random differential equation of geometric Brownian motion with $N = 100$, $t = 1$, $\sigma^2 = 0.05$ and $\mu = 0.05$ parameters

$$\mathrm{Var}(S(t)) = s_0^2 e^{2\mu t + \sigma^2 t}\left(e^{\sigma^2 t} - 1\right) \tag{4}$$

[10].

Later, geometric Brownian motion for values $N = 100$, $t = 1$, $\sigma^2 = 0.05$, $\mu = 0.05$ can be simulated by MCMC method, and the corresponding chart is shown in Fig. 1.

The theoretical basics and initial concepts of risk

Risk

Webster dictionary defines—Value at Risk (VAR) and investment dictionary defines it as potential loss of investment as calculated.

Accident

An accident is a sudden event (such as a crash) that is not planned or intended and that causes damage or injury (Merriam-Webster Dictionary). The adverse accident occurrence probability is the uncertainty regarding the future condition of a phenomenon, and unpredictability and probability of damages can be considered as main elements in the definitions. It seems that common concept in these elements is the one of uncertainty.

Uncertainty

Uncertainty is something that is doubtful or unknown: something that is uncertain (Merriam-Webster Dictionary).

Risk dimensions

Productive and unproductive risk

Productive risk is the one assuring value added, but unproductive risk is without value added (investment in mine exploration project includes productive risk of economic type, but attending gambling, for example, where risking one's property to achieve much greater money, is called unproductive risk).

Controllable and uncontrollable risk

Controllable risk can be controlled by decision maker, while in uncontrollable risk, the decision maker has no control on risk (controllable risk is called reactive risk, and uncontrollable risk is called chance risk).

Profit-and-loss risk

Risk is divided into real (net) risk and involves winning and losing. The real risk includes loss (as car owner incurs damage in case of accident; otherwise it is not changed), and speculative risk includes loss and profit (its good example is the ownership of a factory or company). Simply put, risk can include loss (negative risk) and profit (positive risk) [1].

Different types of risks

There are different types of risks, and some of them are explained briefly.

Market risks

Market risk refers to the risk of losses in the bank's trading book due to changes in equity prices, interest rates, credit spreads, foreign-exchange rates, commodity prices, and other indicators values of which are determined by variable factors in a public market [7].

Exchange rate risk

The exchange rate risk is defined as the variability of a firm's value due to uncertain changes in the rate of exchange [4].

Inflation risk

Unexpected inflation is a change in prices that differs from the consensus view of what inflation is expected to be. When we talk about inflation risk, we are commonly talking about unexpected inflation. While expected inflation must be planned for in retirement budgeting, unexpected inflation causes uncertainty in prices, because it causes an element of surprise to bear upon the market. Since it cannot be predicted, managing the risk of unexpected inflation is critical [8].

Value at risk

Value at risk (VAR) as a statistical criterion defines the maximum expected loss of keeping assets in definite period and at definite confidence level. Assume that the daily value of an asset is 200 million Toman and at probability of 99 %, it is possible that the maximum reduction of this asset on the next day be 10 million Toman. Thus, VAR of this asset in a period of one day at a confidence interval of 99 % is 190 million Toman. We can say that by confidence interval of 99 %, the value of this asset on the next day will not be not less than 190 million Toman.

From math issues, VAR is shown as

$$\text{Prob}\{\delta V \leq -\text{VaR}\} \leq \alpha, \tag{5}$$

where δV is the change of asset value in a definite period

The above equation states that the probability of the asset loss in future period being less than VAR is $1 - \alpha$ [11] (Fig. 2).

In financial sciences, it is assumed that random variables price follows the path depending on Brownian motion as stock price, and one of its common models is geometric Brownian motion.

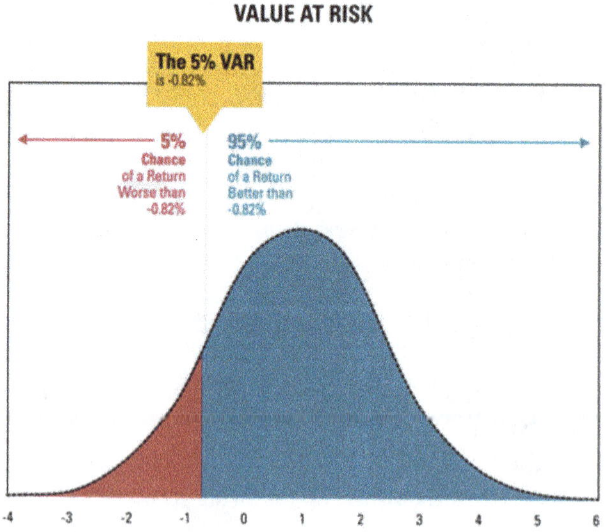

Fig. 2 Value at risk (risk analytics site)

Monte Carlo Method

For a given population L, a parameter, e.g., θ, is estimated. In the Monte Carlo method, an estimate detector $S(x)$ is first determined, in which x is a random variable with density function $f_x(x)$.

The estimate detector should satisfy the following two conditions:

(A) The estimate detector should be unbiased.

$E[S(x)] = \theta$

(B) The estimate detector should have definite variance.

$\mathrm{var}(S(x)) = \sigma^2$

Regarding the random samples $X_1 \ldots N_N$ of the function, density of $f_x(x)$ is used.

$$\hat{\theta}_N(X_1 \ldots X_N) = \frac{1}{N}\sum_{n=1}^{N} S(X_n) \tag{6}$$

$$\mathrm{var}(\hat{\theta}_N) = \frac{\sigma^2}{N} < \infty, \qquad E\left(\hat{\theta}_N\right) = \theta$$

We assume estimator number [6] as Monte Carlo estimator [6].

Monte Carlo simulation and stochastic differential equation

In this simulation, we present the expected value $E[g(X(T))]$ for a solution, X, of a known stochastic differential equation with a known function of g. In general, bipartite approximation error contains two parts: random error and time discretization error. Statistical error estimate is based on the central limit theorem. Error estimation for the time-discretization error of the Euler method directly measures with one remained phrase the accuracy of $\frac{1}{2}$ robust approximation.

Consider the following stochastic differential equation:

$$\mathrm{d}X(t) = a(t, X(t)) + b(t, X(t))\mathrm{d}W(t).$$

How can the value of $E[g(X(T))]$ be calculated on $t_0 \le t \le T$. Monte Carlo method is based on the approximation of

$$E[g(X(T))] \cong \sum_{j=1}^{N} \frac{g(\bar{X}(T;\omega_j))}{N}$$

where $\overline{X}$ is an approximation of X; according to Euler method, the error in the Monte Carlo method is

$$E[g(X(T))] - \sum_{j=1}^{N} \frac{g(\bar{X}(T;\omega_j))}{N}$$
$$= E[g(X(T)) - g(\bar{X}(T))] - \sum_{j=1}^{N} \frac{g(\bar{X}(T;\omega_j)) - E(g(\bar{X}(T)))}{N}.$$

Quasi-Monte Carlo (QMC)

The basic concept of quasi-Monte Carlo method is based on moving the random sample in Monte Carlo method with definite points accurately. The criterion of selection of definite points is that the sequence in [0,1)s has better uniformity than a random sequence. Indeed, these points should be such that [0,1)s is covered uniformly. To measure uniformity, a different concept is used as explained in the following definitions.

Definition 7.1 (*estimation of quasi -Monte Carlo*) Assume $X_1, \ldots, X_N$ is selected of $[\circ, 1)^s$ space, estimation of quasi-Monte Carlo is done as per the formula: $\bar{I}_{\mathrm{QMC}} = \frac{1}{N}\sum_{i=1}^{N} f(X_i)$.

In an ideal model, we replace the set of $x_1, \ldots, x_n$ points with infinite sequence $x_1, x_2, \ldots$ in $[0,1)^s$. A basic condition for this sequence is that the term $\lim_{N\to\infty} \frac{1}{N}\sum_{n=1}^{N} f(X_n) = \int_{[\circ,1)^s} f(x)\mathrm{d}x$ is satisfied.

The satisfaction of this term is achieved as sequence $x_1, x_2, \ldots, x_n$ is distributed uniformly in $[0,1)^s$. The difference, deviation scale of uniformity, is a sequence of points in $[0,1)^s$.

Definition 7.2 (*uniform distribution in* $[\circ, 1)^s$) $\{X_n\}_{n\in N}$ sequence is distributed uniformly in $[\circ, 1)^s$ if for each $x \in \delta[\circ, 1)^s$, we have

$$\lim_{N\to\infty} \frac{1}{N}\sum_{n=1}^{N} f(X_n) = \int_{[\circ,1)^s} f(x)\mathrm{d}x = I$$

Definition 7.3 (*general discrepancy*) Assume p is a set of points $\{X_1, \ldots, X_N\}$, and B is the family of subsets $B \in [\circ, 1)^s$. Then, the general discrepancy of set of points $p = \{X_1, \ldots, X_N\}$ in interval $[\circ, 1)^s$ is as follows:

$$D_N(B, p) = \sup_B \left| \frac{\sum_{n=1}^{N} C_B(X_n)}{N} - \lambda_s(B) \right| \tag{7}$$

where $D_N(B, p)$ is always between [0, 1], and C_B isan attribute function of B. Thus, $\sum_{n=1}^{N} C_B(X_n)$ shows the number of points $1 \le n \le N$ and x_n ε B.

Table 1 Calculation of ten first points of Van der corput sequence in basis 3

n	1	2	3	4	5	6	7	8	9	10
$\varphi_3(n)$	0.333	0.666	0.111	0.444	0.777	0.222	0.555	0.888	0.037	0.370

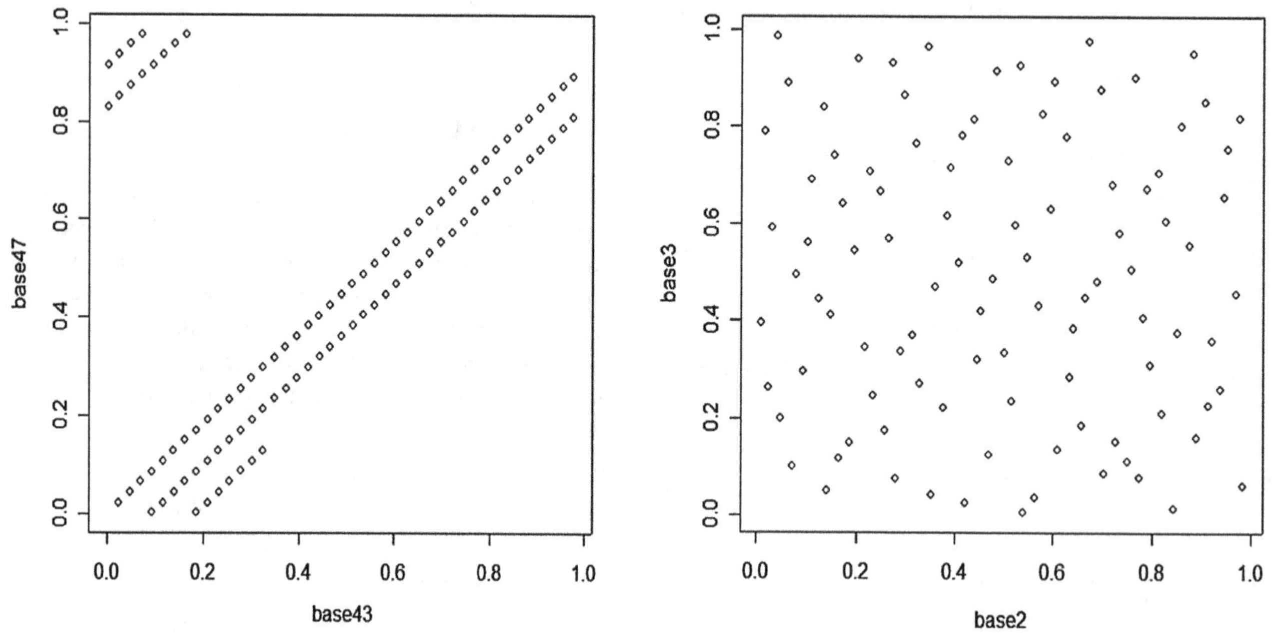

Fig. 3 Halton sequence for 1, 2 and 14, 15 dimensions

Definition 7.4 (*Star discrepancy*) Assume J^* is the family of all sub-intervals $[\circ, 1)^s$as $\prod_{i=1}^{s}[\circ, u_i)^s, \circ < u_i \le 1$. The star discrepancy of the set of $p = \{X_1, \ldots, X_N\}$ on substituting with J^* in place of B in Eq. [6] is as follows:

$$D_N^* = D_N(J^*, P) = \sup_{B_\varepsilon J^*}\left|\frac{\sum_{n=1}^{N} C_B(X_n)}{N} - \lambda(B)\right|. \tag{8}$$

Theorem 7.1 (Koksma Inequality) *If f has bounded changes $v(f)$ in $[0, 1]$, then for each set of $p = \{X_1, \ldots, X_N\}$ of $[0, 1]$, we have*

$$\left|\frac{\sum_{n=1}^{N} f(X_n)}{N} - \int_\circ^1 f(x)dx\right| \le v(f)D_N^*(p)$$

This inequality states that sequences with low discrepancy lead to low error [9].

In Monte Carlo method, there was an aggregation of points as the points were independent, and they had no awareness of each other, and there was little chance that they were very close to each other. Quasi-random sequences had better uniformity and rapid convergence compared with pseudo-random sequence. Some of the quasi-random sequences are mentioned later, and they are calculated by means of MATLAB software.

Table 2 Calculation if first five points of Halton sequence for first four dimensions

n	$\varphi_1(n)$ (base 2)	$\varphi_2(n)$ (base 3)	$\varphi_3(n)$ (base 5)	$\varphi_4(n)$ (base 7)
1	0.500	0.333	0.200	0.142
2	0.250	0.666	0.400	0.285
3	0.750	0.111	0.600	0.428
4	0.125	0.444	0.800	0.571
5	0.625	0.777	0.040	0.714

Van der corput sequence

This is the first sequence by which low discrepancy was formed.

To obtain the nth point of this sequence, at first, n on base b is defined as $n = \sum_{j=\circ}^{m} a_j(n)b^j$ where coefficients $a_j(n)$ include $\{\circ, 1, \ldots, b-1\}$values. We use these coefficients to achieve quasi-random values as $X_n = \phi_b(n) = \sum_{j=\circ}^{m} a_j(n)\frac{1}{b^{j+1}}$. Some of $a_j(n)$ values are nonzero. M is the smallest integer for each $j > m$ as $a_j(n) = 0$ (Table 1).

Van der corput sequence is a unidimensional sequence and generates random data of this sequence; in high dimensions, they can lose their random state and follow a linear function.

Halton sequence

Halton sequence was proposed in 1960, and it is similar to Van der corput sequence. First dimension of Halton sequence is a Van der corput sequence in basis 2, and second dimension is a Van der corput sequence in basis 3. Indeed, we can say Halton sequence is the same as Van der corput sequence with basis value as the nth primary value for the nth dimension of Halton sequence. Halton sequence is a s-dimensional sequence in cubic $[0, 1]^s$. Nth element of Halton sequence in $[0, 1]^s$ is defined as (Table 2)

$$x_n = \left(\phi_{b_1}(n), \phi_{b_2}(n), \ldots, \phi_{b_s}(n)\right) \qquad n = \circ, 1, 2, \ldots \tag{9}$$

Figure 3 shows 100 Halton sequences in bases 2, 3 and dimensions 1, 2 (right figure); and bases 43, 47 and dimensions 14, 15 (left figure). As shown, in low dimension, Halton sequence has suitable dispersion, but by increasing the dimension, convergence is revealed as in high dimensions, its random trend is lost and hypercube is not covered uniformly.

Sobol sequence

Sobol sequence was proposed in 1967. Constant value of basis 2 is used in Sobol sequence for all dimensions. Thus, Sobol sequence is rapid and simpler. This feature generates random numbers with low convergence in high dimensions.

To make this sequence, at first, we write n for basis 2 according to $n = \sum_{i=\circ}^{M} a_i 2^i$ as M is the smallest value

Table 3 Calculation of first 5 points of Sobol sequence for first 5 dimensions

n	Dim = 1	Dim = 2	Dim = 3	Dim = 4	Dim = 5
1	0.500	0.500	0.500	0.500	0.500
2	0.250	0.750	0.250	0.750	0.250
3	0.750	0.250	0.750	0.250	0.750
4	0.125	0.625	0.875	0.875	0.625
5	0.625	0.125	0.375	0.375	0.125

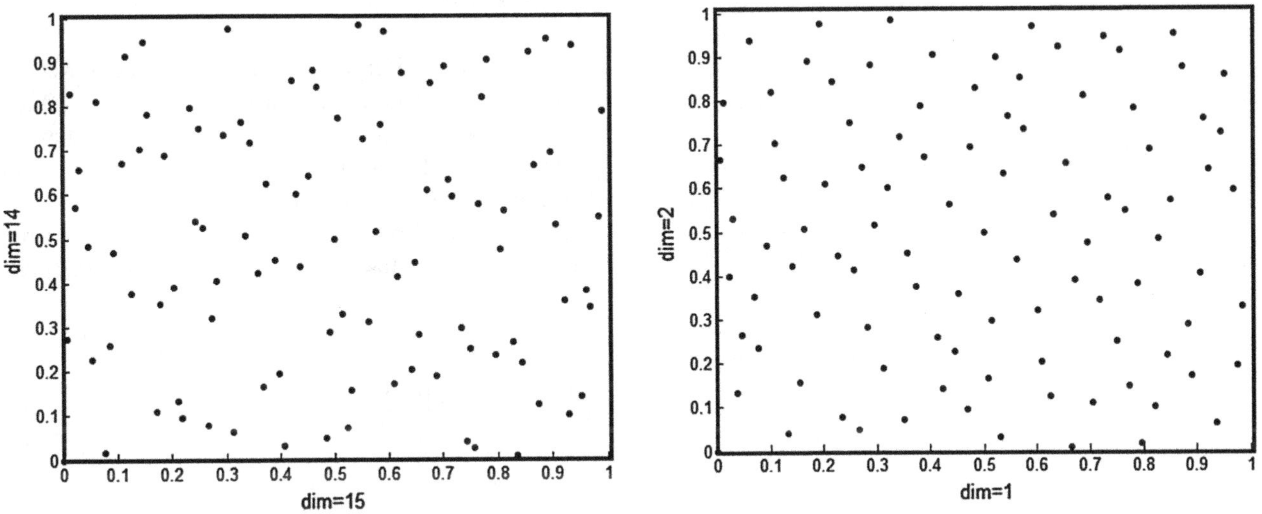

Fig. 4 Sobol sequence for 1, 2 and 14, 15 dimensions

Table 4 The comparison of quasi-random sequences and pseudo-random sequences in integral evaluation (N = 100)

	Pseudo-random sequences	Halton sequence in dimension 1 and basis 2	Sobol sequence in dimension 1 and basis 2	Halton sequence in dimension 10 and basis 29	Sobol sequence in dimension 10 and basis 2
Approximate value integral $\int_0^1 \exp(x)dx$	1.7696	1.7195	1.7195	1.7474	1.7044
Integrating error	0.0513	0.0012	0.0012	0.0291	−0.0139

bigger or equal to $\log_2^n$ and a_is values are zero or one, respectively.

The primary polynomial rank q is assumed as $p = x^q + c_1 x^{q-1} + \cdots + c_{q-1}x + 1$ in which c_is values are zero and one. m_i by coefficients c_i are generated as follows:

$$m_i = 2c_1 m_{i-1} \oplus 2^2 c_2 m_{i-2} \oplus \cdots \oplus 2^{q-1} c_{q-1} m_{i-q+1} \oplus 2^q m_{i-q} \oplus m_{i-q} \quad (10)$$

As $\oplus$ is an operator in computer represented as $1 \oplus \circ = \circ \oplus 1 = 1 \quad 1 \oplus 1 = \circ \oplus \circ = \circ$ m_i values are odd integer values in interval $[1, 2^i - 1]$. $V(i)$s are generated according to $V(i) = \frac{m_i}{2^i}$.

Thus, nth element of Sobol sequence is generated according to $\phi(n) = a{\cdot}v(1) \oplus a_1 v(2) \oplus \cdots \oplus a_{i-1} v(i)$.

To facilitate Sobol sequence generation, Grey code coding is used, and its algorithm is given by $\phi(n) = n \oplus \frac{n}{2}$, and now the adjusted model is given by $\phi(n) = n \oplus \frac{n}{2}$ (Table 3).

Figure 4 shows 100 points of Sobol sequences in dimensions 1, 2 (right figure) and dimensions 14, 15 (left figure). As shown, in high dimensions, this sequence has good uniformity and it makes its advantage compared to other sequences.

Table 5 Comparison of quasi-random sequences and pseudo-random sequences with uniform distribution [0, 1] ($N = 500$)

	Uniform distribution [0, 1]	Pseudo-random sequence	Halton sequence in dimension 3 and basis 5	Sobol sequence in dimension 30 and basis 2
Mean	0.5000	0.4942	0.4984	0.5000
Variance	0.0833	0.0795	0.0835	0.0833
Skewness[a]	0.0000	0.0401	0.0000	0.0000
Kurtosis[b]	1.8000	1.8283	1.7999	1.8005
Min	0.0000	0.0012	0.0000	0.0000
Max	1.0000	0.9991	0.9968	1.0000

[a] $g1 = \frac{\sum_{i=1}^{N} (yi - \bar{y})^3 / N}{s^3}$

[b] $g2 = \frac{\sum_{i=1}^{N} (yi - \bar{y})^4 / N}{s^4} - 3$

Comparison of quasi-random and pseudo-random sequences

Integral evaluation

To show which method is effective on integral evaluation, $\int_0^1 \exp(x)\,dx$ is considered. The exact answer of this integral is 1.7183. Table 4 shows the calculation of this integral by Monte Carlo and quasi-Monte Carlo method, and the integrating error is calculated.

As shown in Table 4, quasi-random sequences in integral evaluation perform better than pseudo-random sequences, and their estimation error is low. As was stated earlier, Sobol sequence in high dimensions is more efficient than Halton sequence.

Uniformity

The higher the uniformity of sequences, the lower the calculation error, and the sequence covered interval 0 to 1 very well. Table 5 shows a series of statistical features of Monte Carlo and quasi-Monte Carlo sequences in comparison with Uniform distribution.

As shown in Table 5, statistical features of quasi-random sequences are closer to uniform distribution [0, 1] than

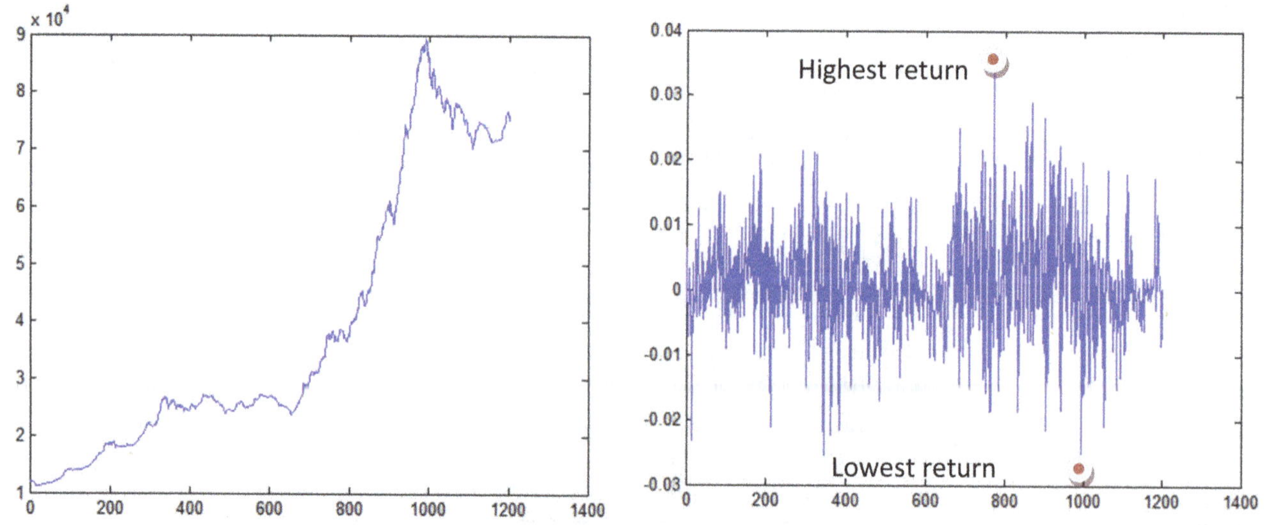

Fig. 5 The daily return of total stock index from Nov 2009 to Nov 2014 and the changes of total stock index from Nov 2009 to Nov 2014

pseudo-random sequence. It means that quasi-random sequences are mostly representing $U[0, 1]$ than pseudo-random sequence. As was said before, these sequences are associated with low discrepancy, their interval [0, 1] coverage is better, and are much more uniform. Also, Sobol sequence is more uniform than Halton sequence.

Nonconference of sequence in high dimensions

As shown in Fig. 3, by increasing dimension, convergence of Halton sequence is increased, and its trend of randomness is lost. Halton sequence has low convergence compared with Van der corput sequence in high dimensions. Sobol sequence as shown in Fig. 4 performs better than other quasi-random sequences as it applies basis 2 for all dimensions, and it leads to low convergence and high speed.

To predict total stock market index and calculation of VAR, a total index of 5 continuous years (Nov 2009–Nov 2014) was extracted from stock organization site, and it is equal to 1204 as the number of working days in these 5 years in Iran (Fig. 5).

As shown in Table 6, kurtosis of return is high, and it indicates that the total stock index distribution of TSE is far from normal value, and because of this, we applied Monte Carlo and quasi-Monte Carlo simulation methods, and these methods do not require normal distribution.

To obtain the prediction values of total TSE index and calculation of VAR, the following stages are considered:

1. Determining time interval as one day ($dt = 1$).
2. Generation of random numbers by Monte Carlo and quasi-Monte Carlo method and putting them in random differential model of geometric Brownian motion ($Si = Si - 1 + Si - 1\left(r\,dt + \sigma\,\varepsilon\sqrt{dt}\right)$) to predict total index.

Table 6 The statistical features of total stock index of TSE

Min	Max	Kurtosis	Skewness	Variance	Mean
−0.0276	0.0344	4.0900	0.1444	0.00005	0.0015

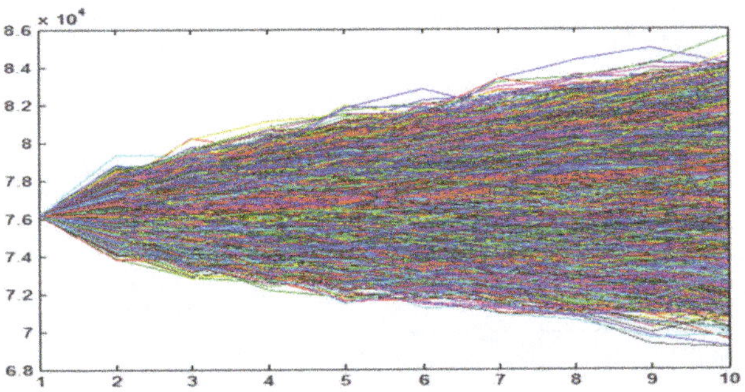

Fig. 6 100,000 simulation paths of total TSE index for 10 days by Monte Carlo method

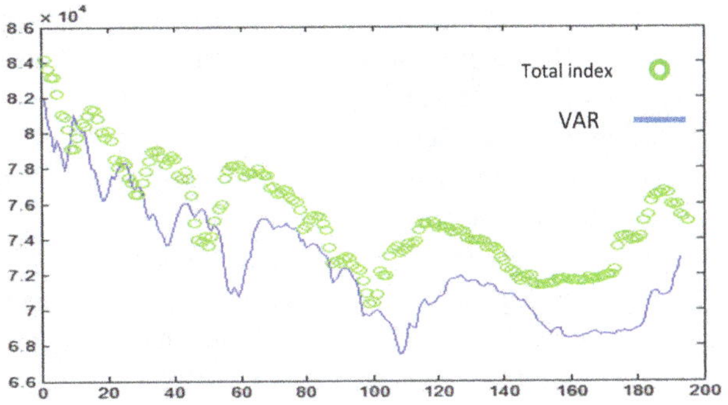

Fig. 7 Total TSE index and VAR calculated by Monte Carlo method

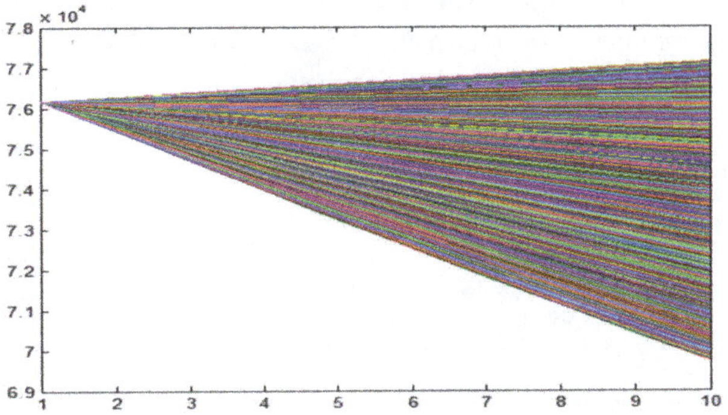

Fig. 8 100,000 simulation paths of total TSE index for 10 days by quasi-Monte Carlo

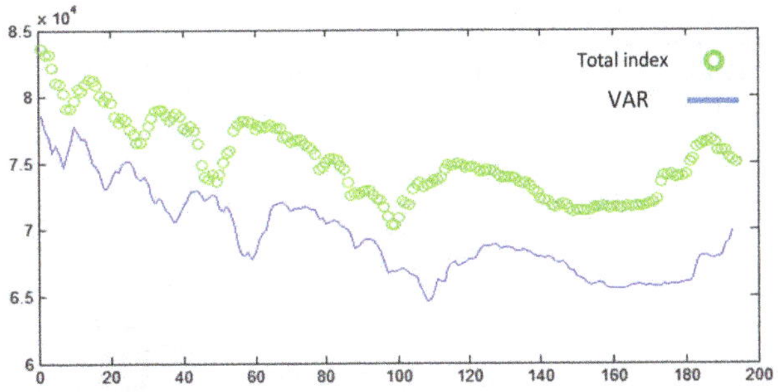

Fig. 9 Total TSE index and VAR calculated by quasi-Monte Carlo method (Sobol sequence)

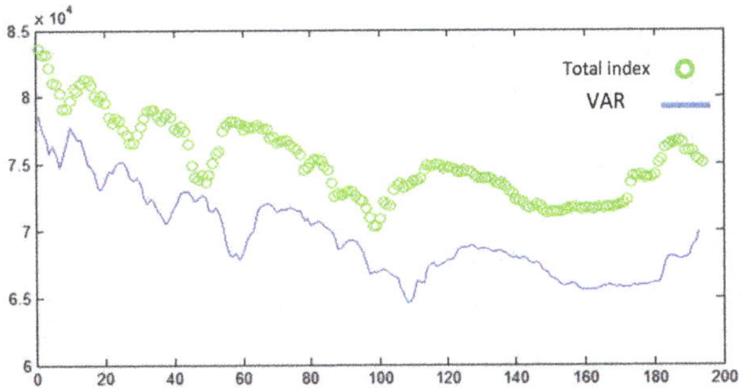

Fig. 10 Total TSE index and VAR calculated by quasi-Monte Carlo method (Halton sequence)

For Stage 2 for a period of 10 days (according to Bazel committee in Swiss), simulation is performed M times, and this value is equal to 100,000 in this study.

3. Sorting total predicted indices of small to big and calculation of its first percentile to determine the VAR at confidence interval of 99 %.
4. Stages 2, 3 are performed L times. As shown in this study, moving window approach is used. This means that first window includes 1000 first data, and the model applies these data to predict 1001 data. In the next stage, the window considers data 2 to 1001, and the model applies these data for prediction of 1002 data, and this window moves forward ($L = 193$) to include 1000 final data. As some(same) values are predicted by this method and as we have their real values, we can evaluate the validation of the model by determining prediction error (Fig. 6).

Table 7 VAR calculation error ratio for TSE total index at confidence interval 99 %

	Monte Carlo	Halton sequence	Sobol sequence
Error ratio	0.05	0.00	0.00

Table 8 Calculation of VAR in the final window for TSE total index at confidence interval 99 %

VAR(193,1) by Monte Carlo method	VAR(193,1) by Sobol sequence
72,967	69,994

As shown in Fig. 7, in some cases, VAR is not predicted precisely, and TSE total index is lower than the maximum loss (Fig. 8).

As shown in Figs. 9 and 10, VAR is predicted accurately by quasi-Monte Carlo method, and the two sequences showed good results as one dimension is used for prediction.

As shown in Table 7, VAR error ratio calculated by Monte Carlo method is 0.05; with higher than 0.01 as the probable error level, this method estimates low values of risk, and this leads to inadequacy of capital to resist against risk.

The results of Table 8 show that there is only one percent of chance that TSE total index under normal market conditions on Nov 18, 2014 achieves by quasi-Monte Carlo method a value of 69,994 or less, and by Monte Carlo method a value of 72,967 or below.

To investigate the adequacy of simulation model, simulated values are compared to real values of TSE site, and prediction values are calculated and shown in Table 9.

Conclusion

In this study, using TSE total index values during the 5 years from Nov 2009 to Nov 2014 and quasi-Monte Carlo and Monte Carlo methods, TSE total index and VAR are simulated and calculated, and it is shown that quasi-Monte Carlo method and Sobol sequence were better than Monte Carlo method with less error, and by confidence interval 99 %, we can invest in TSE, as for one percent of probability, TSE total index is lower than the predicted VAR.

Table 9 The calculation of prediction error of TSE total index at confidence interval 99 %

Date	Main value Total index	Predicted total index by Monte Carlo method	Prediction error of total index by Monte Carlo method	Predicted total index by Sobol sequence	Prediction error of total index by Sobol sequence
Nov 5, 2014	76,167	76,167	0	76,167	0
Nov 8, 2014	76,421	75,850	571	76,130	291
Nov 9, 2014	76,553	76,415	138	76,094	459
Nov 10, 2014	76,561	75,666	895	76,057	504
Nov 11, 2014	76,708	75,642	1066	76,021	687
Nov 12, 2014	76,589	75,291	1298	75,985	604
Nov 15, 2014	75,947	75,726	221	75,948	−1
Nov 16, 2014	75,931	76,078	−147	75,912	19
Nov 17, 2014	75,966	76,712	−746	75,871	90
Nov 18, 2014	75,409	77,430	−2021	75,839	−430

References

1. Akbarian, R., Dyanati, M.: Risk management in Islamic Banking, Islamic economy quarterly (2004)
2. Giles, M., et al.: Quasi-Monte Carlo for finance applications. Anziam Journal, Australia (2008)
3. Huang, A.Y.: An optimization process in Value-at-Risk Estimation". Rev. Finan. Econ. **19**(3), 109–116 (2010)
4. Habibnia, A.: Exchange rate risk measurement and management, LSE Risk and Stochastic Group (2013)
5. Jorion, P.: Value at Risk: The New Benchmark for Managing Financial Risk, 2nd ed. McGraw-Hill (2000)
6. Tan, K.S.: Quasi-Monte Carlo methods, Applications in Finance and Actuarial Science (1998)
7. Mehta. A., et al.: Managing market risk: today and tomorrow. Mckinsey and company (2012)
8. Mitchem. K., Oliver, T.A.: New View on managing the risks of inflation and diversification, state street global advisors (2015)
9. Niederreiter, H.: Random number generation and quasi-Monte Carlo methods. Austrian Academy of Sciences (1992)
10. Sigman, K.: Stationary marked point processes. Springer (2006)
11. Willmott, P.: Quantitative Finance, England, 2nd ed (2006)
12. http://www.merriam-webster.com
13. http://www.tse.ir

20

Suzuki type unique common fixed point theorem in partial metric spaces using (C)-condition

V. M. L. Hima Bindu[1] · G. N. V. Kishore[1] · K. P. R. Rao[2] · Y. Phani[3]

Abstract In this paper, we obtain a Suzuki type unique common fixed point theorem using C-condition in partial metric spaces. In addition, we give an example which supports our main theorem.

Keywords Partial metric · Weakly compatible maps · Suzuki-type contraction · C-condition

Mathematics Subject Classification 54H25 · 47H10 · 54E50

Introduction

The notion of a partial metric space was introduced by Matthews [12] as a part of the study of denotational semantics of data flow networks. In fact, it is widely recognized that partial metric spaces play an important role in constructing models in the theory of computation and domain theory in computer science (see [6]).

Matthews [12] and Romaguera [16] and Altun et al. [2] proved some fixed point theorems in partial metric spaces for a single map. For more works on fixed, common fixed point theorems in partial metric spaces, we refer [1, 3–5, 7–11, 13–15, 17–19]).

The aim of this paper is to prove a Suzuki type unique common fixed point theorem for four maps using (C)-condition in partial metric spaces.

First, we give the following theorem of Suzuki [18].

Theorem 1.1 (See [18]) *Let (X, d) be a complete metric space and let T be a mapping on X. Define a non-increasing function* $\theta : [0, 1) \to (\frac{1}{2}, 1]$ *by*

$$\theta(r) = \begin{cases} 1 & \text{if } 0 \le r \le \frac{(\sqrt{5}-1)}{2}, \\ (1-r)r^{-2} & \text{if } \frac{(\sqrt{5}-1)}{2} \le r \le 2^{-\frac{1}{2}}, \\ (1+r)^{-1} & \text{if } 2^{-\frac{1}{2}} \le r < 1. \end{cases}$$

Assume that there exists $r \in [0, 1)$, *such that*

$$\theta(r)d(x, Tx) \le d(x, y) \Rightarrow d(Tx, Ty) \le rd(x, y)$$

for all $x, y \in X$. *Then, there exists a unique fixed point* z *of* T. *Moreover,* $\lim_n T^n x = z$ *for all* $x \in X$.

Definition 1.2 (See [11]) A mapping T on a metric space (X, d) is called a non-expensive mapping if

$$d(Tx, Ty) \le d(x, y), \quad \forall x, y \in X.$$

Definition 1.3 (See [11]) A mapping T on a metric space (X, d) satisfies the C-condition if

✉ V. M. L. Hima Bindu
v.m.l.himabindu@gmail.com

G. N. V. Kishore
gnvkishore@kluniversity.in; kishore.apr2@gmail.com

K. P. R. Rao
kprrao2004@yahoo.com

Y. Phani
phaniyedlapalli23@gmail.com

1 Department of Mathematics, K L University, Vaddeswaram, Guntur 522 502, Andhra Pradesh, India

2 Department of Mathematics, Acharya Nagarjuna University, Nagarjuna Nagar, Guntur 522 510, Andhra Pradesh, India

3 Department of Basic Sciences, Shri Vishnu Engineering College for Women, Bhimavaram, West Godavari 534 202, Andhra Pradesh, India

$$\frac{1}{2}d(x,Tx) \leq d(x,y) \Rightarrow d(Tx,Ty) \leq d(x,y), \quad \forall x,y \in X.$$

First, we recall some basic definitions and lemmas which play crucial role in the theory of partial metric spaces.

Definition 1.4 (See [12]) A partial metric on a nonempty set X is a function $p: X \times X \to R^+$, such that for all $x,y,z \in X$:

(p_1) $x = y \Leftrightarrow p(x,x) = p(x,y) = p(y,y)$,
(p_2) $p(x,x) \leq p(x,y), p(y,y) \leq p(x,y)$,
(p_3) $p(x,y) = p(y,x)$,
(p_4) $p(x,y) \leq p(x,z) + p(z,y) - p(z,z)$.

The pair (X, p) is called a partial metric space (PMS).

If p is a partial metric on X, then the function $p^s : X \times X \to \mathbb{R}^+$ given by

$$p^s(x,y) = 2p(x,y) - p(x,x) - p(y,y), \tag{1}$$

is a metric on X.

Example 1.5 (See [1, 9, 12]) Consider $X = [0,\infty)$ with $p(x,y) = \max\{x,y\}$. Then, (X, p) is a partial metric space. It is clear that p is not a (usual) metric. Note that in this case, $p^s(x,y) = |x-y|$.

Example 1.6 (See [7]) Let $X = \{[a,b] : a,b, \in \mathbb{R},\ a \leq b\}$ and define $p([a,b],[c,d]) = \max\{b,d\} - \min\{a,c\}$. Then, (X, p) is a partial metric space.

We now state some basic topological notions (such as convergence, completeness, and continuity) on partial metric spaces (see [1, 2, 9, 10, 12].)

Definition 1.7

(i) A sequence $\{x_n\}$ in the PMS (X, p) converges to the limit x if and only if $p(x,x) = \lim_{n\to\infty} p(x,x_n)$.
(ii) A sequence $\{x_n\}$ in the PMS (X, p) is called a Cauchy sequence if $\lim_{n,m\to\infty} p(x_n,x_m)$ exists and is finite.
(iii) A PMS (X, p) is called complete if every Cauchy sequence $\{x_n\}$ in X converges with respect to τ_p, to a point $x \in X$, such that $p(x,x) = \lim_{n,m\to\infty} p(x_n,x_m)$.

The following lemma is one of the basic results in PMS ([1, 2, 9, 10, 12]).

Lemma 1.8

(i) *A sequence $\{x_n\}$ is a Cauchy sequence in the PMS (X, p) if and only if it is a Cauchy sequence in the metric space (X,p^s).*
(ii) *A PMS (X, p) is complete if and only if the metric space (X,p^s) is complete. Moreover*

$$\lim_{n\to\infty} p^s(x,x_n) = 0 \Leftrightarrow p(x,x) = \lim_{n\to\infty} p(x,x_n) = \lim_{n,m\to\infty} p(x_n,x_m).$$

Next, we give two simple lemmas which will be used in the proof of our main result. For the proofs, we refer to [1].

Lemma 1.9 *Assume $x_n \to z$ as $n \to \infty$ in a PMS (X, p), such that $p(z,z) = 0$. Then, $\lim_{n\to\infty} p(x_n,y) = p(z,y)$ for every $y \in X$.*

Lemma 1.10 *Let (X, p) be a PMS. Then*

(A) *If $p(x,y) = 0$, then $x = y$.*
(B) *If $x \neq y$, then $p(x,y) > 0$.*

Remark 1.11 If $x = y$, $p(x, y)$ may not be 0.

Definition 1.12 A pair (T, g) is called weakly compatible pair if they commute at coincidence points.

Now, we prove our main result.

Main result

Theorem 2.1 *Let (X, p) be a partial metric space and let $S,T,f,g : X \to X$ be mappings satisfying*

(2.1.1) $\frac{1}{2}\min\{p(fx,Sx),p(gy,Ty)\} \leq p(fx,gy)$ *implies that* $\psi(p(Sx,Ty)) \leq \alpha(M(x,y)) - \beta(M(x,y))$, *for all x, y in X, where* $\psi,\alpha,\beta : [0,\infty) \to [0,\infty)$ *are such that* ψ *is an altering distance function,* α *is continuous, and* β *is lower semi continuous,* $\alpha(0) = \beta(0) = 0$ *and* $\psi(t) - \alpha(t) + \beta(t) > 0$, *for all* $t > 0$ *and*

$$M(x,y) = \max\left\{\begin{array}{l} p(fx,gy), p(fx,Sx), p(gy,Ty), \\ \frac{1}{2}[p(fx,Ty) + p(gy,Sx)] \end{array}\right\},$$

(2.1.2) $S(X) \subseteq g(X), T(X) \subseteq f(X)$,
(2.1.3) *either f(X) or g(X) is a complete subspace of X,*
(2.1.4) *the pairs (f, S) and (g, T) are weakly compatible.*

Then, S, T, f and g have a unique common fixed point in X.

Proof Let $x_0 \in X$ be arbitrary point in X. From (2.1.2), there exist sequences of $\{x_n\}$ and $\{y_n\}$ in X, such that $Sx_{2n} = gx_{2n+1} = y_{2n}$, $Tx_{2n+1} = fx_{2n+2} = y_{2n+1}$, $n = 0,1,2,\ldots$.

Case (i): Assume that $y_n \neq y_{n+1}$ for all n.
Denote $p_n = p(y_n,y_{n+1})$.
We show that $p_n \leq p_{n-1}$, $n = 1,2,3,\ldots$
Now

$$\frac{1}{2}\min\{p(fx_{2n}, Sx_{2n}), p(gx_{2n+1}, Tx_{2n+1})\} \le p(fx_{2n}, Sx_{2n}) = p(fx_{2n}, gx_{2n+1}).$$

From (2.1.1), we get

$$\psi(p(Sx_{2n}, Tx_{2n+1})) \le \alpha(M(x_{2n}, x_{2n+1})) - \beta(M(x_{2n}, x_{2n+1})).$$

$$M(x_{2n},x_{2n+1}) = \max\left\{\begin{array}{c} p(y_{2n-1},y_{2n}), p(y_{2n-1},y_{2n}), p(y_{2n},y_{2n+1}), \\ \frac{1}{2}[p(y_{2n-1},y_{2n+1}) + p(y_{2n},y_{2n})] \end{array}\right\}$$
$$= \max\{p_{2n-1}, p_{2n}\}, \text{ from } (p_4).$$

Hence,
$\psi(p_{2n}) \le \alpha(\max\{p_{2n-1}, p_{2n}\}) - \beta(\max\{p_{2n-1}, p_{2n}\})$.

If p_{2n} is maximum, then we have $\psi(p_{2n}) \le \alpha(p_{2n}) - \beta(p_{2n})$, thus $\psi(p_{2n}) - \alpha(p_{2n}) + \beta(p_{2n}) \le 0$, which is a contradiction.

Hence p_{2n-1} is maximum. Thus

$$\psi(p_{2n}) \le \alpha(p_{2n-1}) - \beta(p_{2n-1}) \tag{2}$$
$$< \psi(p_{2n-1}).$$

Since ψ is increasing, we have $p_{2n} \le p_{2n-1}$.

Similarly, we can show that $p_{2n-1} \le p_{2n-2}$.

Thus, $p_n \le p_{n-1}, \quad n = 1, 2, 3, \ldots$

Thus, $\{p_n\}$ is a non-increasing sequence of non-negative real numbers and must converge to a real number, say, $l \ge 0$. Suppose $l > 0$.

Letting $n \to \infty$ in (2), we get $\psi(l) \le \alpha(l) - \beta(l)$.

Thus, $\psi(l) - \alpha(l) + \beta(l) \le 0$, which is a contradiction. Hence, $l = 0$.

Thus

$$\lim_{n\to\infty} p(y_n, y_{n+1}) = 0. \tag{3}$$

Hence, from (p_2), we get

$$\lim_{n\to\infty} p(y_n, y_n) = 0. \tag{4}$$

By definition of p^s, (3), and (4), we get

$$\lim_{n\to\infty} p^s(y_n, y_{n+1}) = 0. \tag{5}$$

Now, we prove that $\{y_{2n}\}$ is a Cauchy sequence in (X, p^s). On contrary, suppose that $\{y_{2n}\}$ is not Cauchy.

There exist $\epsilon > 0$ and monotone increasing sequences of natural numbers $\{2m_k\}$ and $\{2n_k\}$, such that $n_k > m_k$,

$$p^s(y_{2m_k}, y_{2n_k}) \ge \epsilon \tag{6}$$

and

$$p^s(y_{2m_k}, y_{2n_k-2}) < \epsilon. \tag{7}$$

From (6) and (7), we obtain

$$\epsilon \le p^s(y_{2m_k}, y_{2n_k})$$
$$\le p^s(y_{2m_k}, y_{2n_k-2}) + p^s(y_{2n_k-2}, y_{2n_k-1}) + p^s(y_{2n_k-1}, y_{2n_k})$$
$$\le \epsilon + p^s(y_{2n_k-2}, y_{2n_k-1}) + p^s(y_{2n_k-1}, y_{2n_k}).$$

Letting $k \to \infty$ and then using (5), we get

$$\lim_{k\to\infty} p^s(y_{2m_k}, y_{2n_k}) = \epsilon. \tag{8}$$

Hence, from definition of p^s and (4), we have

$$\lim_{k\to\infty} p(y_{2m_k}, y_{2n_k}) = \frac{\epsilon}{2}. \tag{9}$$

Letting $k \to \infty$ and then using (8) and (5) in $|p^s(y_{2n_k+1}, y_{2m_k}) - p^s(y_{2n_k}, y_{2m_k})| \le p^s(y_{2n_k+1}, y_{2n_k})$ we obtain

$$\lim_{k\to\infty} p^s(y_{2n_k+1}, y_{2m_k}) = \epsilon. \tag{10}$$

Hence, we have

$$\lim_{k\to\infty} p(y_{2n_k+1}, y_{2m_k}) = \frac{\epsilon}{2}. \tag{11}$$

Letting $k \to \infty$ and then using (8) and (5) in $|p^s(y_{2n_k}, y_{2m_k-1}) - p^s(y_{2n_k}, y_{2m_k})| \le p^s(y_{2m_k-1}, y_{2m_k})$, we get

$$\lim_{k\to\infty} p^s(y_{2n_k}, y_{2m_k-1}) = \epsilon. \tag{12}$$

Hence, we have

$$\lim_{k\to\infty} p(y_{2n_k}, y_{2m_k-1}) = \frac{\epsilon}{2}. \tag{13}$$

Letting $k \to \infty$ and then using (12) and (5) in $|p^s(y_{2m_k-1}, y_{2n_k+1}) - p^s(y_{2m_k-1}, y_{2n_k})| \le p^s(y_{2n_k+1}, y_{2n_k})$ we obtain

$$\lim_{k\to\infty} p^s(y_{2m_k-1}, y_{2n_k+1}) = \epsilon. \tag{14}$$

Hence, we get

$$\lim_{k\to\infty} p(y_{2m_k-1}, y_{2n_k+1}) = \frac{\epsilon}{2}. \tag{15}$$

If

$$\frac{1}{2}\min\{p(y_{2m_k-1}, y_{2m_k}), p(y_{2n_k}, y_{2n_k+1})\} > p(y_{2m_k-1}, y_{2n_k}),$$

then letting $k \to \infty$, we get $0 \ge \frac{\epsilon}{2}$ from (3) and (13).

It is a contradiction. Hence $\frac{1}{2}\min\{p(y_{2m_k-1}, y_{2m_k}), p(y_{2n_k}, y_{2n_k+1})\} \le p(y_{2m_k-1}, y_{2n_k}) = p(fx_{2m_k}, gx_{2n_k+1})$.

From (2.1.1), we have

$$\begin{aligned}
&\psi(p(y_{2m_k},y_{2n_k+1}))\\
&=\psi(p(Sx_{2m_k},Tx_{2n_k+1}))\\
&\le\alpha\left(\max\left\{\begin{array}{c}p(y_{2m_k-1},y_{2n_k}),p(y_{2m_k-1},y_{2m_k}),p(y_{2n_k},y_{2n_k+1}),\\ \frac{1}{2}[p(y_{2m_k-1},y_{2n_k+1})+p(y_{2n_k},y_{2m_k})]\end{array}\right\}\right)\\
&-\beta\left(\max\left\{\begin{array}{c}p(y_{2m_k-1},y_{2n_k}),p(y_{2m_k-1},y_{2m_k}),p(y_{2n_k},y_{2n_k+1}),\\ \frac{1}{2}[p(y_{2m_k-1},y_{2n_k+1})+p(y_{2n_k},y_{2m_k})]\end{array}\right\}\right).
\end{aligned}$$

Letting $k\to\infty$ and then using (11), (13), (3), (15), and (9), we have

$$\begin{aligned}
\psi\left(\frac{\epsilon}{2}\right)&\le\alpha\left(\max\left\{\frac{\epsilon}{2},0,0,\frac{1}{2}\left[\frac{\epsilon}{2}+\frac{\epsilon}{2}\right]\right\}\right)\\
&\quad-\beta\left(\max\left\{\frac{\epsilon}{2},0,0,\frac{1}{2}\left[\frac{\epsilon}{2}+\frac{\epsilon}{2}\right]\right\}\right)\\
&=\alpha\left(\frac{\epsilon}{2}\right)-\beta\left(\frac{\epsilon}{2}\right)\\
&<\psi\left(\frac{\epsilon}{2}\right),
\end{aligned}$$

which is a contradiction. Hence, $\{y_{2n}\}$ is Cauchy.

In addition, $|p^s(y_{2n+1},y_{2m+1})-p^s(y_{2n},y_{2m})|\le p^s(y_{2n+1},y_{2n})+p^s(y_{2m},y_{2m+1})$.

Letting $n,m\to\infty$, we have

$$\lim_{n,m\to\infty}p^s(y_{2n+1},y_{2m+1})=0.$$

Hence, $\{y_{2n+1}\}$ is Cauchy. Thus $\{y_n\}$ is a Cauchy sequence in (X,p^s).

Hence, we have $\lim_{n,m\to\infty}p^s(y_n,y_m)=0$.

Now, from the definition of p^s and from (4), we obtain

$$\lim_{n,m\to\infty}p(y_n,y_m)=0. \tag{16}$$

Therefore, $\{y_n\}$ is Cauchy sequence in X.

Suppose $g(X)$ is complete.

Since $y_{2n}=Sx_{2n}=gx_{2n+1}$, it follows $\{y_{2n}\}\subseteq g(X)$ is a Cauchy sequence in the complete metric space $(g(X),p^s)$, it follows that $\{y_{2n}\}$ converges in $(g(X),p^s)$.

Thus, $\lim_{n\to\infty}p^s(y_{2n},u)=0$ for some $u\in g(X)$.

That is, $y_{2n}\to u=gt\in g(X)$ for some $t\in X$.

Since $\{y_n\}$ is Cauchy in X and $\{y_{2n}\}\to u$, it follows that $\{y_{2n+1}\}\to u$.

From Lemma (1.2.5), we get

$$p(u,u)=\lim_{n\to\infty}p(y_{2n+1},u)=\lim_{n\to\infty}p(y_{2n},u)=\lim_{n,m\to\infty}p(y_n,y_m). \tag{17}$$

From (16) and (17), we obtain

$$p(u,u)=\lim_{n\to\infty}p(y_{2n+1},u)=\lim_{n\to\infty}p(y_{2n},u)=0. \tag{18}$$

Now, we claim that, for each $n\ge 1$, at least, one of the following assertions holds:

$$\frac{1}{2}p(y_{2n-1},y_{2n})\le p(y_{2n-1},u)\ \text{or}\ \frac{1}{2}p(y_{2n},y_{2n+1})\le p(y_{2n},u).$$

On the contrary, suppose that

$$\frac{1}{2}p(y_{2n-1},y_{2n})>p(y_{2n-1},u)\ \text{and}\ \frac{1}{2}p(y_{2n},y_{2n+1})>p(y_{2n},u)$$

for some $n\ge 1$.

Then we have

$$\begin{aligned}
p_{2n-1}=p(y_{2n-1},y_{2n})&\le p(y_{2n-1},u)+p(u,y_{2n})-p(u,u)\\
&<\frac{1}{2}[p(y_{2n-1},y_{2n})+p(y_{2n},y_{2n+1})]\\
&\le\frac{1}{2}[p_{2n-1}+p_{2n}]\\
&\le p_{2n-1},
\end{aligned}$$

which is a contradiction, and so, the claim holds.

Sub case(a) : Suppose $\frac{1}{2}p(y_{2n-1},y_{2n})\le p(y_{2n-1},u)$.

Suppose $Tt\ne u$.

We have

$$\begin{aligned}
\frac{1}{2}\min\{p(fx_{2n},Sx_{2n}),p(gt,Tt)\}&\le\frac{1}{2}p(fx_{2n},Sx_{2n})\\
&=\frac{1}{2}p(y_{2n-1},y_{2n})\\
&\le p(y_{2n-1},u)\\
&=p(fx_{2n},gt).
\end{aligned}$$

From (2.1.1), we get

$$\begin{aligned}
&\psi(p(Sx_{2n},Tt))\le\alpha(M(x_{2n},t))-\beta(M(x_{2n},t))\\
&\le\alpha\left(\max\left\{\begin{array}{c}p(fx_{2n},gt),p(fx_{2n},Sx_{2n}),p(gt,Tt),\\ \frac{1}{2}[p(fx_{2n},Tt)+p(gt,Sx_{2n})]\end{array}\right\}\right)\\
&-\beta\left(\max\left\{\begin{array}{c}p(fx_{2n},gt),p(fx_{2n},Sx_{2n}),p(gt,Tt),\\ \frac{1}{2}[p(fx_{2n},Tt)+p(gt,Sx_{2n})]\end{array}\right\}\right).
\end{aligned}$$

Letting $n\to\infty$ and using (17), (18), we get

$$\begin{aligned}
\psi(p(u,Tt))&\le\alpha\left(\max\left\{\begin{array}{c}p(u,gt),p(u,u),p(gt,Tt),\\ \frac{1}{2}[p(u,Tt)+p(gt,u)]\end{array}\right\}\right)\\
&-\beta\left(\max\left\{\begin{array}{c}p(u,gt),p(u,u),p(gt,Tt),\\ \frac{1}{2}[p(u,Tt)+p(gt,u)]\end{array}\right\}\right)\\
&=\alpha\left(\max\left\{\begin{array}{c}p(u,u),p(u,u),p(u,Tt),\\ \frac{1}{2}[p(u,Tt)+p(u,u)]\end{array}\right\}\right)\\
&-\beta\left(\max\left\{\begin{array}{c}p(u,u),p(u,u),p(u,Tt),\\ \frac{1}{2}[p(u,Tt)+p(u,u)]\end{array}\right\}\right)\\
&\le\alpha(p(u,Tt))-\beta(p(u,Tt))<\psi(p(u,Tt)).
\end{aligned}$$

It is a contradiction. Hence, $Tt = u = gt$.

Since the pair (g, T) is weakly compatible, we have $gu = Tu$.

Suppose $Tu \neq u$.

Since $\frac{1}{2}\min\{p(fx_{2n}, Sx_{2n}), p(gu, Tu)\} \leq p(fx_{2n}, gu)$, from (2.1.1), we get

$$\psi(p(Sx_{2n}, Tu)) \leq \alpha\left(\max\left\{\begin{array}{l} p(fx_{2n}, gu), p(fx_{2n}, Sx_{2n}), p(gu, Tu), \\ \frac{1}{2}[p(fx_{2n}, Tu) + p(gu, Sx_{2n})] \end{array}\right\}\right) - \beta\left(\max\left\{\begin{array}{l} p(fx_{2n}, gu), p(fx_{2n}, Sx_{2n}), p(gu, Tu), \\ \frac{1}{2}[p(fx_{2n}, Tu) + p(gu, Sx_{2n})] \end{array}\right\}\right).$$

Letting $n \to \infty$, we have

$$\begin{aligned} \psi(p(u, Tu)) &\leq \alpha\left(\max\left\{\begin{array}{l} p(u, gu), p(u, u), p(gu, Tu), \\ \frac{1}{2}[p(u, Tu) + p(gu, u)] \end{array}\right\}\right) \\ &\quad - \beta\left(\max\left\{\begin{array}{l} p(u, gu), p(u, u), p(gu, Tu), \\ \frac{1}{2}[p(u, Tu) + p(gu, u)] \end{array}\right\}\right) \\ &\leq \alpha(p(u, Tu)) - \beta(p(u, Tu)) \\ &< \psi(p(u, Tu)), \end{aligned}$$

which is a contradiction.

Hence, $Tu = u$.

Therefore, $u = Tu = gu$.

Since $T(X) \subseteq f(X)$, then there exists $v \in X$, such that $Tu = fv = u$.

Suppose $Sv \neq fv$.

Since $\frac{1}{2}\min\{p(fv, Sv), p(gu, Tu)\} \leq p(fv, gu)$, from (2.1.1), we have

$$\begin{aligned} \psi(p(Sv, fv)) &= \psi(p(Sv, Tu)) \\ &\leq \alpha(M(v, u)) - \beta(M(v, u)) \\ &\leq \alpha\left(\max\left\{\begin{array}{l} p(fv, gu), p(fv, Sv), p(gu, Tu), \\ \frac{1}{2}[p(fv, Tu) + p(gu, Sv)] \end{array}\right\}\right) \\ &\quad - \beta\left(\max\left\{\begin{array}{l} p(fv, gu), p(fv, Sv), p(gu, Tu), \\ \frac{1}{2}[p(fv, Tu) + p(gu, Sv)] \end{array}\right\}\right) \\ &\leq \alpha(p(Sv, Tu)) - \beta(p(Sv, Tu)) \\ &< \psi(p(Sv, Tu)) \\ &= \psi(p(Sv, fv)). \end{aligned}$$

Hence, $Sv = fv = u$.

Since the pair (f, S) is weakly compatible, we have $fu = Su$.

Suppose $Su \neq u$.

Since $\frac{1}{2}\min\{p(fu, Su), p(gt, Tt)\} \leq p(fu, gt)$, from (2.1.1), we have

$$\begin{aligned} \psi(p(Su, u)) &= \psi(p(Su, Tt)) \\ &\leq \alpha\left(\max\left\{\begin{array}{l} p(fu, gt), p(fu, Su), p(gt, Tt), \\ \frac{1}{2}[p(fu, Tt) + p(gt, Su)] \end{array}\right\}\right) \\ &\quad - \beta\left(\max\left\{\begin{array}{l} p(fu, gt), p(fu, Su), p(gt, Tt), \\ \frac{1}{2}[p(fu, Tt) + p(gt, Su)] \end{array}\right\}\right) \\ &\leq \alpha(p(Su, Tt)) - \beta(p(Su, Tt)) < \psi(p(Su, u)). \end{aligned}$$

Is a contradiction. Hence, $u = Su = fu$.

Thus, $Tu = gu = Su = fu = u$.

Hence, u is a common fixed point of S, T, f and g.

Let w be another common fixed point of S, T, f and g.

Since $\frac{1}{2}\min\{p(fu, Su), p(gw, Tw)\} \leq p(fu, gw)$, from (2.1.1), we obtain

$$\begin{aligned} \psi(p(u, w)) &= \psi(p(Su, Tw)) \\ &\leq \alpha\left(\max\left\{\begin{array}{l} p(fu, gw), p(fu, Su), p(gw, Tw), \\ \frac{1}{2}[p(fu, Tw) + p(gw, Su)] \end{array}\right\}\right) \\ &\quad - \beta\left(\max\left\{\begin{array}{l} p(fu, gw), p(fu, Su), p(gw, Tw), \\ \frac{1}{2}[p(fu, Tw) + p(gw, Su)] \end{array}\right\}\right) \\ &\leq \alpha\left(\max\left\{\begin{array}{l} p(u, w), p(u, u), p(w, w), \\ \frac{1}{2}[p(u, w) + p(w, u)] \end{array}\right\}\right) \\ &\quad - \beta\left(\max\left\{\begin{array}{l} p(u, w), p(u, u), p(w, w), \\ \frac{1}{2}[p(u, w) + p(w, u)] \end{array}\right\}\right) \\ &\leq \alpha(p(u, w)) - \beta(p(u, w)) \\ &< \psi(p(u, w)), \end{aligned}$$

which is a contradiction. Hence, $u = w$.

Thus, u is the unique common fixed point of S, T, f and g.

Sub case(b) : Suppose $\frac{1}{2}p(y_{2n}, y_{2n+1}) \leq p(y_{2n}, u)$.

In this case, also, we can prove that u is the unique common fixed point of S, T, f and g by proceeding as in Subcase(a).

Case(ii): Suppose $y_{2m} = y_{2m+1}$ for some m.

Assume that $y_{2m+1} \neq y_{2m+2}$.

$$M(x_{2m+2},x_{2m+1}) = \max\left\{\begin{array}{c} p(y_{2m+1},y_{2m}),p(y_{2m+1},y_{2m+2}),p(y_{2m},y_{2m+1}), \\ \frac{1}{2}[p(y_{2m+1},y_{2m+1})+p(y_{2m},y_{2m+2})] \end{array}\right\}.$$

However, $p(y_{2m+1},y_{2m}) = p(y_{2m+1},y_{2m+1}) \le p(y_{2m+1},y_{2m+2})$, from (p_2) and

$$\begin{aligned} &\frac{1}{2}[p(y_{2m+1},y_{2m+1}) + p(y_{2m},y_{2m+2})] \\ &\le \frac{1}{2}[p(y_{2m},y_{2m+1}) + p(y_{2m+1},y_{2m+2})], \text{ from } (p_4) \\ &\le \frac{1}{2}[p(y_{2m+1},y_{2m+2}) + p(y_{2m+1},y_{2m+2})] \\ &= p(y_{2m+1},y_{2m+2}). \end{aligned}$$

Hence, $M(x_{2m+2},x_{2m+1}) = p(y_{2m+1},y_{2m+2})$.

$$\begin{aligned} \text{Since } &\frac{1}{2}\min\{p(fx_{2m+2},Sx_{2m+2}),p(gx_{2m+1},Tx_{2m+1})\} \\ &\le p(gx_{2m+1},Tx_{2m+1}) \\ &= p(fx_{2m+2},gx_{2m+1}), \end{aligned}$$

from (2.1.1), we get

$$\begin{aligned} \psi(p(y_{2m+2},y_{2m+1})) &= \psi(p(Sx_{2m+2},Tx_{2m+1})) \\ &\le \alpha(M(x_{2m+2},x_{2m+1})) - \beta(M(x_{2m+2},x_{2m+1})) \\ &= \alpha(p(y_{2m+2},y_{2m+1})) - \beta(p(y_{2m+2},y_{2m+1})) \\ &< \psi(p(y_{2m+2},y_{2m+1})). \end{aligned}$$

It is a contradiction. Hence, $y_{2m+2} = y_{2m+1}$.

Continuing in this way, we can conclude that $y_n = y_{n+k}$ for all $k > 0$.

Thus, $\{y_n\}$ is a Cauchy sequence.

The rest of the proof follows as in Case(i). □

The following example illustrates our Theorem 2.1

Example 2.2 Let $X = [0,1]$ and $p(x,y) = \max\{x,y\}$ for all $x,y \in X$. Let $f,g,S,T : X \to X, f(x) = \frac{x}{2}, g(x) = \frac{x}{3}, S(x) = \frac{x}{4}$ and $T(x) = \frac{x}{6}$, Let $\psi,\alpha,\beta : [0,\infty) \to [0,\infty)$ be defined by $\psi(t) = 4t$, $\alpha(t) = 7t$ and $\beta(t) = \frac{7t}{2}$. Clearly, ψ is an altering distance function and α is continuous and β is lower semi continuous, $\alpha(0) = \beta(0) = 0$ and $\psi(t) - \alpha(t) + \beta(t) = \frac{t}{2} > 0$, for all $t > 0$.

Now

$$\begin{aligned} \frac{1}{2}\min\{p(fx,Sx),p(gy,Ty)\} &= \frac{1}{2}\min\{\max\{fx,Sx\},\max\{gy,Ty\}\} \\ &= \frac{1}{2}\min\left\{\max\left\{\frac{x}{2},\frac{x}{4}\right\},\max\left\{\frac{y}{3},\frac{y}{6}\right\}\right\} \\ &= \frac{1}{2}\min\left\{\frac{x}{2},\frac{y}{3}\right\} \\ &\le \frac{1}{2}\max\left\{\frac{x}{2},\frac{y}{3}\right\} \\ &\le p(fx,gy). \end{aligned}$$

$$\begin{aligned} \psi(p(Sx,Ty)) &= 4p(Sx,Ty) \\ &= 4\max\left\{\frac{x}{4},\frac{y}{6}\right\} \\ &= 4 \times \frac{1}{2}\max\left\{\frac{x}{2},\frac{y}{3}\right\} \\ &= 2p(fx,gy) \\ &\le 2M(x,y) \\ &\le 7M(x,y) - \frac{7}{2}M(x,y). \end{aligned}$$

So

$$\psi(p(Sx,Ty)) \le \alpha(M(x,y)) - \beta(M(x,y)).$$

Therefore, all of the conditions of Theorem 2.1 are satisfied and 0 is the unique common fixed point of S, T, f and g.

Acknowledgements The authors are thankful to referee for his valuable suggestions.

References

1. Abdeljawad, T., Karapınar, E., Tas, K.: Existence and uniqueness of a common fixed point on partial metric spaces. Appl. Math. Lett. **24**(11), 1900–1904 (2011)
2. Altun, I., Sola, F., Simsek, H.: Generalized contractions on partial metric spaces. Topol. Appl. **157**(18), 2778–2785 (2010)
3. Altun, I., Erduran, A.: Fixed point theorems for monotone mappings on partial metric spaces. Fixed Point Theory and Applications Article ID 508730, p. 10 (2011). doi:10.1155/2011/508730
4. Aydi, H.: Fixed point results for weakly contractive mappings in ordered partial metric spaces. J. Adv. Math. Stud. **4**(2), 01–12 (2011)
5. Fadail, Z.M., Ahmed, A.G.B., Ansar, A.H., Radenovic, S., Rajovic, M.: Some common fixed point results of mappings in 0-s-complete metric-like spaces via new functions. Appl. Math. Sci. **9**(83), 5009–5027 (2015)
6. Heckmann, R.: Approximation of metric spaces by partial metric spaces. Appl. Categ. Struct. **7**(1–2), 71–83 (1999)
7. Ilić, D., Pavlović, V., Rakočević, V.: Some new extensions of Banach's contraction principle to partial metric spaces. Appl. Math. Lett. **24**(8), 1326–1330 (2011)
8. Kadelburg, Z., Radenovic, S.: Fixed points under $\alpha - \beta$ conditions in ordered partial metric spaces. Int. J. Anal. Appl. **5**(1), 91–101 (2014)
9. Karapınar, E., Erhan, I.M.: Fixed point theorems for operators on partial metric spaces. Appl. Math. Lett. **24**(11), 1894–1899 (2011)
10. Karapınar, E.: Generalizations of Caristi Kirk's theorem on partial metric spaces. Fixed Point Theory Appl. **2011**, 4 (2011)
11. Karapinar, E., Erhan, I.M., Aksoy, U.: Weak ψ-contractions on partially ordered metric spaces and applications to boundary value problems. Bound. Value Probl. **2014**, 149 (2014). doi:10.1186/s13661-014-0149-8
12. Matthews, S.G.: Partial metric topology. In: Proceedings of the 8th Summer Conference on General Topology and Applications 1994. vol. 728, pp. 183–197. Annals of the New York Academy of Sciences (1994)
13. Mustafa, Z., Huang, H., Radenovic, S.: Some remarks on the paper "Some fixed point generalizations are not real generalizations". J. Adv. Math. Stud. 05/11/2015 (to appear)

14. Radenovic, S.: Classical fixed point results in 0-complete partial metric spaces via cyclic-type extension. Allahabad Math. Soc. **31**(1), 39–55 (2016)
15. Rao, K.P.R., Kishore, G.N.V.: A unique common fixed point theorem for four maps under $\psi - \phi$ contractive condition in partial metric spaces. Bull. Math. Anal. Appl. **3**(3), 56–63 (2011)
16. Romaguera, S.: A Kirk type characterization of completeness for partial metric spaces. Fixed Point Theory Appl. Article ID 493298, p. 6 (2010)
17. Shukla, S., Radenovic, S., Vetro, C.: Set-valued Hardy–Rogers type ontraction in 0-complete partial metric spaces. Int. J. Math. Math. Sci. Article ID 652925, p. 9 (2014)
18. Suzuki, T.: A generalized Banach contraction principle which characterizes metric completeness. Proc. Am. Math. Soc. **136**, 1861–1869 (2008)
19. Valero, O.: On Banach fixed point theorems for partial metric spaces. Appl. Gen. Topol. **6**(2), 229–240 (2005)

21

On a p-Kirchhoff-type problem arising in ecosystems

S. H. Rasouli[1] · Z. Firouzjahi[1]

Abstract In this article, we discuss the existence of positive solutions for an ecological model of the form:

$$\begin{cases} -M(\int_\Omega \mid \nabla u \mid^p \, dx)\Delta_p u = \dfrac{au^{p-1} - bu^{\gamma-1} - c}{u^\alpha}, & x \in \Omega, \\ u = 0, & x \in \partial\Omega, \end{cases}$$

where Ω is a bounded domain with smooth boundary, $\Delta_p u = \text{div}(|\nabla u|^{p-2}\nabla u)$, $1<p<\gamma$, $M : [0,\infty) \longrightarrow (0,\infty)$ is a continuous and increasing function, $a > 0$, $b > 0$, $c \geq 0$, and $\alpha \in (0,1)$. This model describes the steady states of a logistic growth model with grazing and constant yield harvesting. It also describes the dynamics of the fish population with natural predation and constant yield harvesting. We discuss the existence of a positive solution for given a, b, γ and small values of c.

Keywords Positive solutions · Sub-supersolutions · p-Kirchhoff-type problems

Mathematics Subject Classification 35J55 · 35J65

Introduction

In this paper, we are interested in the existence of positive solutions for the p-Kirchhoff-type problems

$$\begin{cases} -M(\int_\Omega \mid \nabla u \mid^p \, dx)\Delta_p u = \dfrac{au^{p-1} - bu^{\gamma-1} - c}{u^\alpha}, & x \in \Omega, \\ u = 0, & x \in \partial\Omega, \end{cases} \tag{1}$$

where $M : [0,\infty) \longrightarrow (0,\infty)$ is a continuous and increasing function, $c \geq 0$, $a, b > 0$, Ω is a bounded domain with smooth boundary, Δ_p denotes the p-Laplacian operator defined by $\Delta_p z = \text{div}\,(|\nabla z|^{p-2}\nabla z)$, $1<p<\gamma$ and $\alpha \in (0,1)$.

Here u is the population density and $\frac{au^{p-1}-bu^{\gamma-1}}{u^\alpha}$ represents logistics growth. This model describes grazing of a fixed number of grazers on a logistically growing species (see [11]). The herbivore density is assumed to be a constant which is a valid assumption for managed grazing systems and the rate of grazing is given by $\frac{c}{u^\alpha}$. At high levels of vegetation density this term saturates to c as the grazing population is a constant. This model has also been applied to describe the dynamics of fish populations (see [15]). In the case of the fish population the term $\frac{c}{u^\alpha}$ corresponds to natural predation.

In recent years, problems involving Kirchhoff-type operators have been studied in many papers, we refer to [3, 4, 6, 10, 14] in which the authors have used the variational and topological methods to get the existence of solutions. In this article, we are motivated by the ideas introduced in [7, 12, 13] and properties of Kirchhoff-type operators in [3, 4, 6], we study problem (1) in semipositone case (i.e., $\lim_{s\longrightarrow 0^+} f(s) = -\infty$; $f(s) = \frac{as^{p-1}-bs^{\gamma-1}-c}{s^\alpha}$); see [5, 7–9]). Using sub-supersolution techniques, we prove the existence of a positive solution for the problem.

To precisely state our existence result we consider the eigenvalue problem

$$\begin{cases} -\Delta_p \phi = \lambda \mid \phi \mid^{p-2} \phi, & x \in \Omega, \\ \phi = 0, & x \in \partial\Omega. \end{cases} \tag{2}$$

Let ϕ be the eigenfunction corresponding to the first eigenvalue λ_1 of (3) such that $\phi(x) > 0$ in Ω and $||\phi||_\infty = 1$. It can be shown that $\frac{\partial\phi}{\partial n} < 0$ on $\partial\Omega$. Here n is the outward normal. Let $m, \delta > 0$ and $\mu > 0$ be such that:

✉ S. H. Rasouli
s.h.rasouli@nit.ac.ir

[1] Department of Mathematics, Faculty of Basic Sciences, Babol University of Technology, Babol, Iran

$$\mu \leq \phi \leq 1, \quad x \in \Omega - \overline{\Omega_\delta}, \tag{3}$$

$$|\nabla \phi|^p \geq m, \quad x \in \overline{\Omega_\delta}, \tag{4}$$

with $\overline{\Omega_\delta} := \{x \in \Omega | d(x, \partial\Omega) \leq \delta\}$. This is possible since $|\nabla\phi|^p \neq 0$ on $\partial\Omega$ while $\phi = 0$ on $\partial\Omega$. We will also consider the unique solution $e \in W_0^{1,p}(\Omega)$ of the boundary value problem

$$\begin{cases} -\Delta_p e = 1, & x \in \Omega, \\ e = 0, & x \in \partial\Omega, \end{cases}$$

to discuss our existence result, it is known that $e > 0$ in Ω and $\frac{\partial e}{\partial n} < 0$ on $\partial\Omega$.

Existence results

In this section, we shall establish our existence result via the method of sub-supersolution. A function ψ is said to be a subsolution of (1), if it is in $W_0^{1,p}(\Omega)$ such that

$$- M\left(\int_\Omega |\nabla\psi|^p \,dx\right) \int_\Omega |\nabla\psi|^{p-2} \nabla\psi \cdot \nabla w dx \leq \int_\Omega \left[\frac{a\psi^{p-1} - b\psi^{\gamma-1} - c}{\psi^\alpha}\right] w dx,$$

and z is said supersolution of (1), if it is in $W_0^{1,p}(\Omega)$ such that

$$- M\left(\int_\Omega |\nabla z|^p \,dx\right) \int_\Omega |\nabla z|^{p-2} \nabla z \cdot \nabla w dx \geq \int_\Omega \left[\frac{az^{p-1} - bz^{\gamma-1} - c}{z^\alpha}\right] w dx,$$

for all $w \in W = \{w \in C_0^\infty(\Omega) | w \geq 0, x \in \Omega\}$. Then the following result holds:

Then the following result holds:

Lemma 2.1 (See [1, 2, 8]) Suppose there exist sub and supersolutions ψ and z respectively of (1) such that $\psi \leq z$. Then (1) has a solution u such that $\psi \leq u \leq z$

Now we state our main result.

Theorem 2.2 Let there exist constants $M_0 > 0$ and $M_\infty \geq 0$ such that $M_0 \leq M(t) \leq M_\infty$ for all $t \in [0, \infty)$. Given $a, b > 0$, $1 < p < \gamma$, and $\alpha \in (0, 1)$, there exists a constant $c_1 = c_1(a, b, \alpha, \gamma, p, \Omega) > 0$ such that for $c < c_1$, (1) has a positive solution.

Remark 2.3 In the nonsingular case ($\alpha = 0$), positive solutions exist only when $a > \lambda_1$(the principle eigenvalue) (see [12, 13]). But in the singular case, we establish the existence of a positive solution for any $a > 0$.

Proof of Theorem 2.2 We start with the construction of a positive subsolution for (1). Fix $\beta \in (1, \frac{p}{p-1+\alpha})$. Define $\psi = k\phi^\beta$, where $k > 0$ is such that $a \geq 2bk^{\gamma-p} + M_\infty \beta^{p-1}\lambda_1 k^\alpha$. Define

$$c_1 := \min\left\{M_\infty k^{p-1+\alpha}\beta^{p-1}(\beta-1)(p-1)m^p, \frac{1}{2}M_\infty k^{p-1}\mu^{\beta(p-1)}\left(a - \beta^{p-1}\lambda_1 k^\alpha\right)\right\}.$$

Note that $c_1 > 0$ by the choice of k and β. A calculation shows that

$$\nabla\psi = k\beta\phi^{\beta-1},$$

and

$$\begin{aligned}
&-M\left(\int_\Omega |\nabla\psi|^p \,dx\right)\Delta_p\psi \\
&= M\left(\int_\Omega |\nabla\psi|^p \,dx\right)\int_\Omega |\nabla\psi|^{p-2}\nabla\psi \cdot \nabla w dx \\
&= k^{p-1}\beta^{p-1}M\left(\int_\Omega |\nabla\psi|^p \,dx\right) \\
&\quad \times \int_\Omega \phi^{(\beta-1)(p-1)}|\nabla\phi|^{p-2}\nabla\phi\nabla w dx \\
&= k^{p-1}\beta^{p-1}M\left(\int_\Omega |\nabla\psi|^p \,dx\right) \\
&\quad \times \int_\Omega |\nabla\phi|^{p-2}\nabla\phi\left[\nabla(\phi^{(\beta-1)(p-1)}w) - w\nabla(\phi^{(\beta-1)(p-1)})\right]dx \\
&= k^{p-1}\beta^{p-1}M\left(\int_\Omega |\nabla\psi|^p \,dx\right) \\
&\quad \times \int_\Omega |\nabla\phi|^{p-2}\nabla\phi\nabla(\phi^{(\beta-1)(p-1)}w)dx \\
&\quad - k^{p-1}\beta^{p-1}M\left(\int_\Omega |\nabla\psi|^p \,dx\right) \\
&\quad \times \int_\Omega |\nabla\phi|^{p-2}\nabla\phi\nabla(\phi^{(\beta-1)(p-1)})w dx \\
&= k^{p-1}\beta^{p-1}M\left(\int_\Omega |\nabla\psi|^p \,dx\right) \\
&\quad \times \int_\Omega \left[\lambda_1\phi^{\beta(p-1)} - (\beta-1)(p-1)|\nabla\phi|^p\phi^{-p+\beta(p-1)}\right]w dx \\
&\leq M_\infty k^{p-1}\beta^{p-1}\int_\Omega [\lambda_1\phi^{\beta(p-1)} \\
&\quad - (\beta-1)(p-1)|\nabla\phi|^p\phi^{-p+\beta(p-1)}]w dx.
\end{aligned}$$

Thus ψ is a subsolution of (1) if

$$\begin{aligned}
&M_\infty k^{p-1}\beta^{p-1}\left[\lambda_1\phi^{\beta(p-1)} - (\beta-1)(p-1)|\nabla\phi|^p\phi^{-p+\beta(p-1)}\right] \\
&\leq ak^{p-1-\alpha}\phi^{\beta(p-1-\alpha)} - bk^{\gamma-1-\alpha}\phi^{\beta(\gamma-1-\alpha)} - \frac{c}{k^\alpha\phi^{\alpha\beta}}.
\end{aligned}$$

For this, we have to show the following three inequalities:

$$-k^{p-1-\alpha}\phi^{\beta(p-1-\alpha)}\left(a-M_\infty k^\alpha\beta^{p-1}\lambda_1\phi^{\alpha\beta}\right)$$
$$\leq -2bk^{\gamma-1-\alpha}\phi^{\beta(\gamma-1-\alpha)}, \quad x\in\Omega,$$

$$-\frac{1}{2}k^{p-1-\alpha}\phi^{\beta(p-1-\alpha)}(a-M_\infty k^\alpha\beta^{p-1}\lambda_1\phi^{\alpha\beta})$$
$$\leq -\frac{c}{k^\alpha\phi^{\alpha\beta}}, \quad x\in\Omega-\overline{\Omega_\delta},$$

$$-M_\infty k^{p-1}\beta^{p-1}(p-1)(\beta-1)\frac{|\nabla\phi|^p}{\phi^{p-\beta(p-1)}}$$
$$\leq -\frac{c}{k^\alpha\phi^{\alpha\beta}}, \quad x\in\overline{\Omega_\delta},$$

by the choice of k, we have:

$$-\frac{1}{2}k^{p-1-\alpha}\phi^{\beta(p-1-\alpha)}(a-M_\infty k^\alpha\beta^{p-1}\lambda_1\phi^{\alpha\beta})$$
$$\leq -bk^{\gamma-1-\alpha}\phi^{\beta(p-1-\alpha)}$$
$$\leq -bk^{\gamma-1-\alpha}\phi^{\beta(\gamma-1-\alpha)}. \tag{5}$$

Now, we have in $\overline{\Omega_\delta}$, $|\nabla\phi|^p\geq m$, and $c<M_\infty k^{p-1+\alpha}\beta^{p-1}(\beta-1)(p-1)m^p$, then the following inequalities hold:

$$-M_\infty k^{p-1}\beta^{p-1}(p-1)(\beta-1)\frac{|\nabla\phi|^p}{\phi^{p-\beta(p-1)}}$$
$$\leq\frac{-M_\infty k^{p-1+\alpha}\beta^{p-1}(\beta-1)(p-1)m^p}{k^\alpha\phi^{\alpha\beta}\phi^{p-\beta(p-1)-\alpha\beta}}$$
$$\leq -\frac{c}{k^\alpha\phi^{\alpha\beta}\phi^{p-\beta(p-1)-\alpha\beta}}.$$

On the other hand, since $p-\beta(p-1+\alpha)>0$,

$$-\frac{c}{k^\alpha\phi^{\alpha\beta}\phi^{p-\beta(p-1)-\alpha\beta}}\leq -\frac{c}{k^\alpha\phi^{\alpha\beta}}.$$

Hence

$$-M_\infty k^{p-1}\beta^{p-1}(p-1)(\beta-1)\frac{|\nabla\phi|^p}{\phi^{p-\beta(p-1)}}\leq -\frac{c}{k^\alpha\phi^{\alpha\beta}}. \tag{6}$$

Finally, in $\Omega-\overline{\Omega_\delta}$ using $\phi\geq\mu$ and $c<\frac{1}{2}M_\infty k^{p-1}\mu^{\beta(p-1)}(a-\beta^{p-1}\lambda_1 k^\alpha)$, we have:

$$-\frac{1}{2}k^{p-1-\alpha}\phi^{\beta(p-1-\alpha)}(a-M_\infty k^\alpha\beta^{p-1}\lambda_1\phi^{\alpha\beta})$$
$$\leq\frac{-k^{p-1}\phi^{\beta(p-1)}(a-M_\infty k^\alpha\beta^{p-1}\lambda_1)}{2k^\alpha\phi^{\alpha\beta}}$$
$$\leq -\frac{c}{k^\alpha\phi^{\alpha\beta}}. \tag{7}$$

For $c<c_1$, by (6) and (7) the Eq. (5) holds. Thus ψ is a subsolution of (1).

Now for a supersolution choose $z:=Ne$, where $N>0$ is such that $Ne\geq\psi$ and

$$\frac{au^{p-1}-bu^{\gamma-1}-c}{M_0u^\alpha}\leq N^{p-1},$$

for all $u>0$. We have

$$-M\left(\int_\Omega|\nabla z|^p\,dx\right)\triangle_p z=M\left(\int_\Omega|\nabla z|^p\,dx\right)N^{p-1}$$
$$\geq M_0N^{p-1}$$
$$\geq\left[\frac{az^{p-1}-bz^{\gamma-1}-c}{z^\alpha}\right].$$

i.e., z is a supersolution of (1) with $z\geq\psi$ for N large (note $|\nabla e|\neq 0;\partial\Omega$). Thus, there exists a positive solution u of (1) such that $\psi\leq u\leq z$. This completes the proof of Theorem 2.2. □

Acknowledgments The authors thank the referees for their appreciation, valuable comments and suggestions.

References

1. Afrouzi, G.A., Chung, N.T., Shakeri, S.: Existence of positive solutions for Kirchhoff type equations. Electron. J. Diff. Eqs. **180**, 1–8 (2013)
2. Cui, S.: Existence and nonexistence of positive solution for singular semilinear elliptic boundary value problems, Nonlinear Anal. pp. 149–176 (2000)
3. Dai, G.: Three solutions for a nonlocal Dirichlet boundary value problems involving the p(x)-Laplacian. Appl. Anal. **92**(1), 191–210 (2013)
4. Dai, G., Ma, R.: Solutions for a p(x)-Kirchhoff type equation with Neumann boundary data. Nonlinear Anal. Real World Appl. **12**, 2666–2680 (2011)
5. Goddard ll, J., Lee, E.K., Sankar, L., Shivaji, R.: Existence results for classses of infinite semipositone problems. Boundary Value Problems, 2013:97, 1–9 (2013)
6. Han, X., Dai, G.: On the sub-supersolution method for a p(x)-Kirchhoff type equations. J. Integr. Appl. **2012**, 283 (2012)
7. Lee, E.K., Shivaji, R., Ye, J.: Positive solutions for infinite semipositone problems with falling zeros. Nonlinear Anal. **72**, 4475–4479 (2010)
8. Lee, E.K., Shivaji, R., Ye, J.: Classes of infinite semipositone systems. Proc. R. Soc. Edinb. **139A**, 853–865 (2009)
9. Lee, E.K., Shivaji, R., Ye, J.: Classes of infinite semipositone $n\times n$ systems. Differ. Integr. Eqs. **24**(3–4), 361–370 (2011)
10. Ma, T.F.: Remarks on an elliptic equation of Kirchhoff type. Nonlinear Anal. **63**, 1967–1977 (2005)
11. Noy-Meir, I.: Stability of grazing systems an application of predator-prey graphs. J. Ecol. **63**, 459–482 (1975)
12. Oruganti, S., Shi, J., Shivaji, R.: Diffusive logistic equation with constant yield harvesting, I: steady states. Trans. Am. Math. Soc. **345**(9), 3601–3619 (2002)
13. Oruganti, S., Shi, J., Shivaji, R.: Logistic equation with the p-Laplacian and constant yeild harvesting. Abstr. Appl. Anal. **9**, 723–727 (2004)
14. Ricceri, B.: On an elliptic Kirchhoff type problems depending on two parameter. J. Global Optim. **46**(4), 543–549 (2010)
15. Steele, J.H., Henderson, E.W.: Modelling long term fluctuations in fish stocks. Science **224**, 985–987 (1984)

22

Symmetric-periodic solutions for some types of generalized neutral equations

Rabha W. Ibrahim[1] · **M. Z. Ahmad**[2] · **M. Jasim Mohammed**[2]

Abstract The existence of symmetric-periodic outcomes for a class of fractional differential equations has been increasingly studied. Such study has used various methods such as fixed point theory, critical point theory, and approximation theory. In this work, we study the m-pseudo almost automorphic (m-PΛΛ) outcomes for a category of fractional neutral differential equations. To satisfy this aim, we introduce composition results under suitable conditions and employ them to establish some extant outcomes using interpolation theory mixed with fixed point technique. Examples are illustrated.

Keywords Fractional calculus · Fractional differential equations · Periodic symmetric solution

Mathematics Subject Classification 34A08

Introduction

The symmetry in the field of differential equations is a transformation that preserves its domestic of results invariant. Symmetry study can be utilized to resolve some classes of ordinary, partial, fractional differential equations, though defining the symmetries can be computationally concentrated like other mathematical methods. The best method for symmetry is by finding the periodic solution of the differential equation.

In 1962, Bochner [1] introduced the concept of almost automorphy, which is an important generalization of almost periodicity. The concept of almost periodic functions was introduced by Bohr [2]. It was named as PAA functions because they originally presented themselves, in their work in differential geometry, as scalars and tensors on manifolds with (discrete) groups of automorphisms [3]. Later, PΛΛ function has become one of the most attractive topics in the qualitative theory of evolution equations, and there have been several interesting, natural and powerful generalizations of the classical PΛΛ functions [4–9]. Recently, Digana et al., studied the concept for different classes of ODF and PDE (see [10–14]). Xiao et al. [15] introduced the concept of PΛΛ functions for a natural and a significant extension of PΛΛ functions. Moreover, they proved that the space of PΛΛ functions is complete; so they solve a key fundamental problem on this issue and pave the road to further study the applications of PΛΛ functions. They investigated the existence of PΛΛ to

$$u'(t) = \Lambda(t)u(t) + f(t)$$

and

$$u'(t) = \Lambda(t)u(t) + f(t, u(t))$$

in a Banach space. Chang and Luos [16] presented a composition theorem for m-PΛΛ function, which was proved under appropriate conditions. They applied this theorem to investigate whether the m-PΛΛ solutions exist in the neutral differential equation as follows:

✉ Rabha W. Ibrahim
rabhaibrahim@yahoo.com

M. Z. Ahmad
mzaini@unimap.edu.my

M. Jasim Mohammed
mohdmath87@gmail.com

1 Faculty of Computer Science and Information Technology, University, Malaya, 50603 Kuala Lumpur, Malaysia

2 Institute of Engineering Mathematics, Universiti Malaysia Perlis, 02600 Arau, Perlis, Malaysia

$$\frac{\mathrm{d}}{\mathrm{d}t}[u(t)+f(t,u(t))] = \Lambda u(t) + g(t,u(t)), \quad t \in \mathbb{R}.$$

Periodic motion is a very important and special phenomena not only in natural science, but also in social science, such as climate, food supplement, insecticide population and sustainable development. Periodic solutions are desired property in differential equations, constituting one of the most important research directions in the theory of differential equations. The existence of periodic solutions is often a desired property in dynamical systems, constituting one of the most important research directions in the theory of dynamical systems, with applications ranging from celestial mechanics to biology and finance. Fractional differential equations (FDEs) are the most important generalizations of the field of ODE [17–21]. Recent investigations in physics, engineering, biological sciences and other fields have demonstrated that the dynamics of many systems are described more accurately using FDEs, and that FDE with delay are often more realistic to describe natural phenomena than those without delay. Periodic solution fractional differential equations have been studied by many researchers. They studied periodic solutions of the equation (see [17, 18])

$$D^{\alpha}u(t) + BD^{\beta}u(t)\Lambda u(t) = f(t), \quad 0 \leq t \leq 2\pi,$$

where A and B are closed linear operators defined on a complex Banach space X with domains $D(A)$ and $D(B)$, respectively, $0 \leq \beta < \alpha \leq 2$.

The aim of this paper is to study the existence of periodic solutions for the following FDE:

$$D^{\mu}\Big(\upsilon(t) + \varphi(t,\upsilon(t))\Big) = \Lambda\upsilon(t) + \vartheta(t,\upsilon(t)), \quad t \in \Re \qquad (1)$$

$\forall\mu \in (0,1]$, where $\Lambda : dom(\Lambda) \subset \chi \to \chi$ is considered the operator of a hyperbolic analytic semigroup $T(t)_{t\geq 0}$, and $\varphi : \Re \times \chi \to \chi_{\delta}(<\lambda<\delta<), \vartheta \;\; \Re \times \chi \to \chi$ are appropriate continuous functions; χ_{δ} refers to the appropriate interpolation space and D^{μ} is the Riemann–Liouville fractional differential operator (R–L operator).

This paper is classified as follows. In "Setting", we present some basic definitions, lemmas, and setting results which will be used in this study. In "Findings", we introduce some existence results of almost-periodic and mild solutions of the fractional neutral differential equation. Examples are illustrated in the sequel.

Setting

The researchers allocated this section to investigate some results required in the sequel. In this paper, the notations $(\chi, \|\bullet\|)$ and $(\Upsilon, \|\bullet\|_{\Upsilon})$ denote the two Banach spaces, whereas $BC(\Re,\chi)$ refers to the Banach space of all bounded continuous functions from $\Re$ to χ, qualified with the supremum norm $\|\varphi\|_{\infty} = sup_{t\in\Re}\|\varphi(t)\|$. Let χ_{λ} be a space mediated between $dom(\Lambda)$ and $\chi.B(\Re,\chi_{\lambda})$ for $\lambda \in (0,1)$ refers to Banach space of all bounded continuous functions $\sigma : \Re \to \chi_{\lambda}$ when supported with the $\lambda - sup$ norm:

$$\|\sigma\|_{\lambda,\infty} := sup_{t\in\Re}\|\sigma(t)\|_{\lambda}$$

for $\sigma \in BC(\Re,\chi_{\lambda})$. Throughout this paper, $\wp$ denotes the Lebesgue field of $\Re$ and $\aleph$ the set of all positive measures m on $\wp$ satisfying $m(\Re) = +\infty$ and $m([a,b]) < +\infty$, for all $a,b \in \Re(\mathfrak{a}<\mathfrak{b})$.

Definition 2.1 [3] A continuous function $\varphi : \Re \to \chi$ is referred to as automorphic in the case that every sequence of real numbers $(\varsigma_{\eta})_{\eta\in N}$ has a subsequence $(\varsigma'_{\eta})_{\eta\in N} \subset (\varsigma_{\eta})_{\eta\in N}$, such that

$$\lim_{\eta,n\to\infty}\|\varphi(t+\varsigma_{\eta}-\varsigma_{n})-\varphi(t)\| = 0.$$

Define

$$\rho\Lambda\Lambda_0(\Re,\chi) = \left\{\Phi \in \mathfrak{BC}(\Re,\chi) \;\; \lim_{\mathfrak{T}\to\infty}\frac{}{\mathfrak{T}}\int_{-\mathfrak{T}}^{\mathfrak{T}}\|\varphi(\tau)\|\mathrm{d}\tau = \right\}.$$

Likely, $\rho\Lambda\Lambda_0(\Re\times\chi,\chi)$ is defined as the gathering of combined continuous functions $\varphi:\Re\times\chi\to\chi$ which belong to $BC(\Re\times\chi,\chi)$ and satisfy

$$\lim_{T\to\infty}\frac{1}{2T}\int_{-T}^{T}\|\varphi(\tau,x)\|\mathrm{d}\tau = 0$$

uniformly in a compact subset of χ.

Definition 2.2 [15] A continuous function $\varphi : \Re \to \chi$ (respectively, $\Re\times\chi\to\chi$) represents pseudo automorphic when decomposed as $\varphi = \vartheta + \Phi$, where $\vartheta \in \Lambda\Lambda(\Re,\chi)$ (respectively, $\Lambda\Lambda(\Re\times\chi,\chi)$) and $\Phi \in \rho\Lambda\Lambda_0(\Re,\chi)$(respectively, $\rho\Lambda\Lambda_0(\Re\times\chi,\chi)$)). Denote by $\rho\Lambda\Lambda(\Re,\chi)$ (respectively, $\rho\Lambda\Lambda(\Re\times\chi,\chi)$) the set of all such functions.

Definition 2.3 [19] Let $m \in \aleph$. A bounded continuous function $\varphi : \Re \to \chi$ is referred to as m-ergodic if φ is ergodic with respect to m (measure) i.e.

$$\lim_{r\to\infty}\frac{1}{m([-r,r])}\int_{[-r,r]}\|\varphi(t)\|\mathrm{d}m(t) = 0.$$

The space of all such functions is denoted as $\zeta(\Re,\chi,\mathfrak{m})$ and $(\zeta(\Re,\chi,\mathfrak{m}),\|\bullet\|_{\infty})$ is a Banach space (see [19], Proposition 2.13]). The word ergodic (work) is employed to explain the dynamical system which has the same behavior averaged during the item time as averaged over the phase space.

Definition 2.4 [19] Let $m \in \aleph$. A continuous function $\varphi : \Re \to \chi$ is stated to be m-P$\Lambda\Lambda$ if φ comes in the form

$\varphi = \vartheta + \Phi$, where $\vartheta \in \Lambda\Lambda(\mathfrak{R}, \chi)$ and $\Phi \in \zeta(\mathfrak{R}, \chi, \mathfrak{m})$. So, all such functions have a space denoted by $\rho\Lambda\Lambda(\mathfrak{R}, \chi, \mathfrak{m})$. Most clearly, we have $\Lambda\Lambda(\mathfrak{R}, \chi) \subset \rho\Lambda\Lambda(\mathfrak{R}, \chi, \mathfrak{m}) \subset BC(\mathfrak{R}, \chi)$.

Lemma 2.5 [20, Theorem 2.2.6] *If $\varphi : \mathfrak{R} \times \chi \mapsto \chi$ is P$\Lambda\Lambda$, and assume that $\varphi(t, \bullet)$ is uniformly continuous on each bounded subset $\kappa \subset \chi$ uniformly for $t \in \mathfrak{R}$, that is for any $\zeta > 0$, there exists $S > 0$ such that $x, y \in \kappa$ and $\|x - y\| < S$ imply that $\|\varphi(t, x) - \varphi(t, y)\| < \zeta$ for all $t \in \mathfrak{R}$. Let $\Phi : \mathfrak{R} \mapsto \chi$ be P $\Lambda\Lambda$. Then the function $F : \mathfrak{R} \mapsto \chi$ defined by $F(t) = \varphi(t, \Phi(t))$ is P$\Lambda\Lambda$.*

Lemma 2.6 [19, Theorem 4.1] *Let $m \in \aleph$ and $\varphi \in \rho\Lambda\Lambda(\mathfrak{R}, \chi, \mathfrak{m})$ be such that $\varphi = \vartheta + \Phi$, where $\vartheta \in \Lambda\Lambda(\mathfrak{R}, \chi)$ and $\Phi \in \zeta(\mathfrak{R}, \chi, \mathfrak{m})$. If $\rho\Lambda\Lambda(\mathfrak{R}, \chi, \mathfrak{m})$ is translation invariant, then*

$$\{\vartheta(t) : t \in \mathfrak{R}\} \subset \overline{\{\varphi(\mathrm{t}) \;\; \mathrm{t} \in \mathfrak{R}\}}$$

(the closure of the range of φ).

Lemma 2.7 [19, Theorem 2.14] *Let $m \in \aleph$ and I be the bounded interval (eventually $I = \emptyset$). Suppose that $\varphi \in BC(\mathfrak{R}, \chi)$. The assertions indicated as following are equivalent.*

(i) $\varphi \in \zeta(\mathfrak{R}, \chi, \mathfrak{m})$;

(ii) $\lim_{r \to +\infty} \frac{1}{m([-r,r] \setminus I)} \int_{[-r,r] \setminus I} \|\varphi(t)\| \mathrm{d}m(t) = 0$;

(iii) For any $\zeta > 0$, $\lim_{r \to +\infty} \frac{m(\{t \in [-r,r] \setminus I : \|\varphi(t)\| > \zeta\})}{m([-r,r] \setminus I)} = 0$.

In the sequel, we need some notions and properties of intermediate spaces and hyperbolic semi groups. Let χ and Z be Banach spaces, with norms $\| \bullet \|_{\chi}, \| \bullet \|_Z$, respectively, and assume that Z is continuously embedded in χ, that is, $Z \hookrightarrow \chi$.

Definition 2.8 The Riemann–Liouville fractional integral is defined as follows:

$$I^{\mu}u(t) = \frac{1}{\Gamma(\mu)} \int_0^t (t - \varsigma)^{\mu - 1} u(\varsigma) \mathrm{d}\varsigma,$$

where Γ denotes the gamma function (see [22, 23]).

Definition 2.9 The Riemann–Liouville fractional derivative is defined as follows:

$$D^{\mu}u(t) = \frac{1}{\Gamma(1 - \mu)} \frac{\mathrm{d}}{\mathrm{d}t} \int_0^t (t - \varsigma)^{-\mu} u(\varsigma) \mathrm{d}\varsigma, \quad 0 < t < \infty.$$

Definition 2.10 [8, *Definition* 2.5] A semi group $(T(t))_{t \geq 0}$ on χ is stated to be hyperbolic if there is a projection ρ and constants $\aleph, S > 0$ such that each $T(t)$ commutes with $\rho, Ker\rho$ is invariant with respect to $T(t), T(t) : ImQ \to ImQ$ is invertible and for every $x \in \chi$

$$\|T(t)\rho x\| \leq M\varrho^{-st}\|x\|, \quad fort \geq 0; \tag{2}$$

$$\|T(t)Qx\| \leq M\varrho^{-st}\|x\|, \quad fort \leq 0, \tag{3}$$

where $Q := I - \rho$ and, for $t < 0, T(t) = T(-t)^{-1}$.

Definition 2.11 [14] A linear operator $\Lambda : dom(\Lambda) \subset \chi \to \chi$ (not necessarily densely defined) is referred to as sectorial if the following hold: there exist constants $\mho \in \mathfrak{R}, \theta \in (\frac{\pi}{}, \pi)$ and $M > 0$ such that

$$p(\Lambda) \subset \varsigma_{\theta,\mho} := \{\alpha \in \ell : \alpha \neq \mho, |arg(\alpha - \mho)| < \theta\},$$

$$\|\mathfrak{R}(\alpha, \Lambda)\| \leq \frac{\mathfrak{M}}{|\alpha - \mho|'}, \quad \alpha \in \varsigma_{\theta,\mho}.$$

Definition 2.12 [8, *Definition* 2.7] Let $0 \leq \lambda \leq 1$. A Banach space Υ such that $Z \hookrightarrow \Upsilon \hookrightarrow \chi$ refers to the class J_{λ} between χ and Z if there is a constant $c > 0$ such that

$$\|x\|_{\Upsilon} \leq c\|x\|^{1-\lambda}\|x\|_Z^{\lambda}(x \in Z).$$

In this case, we write $\Upsilon \in J_{\lambda}((X), Z)$.

Definition 2.13 [8, *Definition* 2.8] Let $\Lambda : dom(\Lambda) \subset \chi \to \chi$ be a sectorial operator. A Banach space $(\chi_{\lambda}, \| \bullet \|_{\lambda}), \lambda \in (0, 1)$ is said to be an intermediate space between χ and $dom(\Lambda)$ if $\chi_{\lambda} \in J_{\lambda}$.

For the problem (1), we list the following assumptions:

(H1) If $0 \leq \lambda < \delta < 1$, then we let k_1 be the bound of the embedding $\chi_{\lambda} \hookrightarrow \chi$, that is

$$\|\upsilon\| \leq k_1 \|\upsilon\|_{\lambda} \quad \text{for } \upsilon \in \chi_{\lambda}.$$

(H2) Let $0 \leq \lambda < \delta < 1$ and the function $\varphi : \mathfrak{R} \times \chi \to \chi_{\delta}$ belongs to $\rho\Lambda\Lambda(\mathfrak{R}, \chi_{\delta}, \mathfrak{m})$, while $\vartheta : \mathfrak{R} \times \chi \to \chi$ belongs to $\rho\Lambda\Lambda(\mathfrak{R}, \chi, \mathfrak{m})$. Moreover, the functions φ, ϑ are uniformly Lipschitz in rotation to the second following argument: there exist $K > 0$ such that

$$\|\varphi(t, \upsilon) - \varphi(t, \upsilon)\|_{\delta} \leq K\|\upsilon - v\|$$

and

$$\|\vartheta(t, \upsilon) - \vartheta(t, \upsilon)\| \leq K\|\upsilon - v\|$$

for all $\upsilon, v \in \chi$ and $t \in \mathfrak{R}$.

Findings

In the present section, a composition theorem is proved for m-P$\Lambda\Lambda$ functions under appropriate conditions. Then, we apply this composition theorem to obtain some results regarding Eq. (1).

Auxiliary outcomes

Theorem 3.1 *Let* $m \in \aleph$ *and* $\varphi = \vartheta + \hbar \in \rho\Lambda\Lambda(\mathfrak{R} \times \chi, \chi, \mathfrak{m})$. *Suppose that*

(H3) $\varphi(t, \chi)$ *is uniformly continuous on any bounded subset* $\kappa \subset \chi$ *uniformly in* $t \in \mathfrak{R}$;

(H4) $\vartheta(t, \chi)$ *is uniformly continuous on any bounded subset* $\kappa \subset \chi$ *uniformly in* $t \in \mathfrak{R}$.

If $\Phi \in \rho\Lambda\Lambda(\mathfrak{R}, \chi, \mathfrak{m})$ *then* $F(\bullet) := \varphi(\bullet, \Phi(\bullet)) \in \rho\Lambda\Lambda(\mathfrak{R} \times \chi, \mathfrak{m})$.

Proof Let $\varphi = \vartheta + \hbar$ with $\vartheta \in \Lambda\Lambda(\mathfrak{R} \times \chi, \chi), \mathfrak{h} \in \zeta(\mathfrak{R} \times \chi, \chi, \mathfrak{m})$ and $\Phi = \upsilon + \nu$, with $\upsilon \in \Lambda\Lambda(\mathfrak{R}, \chi)$ and $\nu \in \zeta(\mathfrak{R}, \chi, \mathfrak{m})$. Define a function f as follows:

$$\begin{aligned} f(t) &:= \vartheta(t, \upsilon(t)) + \varphi(t, \Phi(t)) - \vartheta(t, \upsilon(t)) \\ &= \vartheta(t, \upsilon(t)) + \varphi(t, \Phi(t)) - \varphi(t, \upsilon(t)) + \hbar(t, \upsilon(t)). \end{aligned}$$

Let us restate

$$G(t) = \vartheta(t, \upsilon(t)), \phi(t) = \varphi(t, \Phi(t)) - \varphi(t, \upsilon(t)), H(t) = \hbar(t, \upsilon(t)).$$

Therefore, we obtain $f(t) = G(t) + \phi(t) + H(t)$. By Lemma 2.5, we conclude that $G(t) \in \Lambda\Lambda(\mathfrak{R}, \chi)$ and obviously $\phi(t) \in BC(\mathfrak{R}, \chi)$. We proceed to show that $\phi(t) \in \zeta(\mathfrak{R}, \chi, \mathfrak{m})$. It suffices to prove that

$$\mathrm{Lim}_{r\to\infty} \frac{1}{m([-r, r])} \int_{[-r,r]} \|\phi(t)\| \mathrm{d}m(t) = 0.$$

By Lemma 2.6 , $\upsilon(\mathfrak{R}) \subset \overline{\Phi(\mathfrak{R})}$ which is a bounded set. From the hypothesis (*H*3) with $\kappa = \overline{\Phi(\mathfrak{R})}$ yields that for each $\zeta > 0$, there is an existence of a constant $S > 0$ such that for all $t \in \mathfrak{R}$,

$$\|\Phi - \upsilon\| \leq S \Rightarrow \|\varphi(t, \Phi(t)) - \varphi(t, \upsilon(t))\| \leq \zeta.$$

From the following set: $\Lambda_{r,\zeta} = \{t \in [-r, r] : \|\varphi(t)\| > \zeta\}$, we get

$$\begin{aligned} \Lambda_{r,\zeta}(\phi) &= \Lambda_{r,\zeta}(\varphi(t, \Phi(t)) - \varphi(t, u(t))) \\ &\subset \Lambda_{r,S}(\Phi(t) - \upsilon(t)) = \Lambda_{r,S}(\nu). \end{aligned}$$

Therefore, the following inequality carries:

$$\begin{aligned} &\frac{m(\{t \in [-r, r] : \|\varphi(t, \Phi(t)) - \varphi(t, \upsilon(t))\| > \zeta\})}{m([-r, r])} \\ &\leq \frac{m(\{t \in [-r, r] : \|\Phi(t) - \upsilon(t)\| > S\})}{m([-r, r])}. \end{aligned}$$

Since $\Phi(t) = \upsilon(t) + \nu(t)$ and $\nu \in \zeta(\mathfrak{R}, \chi, \mathfrak{m})$, Lemma 2.7 states that for the above-mentioned S, we have

$$\lim_{r\to\infty} \frac{m(\{t \in [-r, r] : \|\varphi(t - \upsilon(t)\| > \zeta\})}{m([-r, r])} = 0,$$

and then we get

$$\lim_{r\to\infty} \frac{m(\{t \in [-r, r] : \|\varphi(t, \Phi(t)) - \varphi(t, \upsilon(t))\| > \zeta\})}{m([-r, r])} = 0. \tag{4}$$

Again, in view of Lemma 2.7 and Eq. (8), we attain $\phi(t) \in \zeta(\mathfrak{R}, \chi, \mathfrak{m})$. Finally, we have to prove $H(t) = \hbar(t, \upsilon(t)) \in \zeta(\mathfrak{R}, \chi, \mathfrak{m})$. Since υ is continuous on $\mathfrak{R}$ as PΛΛ function, the set $\upsilon([-r, r])$ can be taken as compact. Therefore, the function $\vartheta \in \Lambda\Lambda(\mathfrak{R} \times \chi, \chi)$, and ϑ is uniformly continuous on $[-r, r] \times \upsilon([-r, r])$. Then, (H3) implies that $\hbar(t, \chi)$ is uniformly continuous with $X \in \upsilon([-r, r])$ uniformly in $t \in ([-r, r])$. Thus for any $\zeta > 0$, a constant $S > 0$ exists such that for $X_1, X_2 \in \upsilon([-r, r])$ with $\|X_1 - X_2\| < S$, we have

$$\|\hbar(t, X_1) - \hbar(t, X_2)\| < \frac{\zeta}{2}, \quad \forall t \in [-r, r]. \tag{5}$$

By the compactness of the set $\upsilon([-r, r])$, we conclude that there is an existence of finite balls Θ_K with $\delta_K \in \upsilon([-r, r]), K = 1, \ldots, n$ and radius S indicated above, such that $\upsilon([-r, r]) \subset \cup_n^K \Theta_K$.

Then the sets $U_K := \{t \in [-r, r] : \upsilon(t) \in \Theta_K\}, K = 1, \ldots, n$ are open in $[-r, r] = \cup_{K=1}^n U_K$. Define V_K by

$$V_1 = U_1, V_K = U_K - \cup_{i=1}^{K-1} U_i, \quad 2 \leq K \leq n.$$

Then it is clear that $V_i \cap V_j = \emptyset$, if $i \neq j, 1 \leq i, j \leq n$. So, we obtain

$$\begin{aligned} \wedge &:= \{t \in [-r, r] : \|H(t)\| \geq \zeta\} = \{t \in [-r, r] : \|\hbar(t, \upsilon(T))\| \geq \zeta\} \\ &\subset \cup_{K=1}^n \{t \in V_K : \|\hbar(t, \upsilon(t)) - \hbar(t, \delta_K)\| + \|\hbar(t, \delta_K)\| \geq \zeta\} \\ &\subset \cup_{K=1}^n \left(\left\{ t \in V_K : \|\hbar(t, \upsilon(t)) - \hbar(t, \delta_K)\| \geq \frac{\zeta}{2} \right\} \right. \\ &\left. \cup \left\{ t \in V_K : \|\hbar(t, \delta_K)\| \geq \frac{\zeta}{2} \right\} \right). \end{aligned}$$

By (9), we obtain

$$\left\{ t \in V_K : \|\hbar(t, \upsilon(t)) - \hbar(t, \delta_K)\| \geq \frac{\zeta}{2} \right\} = \emptyset, \quad K = 1, \ldots, n.$$

Thus , if we set $\Lambda_{r,\frac{\zeta}{2}}(\hbar_K) := \Lambda_{r,\frac{\zeta}{2}}(\hbar(t, \delta_K))$, then $\Lambda_{r,\zeta}(H) \subset U_{K=1}^n \Lambda_{r,\frac{\zeta}{2}}(\hbar_K)$ and

$$\frac{1}{m([-r, r])} \int_{[-r,r]} \|H(t)\| \mathrm{d}m(t) \leq \sum_{K=1}^n \frac{1}{m([r, -r])} \int_{[r,-r]} \|H(t)\| \mathrm{d}m(t).$$

Since $\hbar \in \zeta(\mathfrak{R} \times \chi, \chi, \mathfrak{m})$, we have

$$\lim_{r\to\infty} \frac{1}{m([r, -r])} \int_{[r,-r]} \|\hbar_K(t)\| \mathrm{d}m(t) = 0, \quad K = 1, \ldots, n.$$

It indicates that $\lim_{r\to\infty} \frac{1}{m([r,-r])} \int_{[r,-r]} \|H(t)\| \mathrm{d}m(t) = 0$. According to Lemma 2.7, we impose

$H(t) = \hbar(t, \upsilon(t)) \in \zeta(\Re, \chi, \mathfrak{m})$.

This ends the proof. □

Throughout the remaining parts of this paper, it is proposed that there is existence of two real numbers λ, δ such that $0<\lambda<\delta<1$ and $2\delta > \lambda + 1$. Moreover, we define the following fractional operators $\gamma_1^\mu, \gamma_2^\mu, \gamma_3^\mu$, and γ_4^μ by

$$(\gamma_1^\mu(\upsilon)(t)) := \int_{-\infty}^{t} \frac{\Lambda T^\mu(t-\varsigma)\rho}{\Gamma(\mu)} \varphi(\varsigma, \upsilon(\varsigma))d\varsigma,$$

$$(\gamma_2^\mu(\upsilon)(t)) := \int_{t}^{\infty} \frac{\Lambda T^\mu(t-\varsigma)Q}{\Gamma(\mu)} \varphi(\varsigma, \upsilon(\varsigma))d\varsigma,$$

$$(\gamma_3^\mu(\upsilon)(t)) := \int_{-\infty}^{t} \frac{\Lambda T^\mu(t-\varsigma)\rho}{\Gamma(\mu)} \vartheta(\varsigma, \upsilon(\varsigma))d\varsigma,$$

$$(\gamma_4^\mu(\upsilon)(t)) := \int_{t}^{\infty} \frac{\Lambda T^\mu(t-\varsigma)Q}{\Gamma(\mu)} \vartheta(\varsigma, \upsilon(\varsigma))d\varsigma,$$

where $\mu \in (0, 1]$ and $T(t)_{t\geq 0}$ is the analytic semigroup. It is clear that, if $T^\mu = (t-\varsigma)^{\mu-1}$, then we obtain the RL-integral operator.

Lemma 3.2 *Let $m \in \aleph$, $\upsilon \in \rho\Lambda\Lambda(\Re, \chi_\lambda, \mathfrak{m})$ and (H1)–(H2) hold. Then,*

$$\gamma_3^\mu, \gamma_4^\mu : \rho\Lambda\Lambda(\Re, \chi_\lambda, \mathfrak{m}) \longrightarrow \rho\Lambda\Lambda(\Re, \chi_\lambda, \mathfrak{m}).$$

Proof Let $\upsilon \in \rho\Lambda\Lambda(\Re, \chi_\lambda, \mathfrak{m})$. Putting $\hbar(t) = \vartheta(t, \upsilon(t))$ and by Theorem 3.1, it indicates that $\hbar \in \rho\Lambda\Lambda(\Re, \chi, \mathfrak{m})$ for each $\upsilon \in \rho\Lambda\Lambda(\Re, \chi_\lambda, \mathfrak{m})$. Setting $\hbar = \Phi + \xi$, where $\Phi \in \Lambda\Lambda(\Re, \chi)$ and $\xi \in \zeta(\Re, \chi, \mathfrak{m})$. Therefore, $\gamma_3^\mu \upsilon$ can be read as

$$(\gamma_3^\mu(\upsilon)(t)) := \int_{-\infty}^{t} \frac{T^\mu(t-\varsigma)\rho}{\Gamma(\mu)} \Phi(\varsigma)d\varsigma + \int_{-\infty}^{t} \frac{T^\mu(t-\varsigma)\rho}{\Gamma(\mu)} \xi(\varsigma)d\varsigma.$$

Let

$$\phi(t) = \int_{-\infty}^{t} \frac{T^\mu(t-\varsigma)\rho}{\Gamma(\mu)} \Phi(\varsigma)d\varsigma$$

and

$$\psi(t) = \int_{-\infty}^{t} \frac{T^\mu(t-\varsigma)\rho}{\Gamma(\mu)} \xi(\varsigma)d\varsigma,$$

for each $t \in \Re$. It can be realized that $\phi \in \Lambda\Lambda(\Re, \chi_\lambda)$. Consider a sequence $(\varsigma_\eta')_{\eta\in N}$, then there is a subsequence $(\varsigma_\eta)_{\eta\in N}$ such that

$$\lim_{\eta,n\to\infty} \|\Phi(t+\varsigma_\eta-\varsigma_n)-\Phi(t)\| = 0. \tag{6}$$

Moreover, we have

$$\phi(t+\varsigma_\eta-\varsigma_n)-\phi(t) = \int_{-\infty}^{t+\varsigma_\eta-\varsigma_n} \frac{T^\mu(t+\varsigma_\eta-\varsigma_n-\varsigma)\rho}{\Gamma(\mu)} \Phi(\varsigma)d\varsigma - \int_{-\infty}^{t} \frac{T^\mu(t-\varsigma)\rho}{\Gamma(\mu)} \Phi(\varsigma)d\varsigma$$
$$= \int_{-\infty}^{0} \frac{T^\mu(-\varsigma)\rho}{\Gamma(\mu)} [\Phi(\varsigma+t+\varsigma_\eta-\varsigma_n) - \Phi(\varsigma+t)]d\varsigma.$$

Then, we get

$$\|\phi(t+\varsigma_\eta-\varsigma_n)-\phi(t)\|_\lambda \leq \int_{-\infty}^{0} \left\| \frac{T^\mu(-\varsigma)\rho}{\Gamma(\mu)} [\Phi(\varsigma+t+\varsigma_\eta-\varsigma_n) - \Phi(\varsigma+t)] \right\|_\lambda d\varsigma.$$

Hence, by (4) and the fact that $\|T^\mu\|_\lambda \leq \|T\|_\lambda$, we conclude

$$\|\phi(t+\varsigma_\eta-\varsigma_n)-\phi(t)\|_\lambda \leq \int_{-\infty}^{0} \frac{M(\lambda)\varsigma^{-\lambda}\varrho^{-\epsilon\varsigma}}{\Gamma(\mu)} \|\Phi(\varsigma+t+\varsigma_\eta-\varsigma_n) - \Phi(\varsigma+t)\|d\varsigma.$$

The outcome obtains from (10) and the Lebesgue dominated convergence theorem. Lastly, we aim to show that $\psi(t) \in \zeta(\Re, \chi_\lambda, \mathfrak{m})$. A computation yields

$$\frac{1}{m([-r,r])}\int_{[-r,r]} \|\psi(t)\|_\lambda dm(t) = \frac{1}{m([-r,r])}\int_{[-r,r]} \left\| \int_{-\infty}^{t} \frac{T^\mu(t-\varsigma)\rho}{\Gamma(\mu)} \xi(\varsigma)d\varsigma \right\|_\lambda dm(t)$$
$$\leq \frac{1}{m([-r,r])}\int_{[-r,r]} \int_{-\infty}^{t} \left\| \frac{T^\mu(t-\varsigma)\rho}{\Gamma(\mu)} \xi(\varsigma) \right\| d\varsigma dm(t)$$
$$\leq \frac{1}{m([-r,r])}\int_{[-r,r]} \int_{-\infty}^{t} \frac{M(\lambda)(t-\varsigma)^{-\mu\lambda}\varrho^{-\epsilon(t-\varsigma)}}{\Gamma(\mu)} \|\xi(\varsigma)\| d\varsigma dm(t)$$
$$\leq \frac{M(\lambda)}{\Gamma(\mu)} \int_{0}^{\infty} \varsigma^{-\mu\lambda}\varrho^{-\epsilon\varsigma} \left(\frac{1}{m([-r,r])}\int_{[-r,r]} \|\xi(t-\varsigma)\| dm(t) \right) d\varsigma.$$

In fact, the space $\zeta(\Re, \chi, \mathfrak{m})$ is invariant (preserved by some function); it shows that $t \mapsto \xi(t-\varsigma)$ belongs to $\zeta(\Re, \chi, \mathfrak{m})$ for each $\varsigma \in \Re$ and hence

$$\lim_{r\to\infty} \frac{1}{m([-r,r])}\int_{[-r,r]} \|\xi(t-\varsigma)\| dm(t) = 0.$$

Consequently, by utilizing the Lebesgue dominated convergence theorem, we have

$$\lim_{r\to\infty} \frac{M(\lambda)}{\Gamma(\mu)} \int_{0}^{\infty} \varsigma^{-\lambda}\varrho^{-\epsilon\varsigma} \left(\frac{1}{m([-r,r])}\int_{[-r,r]} \|\xi(t-\varsigma)\| dm(t) \right) d\varsigma = 0,$$

similarly, by applying (5) to $\gamma_4^\mu \upsilon$. This completes the proof. □

Lemma 3.3 *Let $m \in \aleph$, and $\upsilon \in \rho\Lambda\Lambda(\Re, \chi, \mathfrak{m})$. If (H1)–(H2) are satisfied, then*

$$\gamma_1^\mu, \gamma_2^\mu : \rho\Lambda\Lambda(\Re, \chi_\delta, \mathfrak{m}) \longrightarrow \rho\Lambda\Lambda(\Re, \chi_\delta, \mathfrak{m}).$$

Proof Let $\upsilon \in \rho\Lambda\Lambda(\Re, \chi, \mathfrak{m})$ and $\hbar(t) = \varphi(t, \chi, \upsilon(t))$. Then in view of Theorem 3.1, it implies that $\hbar \in \rho\Lambda\Lambda(\Re, \chi_\delta, \mathfrak{m})$ whenever $\upsilon \in \rho\Lambda\Lambda(\Re, \chi_\lambda, \mathfrak{m})$. In particular,

$$\|\hbar\|_{\infty,\delta} = \sup_{t\in\Re}\|\varphi(t, \upsilon(t))\|_\delta < \infty.$$

Now, we write $\hbar = \Phi + \Psi$, where $\Phi \in \Lambda\Lambda(\Re, \chi_\delta), \Psi \in \zeta(\Re, \chi_\delta, \mathfrak{m})$, that is, $\gamma_1^\mu\hbar = E\Phi + E\Psi$ where

$$E\Phi(t) := \int_{-\infty}^{t} \frac{\Lambda T^\mu(t-\varsigma)\rho}{\Gamma(\mu)}\Phi(\varsigma)d\varsigma,$$

$$E\Psi(t) := \int_{-\infty}^{t} \frac{\Lambda T^\mu(t-\varsigma)\rho}{\Gamma(\mu)}\Psi(\varsigma)d\varsigma.$$

First, we need to show that $E\Phi(t) \in \Lambda\Lambda(\Re, \chi_\lambda)$. Consider a sequence $(\varsigma'_\eta)_{\eta\in N}$ in $t \in \Re$, since $\Phi(t) \in \Lambda\Lambda(\Re, \chi_\delta)$, a subsequence $(S_\eta)_{\eta\in N}$ exists such that

$$\lim_{\eta,n\to\infty} \|\Phi(t, +\varsigma_\eta - \varsigma_n) - \Phi(t)\|_\delta = 0. \tag{7}$$

In addition, since

$$\begin{aligned} &E\Phi(t + \varsigma_\eta - \varsigma_n) - E\Phi(t) \\ &= \int_{-\infty}^{t+\varsigma_\eta-\varsigma_n} \frac{\Lambda T^\mu(t + \varsigma_\eta - \varsigma_n - \varsigma)\rho}{\Gamma(\mu)}\Phi(\varsigma)d\varsigma \\ &- \int_{-\infty}^{t} \frac{\Lambda T^\mu(t-\varsigma)\rho}{\Gamma(\mu)}\Phi(\varsigma)d\varsigma \\ &= \int_{-\infty}^{0} \frac{AT^\mu(-\varsigma)\rho[\Phi(\varsigma + t + \varsigma_\eta - \varsigma_n) - \Phi(\varsigma + t)]}{\Gamma(\mu)}d\varsigma. \end{aligned}$$

Then, a computation implies

$$\begin{aligned} &\|E\Phi(t + \varsigma_\eta - \varsigma_n) - E\Phi(t)\|_\lambda \\ &\le \int_{-\infty}^{0} \left\| \frac{\Lambda T^\mu(-\varsigma)\rho[\Phi(\varsigma + t + \varsigma_\eta - \varsigma_n) - \Phi(\varsigma + t)]}{\Gamma(\mu)} \right\|_\lambda d\varsigma. \end{aligned}$$

Hence, by (6) and the fact that $\|\Lambda(T^\mu)\| \le \|\Lambda(T)\|$, we receive

$$\begin{aligned} &\|E\Phi(t + \varsigma_\eta - \varsigma_n) - E\Phi(t)\|_\lambda \\ &\le \int_{-\infty}^{0} \frac{c\varsigma^{\delta-\lambda-1}\varrho^{-\epsilon\varsigma}}{\Gamma(\mu)}\|\Phi(\varsigma + t + \varsigma_\eta + \varsigma_n) - \Phi(\varsigma + t)\|_\delta d\varsigma. \end{aligned}$$

The result comes from Eq. (11) and the Lebesgue's dominated theorem. Finally, we reveal that $E\Psi(t) \in \zeta(\Re, \chi_\lambda, \mathfrak{m})$. We have

$$\begin{aligned} &\frac{1}{m([-r,r])}\int_{[-r,r]} \|E\Psi(t)\|_\lambda dm(t) \\ &= \frac{1}{m([-r,r])}\int_{[-r,r]} \left\| \int_{-\infty}^{t} \frac{AT^\mu(t-\varsigma)\rho}{\Gamma(\mu)}\Psi(\varsigma)d\varsigma \right\|_\lambda dm(t) \\ &\le \frac{1}{m([-r,r])}\int_{[-r,r]} \int_{-\infty}^{t} \left\| \frac{AT^\mu(t-\varsigma)\rho}{\Gamma(\mu)}\Psi(\varsigma) \right\|_\lambda d\varsigma dm(t) \\ &\le \frac{1}{m([-r,r])}\int_{[-r,r]} \int_{-\infty}^{t} \frac{c(t-\varsigma)^{\delta-\lambda-1}\varrho^{-\epsilon(t-\varsigma)}}{\Gamma(\mu)}\|\Psi(\varsigma)\|_\delta d\varsigma dm(t) \\ &\le \frac{c}{\Gamma(\mu)}\int_{0}^{\infty} \varsigma^{\delta-\lambda-1}\varrho^{-\epsilon\varsigma}\left(\frac{1}{m([-r,r])}\int_{[-r,r]} \|\Psi(t-\varsigma)\|_\delta dm(t)\right)d\varsigma. \end{aligned}$$

Therefore, we obtain

$$\lim_{r\to\infty} \frac{1}{([-r,r])}\int_{[-r,r]} \|\Psi(t-\varsigma)\|_\delta dm(t) = 0$$

as

$$\varsigma \longrightarrow \Psi(t-\varsigma) \in \zeta(\Re, \chi_\delta, \mathfrak{m})$$

for every $\varsigma \in \Re$. The proof is completed by applying the Lebesgue's dominated convergence theorem and similarly for $\gamma_2^\mu\upsilon$ using (7). □

m−PΛΛM outcomes

The rest of this section is conducted to find the existence of m-PΛΛ mild solutions (m-PΛΛM) of Eq. (1). Recently, Ibrahim et al. studied the mild solution of a class of FDE, by utilizing the fractional resolvent concept (see [24, 25]).

Definition 3.4 Let $\lambda \in (0, 1)$. A bounded continuous function $\upsilon : \Re \to \chi_\lambda$ is stated to be a mild solution to (1) indicate that the function $\varsigma \to \frac{\Lambda T^\mu(t-\varsigma)\rho}{\Gamma(\mu)}\varphi(\varsigma, \upsilon(\varsigma))$ is integrable on $(-\infty, t), \varsigma \to AT^\mu(t-\varsigma)Q\varphi(\varsigma, \upsilon(\varsigma))$ is integrable on (t, ∞) and

$$\begin{aligned} \upsilon(t) = &-\varphi(t, \upsilon(t)) - \int_{-\infty}^{t} \frac{\Lambda T^\mu(t,\varsigma)\rho}{\Gamma(\mu)}\varphi(\varsigma, \upsilon(\varsigma))d\varsigma \\ &+ \int_{t}^{\infty} \frac{\Lambda T^\mu(t,\varsigma)Q}{\Gamma(\mu)}\varphi(\varsigma, \upsilon(\varsigma))d\varsigma + \int_{-\infty}^{t} \frac{T^\mu(t,\varsigma)\rho}{\Gamma(\mu)}\vartheta(\varsigma, \upsilon(\varsigma))d\varsigma \\ &- \int_{t}^{\infty} \frac{\Lambda T^\mu(t,\varsigma)Q}{\Gamma(\mu)}\vartheta(\varsigma, \upsilon(\varsigma))d\varsigma \end{aligned}$$

for each $t \in \Re$.

Theorem 3.5 *Let $m \in \aleph$. Under the assumptions (H1) and (H2), Eq. (1) admits a unique m-PΛΛM solution for some constants $K > 0$.*

Proof Consider the fractional integral operator $\wedge : \rho\Lambda\Lambda(\Re, \chi_\lambda, \mathfrak{m}) \longrightarrow \rho\Lambda\Lambda(\Re, \chi_\lambda, \mathfrak{m})$ such that

$$\wedge\upsilon(t) := -\varphi(t,\upsilon(t)) - \int_{-\infty}^{t}\frac{\Lambda T^{\mu}(t,\varsigma)\rho}{\Gamma(\mu)}\varphi(\varsigma,\upsilon(\varsigma))d\varsigma + \int_{t}^{\infty}\frac{\Lambda T^{\mu}(t,\varsigma)Q}{\Gamma(\mu)}\varphi(\varsigma,\upsilon(\varsigma))d\varsigma + \int_{-\infty}^{t}\frac{T^{\mu}(t,\varsigma)\rho}{\Gamma(\mu)}\vartheta(\varsigma,\upsilon(\varsigma))d\varsigma - \int_{t}^{\infty}\frac{AT^{\mu}(t,\varsigma)Q}{\Gamma(\mu)}\vartheta(\varsigma,\upsilon(\varsigma))d\varsigma.$$

It has been formerly shown that for every $\upsilon \in \rho\Lambda\Lambda(\mathfrak{R},\chi_\lambda,\mathfrak{m})$, $\varphi(\bullet,\upsilon(\bullet)) \in \rho\Lambda\Lambda(\mathfrak{R},\chi_\lambda,\mathfrak{m})$ (see Theorem 3.1). In view of Lemmas 3.2 and 3.3, it documents that $\wedge : \rho\Lambda\Lambda(\mathfrak{R},\chi_\lambda,\mathfrak{m}) \longrightarrow \rho\Lambda\Lambda(\mathfrak{R},\chi_\lambda,\mathfrak{m})$. Our aim is to show that $\wedge$ has a unique fixed point. For this purpose, we employ Lemmas 2.14 and 2.15. Let $v,\varpi \in \rho\Lambda\Lambda(\mathfrak{R},\chi_\lambda,\mathfrak{m})$, then for γ_1^{μ}, we conclude

$$\begin{aligned}\|\gamma_1^{\mu}(v)(t)-\gamma_1^{\mu}(\varpi)(t)\|_\lambda &\le \int_{-\infty}^{t}\left\|\frac{\Lambda T^{\mu}(t-\varsigma)\rho}{\Gamma(\mu)}[\varphi(\varsigma,v(\varsigma))-\varphi(\varsigma,\varpi(\varsigma))]\right\|_\lambda d\varsigma\\ &\le \int_{-\infty}^{t}\left\|\frac{\Lambda T(t-\varsigma)\rho}{\Gamma(\mu)}[\varphi(\varsigma,v(\varsigma))-\varphi(\varsigma,\varpi(\varsigma))]\right\|_\lambda d\varsigma\\ &\le \int_{-\infty}^{t}\frac{c(t-\varsigma)^{\delta-\lambda-1}\varrho^{-\epsilon(t-\varsigma)}}{\Gamma(\mu)}\|\varphi(\varsigma,v(\varsigma)) - \varphi(\varsigma,\varpi(\varsigma))\|_\delta d\varsigma \le \frac{\kappa_1 K}{\Gamma(\mu)}\|v-\varpi\|_{\lambda,\infty}.\end{aligned}$$

Now, for γ_2^{μ}, we obtain

$$\begin{aligned}\|\gamma_2^{\mu}(v)(t)-\gamma_2^{\mu}(\varpi)(t)\|_\lambda &\le \int_{t}^{\infty}\left\|\frac{\Lambda T^{\mu}(t-\varsigma)Q}{\Gamma(\mu)}[\varphi(\varsigma,v(\varsigma))-\varphi(\varsigma,\varpi(\varsigma))]\right\|_\lambda d\varsigma\\ &\le \int_{t}^{\infty}\left\|\frac{\Lambda T(t-\varsigma)Q}{\Gamma(\mu)}[\varphi(\varsigma,v(\varsigma))-\varphi(\varsigma,\varpi(\varsigma))]\right\|_\lambda d\varsigma\\ &\le \int_{t}^{\infty}\frac{c\varrho^{s(t-\varsigma)}}{\Gamma(\mu)}\|\varphi(\varsigma,v(\varsigma))-\varphi(\varsigma,\varpi(\varsigma))\|_\delta d\varsigma\\ &\le \frac{\kappa_2 K}{\Gamma(\mu)}\|v-\varpi\|_{\lambda,\infty}.\end{aligned}$$

Now, for γ_3^{μ} and γ_4^{μ}, the following approximations can be given:

$$\begin{aligned}\|\gamma_3^{\mu}(v)(t)-\gamma_3^{\mu}(\varpi)(t)\|_\lambda &\le \int_{-\infty}^{t}\left\|\frac{T^{\mu}(t-\varsigma)\rho}{\Gamma(\mu)}[\vartheta(\varsigma,v(\varsigma))-\vartheta(\varsigma,\varpi(\varsigma))]\right\|_\lambda d\varsigma\\ &\le \int_{-\infty}^{t}\left\|\frac{T(t-\varsigma)\rho}{\Gamma(\mu)}[\vartheta(\varsigma,v(\varsigma))-\vartheta(\varsigma,\varpi(\varsigma))]\right\|_\lambda d\varsigma\\ &\le \int_{-\infty}^{t}\frac{M(\lambda)(t-\varsigma)^{-\lambda}\varrho^{-\epsilon(t-\varsigma)}}{\Gamma(\mu)}\|\vartheta(\varsigma,v(\varsigma)) - \vartheta(\varsigma,\varpi(\varsigma))\|d\varsigma \le \frac{\kappa_3 K}{\Gamma(\mu)}\|v-\varpi\|_{\lambda,\infty}.\end{aligned}$$

Also, we attain

$$\begin{aligned}\|\gamma_4^{\mu}(v)(t)-\gamma_4^{\mu}(\varpi)(t)\|_\lambda &\le \int_{t}^{\infty}\left\|\frac{T^{\mu}(t-\varsigma)Q}{\Gamma(\mu)}[\vartheta(\varsigma,v(\varsigma))-\vartheta(\varsigma,\varpi(\varsigma))]\right\|_\lambda d\varsigma\\ &< \int_{t}^{\infty}\left\|\frac{T(t-\varsigma)Q}{\Gamma(\mu)}[\vartheta(\varsigma,v(\varsigma))-\vartheta(\varsigma,\varpi(\varsigma))]\right\|_\lambda d\varsigma\\ &\le \int_{t}^{\infty}\frac{C(\lambda)\varrho^{s(t-\varsigma)}}{\Gamma(\mu)}\|\vartheta(\varsigma,v(\varsigma))-\vartheta(\varsigma,\varpi(\varsigma))\|d\varsigma\\ &\le \frac{\kappa_4 K}{\Gamma(\mu)}\|v-\varpi\|_{\lambda,\infty}.\end{aligned}$$

Joining the above inequalities yields $\|\wedge v - \wedge\varpi\|_{\lambda,\infty} \le K\Xi$, where

$$\Xi := \frac{\kappa_1+\kappa_2+\kappa_3+\kappa_4}{\Gamma(\mu)}.$$

Therefore, if $K<\Xi^{-1}$, then in view of the Banach fixed point theorem, Eq. (1) has a unique solution, which clearly is the only m-P$\Lambda\Lambda$M solution. This completes the proof. □

Example 3.6 Consider the equation

$$D^{\mu}\Big((1+t)\upsilon(t)\Big) = \Big((1+t)\upsilon(t)\Big), \quad t\in[0,1],\ \mu\in(0,1]. \tag{8}$$

Let $\mu = 0.15$. It is clear that $\varphi(t,\upsilon(t)) = \vartheta(t,\upsilon(t)) = t\upsilon(t)$. Thus, they are Lipschitz with $K = 1$, $t\in[0,1]$. Moreover, $\kappa_1 = \kappa_2 = \kappa_3 = \kappa_4 = 2.7/2$, this implies that $\frac{\kappa_1+\kappa_2+\kappa_3+\kappa_4}{\Gamma(0.15)} = \frac{4\times 1.11}{6.22} = 0.874 \rightarrow \Xi^{-1} = 1.14 > K = 1$, $k_1 = 1$ such that $\lambda = 1/2$ and $\|\upsilon\| \le k_1\|\upsilon\|_\lambda$. In addition, the functions φ and ϑ are bounded and uniformly continuous for all $t\in[0,1]$. Hence, all the conditions of Theorem 3.5 are achieved; therefore, Eq. (12) has a unique periodic solution. Note that if $\mu = 0.5$, we obtain $\frac{\kappa_1+\kappa_2+\kappa_3+\kappa_4}{\Gamma(0.5)} = \frac{4\times 1.11}{1.77} = 3.0 \rightarrow \Xi^{-1} = 0.3 < K = 1$, then Theorem 3.5 is field. From the above computation, Eq. (12) has a unique periodic solution when $0<\mu<0.4$.

Example 3.7 Consider the equation

$$D^{\mu}\Big((1/4+\sin t)\upsilon(t)\Big) = \Big((1/2+\cos t)\upsilon(t)\Big), \quad t\in[0,2\pi],\ \mu\in(0,1]. \tag{9}$$

Obviously, $\varphi(t,\upsilon(t)) = \sin t\,\upsilon(t)$, $\vartheta(t,\upsilon(t)) = \cos t\,\upsilon(t)$. Thus, they are Lipschitz with $K = \max_{t\in[0,2\pi]}\{\sin(t),\cos(t)\} = 1$, $\kappa_1 = \kappa_2 = \kappa_3 = \kappa_4 = 2.7/4$; this implies that $\frac{\kappa_1+\kappa_2+\kappa_3+\kappa_4}{\Gamma(0.15)} = 0.4 \rightarrow \Xi^{-1} = 2.5 > K = 1$. Moreover, $k_1 = 1/2$ such that $\lambda = 1/4$ and $\|\upsilon\| \le k_1\|\upsilon\|_\lambda$. In addition, the functions φ and ϑ are bounded and uniformly continuous for all $t\in[0,2\pi]$. Hence, all the conditions of Theorem 3.5 are achieved; therefore, Eq. (13) admits a unique periodic solution. If $\mu = 0.5$, then we have $\Xi^{-1} = 0.66 < K = 1$ and hence Theorem 3.5 is true when $0<\mu<0.4$.

Conclusions

In the current study, we suggested symmetry of a class of fractional differential equations (the fractional calculus depends on the RL fractional operators) by utilizing its periodic solutions. We studied a special class of FDEs. This class is a generalization of the neutral equation. We proved a composition theorem for m-P$\Lambda\Lambda$ functions under appropriate conditions. Our technique is based on interpolation theory and Banach's fixed point theorem. Therefore, the solution, in this case, is unique. Moreover, we investigated the mild solution, for such a class by illustrating a new fractional resolvent concept. This functional is constructed to keep the periodicity of the solution and consequently its symmetry.

Author contribution The authors jointly worked on deriving the results and approved the final manuscript. There is no conflict of interests regarding the publication of this article.

Acknowledgements The authors would like to thank the referees for their useful suggestions for improving the work.

Compliance with ethical standards

Conflict of interest The authors declare that they have no competing interests

References

1. Bochner, S.: A new approach to almost periodicity. Proc. Ncad. Sci. USA **48**, 2039–2043 (1962)
2. Bohr, H.: Zur Theorie der fastperiodischen funktionen I. Acta Math **45**, 29–127 (1925)
3. Bochner, S.: Continuous mappings of almost automorphic and almost periodic functions. Proc. Acad. Sci. USA **52**, 907–910 (1964)
4. NGuerekata G.M.: Almost auto morphic and almost periodic functions in abstract spaces. Kluwer Academic, Plenum Publishers, New York, Boston, Moscow, London (2001)
5. NGuerekata G.M.: Topics in almost automorphy. Springer, New York, Boston, Dordrecht, London Moscow (2005)
6. Ding, H.S., et al.: Almost automorphic solutions of nonautonomous evolution equations. Nonlinear Anal. TMA **70**, 4158–4164 (2009)
7. Ding, H.S., et al.: Almost automorphic solutions to nonautonomous semilinear evolution equations in Banach spaces. Nonlinear Anal. TMA **73**, 1426–1438 (2010)
8. Bruno, A., et al.: Almost automorphic solutions of hypersolic evolution equations. Banach J. Math. Anal. **6**, 190–200 (2012)
9. Abbas, S., Xia, Y.: Existence and attractivity of k-almost automorphic solutions of a model of cellular neural network with delay. Acta Math. Sci. **1**, 290–302 (2013)
10. Diagana, T.: Pseudo almost periodic functioins in Banach spaces. Nova Science Publishers, New York (2007)
11. Diagana T.: Existence of almost automorphic solutions to some neutral functional differential equations with infinite delay. Electron. J. Differ. Equ. **2008**, 114 (2008)
12. Diagana, T., et al.: Almost automorphic mild solutions to some partial neutral functional-differential equations and applications. Nonlinear Anal. Theor. Methods Appl. **69**(5), 1485–1493 (2008)
13. Diagana, T.: Pseudo-almost periodic solutions for some classes of nonautonomous partial evolution equations. J. Franklin Inst. **348**, 2082–2098 (2011)
14. Bezandry, P.H., Diagana, T.: Almost periodic Stochastic processes. Springer, New York, Dordrecht, Heidelberg, London (2011)
15. Xiao, T.J., et al.: Pseudo almost automorphic mild solutions to nonautonomous differential equations and applications. Nonlinear Anal. TMA **70**, 4079–4085 (2009)
16. Chang, Yong-Kui, Luo, Xiao-Xia: Existence of −pseudo almost automorphic solutions to a neutral differential equation by interpolation theory. Filomat **28**(3), 603–614 (2014)
17. Lizama C., Poblete V.: Periodic solutions of fractional differential equations with delay. J. Evol. Equ. **11**(1), 57–70 (2011)
18. Keyantuo, V., Lizama, C.: A characterization of periodic solutions for time-fractional differential equations in UMD spaces and applications. Math. Nachrichten **284**(4), 494–506 (2011)
19. Blot, J., et al.: Measure theory and pseudo almost automorphic functions: new developments and aplications. Nonlinear Anal. **75**, 2426–2447 (2012)
20. Tunc, C.: On the existence of periodic solutions of functional differential equations of the third order. Appl. Comput. Math. **15**(2), 189–199 (2016)
21. Ibrahim R.W., et al.: Periodicity and positivity of a class of fractional differential equations. SpringerPlus, 1;**5**(1), 1–0 (2016)
22. Podlubny, I.: Fractional differential equations. Academic Press, San Diego, Boston, NewYork, London, Tokyo, Toronto (1999)
23. Kilbas A.A., et al.: Theory and applications of fractional differential equations. vol. 204. Elsevier Science Limited (2006)
24. Ibrahim, R.W., et al.: Existence and uniqueness for a class of iterative fractional differential equations. Adv. Differ. Equ. **78**, 1–13 (2015)
25. Ibrahim, R.W., et al.: Existence results for a family of equations of fractional resolvent. Sains Malay **44**(2), 295–300 (2015)

23

Common fixed point theorems for infinite families of contractive maps

Reza Allahyari[2] · Reza Arab[1] · Ali Shole Haghighi[2]

Abstract In this paper, we prove some fixed point theorems for infinite families of self-mappings of a complete metric space satisfying some new conditions of common contractivity. An example is presented to show the effectiveness of our results.

Keywords Fixed point · Family of contractive maps

Mathematics Subject Classification 47H10 · 54H25

Introduction and preliminaries

Fixed point theory constitutes an important and the core part of the subject of nonlinear functional analysis and is useful for proving the existence theorems for nonlinear differential and integral equations. The Banach contraction principle [3] is the simplest and one of the most versatile elementary results in fixed point theory, which is a very popular tool for solving existence problems in many branches of mathematical analysis. Several authors have extended the Banach's fixed point theorem in various ways. The family of contraction mappings was introduced and studied by Ćirić [7] and Tasković [11]. Also in the process, the study of existence of common fixed point for finite and infinite family of self-mapping has been carried out by many authors. For example, one may refer [1, 2, 4–6, 12–14].

Recently, some new results for the existence and uniqueness of fixed points were presented for the cases of partially ordered metric spaces, cone metric spaces and fuzzy metric spaces (for example, see [1, 15–18]). Also, the study of common fixed points for a family of contractive type maps has been paid attention, and many interesting fixed point results have been obtained (for example, see [2, 7–11]).

The aim of this paper is to define some new conditions of common contractivity for an infinite family of mappings and give some new results on the existence and uniqueness of common fixed points in the setting of complete metric space.

Here, we state some known definitions and facts. We refer for more details to [1, 7].

Definition 1 Let X be a nonempty set and let $\{T_n\}$ be a family of self-mappings on X. A point $x_0 \in X$ is called a common fixed point for this family iff $T_n(x_0) = x_0$, for each $n \in N$.

The following interesting theorem was given by Ćirić [7] for a family of generalized contractions.

Theorem 1 *Let (X, d) be a complete metric space and let $\{T_\alpha\}_{\alpha \in J}$ be a family of self-mappings of X. If there exists fixed $\beta \in J$ such that for each $\alpha \in J$:*

$$d(T_\alpha x, T_\beta y) \le \lambda \max\{d(x,y), d(x,T_\alpha x), d(y,T_\beta y), \frac{1}{2}[d(x,T_\beta y) + d(y,T_\alpha x)]\}, \tag{1}$$

✉ Reza Arab
mathreza.arab@iausari.ac.ir

Reza Allahyari
rezaallahyari@mshdiau.ac.ir

Ali Shole Haghighi
ali.sholehaghighi@gmail.com

1 Department of Mathematics, Sari Branch, Islamic Azad University, Sari, Iran

2 Department of Mathematics, Mashhad Branch, Islamic Azad University, Mashhad, Iran

for some $\lambda = \lambda(\alpha) \in (0,1)$ *and all* $x,\ y \in X$, *then all* T_α *have a unique common fixed point, which is a unique fixed point of each* T_α, $\alpha \in J$.

Common fixed point theorems for a family of mappings

In this section, we prove existence of a unique common fixed point for a family of contractive type self-maps on a complete metric space.

Theorem 2 *Let (X, d) be a complete metric space and* $0 \le a_{i,j} (i, j = 1, 2, \ldots)$ *satisfy*

(i) for each j, $\overline{\lim}_{i \to \infty} a_{i,j} < 1$
(ii) $\sum_{n=1}^{\infty} A_n < \infty$ *where* $A_n = \prod_{i=1}^{n} \frac{a_{i,i+1}}{1-a_{i,i+1}}$.

If $\{T_n\}$ *is a sequence of self-maps on X satisfying*

$$d(T_i x, T_j y) \le a_{i,j}[d(x, T_j y) + d(y, T_i x)], \tag{2}$$

for $x, y \in X; i, j = 1, 2, \ldots$ with $x \neq y$ and $i \neq j$ then all T_ns have a unique common fixed point in X.

Proof For any $x_0 \in X$, let $x_n = T_n(x_{n-1}), n = 1, 2, \ldots$, then using (2.1) we get

$$\begin{aligned} d(x_1, x_2) = d(T_1(x_0), T_2(x_1)) &\le a_{1,2}[d(x_0, T_2(x_1)) + d(x_1, T_1(x_0))] \\ &\le a_{1,2}[d(x_0, x_2) + d(x_1, x_1)] \\ &= a_{1,2} d(x_0, x_2) \\ &\le a_{1,2}[d(x_0, x_1) + d(x_1, x_2)] \end{aligned}$$

which implies

$$(1 - a_{1,2}) d(x_1, x_2) \le a_{1,2} d(x_0, x_1).$$

So

$$d(x_1, x_2) \le \frac{a_{1,2}}{1 - a_{1,2}} d(x_0, x_1).$$

Also we have,

$$\begin{aligned} d(x_2, x_3) = d(T_2(x_1), T_3(x_2)) &\le a_{2,3}[d(x_1, T_3(x_2)) + d(x_2, T_2(x_1))] \\ &\le a_{2,3}[d(x_1, x_3) + d(x_2, x_2)] \\ &= a_{2,3} d(x_1, x_3) \\ &\le a_{2,3}[d(x_1, x_2) + d(x_2, x_3)] \end{aligned}$$

implies

$$(1 - a_{2,3}) d(x_2, x_3) \le a_{2,3} d(x_1, x_2).$$

So

$$\begin{aligned} d(x_2, x_3) &\le \frac{a_{2,3}}{1 - a_{2,3}} d(x_1, x_2) \\ &\le \frac{a_{1,2}}{1 - a_{1,2}} \cdot \frac{a_{2,3}}{1 - a_{2,3}} d(x_0, x_1). \end{aligned}$$

In general, we get

$$d(x_n, x_{n+1}) \le \prod_{i=1}^{n} \frac{a_{i,i+1}}{1 - a_{i,i+1}} d(x_0, x_1) = A_n d(x_0, x_1). \tag{3}$$

Therefore, for $m, n \in N,\ m \ge n$, and using (2.2)

$$\begin{aligned} d(x_n, x_m) &\le \sum_{k=n}^{m-1} d(x_k, x_{k+1}) \\ &\le \sum_{k=n}^{m-1} \prod_{i=1}^{k} \frac{a_{i,i+1}}{1 - a_{i,i+1}} d(x_0, x_1) \\ &= \sum_{k=n}^{m-1} A_k d(x_0, x_1). \end{aligned}$$

Thus $\{x_n\}$ is a Cauchy sequence and by completeness of X, $\{x_n\}$ converges to x (say) in X.

So using (2.1), for any positive integer m we have

$$\begin{aligned} d(x, T_m x) &\le d(x, x_n) + d(x_n, T_m x) \\ &= d(x, x_n) + d(T_n x_{n-1}, T_m x) \\ &\le d(x, x_n) + a_{n,m}[d(x_{n-1}, T_m x) + d(x, T_n x_{n-1})]. \end{aligned}$$

Taking $\overline{\lim}$ as $n \longrightarrow \infty$, we get

$$d(x, T_m x) \le \overline{\lim} a_{n,m} d(x, T_m x),$$

and it follows that $d(x, T_m x) = 0$ which shows that x is a common fixed of $\{T_m\}$.

Now we prove the uniqueness of the common fixed point x. Suppose that y be another common fixed point of $\{T_k\}$. Since $\sum_{n=1}^{\infty} A_n < \infty$ so $\lim_{n \to \infty} A_n = 0$ and therefore there exists an $i_0 \in N$ such that $a_{i_0, i_0+1} < \frac{1}{2}$. Thus, from (2.1) we have

$$\begin{aligned} d(x, y) &= d(T_{i_0} x, T_{i_0+1} y) \\ &\le a_{i_0, i_0+1}[d(x, T_{i_0+1} y) + d(y, T_{i_0} x)] \\ &= a_{i_0, i_0+1}[d(x, y) + d(y, x)] \\ &< d(x, y) \end{aligned}$$

which implies that $x = y$. So the uniqueness is proved and the proof is complete.

Corollary 1 *In addition to hypotheses of Theorem 2, suppose that for every* $n \in N$, *there exists a* $k_n \in N$ *such that* $a_{n,k_n} < \frac{1}{2}$, *then every* T_n *has a unique fixed point in X.*

Proof Following the proof of Theorem 2, $\{T_n\}$ have a unique common fixed point $x \in X$. If y is another fixed point of a T_m then

$$\begin{aligned} d(x, y) = d(T_m y, T_{k_m} x) &\le a_{k_m, m}[d(x, T_m y) + d(y, T_{k_m} x)] \\ &= a_{k_m, m}[d(x, y) + d(y, x)] \\ &< d(x, y), \end{aligned}$$

which implies $d(x,y)=0$. Therefore, $x=y$, which gives the desired result.

Example 1 Let $X=[0,1]$ be a complete metric space with the distance $d(x,y)=|x-y|$, $x,y\in X$, and $T_n: X\longrightarrow X$ be defined by

$$T_n(x)=\begin{cases}1, & 0<x\leq 1,\\ \frac{2}{3}+\frac{1}{n+2}, & x=0.\end{cases}$$

Let $a_{i,j}=\frac{1}{3}+\frac{1}{|i-j|+6}, i\neq j$, then for each j, $\overline{\lim}_{i\longrightarrow\infty}a_{i,j}<1$ and $A_n=\Pi_{i=1}^{n}\frac{a_{i,i+1}}{1-a_{i,i+1}}=\left(\frac{10}{11}\right)^n$, therefore $\sum_{n=1}^{\infty}\left(\frac{10}{11}\right)^n<\infty$. Now we prove that for each $x,y\in X$,

$$d(T_ix,T_jy)\leq a_{i,j}[d(x,T_jy)+d(y,T_ix)].$$

There are three possible cases:

1. $x\in(0,1]$, $y\in(0,1]$. Then

$$\begin{aligned}d(T_ix,T_jy)&=|T_ix-T_jy|=0\\&\leq a_{i,j}(|x-1|+|y-1|)\\&=a_{i,j}(d(x,T_jy)+d(y,T_ix)).\end{aligned}$$

2. $x\in(0,1]$, $y=0$. Then

$$\begin{aligned}d(T_ix,T_jy)&=|T_ix-T_j(0)|=\left|\frac{1}{3}-\frac{1}{j+2}\right|\leq\frac{1}{3}\\&\leq\left(\frac{1}{3}+\frac{1}{|i-j|+6}\right)\left(\left|x-\frac{2}{3}-\frac{1}{j+2}\right|+|0-1|\right)\\&=a_{i,j}[d(x,T_jy)+d(y,T_ix)].\end{aligned}$$

3. $x=y=0$, $i<j$. Then

$$\begin{aligned}d(T_ix,T_jy)&=|T_i(0)-T_j(0)|=\left|\frac{2}{3}+\frac{1}{i+2}-\frac{2}{3}-\frac{1}{j+2}\right|\\&=\left|\frac{1}{i+2}-\frac{1}{j+2}\right|\\&\leq\frac{1}{i+2}\\&\leq\left(\frac{1}{3}+\frac{1}{|i-j|+6}\right)\left(\left|\frac{2}{3}+\frac{1}{j+2}\right|+\left|\frac{2}{3}+\frac{1}{i+2}\right|\right)\\&=a_{i,j}[d(x,T_jy)+d(y,T_ix)].\end{aligned}$$

So all the conditions of Theorem 2 are satisfied and note that $x=1$ is the only fixed point for all T_n.

Theorem 3 *Let (X, d) be a complete metric space and $0\leq a_{i,j}(i,j=1,2,...)$, satisfy*

(i) *for each j, $\overline{\lim}_{i\longrightarrow\infty}a_{i,j}<1$,*
(ii) *$\sum_{n=1}^{\infty}A_n<\infty$ where $A_n=\Pi_{i=1}^{n}\frac{a_{i,i+1}}{1-a_{i,i+1}}$.*

If $\{T_n\}$ is a sequence of self-maps on X satisfying

$$d(T_ix,T_jy)\leq a_{i,j}\max\{d(x,y),d(x,T_ix),d(y,T_jy),d(x,T_jy),d(y,T_ix)\},\tag{4}$$

for all $x,y\in X, i,j=1,2,...$ with $x\neq y$ and $i\neq j$ then all T_n,s have a unique common fixed point in X. Further, if $x\in X$ be unique common fixed point of $\{T_n\}$,s then x is a unique fixed point for all T_n, s.

Proof For any $x_0\in X$, let $x_n=T_n(x_{n-1}), n=1,2,...$, then using (2.3) we obtain

$$\begin{aligned}d(x_1,x_2)&=d(T_1(x_0),T_2(x_1))\\&\leq a_{1,2}\max\{d(x_0,x_1),d(x_0,T_1x_0),d(x_1,T_2x_1),\\&\quad d(x_0,T_2(x_1)),d(x_1,T_1(x_0))\}\\&=a_{1,2}\max\{d(x_0,x_1),d(x_0,x_1),d(x_1,x_2),d(x_0,x_2),\\&\quad d(x_1,x_1)\}\\&=a_{1,2}\max\{d(x_0,x_1),d(x_1,x_2),d(x_0,x_2)\}\\&\leq a_{1,2}(d(x_0,x_1)+d(x_1,x_2)).\end{aligned}$$

Therefore,

$$d(x_1,x_2)\leq\frac{a_{1,2}}{1-a_{1,2}}d(x_0,x_1).$$

Also

$$\begin{aligned}d(x_2,x_3)&=(T_2(x_1),T_3(x_2))\\&\leq a_{2,3}\max\{d(x_1,x_2),d(x_1,T_2x_1),d(x_2,T_3x_2),\\&\quad d(x_1,T_3(x_2)),d(x_2,T_2(x_1))\}\\&\leq a_{2,3}\max\{d(x_1,x_2),d(x_2,x_3),d(x_1,x_3)\},\end{aligned}$$

which similar to the previous case we get

$$d(x_2,x_3)\leq\frac{a_{2,3}}{1-a_{2,3}}d(x_1,x_2).$$

Hence we have

$$d(x_2,x_3)\leq\frac{a_{1,2}}{1-a_{1,2}}\cdot\frac{a_{2,3}}{1-a_{2,3}}d(x_0,x_1).$$

In general

$$d(x_n,x_{n+1})\leq\prod_{i=1}^{n}\frac{a_{i,i+1}}{1-a_{i,i+1}}d(x_0,x_1).\tag{5}$$

Therefore, for $m,n\in N$, $m\geq n$, and using (2.4)

$$\begin{aligned}d(x_n,x_m)&\leq\sum_{k=n}^{m-1}d(x_k,x_{k+1})\\&\leq\sum_{k=n}^{m-1}\prod_{i=1}^{k}\frac{a_{i,i+1}}{1-a_{i,i+1}}d(x_0,x_1)\\&=\sum_{k=n}^{m-1}A_kd(x_0,x_1).\end{aligned}$$

Thus $\{x_n\}$ is a Cauchy sequence and by completeness of X, $\{x_n\}$ converges to x (say) in X.

So for any positive integer m,

$$\begin{aligned}d(x,T_mx) &\leq d(x,x_n)+d(x_n,T_mx)=d(x,x_n)+d(T_nx_{n-1},T_mx)\\ &\leq d(x,x_n)+a_{n,m}\max\{d(x_{n-1},x),d(x_{n-1},T_nx_{n-1}),\\ &\quad d(x,T_mx),d(x_{n-1},T_mx),d(x,T_nx_{n-1})\}\\ &\leq d(x,x_n)+a_{n,m}\max\{d(x_{n-1},x),d(x_{n-1},x_n),d(x,T_mx),\\ &\quad d(x_{n-1},T_mx),d(x,x_n)\}.\end{aligned}$$

Taking $\overline{\lim}$ as $n\longrightarrow\infty$, we get

$$d(x,T_mx)\leq\overline{\lim}_{n\longrightarrow\infty}a_{n,m}d(x,T_mx).$$

From condition (i), it follows that $d(x,T_mx)=0$ gives x as a common fixed point of $\{T_m\}$.

Let y be another fixed point of $\{T_n\}$, then

$$\begin{aligned}d(x,y)&=d(T_nx,T_my)\\ &\leq a_{n,m}\max\{d(x,y),d(x,T_n(x),d(y,T_my),d(x,T_my),\\ &\quad d(y,T_nx)\}\\ &\leq a_{n,m}d(x,y).\end{aligned}$$

Taking $\overline{\lim}$ as $n\longrightarrow\infty$, we get

$$d(x,y)\leq\overline{\lim}_{n\longrightarrow\infty}a_{n,m}d(x,y),$$

which is possible only when $x=y$. Hence x is the unique common fixed point of $\{T_n\}$. Further, if $y\in X$ is a unique fixed point of T_k, then according to $\overline{\lim}_{i\longrightarrow\infty}a_{i,k}<1$, there exists an $i_k\in N$ such that $a_{i_k,k}<1$. Thus, by (2.3) we have

$$\begin{aligned}d(x,y)&=d(T_{i_k}x,T_ky)\\ &\leq a_{i_k,k}\max\{d(x,y),d(x,T_{i_k}x),d(y,T_ky),d(x,T_ky),d(y,T_{i_k}x)\}\\ &\leq a_{i_k,k}d(x,y),\end{aligned}$$

which implies $d(x,y)=0$ and hence $x=y$.

Theorem 4 *Let (X, d) be a complete metric space and* $0\leq a_{i,j}+b_{i,j}<1(i,j=1,2,...)$, *satisfy*

(i) *for each j,* $\overline{\lim}_{i\longrightarrow\infty}a_{i,j}<1$ *and* $\overline{\lim}_{i\longrightarrow\infty}b_{i,j}<1$,

(ii) $\sum_{n=1}^{\infty}A_n<\infty$ *where* $A_n=\prod_{i=1}^{n}\frac{b_{i,i+1}}{1-a_{i,i+1}}$.

If $\{T_n\}$ *is a sequence of self-maps on X satisfying*

$$d(T_ix,T_jy)\leq a_{i,j}d(y,T_jy)\varphi(d(x,T_ix),d(x,y))+b_{i,j}d(x,y),\tag{6}$$

for all $x,y\in X,i,j=1,2,...$ *with* $x\neq y$ *and* $i\neq j$ *where* $\varphi:[0,\infty)\times[0,\infty)\longrightarrow[0,\infty)$ *is a continuous function such that* $\varphi(t,t)=1$ *for all* $t\in[0,\infty)$ *then, all* T_n *have a unique common fixed point in X.*

Proof For any $x_0\in X$, let $x_n=T_n(x_{n-1})$, $n=1,2,...,$ then from (2.5) we obtain

$$\begin{aligned}d(x_1,x_2)&=d(T_1(x_0),T_2(x_1))\\ &\leq a_{1,2}d(x_1,T_2x_1)\varphi(d(x_0,T_1x_0),d(x_0,x_1))+b_{1,2}d(x_0,x_1)\\ &\leq a_{1,2}d(x_1,x_2)\varphi(d(x_0,x_1),d(x_0,x_1))+b_{1,2}d(x_0,x_1)\\ &\leq a_{1,2}[d(x_1,x_2)+b_{1,2}d(x_0,x_1)],\end{aligned}$$

implies

$$(1-a_{1,2})d(x_1,x_2)\leq b_{1,2}d(x_0,x_1).$$

Hence we have

$$d(x_1,x_2)\leq\frac{b_{1,2}}{1-a_{1,2}}d(x_0,x_1).$$

Also,

$$\begin{aligned}d(x_2,x_3)&=(T_2(x_1),T_3(x_2))\\ &\leq a_{2,3}d(x_2,T_3x_2)\varphi(d(x_1,T_2x_1),d(x_1,x_2))+b_{2,3}d(x_1,x_2)\\ &\leq a_{2,3}d(x_2,x_3)\varphi(d(x_1,x_2),d(x_1,x_2))+b_{2,3}d(x_1,x_2)\\ &\leq a_{2,3}d(x_2,x_3)+b_{2,3}d(x_1,x_2).\end{aligned}$$

Then

$$\begin{aligned}d(x_2,x_3)&\leq\frac{b_{2,3}}{1-a_{2,3}}d(x_1,x_2)\\ &\leq\frac{b_{1,2}}{1-a_{1,2}}\times\frac{b_{2,3}}{1-a_{2,3}}d(x_0,x_1).\end{aligned}$$

Generally we conclude that

$$d(x_n,x_{n+1})\leq\prod_{i=1}^{n}\frac{b_{i,i+1}}{1-a_{i,i+1}}d(x_0,x_1)=A_nd(x_0,x_1).\tag{7}$$

Therefore, for $m,n\in N$, $m\geq n$, and using (2.6) we get

$$\begin{aligned}d(x_n,x_m)&\leq\sum_{k=n}^{m-1}d(x_k,x_{k+1})\\ &\leq\sum_{k=n}^{m-1}A_kd(x_0,x_1).\end{aligned}$$

Now passing to limit $n,m\longrightarrow\infty$, we get $d(x_n,x_m)\longrightarrow 0$.

Thus $\{x_n\}$ is a Cauchy sequence and by completeness of X, $\{x_n\}$ converges to x in X that is $\lim_{n\longrightarrow\infty}x_n=x\in X$.

So, for any positive integer m,

$$\begin{aligned}d(x,T_mx)&\leq d(x,x_n)+d(x_n,T_mx)=d(x,x_n)+d(T_nx_{n-1},T_mx)\\ &\leq d(x,x_n)+a_{n,m}d(x,T_mx)\varphi(d(x_{n-1},T_nx_{n-1}),d(x_{n-1},x))\\ &\quad+b_{n,m}d(x_{n-1},x)\\ &\leq d(x,x_n)+a_{n,m}d(x,T_mx)\varphi(d(x_{n-1},x_n),d(x_{n-1},x_n))\\ &\quad+b_{n,m}d(x_{n-1},x)\\ &\leq d(x,x_n)+a_{n,m}d(x,T_mx)+b_{n,m}d(x_{n-1},x).\end{aligned}$$

Taking $\overline{\lim}$ as $n\longrightarrow\infty$, we get

$$d(x,T_mx)\leq\overline{\lim}a_{n,m}d(x,T_mx)<d(x,T_mx).$$

It follows that $d(x, T_m x) = 0$ gives x as a common fixed of $\{T_m\}$.

Let y be another common fixed point, then

$$\begin{aligned} d(x,y) &= d(T_n x, T_m y) \leq a_{n,m} d(y, T_m y)\varphi(d(x, T_n(x)), d(x,y)) \\ &\quad + b_{n,m} d(x,y) \\ &= a_{n,m} d(y,y)\varphi(d(x,x), d(x,y)) + b_{n,m} d(x,y) \\ &= b_{n,m} d(x,y). \end{aligned}$$

Taking $\overline{lim}$ as $n \longrightarrow \infty$, we get $x = y$. So, the uniqueness is proved. □

Author contributions All authors contributed equally and significantly in writing this paper. All authors read and approved the final manuscript.

Compliance with ethical standards

Conflict of interest The authors declare that they have no competing interests.

References

1. Aghajani, A.: Radenović)-contractive condition in partially ordered metric spaces. Appl. Math. Comput. **218**, 5665–5670 (2012)
2. Alaca, C.: A Common fixed point theorem for a family of selfmappings satisfying a general contractive condition of operator type. Albania. J. Math. **3**(1), 13–17 (2009)
3. Banach, S.: Surles oprations dans les ensembles abstraits et leurs application aux quations intgrales. Fund. Mat. **3**, 133–181 (1922)
4. Buong, N., Thi Quynh Anh, N.: An implicit iteration method for variational inequalities over the set of common fixed points for a finite family of nonexpansive mappings in Hilbert spaces, fixed point theory and applications, (2011), in press
5. Husain, S.A., Sehgal, V.M.: On common fixed points for a family of mappings. Bull. Austral. Math. **13**, 261–267 (1975)
6. Hu, L.G., Wang, J.P.: Strong Convergence of a New Iteration for a Finite Family of Accretive Operators. Fixed Point Theory Appl. (2009). doi:10.1155/2009/491583
7. Ćirić, L. B.: On a family of contractive maps and fixed points, Publ. Inst. Math, **17** (1974), no. 31, 45–51
8. Mai, J.H., Liu, X.H.: Fixed-point theorems for families of weakly non-expansive maps. J. Math. Anal. Appl. **334**, 932–949 (2007)
9. Park, B. S.: Sequences of quasi-contractions and fixed, J. Korean Math. Soc., **14** (1977), no. 1
10. Samet, B.: Common fixed point theorems involving two pairs of weakly compatible mappings in K-metric spaces. Appl. Math. Lett. **24**, 1245–1250 (2011)
11. Taskovic, M.R.: On a family of contractive maps. Bull. Austral. Math. Soc **13**, 301–308 (1975)
12. Tang, Y.C., Peng, J.G.: Approximation of common fixed points for a finite family of uniformly Quasi-Lipschitzian mappings in Banach Spaces. Thai J. Math. **8**(1), 63–71 (2010)
13. Wangkeeree, R.: Implicit iteration process for finite family of nonexpansive nonself-mappings in Banach spaces. Int. J. Math. Sci. Engg. Appls. (IJMSEA), **1** (2007), no. 1, 1–12
14. Zuo, Z.: Iterative approximations for a family of multivalued mappings in Banach Spaces. J. Math. Inequal. **4**(4), 549–560 (2010)
15. Alsulami, H., Karapinar, E., Roldan, A.: A short note on Common fixed point theorems for non-compatible self-maps in generalized metric spaces. J. Inequal Appl. (accepted 29 January 2015)
16. Chauhan, S., Karapinar, E.: Some integral type common fixed point theorems satisfying Ψ-contractive conditions. Bull. Belg. Math. Soc. Simon Stevin **21**(4), 593–612 (2014)
17. Roldan, A., Karapinar, E., Kumam, p.: On unified fixed point theorems in fuzzy metric spaces via common limit range property. J. Inequal. Appl
18. Bilgili, N., Karapinar, E., Turkoglu, D.: A note on common fixed points for (ψ, α, β)-weakly contractive mappings in generalized metric spaces. Fixed Point Theory Appl. **2013**, 287 (2013)

24

Hermite–Hadamard-type inequalities for generalized *s*-convex functions on real linear fractal set $\mathbb{R}^{\alpha}(0<\alpha<1)$

Huixia Mo[1] · Xin Sui[1]

Abstract In this paper, we establish the Hermite–Hadamard-type inequalities for the generalized *s*-convex functions in the second sense on real linear fractal set $\mathbb{R}^{\alpha}(0<\alpha<1)$.

Keywords Fractal set · Local fractional integral · Hermite–Hadamard-type inequality · Generalized *s*-convex function

Introduction

The convex function plays an important role in the class mathematical analysis course and other fields. In [1], Hudzik and Maligranda introduced two kinds of *s*-convex functions in the space of European space $\mathbb{R}$. In addition, many important inequalities are established for the *s*-convex functions in $\mathbb{R}$. For example, the Hermite–Hadamard's inequality is one of the best known results in the literature, see [2–4] and so on.

In recent years, the fractal theory has received significantly remarkable attention [5]. The calculus on fractal set can lead to better comprehension for the various real-world models from the engineering and science [6].

On the fractal set, Mo et al. [7, 8] introduced the definition of the generalized convex function and established Hermite–Hadamard-type inequality. In [9], the authors introduced two kinds of generalized *s*-convex functions on fractal sets $\mathbb{R}^{\alpha}(0<\alpha<1)$.

The definitions of the generalized *s*-convex functions are as follows:

Definition 1.1 [9] Suppose that $\mathbb{R}_+ = [0,\infty)$. If the function $f : \mathbb{R}_+ \to \mathbb{R}^{\alpha}$ satisfies the following inequality:

$$f(\lambda_1 u + \lambda_2 v) \le \lambda_1^{\alpha s} f(u) + \lambda_2^{\alpha s} f(v), \tag{1.1}$$

for all $u, v \in \mathbb{R}_+$ and all $\lambda_1, \lambda_2 \ge 0$ with $\lambda_1^s + \lambda_2^s = 1$ and $0<s<1$, then f is said to be a generalized *s*-convex function in the first sense. We denote this by $f \in GK_s^1$.

Definition 1.2 [9] Suppose that $\mathbb{R}_+ = [0,\infty)$. If the function $f : \mathbb{R}_+ \to \mathbb{R}^{\alpha}$ satisfies the following inequality:

$$f(\lambda_1 u + \lambda_2 v) \le \lambda_1^{\alpha s} f(u) + \lambda_2^{\alpha s} f(v), \tag{1.2}$$

for all $u, v \in \mathbb{R}_+$ and all $\lambda_1, \lambda_2 \ge 0$ with $\lambda_1 + \lambda_2 = 1$ and $0<s<1$, then f is said to be a generalized *s*-convex function in the second sense. We denote this by $f \in GK_s^2$.

Note that the generalized *s*-convex function in both sense is generalized convex function [9] for $s = 1$.

Inspired by [2, 3, 8], in this paper, we will establish the Hermite–Hadamard-type inequalities for the generalized *s*-convex functions.

Preliminaries

Now, let us review the operations with real line number on fractal space. In addition, we will use the Gao–Yang–Kang's idea to describe the definitions of the local fractional derivative and local fractional integral [10–14].

Let a^{α}, b^{α} and c^{α} belong to the set $\mathbb{R}^{\alpha}(0<\alpha<1)$ of real line numbers, then

The research is supported by the NNSF of China (11601035, 11471050).

✉ Huixia Mo
huixiamo@bupt.edu.cn

1 School of Science, Beijing University of Posts and Telecommunications, Beijing 100876, China

1. $a^\alpha b^\alpha$ and $a^\alpha + b^\alpha$ belong to the set $\mathbb{R}^\alpha$;
2. $a^\alpha + b^\alpha = (a+b)^\alpha = b^\alpha + a^\alpha = (b+a)^\alpha$;
3. $a^\alpha + (b^\alpha + c^\alpha) = (a^\alpha + b^\alpha) + c^\alpha$;
4. $a^\alpha b^\alpha = (ab)^\alpha = b^\alpha a^\alpha = (ba)^\alpha$;
5. $a^\alpha (b^\alpha c^\alpha) = (a^\alpha b^\alpha) c^\alpha$;
6. $a^\alpha (b^\alpha + c^\alpha) = a^\alpha b^\alpha + a^\alpha c^\alpha$;
7. $0^\alpha + a^\alpha = a^\alpha + 0^\alpha = a^\alpha$ and $1^\alpha \cdot a^\alpha = a^\alpha \cdot 1^\alpha = a^\alpha$.

Definition 2.1 ([10]) If the function $f : [a,b] \to \mathbb{R}^\alpha$ satisfies the inequality

$$|f(x) - f(y)| < c|x-y|^\alpha, \quad (x, y \in [a,b]),$$

for $c > 0$ and $\alpha (0 < \alpha \leq 1)$, then f is called a Hölder continuous function. In this case, we think that f is in the space $C_\alpha[a,b]$.

Definition 2.2 [10] Let $\Delta^\alpha(f(x) - f(x_0)) \cong \Gamma(1 + a)(f(x) - f(x_0))$. Then, the local fractional derivative of f of order α at $x = x_0$ is defined by

$$f^{(\alpha)}(x_0) = \frac{d^\alpha f(x)}{dx^\alpha}|_{x=x_0} = \lim_{x \to x_0} \frac{\Delta^\alpha(f(x) - f(x_0))}{(x - x_0)^\alpha}.$$

If there exists $f^{((k+1)\alpha)}(x) = \overbrace{D_x^\alpha \ldots D_x^\alpha}^{k+1 \text{ times}} f(x)$ for any $x \in I \subseteq \mathbb{R}$, then we denoted $f \in D_{(k+1)\alpha}(I)$, where $k = 0, 1, 2 \ldots$.

Definition 2.3 [10] For $f \in C_\alpha[a,b]$, the local fractional integral of the function f is defined by

$$\begin{aligned} &{}_aI_b^{(\alpha)} f(x) \\ &= \frac{1}{\Gamma(1+a)} \int_a^b f(t)(dt)^\alpha \\ &= \frac{1}{\Gamma(1+a)} \lim_{\Delta t \to 0} \sum_{j=0}^{N} f(t_j)(\Delta t_j)^\alpha, \end{aligned}$$

where $\Delta t_j = t_{j+1} - t_j$, $\Delta t = \max\{\Delta t_1, \Delta t_2, \Delta t_j, \ldots\}$ and $[t_j, t_j + 1]$, $j = 0, \ldots, N-1$, $t_0 = a$, $t_N = b$, is a partition of the interval $[a, b]$.

Lemma 2.1 [10] *Let* $f \in C_\alpha[g(a), g(b)]$ *and* $g \in C_1[a,b]$, *then*

$${}_{g(a)}I_{g(b)}^{(\alpha)} f(x) = {}_aI_b^{(\alpha)} f(g)(s)[g'(s)]^\alpha.$$

Lemma 2.2 [10]

1. Let $f(x) = g^{(\alpha)}(x) \in C_\alpha[a,b]$, *then we have*

   $${}_aI_b^{(\alpha)} f(x) = g(b) - g(a).$$

2. Let $f(x), g(x) \in D_\alpha[a,b]$ and $f^{(\alpha)}(x), g^{(\alpha)}(x) \in C_\alpha[a,b]$, *then we have*

   $${}_aI_b^\alpha f(x) g^{(\alpha)}(x) = f(x)g(x)\Big|_a^b - {}_aI_b^{(\alpha)} f^{(\alpha)}(x) g(x).$$

Lemma 2.3 [10]

$$\frac{d^\alpha x^{k\alpha}}{dx^\alpha} = \frac{\Gamma(1+k\alpha)}{\Gamma(1+(k-1)\alpha)} x^{(k-1)\alpha}.$$

From the above formula and Lemma 2.2, *we have*

$$\frac{1}{\Gamma(1+\alpha)} \int_a^b x^{k\alpha}(dx)^\alpha = \frac{\Gamma(1+k\alpha)}{\Gamma(1+(k+1)\alpha)} (b^{(k+1)\alpha} - a^{(k+1)\alpha}), k \in R.$$

Lemma 2.4 [10] (Generalized Hölder's inequality) *Let* $f, g \in C_\alpha[a,b]$ *and* $p, q > 1$ *with* $1/p + 1/q = 1$. *Then, it follows that*

$$\frac{1}{\Gamma(1+\alpha)} \int_a^b |f(x)g(x)|(\mathrm{d}x)^\alpha \leq \left(\frac{1}{\Gamma(1+\alpha)} \int_a^b |f(x)|^p (\mathrm{d}x)^\alpha \right)^{1/p} \times \left(\frac{1}{\Gamma(1+\alpha)} \int_a^b |g(x)|^q (\mathrm{d}x)^\alpha \right)^{1/q}.$$

Main results

Theorem 3.1 *Let* $f : \mathbb{R}_+ \to \mathbb{R}^\alpha$ *be a generalized s-convex function in the second sense for* $0 < s < 1$ *and* $a, b \in [0, \infty)$ with $a < b$. *Then, for* $f \in C_\alpha[a,b]$, *the following inequalities hold*:

$$\frac{2^{(s-1)\alpha}}{\Gamma(1+\alpha)} f\left(\frac{a+b}{2}\right) \leq \frac{{}_aI_b^\alpha f(x)}{(b-a)^\alpha} \leq \frac{\Gamma(1+s\alpha)}{\Gamma(1+(s+1)\alpha)} (f(a) + f(b)). \tag{3.1}$$

Proof Let $x = a + b - t$. Then, from Lemma 2.1, we have ${}_{\frac{a+b}{2}}I_b^{(\alpha)} f(x) = {}_aI_{\frac{a+b}{2}}^{(\alpha)} f(a+b-t)$.

Since f is a generalized s-convex function in the second sense, then

$$\begin{aligned} {}_aI_b^{(\alpha)} f(x) &= {}_aI_{\frac{a+b}{2}}^{(\alpha)} [f(x) + f(a+b-x)] \\ &\geq 2^{\alpha s} {}_aI_{\frac{a+b}{2}}^{(\alpha)} f\left(\frac{a+b}{2}\right) \\ &= \frac{2^{(s-1)\alpha}}{\Gamma(1+\alpha)} (b-a)^\alpha f\left(\frac{a+b}{2}\right). \end{aligned}$$

In the other hand, let $x = b - (b-a)t$, $0 \leq t \leq 1$, then we get

$$\begin{aligned} {}_aI_b^{(\alpha)} f(x) &= (b-a)^\alpha {}_0I_1^{(\alpha)} f[ta + (1-t)b] \\ &\leq (b-a)^\alpha {}_0I_1^{(\alpha)} [t^{\alpha s} f(a) + (1-t)^{\alpha s} f(b)] \\ &= (b-a)^\alpha [f(a) {}_0I_1^{(\alpha)} t^{\alpha s} + f(b) {}_0I_1^{(\alpha)} (1-t)^{\alpha s}]. \end{aligned}$$

From Lemma 2.3, it is easy to see that

$${}_0I_1^{(\alpha)} t^{\alpha s} = \frac{\Gamma(1+s\alpha)}{\Gamma(1+(s+1)\alpha)},$$

and

$$ {}_0I_1^{(\alpha)}(1-t)^{\alpha s} = \frac{\Gamma(1+s\alpha)}{\Gamma(1+(s+1)\alpha)}. $$

Therefore

$$ {}_aI_b^{(\alpha)}f(x) \le (b-a)^\alpha \frac{\Gamma(1+s\alpha)}{\Gamma(1+(s+1)\alpha)}(f(a)+f(b)). $$

Combining the above estimates, we obtain

$$ \frac{2^{(s-1)\alpha}}{\Gamma(1+\alpha)} f\left(\frac{a+b}{2}\right) \le \frac{{}_aI_b^\alpha f(x)}{(b-a)^\alpha} \le \frac{\Gamma(1+s\alpha)}{\Gamma(1+(s+1)\alpha)}(f(a)+f(b)). $$

□

Theorem 3.2 *Let $I \subset \mathbb{R}$ be an interval, and I^0 be the interior of I. Suppose that $f: I \to \mathbb{R}^\alpha$ is a differentiable function on I^0 such that $f^{(\alpha)} \in C_\alpha[a,b]$, where $a, b \in I^0$* with $a<b$. If $|f^{(\alpha)}|^q$ *is a generalized s-convex function in the second sense on [a, b] for some fixed $s \in (0,1)$ and $q \ge 1$, then*

$$ \left|\frac{f(a)+f(b)}{2^\alpha} - \frac{\Gamma(1+\alpha)}{(b-a)^\alpha}{}_aI_b^{(\alpha)}f(x)\right| \le \frac{(b-a)^\alpha}{2^\alpha}\left(\frac{\Gamma(1+\alpha)}{\Gamma(1+2\alpha)}\right)^{\frac{q-1}{q}} \left[\frac{\Gamma(1+s\alpha)}{\Gamma(1+(s+1)\alpha)} + \frac{\Gamma(1+\alpha)\Gamma(1+s\alpha)}{\Gamma(1+(s+2)\alpha)}\left(\left(\frac{1}{2}\right)^{\alpha s} - 2^\alpha\right)\right]^{\frac{1}{q}} \times \left[|f^{(\alpha)}(a)|^q + |f^{(\alpha)}(b)|^q\right]^{\frac{1}{q}}. $$

To show Theorem 3.2 is right, we need the following Lemma.

Lemma 3.1 ([8]) *Let $f: I \to \mathbb{R}^\alpha$, $I \subset [0,\infty)$. If $f \in D_\alpha(I^0)$* and $f^{(\alpha)} \in C_\alpha[a,b]$ *for $a, b \in I^0$ with $a<b$, then the following equality holds*:

$$ \frac{f(a)+f(b)}{2^\alpha} - \frac{\Gamma(1+\alpha)}{(b-a)^\alpha}{}_aI_b^{(\alpha)}f(x) = \frac{(b-a)^\alpha}{2^\alpha}\frac{1}{\Gamma(1+\alpha)}\int_0^1 (1-2t)^\alpha f^{(\alpha)}(ta+(1-t)b)(dt)^\alpha. $$

Now, let us give the proof of Theorem 3.2.

Proof From Lemma 3.1, it is obvious that

$$ \left|\frac{f(a)+f(b)}{2^\alpha} - \frac{\Gamma(1+\alpha)}{(b-a)^\alpha}{}_aI_b^{(\alpha)}f(x)\right| \le \frac{(b-a)^\alpha}{2^\alpha}\frac{1}{\Gamma(1+\alpha)}\int_0^1 |1-2t|^\alpha |f^{(\alpha)}(ta+(1-t)b)|(dt)^\alpha. \tag{3.2} $$

Let us estimate

$$ \frac{1}{\Gamma(1+\alpha)}\int_0^1 |1-2t|^\alpha |f^{(\alpha)}(ta+(1-t)b)|(dt)^\alpha, $$

for $q=1$ and $q>1$.

Case I. $q=1$.

Since $|f^{(\alpha)}|$ is generalized sconvex on [a, b] in the second sense, we can know that for any $t \in [0,1]$

$$ |f^{(\alpha)}(ta+(1-t)b)| \le t^{\alpha s}|f^{(\alpha)}(a)| + (1-t)^{\alpha s}|f^{(\alpha)}(b)|. $$

Then, we have

$$ \begin{aligned} &\frac{1}{\Gamma(1+\alpha)}\int_0^1 |1-2t|^\alpha |f^{(\alpha)}(ta+(1-t)b)|(dt)^\alpha \\ &\le \frac{1}{\Gamma(1+\alpha)}\int_0^1 |1-2t|^\alpha \left[t^{\alpha s}|f^{(\alpha)}(a)| + (1-t)^{\alpha s}|f^{(\alpha)}(b)|\right](dt)^\alpha \\ &= \left[|f^{(\alpha)}(a)|\frac{1}{\Gamma(1+\alpha)}\int_0^1 t^{\alpha s}|1-2t|^\alpha(dt)^\alpha + |f^{(\alpha)}(b)| \frac{1}{\Gamma(1+\alpha)}\int_0^1 (1-t)^{\alpha s}|1-2t|^\alpha(dt)^\alpha\right]. \end{aligned} \tag{3.3} $$

From Lemmas 2.2 and 2.3, it is easy to see that

$$ \begin{aligned} &\frac{1}{\Gamma(1+\alpha)}\int_0^1 t^{\alpha s}|1-2t|^\alpha(dt)^\alpha \\ &= \frac{1}{\Gamma(1+\alpha)}\left[\int_0^{\frac{1}{2}} t^{\alpha s}(1-2t)^\alpha(dt)^\alpha + \int_{\frac{1}{2}}^1 t^{\alpha s}(2t-1)^\alpha(dt)^\alpha\right] \\ &= \left[\frac{\Gamma(1+s\alpha)}{\Gamma(1+(s+1)\alpha)} + \frac{\Gamma(1+\alpha)\Gamma(1+s\alpha)}{\Gamma(1+(s+2)\alpha)}\left(\left(\frac{1}{2}\right)^{\alpha s} - 2^\alpha\right)\right]. \end{aligned} \tag{3.4} $$

In addition, let $1-t=x$, then by Lemma 2.1 and (3.4), we have

$$ \begin{aligned} &\frac{1}{\Gamma(1+\alpha)}\int_0^1 (1-t)^{\alpha s}|1-2t|^\alpha(dt)^\alpha \\ &= \frac{1}{\Gamma(1+\alpha)}\int_0^1 x^{\alpha s}|1-2x|^\alpha(dx)^\alpha \\ &= \left[\frac{\Gamma(1+s\alpha)}{\Gamma(1+(s+1)\alpha)} + \frac{\Gamma(1+\alpha)\Gamma(1+s\alpha)}{\Gamma(1+(s+2)\alpha)}\left(\left(\frac{1}{2}\right)^{\alpha s} - 2^\alpha\right)\right]. \end{aligned} \tag{3.5} $$

Thus, substituting (3.4) and (3.5) into (3.3), we have

$$\frac{1}{\Gamma(1+\alpha)}\int_0^1 |1-2t|^\alpha |f^{(\alpha)}(ta+(1-t)b)|(dt)^\alpha \leq \frac{\Gamma(1+\alpha)}{\Gamma(1+2\alpha)}\left[\frac{\Gamma(1+s\alpha)}{\Gamma(1+(s+1)\alpha)}+\frac{\Gamma(1+\alpha)\Gamma(1+s\alpha)}{\Gamma(1+(s+2)\alpha)} \times \left(\left(\frac{1}{2}\right)^{\alpha s}-2^\alpha\right)\right] \times \left[|f^{(\alpha)}(a)|^q+|f^{(\alpha)}(b)|\right]. \tag{3.6}$$

Thus, from (3.2), we obtain

$$\left|\frac{f(a)+f(b)}{2^\alpha}-\frac{\Gamma(1+\alpha)}{(b-a)^\alpha}{}_aI_b^{(\alpha)}f(x)\right| \leq \frac{(b-a)^\alpha}{2^\alpha}\frac{\Gamma(1+\alpha)}{\Gamma(1+2\alpha)}\left[\frac{\Gamma(1+s\alpha)}{\Gamma(1+(s+1)\alpha)} + \frac{\Gamma(1+\alpha)\Gamma(1+s\alpha)}{\Gamma(1+(s+2)\alpha)}\left(\left(\frac{1}{2}\right)^{\alpha s}-2^\alpha\right)\right] \times \left[|f^{(\alpha)}(a)|^q+|f^{(\alpha)}(b)|\right].$$

Case II. $q > 1$.

Using the generalized Hölder's inequality (Lemma 2.4), we obtain

$$\frac{1}{\Gamma(1+\alpha)}\int_0^1 |1-2t|^\alpha |f^{(\alpha)}(ta+(1-t)b)|(dt)^\alpha = \frac{1}{\Gamma(1+\alpha)} \int_0^1 |1-2t|^{\alpha\frac{q-1}{q}}|1-2t|^{\alpha\frac{1}{q}}|f^{(\alpha)}(ta+(1-t)b)|(dt)^\alpha \leq \left(\frac{1}{\Gamma(1+\alpha)}\int_0^1 |1-2t|^\alpha(dt)^\alpha\right)^{\frac{q-1}{q}} \times \left(\frac{1}{\Gamma(1+\alpha)}\int_0^1 |1-2t|^\alpha|f^{(\alpha)}(ta+(1-t)b)|^q(dt)^\alpha\right)^{\frac{1}{q}}. \tag{3.7}$$

It is obvious that

$$\frac{1}{\Gamma(1+\alpha)}\int_0^1 |1-2t|^\alpha(dt)^\alpha = \frac{1}{\Gamma(1+\alpha)}\int_0^{\frac{1}{2}}(1-2t)^\alpha(dt)^\alpha + \frac{1}{\Gamma(1+\alpha)}\int_{\frac{1}{2}}^1(2t-1)^\alpha(dt)^\alpha = \frac{\Gamma(1+\alpha)}{\Gamma(1+2\alpha)}. \tag{3.8}$$

Moreover, since $|f^{(\alpha)}|^q$ is generalized s convex in the second sense on $[a, b]$, then

$$\frac{1}{\Gamma(1+\alpha)}\int_0^1 |1-2t|^\alpha |f^{(\alpha)}(ta+(1-t)b)|^q(dt)^\alpha \leq \frac{1}{\Gamma(1+\alpha)}\int_0^1 |1-2t|^\alpha \left(t^{\alpha s}|f^{(\alpha)}(a)|^q+(1-t)^{\alpha s}|f^{(\alpha)}(b)|^q\right)(dt)^\alpha = |f^{(\alpha)}(a)|^q\frac{1}{\Gamma(1+\alpha)}\int_0^1|1-2t|^\alpha t^{\alpha s}(dt)^\alpha + |f^{(\alpha)}(b)|^q\frac{1}{\Gamma(1+\alpha)}\int_0^1|1-2t|^\alpha(1-t)^{\alpha s}(dt)^\alpha.$$

From (3.3) and (3.4), it is easy to see that

$$\frac{1}{\Gamma(1+\alpha)}\int_0^1|1-2t|^\alpha t^{\alpha s}(dt)^\alpha = \frac{1}{\Gamma(1+\alpha)}\int_0^1|1-2t|^\alpha(1-t)^{\alpha s}(dt)^\alpha = \left[\frac{\Gamma(1+s\alpha)}{\Gamma(1+(s+1)\alpha)}+\frac{\Gamma(1+\alpha)\Gamma(1+s\alpha)}{\Gamma(1+(s+2)\alpha)}\left(\left(\frac{1}{2}\right)^{\alpha s}-2^\alpha\right)\right].$$

Therefore

$$\frac{1}{\Gamma(1+\alpha)}\int_0^1 |1-2t|^\alpha |f^{(\alpha)}(ta+(1-t)b)|^q(dt)^\alpha = \left[\frac{\Gamma(1+s\alpha)}{\Gamma(1+(s+1)\alpha)}+\frac{\Gamma(1+\alpha)\Gamma(1+s\alpha)}{\Gamma(1+(s+2)\alpha)}\left(\left(\frac{1}{2}\right)^{\alpha s}-2^\alpha\right)\right]^{\frac{1}{q}} \left[|f^{(\alpha)}(a)|^q+|f^{(\alpha)}(b)|^q\right]^{\frac{1}{q}}. \tag{3.9}$$

Thus, substituting (3.8) and (3.9) into (3.7), we have

$$\frac{1}{\Gamma(1+\alpha)}\int_0^1 |1-2t|^\alpha |f^{(\alpha)}(ta+(1-t)b)|(dt)^\alpha \leq \left(\frac{\Gamma(1+\alpha)}{\Gamma(1+2\alpha)}\right)^{\frac{q-1}{q}} \left[\frac{\Gamma(1+s\alpha)}{\Gamma(1+(s+1)\alpha)}+\frac{\Gamma(1+\alpha)\Gamma(1+s\alpha)}{\Gamma(1+(s+2)\alpha)}\left(\left(\frac{1}{2}\right)^{\alpha s}-2^\alpha\right)\right]^{\frac{1}{q}} \left[|f^{(\alpha)}(a)|^q+|f^{(\alpha)}(b)|^q\right]^{\frac{1}{q}}.$$

Therefore, from (3.2), it follows that

$$\left|\frac{f(a)+f(b)}{2^\alpha}-\frac{\Gamma(1+\alpha)}{(b-a)^\alpha}{}_aI_b^{(\alpha)}f(x)\right|$$
$$\leq\frac{(b-a)^\alpha}{2^\alpha}\left(\frac{\Gamma(1+\alpha)}{\Gamma(1+2\alpha)}\right)^{\frac{q-1}{q}}$$
$$\left[\frac{\Gamma(1+s\alpha)}{\Gamma(1+(s+1)\alpha)}+\frac{\Gamma(1+\alpha)\Gamma(1+s\alpha)}{\Gamma(1+(s+2)\alpha)}\left(\left(\frac{1}{2}\right)^{\alpha s}-2^\alpha\right)\right]^{\frac{1}{q}}$$
$$\times\left[|f^{(\alpha)}(a)|^q+|f^{(\alpha)}(b)|^q\right]^{\frac{1}{q}}.$$

Thus, we complete the proof of Theorem 3.2. □

Theorem 3.3 *Suppose that* $f\colon I\to\mathbb{R}^\alpha$, $I\subset[0,\infty)$ *is a differentiable function on* I^0, *such that* $f^{(\alpha)}\in C_\alpha[a,b]$, *where* $a,b\in I$ *with* $a<b$. *If* $|f^{(\alpha)}|^q$ *is a generalized s-convex function in the second sense on* $[a,b]$ *for some fixed* $s\in(0,1)$ *and* $q>1$, *then*

$$\left|\frac{f(a)+f(b)}{2^\alpha}-\frac{\Gamma(1+\alpha)}{(b-a)^\alpha}{}_aI_b^{(\alpha)}f(x)\right|$$
$$\leq\frac{(b-a)^\alpha}{2^\alpha}\left[\frac{\Gamma(1+\frac{q}{q-1}\alpha)}{2^\alpha\Gamma(1+\frac{2q-1}{q-1}\alpha)}\right]^{\frac{q-1}{q}}$$
$$\left(\frac{\Gamma(1+s\alpha)}{2^\alpha\Gamma(1+(s+1)\alpha)}\right)^{\frac{1}{q}}$$
$$\times\left[\left(|f^{(\alpha)}(a)|^q+\left|f^{(\alpha)}\left(\frac{a+b}{2}\right)\right|^q\right)^{\frac{1}{q}}\right.$$
$$\left.+\left(\left|f^{(\alpha)}\left(\frac{a+b}{2}\right)\right|^q+|f^{(\alpha)}(b)|^q\right)^{\frac{1}{q}}\right].$$

Proof From Lemma 3.1, we have

$$\left|\frac{f(a)+f(b)}{2^\alpha}-\frac{\Gamma(1+\alpha)}{(b-a)^\alpha}{}_aI_b^{(\alpha)}f(x)\right|$$
$$\leq\frac{(b-a)^\alpha}{2^\alpha}\frac{1}{\Gamma(1+\alpha)}\int_0^1|1-2t|^\alpha|f^{(\alpha)}(ta+(1-t)b)|(dt)^\alpha$$
$$\leq\frac{(b-a)^\alpha}{2^\alpha}\left[\frac{1}{\Gamma(1+\alpha)}\int_0^{\frac{1}{2}}(1-2t)|f^{(\alpha)}(ta+(1-t)b)|(dt)^\alpha\right.$$
$$\left.+\frac{1}{\Gamma(1+\alpha)}\int_{\frac{1}{2}}^1(2t-1)|f^{(\alpha)}(ta+(1-t)b)|(dt)^\alpha\right].\tag{3.10}$$

Let us estimate

$$\frac{1}{\Gamma(1+\alpha)}\int_0^{\frac{1}{2}}(1-2t)|f^{(\alpha)}(ta+(1-t)b)|(dt)^\alpha$$

and

$$\frac{1}{\Gamma(1+\alpha)}\int_{\frac{1}{2}}^1(2t-1)|f^{(\alpha)}(ta+(1-t)b)|(dt)^\alpha,$$

respectively.

Using the generalized Hölder's inequality(Lemma 2.4), we obtain

$$\frac{1}{\Gamma(1+\alpha)}\int_0^{\frac{1}{2}}(1-2t)|f^{(\alpha)}(ta+(1-t)b)|(dt)^\alpha$$
$$\leq\left(\frac{1}{\Gamma(1+\alpha)}\int_0^{\frac{1}{2}}(1-2t)^{\alpha\frac{q}{q-1}}(dt)^\alpha\right)^{\frac{q-1}{q}}\tag{3.11}$$
$$\left(\frac{1}{\Gamma(1+\alpha)}\int_0^{\frac{1}{2}}|f^{(\alpha)}(ta+(1-t)b)|^{\alpha q}(dt)^\alpha\right)^{\frac{1}{q}}.$$

It is easy to see that

$$\frac{1}{\Gamma(1+\alpha)}\int_0^{\frac{1}{2}}(1-2t)^{\alpha\frac{q}{q-1}}(dt)^\alpha=$$
$$\frac{1}{\Gamma(1+\alpha)}\int_{\frac{1}{2}}^1(1-2t)^{\alpha\frac{q}{q-1}}(dt)^\alpha=\frac{\Gamma(1+\frac{q}{q-1}\alpha)}{2^\alpha\Gamma(1+\frac{2q-1}{q-1}\alpha)}.\tag{3.12}$$

Let $|f^{(\alpha)}(ta+(1-t)b)|^q=U(t)$. It is easy to see that $U(t)$ is a generalized *s*convex function in the second sense. Thus, from the right-hand side of (3.1), it follows that

$$\frac{1}{\Gamma(1+\alpha)}\int_0^{\frac{1}{2}}|f^{(\alpha)}(ta+(1-t)b)|^q(dt)^\alpha$$
$$=\frac{1}{\Gamma(1+\alpha)}\int_0^{\frac{1}{2}}U(t)(dt)^\alpha$$
$$\leq\left(\frac{1}{2}-0\right)^\alpha\frac{\Gamma(1+s\alpha)}{\Gamma(1+(s+1)\alpha)}\left(U(0)+U(\tfrac{1}{2})\right)$$
$$=\frac{\Gamma(1+s\alpha)}{2^\alpha\Gamma(1+(s+1)\alpha)}\left(\left|f^{(\alpha)}\left(\frac{a+b}{2}\right)\right|^q+|f^{(\alpha)}(b)|^q\right).\tag{3.13}$$

Thus, substituting (3.12) and (3.13) into (3.11), we get

$$\frac{1}{\Gamma(1+\alpha)}\int_0^{\frac{1}{2}}(1-2t)|f^{(\alpha)}(ta+(1-t)b)|(dt)^\alpha$$
$$\leq\left(\frac{\Gamma(1+\frac{q}{q-1}\alpha)}{2^\alpha\Gamma(1+\frac{2q-1}{q-1}\alpha)}\right)^{\frac{q-1}{q}}\left(\frac{\Gamma(1+s\alpha)}{2^\alpha\Gamma(1+(s+1)\alpha)}\right)^{\frac{1}{q}}$$
$$\left(|f^{(\alpha)}(a)|^q+\left|f^{(\alpha)}\left(\frac{a+b}{2}\right)\right|^q\right)^{\frac{1}{q}}.\tag{3.14}$$

Moreover

$$\frac{1}{\Gamma(1+\alpha)}\int_{\frac{1}{2}}^{1}(2t-1)^{\alpha\frac{q}{q-1}}(dt)^{\alpha}$$

$$=\frac{1}{\Gamma(1+\alpha)}\int_{\frac{1}{2}}^{1}(1-2t)^{\alpha\frac{q}{q-1}}(dt)^{\alpha}=\frac{\Gamma(1+\frac{q}{q-1}\alpha)}{2^{\alpha}\Gamma(1+\frac{2q-1}{q-1}\alpha)}.$$

In addition, similar to the estimate of (3.13), we have

$$\frac{1}{\Gamma(1+\alpha)}\int_{\frac{1}{2}}^{1}|f^{(\alpha)}(ta+(1-t)b)|^{q}(dt)^{\alpha}$$
$$\leq\frac{\Gamma(1+s\alpha)}{2^{\alpha}\Gamma(1+(s+1)\alpha)}\left(\left|f^{(\alpha)}\left(\frac{a+b}{2}\right)\right|^{q}+|f^{(\alpha)}(b)|^{q}\right).$$

Therefore, it is analogues to the estimate of (3.11), we have

$$\frac{1}{\Gamma(1+\alpha)}\int_{0}^{\frac{1}{2}}(1-2t)|f^{(\alpha)}(ta+(1-t)b)|(dt)^{\alpha}$$
$$\leq\left(\frac{1}{\Gamma(1+\alpha)}\int_{\frac{1}{2}}^{1}(2t-1)^{\alpha\frac{q}{q-1}}(dt)^{\alpha}\right)^{\frac{q-1}{q}}$$
$$\left(\frac{1}{\Gamma(1+\alpha)}\int_{\frac{1}{2}}^{1}|f^{(\alpha)}(ta+(1-t)b)|^{q\alpha}(dt)^{\alpha}\right)^{\frac{1}{q}}$$
$$\leq\left(\frac{\Gamma(1+\frac{q}{q-1}\alpha)}{2^{\alpha}\Gamma(1+\frac{2q-1}{q-1}\alpha)}\right)^{\frac{q-1}{q}}\left(\frac{\Gamma(1+s\alpha)}{2^{\alpha}\Gamma(1+(s+1)\alpha)}\right)^{\frac{1}{q}}$$
$$\left(\left|f^{(\alpha)}\left(\frac{a+b}{2}\right)\right|^{q}+|f^{(\alpha)}(b)|^{q})|^{q}\right)^{\frac{1}{q}}. \tag{3.15}$$

Thus, combining (3.10), (3.14), and (3.15), we obtain

$$\left|\frac{f(a)+f(b)}{2^{\alpha}}-\frac{\Gamma(1+\alpha)}{(b-a)^{\alpha}}{}_{a}I_{b}^{(\alpha)}f(x)\right|$$
$$\leq\frac{(b-a)^{\alpha}}{2^{\alpha}}\left[\frac{\Gamma(1+\frac{q}{q-1}\alpha)}{2^{\alpha}\Gamma(1+\frac{2q-1}{q-1}\alpha)}\right]^{\frac{q-1}{q}}\left(\frac{\Gamma(1+s\alpha)}{2^{\alpha}\Gamma(1+(s+1)\alpha)}\right)^{\frac{1}{q}}$$
$$\times\left[\left(|f^{(\alpha)}(a)|^{q}+\left|f^{(\alpha)}\left(\frac{a+b}{2}\right)\right|^{q}\right)^{\frac{1}{q}}\right.$$
$$\left.+\left(\left|f^{(\alpha)}\left(\frac{a+b}{2}\right)\right|^{q}+|f^{(\alpha)}(b)|^{q}\right)^{\frac{1}{q}}\right].$$

□

Therefore, we complete the proof of Theorem 3.3.

Acknowledgements The authors would like to express their gratitude to the Editors and referees for some very valuable suggestion.

References

1. Hudzik, H., Maligranda, L.: Some remarks on s-convex functions. Aequationes Math. **48**(1), 100–111 (1994)
2. Kirmaci, U.S., Bakula, M.K., Özdemir, M.E., Pečarić, J.: Hadamard-type inequalities for s-convex functions. Appl. Math. Comput. **193**(1), 26–35 (2007)
3. Avci, M., Kavurmaci, H., Ozdemir, M.E.: New inequalities of Hermite-Hadamard type via s-convex functions in the second sense with applications. Appl. Math. Comput. **217**, 5171–5176 (2011)
4. Shuang, Y., Yin, H.P., Qi, F.: Hermite-Hadamard type integral inequalities for geometric-arithmetically s-convex functions. Analysis. **33**(2), 197–208 (2013)
5. Kilbas, A.A., Srivastava, H.M., Trujillo, J.J.: Theory and Applications of Fractional Differential Equations. North-Holland Mathematical Studies. Elsevier (North-Holland) Science Publishers, Amsterdam, London and New York (2006)
6. Gao, F., Zhong, W.P., Shen, X.M.: Applications of Yang-Fourier transform to local fractional equations with local fractional derivative and local fractional integral. Adv. Mater. Res. **461**, 306–310 (2012)
7. Mo, H., Sui, X., Yu, D.: Generalized convex functions and some inequalities on fractal sets. Abstr. Appl. Anal. **2014**, Article ID 636751 (2014)
8. Mo, H.: Generalized Hermite-Hadamard Type Inequalities Involving Local Fractional Integrals, arxiv: 1410.1062
9. Mo, H., Sui, X.: Generalized s-convex function on fractal sets. Abstr. Appl. Anal. **2014**, Article ID 254737 (2014)
10. Yang, X.J.: Advanced Local Fractional Calculus and Its Applications. World Science Publisher, New York, NY, USA (2012)
11. Wie, W., Srivastava, H.M., Zhang, Y., Wang, L., Shen, P., Zhang, J.: A local fractional integral inequality on fractal space analogous to andersons inequality. Abstr. Appl. Anal. 2014, Article ID 797561 (2014)
12. Yang, X.J., Baleanu, D., Machado, J.A.T.: Mathematical aspects of heisenberg uncertainty principle within local fractional fourier analysis. Bound. Value Probl. **2013**, 131–146 (2013)
13. Zhao, Y., Cheng, D.F., Yang, X.J.: Approximation solutions for local fractional Schrödinger Equation in the on-dimensional cantorian system. Adv. Math. Phys. 2013, Article ID 291386 (2013)
14. Yang, A.M., Chen, Z.S., Srivastava, H.M., Yang, X.J.: Application of the local fractional series expansion method and the variational iteration method to the Helmholtz equation involving local fractional derivative operators, Abstr. Appl. Anal. **2013**, Article ID 259125 (2013)

25

A Pata-type fixed point theorem

Sriram Balasubramanian

Abstract We prove a fixed point theorem for a Pata-type map defined on a complete (normal) cone metric space. Our results generalize the recent work of M. Chakraborty and S. K. Samanta. An example demonstrating this fact is also presented.

Keywords Fixed point · Kannan · Pata · Contraction · Cone metric spaces

Mathematics Subject Classification 47H10 · 54H25 (Primary) · 37C25 (Secondary)

Introduction

The classical Banach fixed point theorem states that if (X, d) is a complete metric space and $T : X \to X$ is a contraction map, i.e., T satisfies

$$d(Tx, Ty) \leq \alpha d(x, y), \tag{1}$$

for all $x, y \in X$ and some $\alpha \in [0, 1)$, then T has a unique fixed point. i.e., there exists a unique $a \in X$ such that $Ta = a$.

In [4], Pata considered a map $T : X \to X$ on the complete metric space (X, d) that satisfied the condition: for all $x, y \in X$,

$$d(Tx, Ty) \leq (1 - \epsilon)d(x, y) + \Lambda\epsilon^{\alpha}\psi(\epsilon)[1 + \|x\| + \|y\|]^{\beta}, \tag{2}$$

for every $\epsilon \in [0, 1]$, fixed constants $\Lambda \geq 0$, $\alpha \geq 1$ and $\beta \in [0, \alpha]$, a fixed element $x_0 \in X$, $\|z\| = d(z, x_0)$ and an increasing function $\psi : [0, 1] \to [0, \infty)$ which vanishes at and is continuous at 0. He proved that the map T satisfying (2) has a unique fixed point. Moreover, he also demonstrated that if $T : X \to X$ is a contraction map, then T satisfies condition (2), thereby obtaining a generalization of the Banach fixed point theorem.

Another fixed point theorem that is widely popular, is due to Kannan which states that if (X, d) is a complete metric space and the map $T : X \to X$ is a Kannan contraction, i.e., T satisfies

$$d(Tx, Ty) \leq \frac{\gamma}{2}\{d(x, Tx) + d(y, Ty)\} \tag{3}$$

for all $x, y \in X$ and some $\gamma \in [0, 1)$, then T has a unique fixed point.

There are several generalizations of the Kannan fixed point theorem, but the one of particular interest to us is due to Chakraborty and Samanta in [1]. The authors, in [1], consider a map $T : X \to X$ defined on the complete metric space (X, d) that satisfies the condition: for all $x, y \in X$,

$$\begin{aligned} d(Tx, Ty) \leq {} & \frac{1-\epsilon}{2}\{d(x, Tx) + d(y, Ty)\} + \\ & \Lambda\epsilon^{\alpha}\psi(\epsilon)[1 + \|x\| + \|y\| + \|Tx\| + \|Ty\|]^{\beta}, \end{aligned} \tag{4}$$

for every $\epsilon \in [0, 1]$, fixed constants $\Lambda \geq 0$, $\alpha \geq 1$ and $\beta \in [0, \infty)$, a fixed element $x_0 \in X$, $\|z\| = d(z, x_0)$ and a function $\psi : [0, 1] \to [0, \infty)$ which vanishes at and is continuous at 0. The authors prove that the map T has a unique fixed point. Moreover, they also demonstrate that if $T : X \to X$ is a Kannan contraction map, then T satisfies condition (4).

This article contains a generalization of the main result in [1]. The setting considered is that of complete cone

S. Balasubramanian (✉)
Department of Mathematics and Statistics, Indian Institute of Science Education and Research, Kolkata, India
e-mail: bsriram@iiserkol.ac.in

metric spaces, where the underlying cone is normal (see [3]). It is shown that for a map $T : X \to X$ defined on the complete (normal) cone metric space (X, d), which satisfies a Pata-type condition, that is an improved version of (4), there exists a unique fixed point. An example to illustrate the main result is also provided.

Preliminaries and the main result

Let E be a real Banach space. A non-empty closed subset P of E is said to be a *cone* if

(a) $\alpha P + \beta P \subset P$ for all $\alpha, \beta \in [0, \infty)$.
(b) $P \cap (-P) = \{\theta\}$, where $\theta \in E$ is the zero vector.

The cone P is said to be *solid* if the interior of P, which we will denote by $int\, P$, is non-empty.

Examples of solid cones

(1) Let $E = \mathbb{R}$ and $P = [0, \infty)$.
(2) Let $E = \mathbb{R}^2$ and $P = \{(x, y) : x, y \geq 0\}$.
(3) Let $E = \ell^2$ and $P = \{(x_n)_{n \geq 1} : x_n \geq 0\}$.

The norms on E in the examples above are the usual norms.

A cone P in a real Banach space E, induces the following partial order $\preceq$ on E. For $x, y \in E$,

$$x \preceq y \Leftrightarrow y - x \in P.$$

In the case of a solid cone P, we will use the notation $x \ll y$ to denote $y - x \in int\, P$.

A cone P is said to be *normal* if for all $x, y \in P$ such that $x \preceq y$, there exists a constant $\kappa \geq 1$ such that $\|x\| \leq \kappa\|y\|$. The examples (1), (2) and (3) above are normal cones with $\kappa = 1$.

Let X be a nonempty set, E be a real Banach space and $P \subset E$ be a solid normal cone with normal constant $\kappa \geq 1$. A map $d : X \times X \to E$ is said to be a *cone metric* if for all $x, y, z \in E$,

(a) $d(x, y) \succeq \theta$, i.e., $d(x, y) \in P$.
(b) $d(x, y) = \theta$ if and only if $y = x$.
(c) $d(x, y) \preceq d(x, z) + d(z, y)$.

The pair (X, d) is called a *cone metric space*. It is indeed the case that every metric space is a cone metric space.

Examples of cone metric spaces

(1) ([3]) Let E and P be as in example (2) above, $d : \mathbb{R} \times \mathbb{R} \to E$ be the map $d(x, y) = (|x - y|, \alpha|x - y|)$, where $\alpha \geq 0$ is a constant. The pair $(\mathbb{R}, d)$ is a cone metric space.
(2) ([2]) Let (X, ρ) be a metric space, E and P be as in example (3) above, $d : X \times X \to E$ be the map $d(x, y) = \left(\sqrt{2^{-n}\rho(x, y)}\right)_{n \geq 1}$. It can be verified that (X, d) is a cone metric space.

Let (X, d) be a cone metric space. A sequence (x_n) of points in X is said to be *Cauchy* if for any given $c \gg 0$, there exists $N \in \mathbb{N}$ such that $d(x_m, x_n) \ll c$, for all $m, n \geq N$, or equivalently, there exists $M \in \mathbb{N}$ which is independent of p such that $d(x_n, x_{n+p}) \ll c$ for all $n \geq M$.

The sequence (x_n) is said to be *convergent*, if there exists $x \in X$ such that for any given $c \gg 0$, there exists $N \in \mathbb{N}$ such that $d(x_n, x) \ll c$ for all $n \geq N$.

The cone metric space (X, d) is said to be *complete* if every Cauchy sequence in X converges.

The proof of the following lemma can be found in [3].

Lemma 1 Let (X, d) be a cone metric space and (x_n) be a sequence in X.

(1) (x_n) is Cauchy if and only if $\|d(x_m, x_n)\| \to 0$ as $m, n \to \infty$. i.e., Given $\eta > 0$ and $p \in \mathbb{N}$, there exists $N \in \mathbb{N}$ which is independent of p such that $\|d(x_n, x_{n+p})\| < \eta$, for all $n \geq N$.
(2) (x_n) is convergent to x if and only if $\|d(x_n, x)\| \to 0$ as $n \to \infty$.
(3) If (x_n) converges to $x, y \in X$, then $x = y$.

The following is our main result which generalizes Theorem 2.2 in [1].

Theorem 1 Let (X, d) be a complete cone metric space with normal constant $\kappa \geq 1$, $x_0 \in X$, $\Lambda \geq 0$, $\alpha \geq 1$ and $\beta \in [0, \infty)$ be fixed constants and $\psi : [0, 1] \to [0, \infty)$ be such that $\lim_{\epsilon \to 0^+} \psi(\epsilon) = 0$. If for every $x, y \in X$, the map $T : X \to X$ satisfies

$$\|d(Tx, Ty)\| \leq \frac{(1 - \epsilon)}{\kappa} M(x, y) + \Lambda \epsilon^{\alpha} \psi(\epsilon)[1 + \|x\| + \|y\| + \|Tx\| + \|Ty\|]^{\beta}, \tag{5}$$

for every $\epsilon \in [0, 1]$, where $M(x, y) = \max\{\|d(x, Tx)\|, \|d(y, Ty)\|, \frac{1}{2\kappa}\|d(x, y)\|\}$ and $\|z\|$ denotes $\|d(z, x_0)\|$, then T has a unique fixed point.

The proof of the above Theorem is given in Sect. 3. We point out that there is no loss of generality in choosing any such x_0 in (5), simply because a change in x_0 can essentially be absorbed by assigning a different value to Λ, thanks to the triangle inequality of the cone metric d and the sub-additivity of the norm.

Proof of The main result

This section contains a proof of our main result. Although the proof follows a similar pattern as that of Theorem 2.2 in [1], the arguments provided here are different and sometimes simpler.

Proof of Theorem 1 First we prove uniqueness of the fixed point. Suppose that $x, y \in X$ are such that $x \neq y$, $Tx = x$ and $Ty = y$. Letting $\epsilon = 0$ in inequality (5) yields, $\|d(x,y)\| \leq \frac{1}{2\kappa}\|d(x,y)\|$, a contradiction to the fact that $x \neq y$. The uniqueness follows.

To show the existence of a fixed point, consider the sequence $(T^n x_0)$. Without loss of generality, we assume that $T^n x_0 \neq T^{n+1} x_0$, for all $n = 0, 1, 2, \ldots$. Since for each $n \in \mathbb{N}$, and $\epsilon \in [0, 1]$,

$$\begin{aligned}\|d(T^{n+1}x_0, T^n x_0)\| &\leq (1-\epsilon)M(T^n x_0, T^{n-1}x_0)\\ &\quad + \Lambda\epsilon^{\alpha}\psi(\epsilon)[1 + 2\|T^n x_0\| + \|T^{n-1}x_0\|\\ &\quad + \|T^{n+1}x_0\|]^{\beta}, \end{aligned} \tag{6}$$

it follows that there exists no $n \in \mathbb{N}$, for which $M(T^n x_0, T^{n-1}x_0) = \|d(T^{n+1}x_0, T^n x_0)\|$. For otherwise, it would mean that there exists some $m \in \mathbb{N}$ such that for every $\epsilon \in (0, 1]$,

$$\begin{aligned}\|d(T^{m+1}x_0, T^m x_0)\| &\leq \Lambda\epsilon^{\alpha-1}\psi(\epsilon)[1 + 2\|T^m x_0\|\\ &\quad + \|T^{m-1}x_0\| + \|T^{m+1}x_0\|]^{\beta}.\end{aligned}$$

Letting $\epsilon \to 0^+$ yields, $\|d(T^{m+1}x_0, T^m x_0)\| = 0$, i.e., $T^{m+1}x_0 = T^m x_0$, a contradiction. Thus, for each $n \in \mathbb{N}$, letting $\epsilon = 0$ in inequality (6) yields,

$$\|d(T^{n+1}x_0, T^n x_0)\| \leq \|d(T^n x_0, T^{n-1}x_0)\|. \tag{7}$$

Iterating we obtain

$$\|d(T^{n+1}x_0, T^n x_0)\| \leq \|d(Tx_0, x_0)\|, \tag{8}$$

for all $n = 0, 1, 2, \ldots$

Next, we show that the sequence $(\|d(T^n x_0, x_0)\|)$ is bounded above by $c = 2\kappa\|d(x_0, Tx_0)\|$. This is certainly the case when $n = 1$. Assume that $\|d(T^{m-1}x_0, x_0)\| \leq c$. The claim follows from induction, if we show that $\|d(T^m x_0, x_0)\| \leq c$. Using (8), it follows that

$$\begin{aligned}\|d(T^m x_0, x_0)\| &\leq \kappa\{\|d(T^m x_0, Tx_0)\| + \|d(Tx_0, x_0)\|\}\\ &\leq \kappa M(T^{m-1}x_0, x_0) + \frac{c}{2}\\ &= \kappa \max\Big\{\|d(T^{m-1}x_0, T^m x_0)\|, \|d(x_0, Tx_0)\|,\\ &\quad \times \frac{1}{2\kappa}\|d(T^{m-1}x_0, x_0)\|\Big\} + \frac{c}{2}\\ &\leq \kappa \max\Big\{\|d(x_0, Tx_0)\|, \frac{1}{2\kappa}\|d(T^{m-1}x_0, x_0)\|\Big\}\\ &\quad + \frac{c}{2} \leq \kappa\Big\{\frac{c}{2\kappa}\Big\} + \frac{c}{2} = c.\end{aligned}$$

Thus, for all $n \in \mathbb{N}$,

$$\|d(T^n x_0, x_0)\| \leq c. \tag{9}$$

Consider the monotonically decreasing sequence $(\|d(T^n x_0, T^{n+1}x_0)\|)$ [see (7)]. Since it is bounded below by 0, it is convergent. Let $\ell = \lim_{n\to\infty}\|d(T^n x_0, T^{n+1}x_0)\|$. Clearly $\ell \geq 0$. We will in fact, prove that $\ell = 0$. For each $\epsilon \in (0, 1]$, it follows from (6) and (9) that

$$\|d(T^n x_0, T^{n+1}x_0)\| \leq (1-\epsilon)\|d(T^{n-1}x_0, T^n x_0)\| + K\Lambda\epsilon^{\alpha}\psi(\epsilon),$$

where $K = (1 + 4c)^{\beta}$. Rearranging the above inequality and using the fact that the sequence $(\|d(T^n x_0, T^{n+1}x_0)\|)$ is monotonically decreasing yields,

$$\|d(T^n x_0, T^{n+1}x_0)\| \leq \frac{\|d(T^{n-1}x_0, T^n x_0)\|}{(1+\epsilon)} + \frac{K\Lambda\epsilon^{\alpha}\psi(\epsilon)}{(1+\epsilon)}, \tag{10}$$

for each $\epsilon \in (0, 1]$. Fixing such an ϵ and letting $n \to \infty$ in (10), we obtain

$$\ell \leq \frac{\ell}{(1+\epsilon)} + \frac{K\Lambda\epsilon^{\alpha}\psi(\epsilon)}{(1+\epsilon)}$$

Multiplying both sides by $(1+\epsilon)$ and simplifying yields, for each $\epsilon \in (0, 1]$,

$$\ell \leq K\Lambda\epsilon^{\alpha-1}\psi(\epsilon).$$

Letting $\epsilon \to 0^+$ yields, $\ell \leq 0$. Thus, in fact,

$$\ell = 0. \tag{11}$$

Next, we show that the sequence $(T^n x_0)$ is Cauchy. In view of Lemma 1 (i), it suffices to show that, given $\eta > 0$ and $p \in \mathbb{N}$, there exists $M \in \mathbb{N}$ which is independent of p, such that $\|d(T^n x_0, T^{n+p}x_0)\| < \eta$, for all $n \geq M$.

By (11), choose $N \in \mathbb{N}$ such that

$$\|d(T^n x_0, T^{n+1}x_0)\| < \frac{\eta}{2} \tag{12}$$

for all $n \geq N$.

Consider $\|d(T^n x_0, T^{n+p}x_0)\|$ for all $n \geq N + 1$. Letting $\epsilon = 0$ in (5) yields,

$$\|d(T^n x_0, T^{n+p}x_0)\| \leq M(T^{n-1}x_0, T^{n+p-1}x_0).$$

If $M(T^{n-1}x_0, T^{n+p-1}x_0) = \|d(T^{n-1}x_0, T^n x_0)\|$ or $\|d(T^{n+p-1}x_0, T^{n+p}x_0)\|$, then it follows from (12) that

$$\|d(T^n x_0, T^{n+p}x_0)\| < \frac{\eta}{2}. \tag{13}$$

If $M(T^{n-1}x_0, T^{n+p-1}x_0) = \frac{1}{2\kappa}\|d(T^{n-1}x_0, T^{n+p-1}x_0)\|$, then the normality of the underlying cone and an application of the triangle inequality yield,

$$\begin{aligned}\|d(T^n x_0, T^{n+p}x_0)\| &\leq \frac{1}{2\kappa}\|d(T^{n-1}x_0, T^{n+p-1}x_0)\|\\ &\leq \frac{1}{2}\{\|d(T^{n-1}x_0, T^n x_0)\| + \|d(T^n x_0, T^{n+p}x_0)\|\\ &\quad + \|d(T^{n+p}x_0, T^{n+p-1}x_0)\|\}.\end{aligned}$$

From a rearrangement of terms, it follows from (12) that

$$\|d(T^n x_0, T^{n+p} x_0)\| \leq \|d(T^{n-1} x_0, T^n x_0)\| + \|d(T^{n+p} x_0, T^{n+p-1} x_0)\| < \eta. \quad (14)$$

Thus, it follows from (13) and (14) that given $\eta > 0$ and $p \in \mathbb{N}$, by choosing $M = N + 1$, which is independent of p, one has that

$$\|d(T^n x_0, T^{n+p} x_0)\| < \eta,$$

for all $n \geq M$. i.e., the sequence $(T^n x_0)$ is Cauchy in the cone metric space (X, d). By completeness, there exists a unique $u \in X$ such that $T^n x_0 \to u$ as $n \to \infty$.

We complete the proof by showing that this u is a fixed point of T. Let $\eta > 0$ be arbitrary, $B = \Lambda\kappa(1 + \|u\| + \|Tu\| + 2c)^\beta$. Choose $\epsilon_0 \in (0, 1]$ such that

$$B\epsilon_0^{\alpha-1}\psi(\epsilon_0) < \frac{\eta}{2}. \quad (15)$$

Choose $L \in \mathbb{N}$ such that

$$\|d(T^L x_0, u)\| + \|d(T^L x_0, T^{L+1} x_0)\| < \frac{\epsilon_0 \eta}{4\kappa}. \quad (16)$$

Consider $\|d(Tu, u)\|$. From (5), it follows that

$$\begin{aligned}
\|d(Tu, u)\| &\leq \kappa\{\|d(Tu, T^{L+1} x_0)\| + \|d(T^{L+1} x_0, T^L x_0)\| \\
&\quad + \|d(T^L x_0, u)\|\} \\
&\leq (1 - \epsilon_0) M(u, T^L x_0) \\
&\quad + \kappa\Lambda\epsilon_0^\alpha\psi(\epsilon_0)(1 + \|u\| + \|T^L x_0\| + \|Tu\| \\
&\quad + \|T^{L+1} x_0\|)^\beta + \kappa\{\|d(T^{L+1} x_0, T^L x_0)\| \\
&\quad + \|d(T^L x_0, u)\|\} \\
&\leq (1 - \epsilon_0)\{\|d(u, Tu)\| \\
&\quad + \|d(T^{L+1} x_0, T^L x_0)\| + \|d(u, T^L x_0)\|\} \\
&\quad + B\epsilon_0^\alpha\psi(\epsilon_0) + \kappa\{\|d(T^{L+1} x_0, T^L x_0)\| \\
&\quad + \|d(T^L x_0, u)\|\} \\
&\leq (1 - \epsilon_0)\|d(u, Tu)\| \\
&\quad + 2\kappa\{\|d(T^{L+1} x_0, T^L x_0)\| + \|d(u, T^L x_0)\|\} \\
&\quad + B\epsilon_0^\alpha\psi(\epsilon_0).
\end{aligned}$$

Rearranging the terms in the above inequality yields,

$$\|d(u, Tu)\| \leq \frac{2\kappa}{\epsilon_0}\{\|d(T^{L+1} x_0, T^L x_0)\| + \|d(u, T^L x_0)\|\} + B\epsilon_0^{\alpha-1}\psi(\epsilon_0).$$

It follows from (15) and (16) that

$$\|d(u, Tu)\| < \eta.$$

Since $\eta > 0$ is arbitrary, the proof is complete. □

Applications and examples

This section contains applications of Theorem 1, presented as corollaries. Recall the definition of $M(x, y)$ and the notation $\|\cdot\|$ from Theorem 1.

Corollary 1 Let (X, d) be a complete cone metric space with normal constant $\kappa \geq 1$ and $\delta \in \left(0, \frac{1}{\kappa}\right)$. If the map $T : X \to X$ satisfies

$$\|d(Tx, Ty)\| \leq \delta M(x, y)$$

for all $x, y \in X$, then T has a unique fixed point.

Proof It suffices to show that T satisfies condition (5). The result then follows from Theorem 1. Fix $x_0 \in X$. Observe that

$$\begin{aligned}
M(x, y) &= \max\left\{\|d(x, Tx)\|, \|d(y, Ty)\|, \frac{1}{2\kappa}\|d(x, y)\|\right\} \\
&\leq 1 + \|d(x, x_0)\| + \|d(x_0, Tx)\| + \|d(y, x_0)\| \\
&\quad + \|d(x_0, Ty)\| \\
&= 1 + \|x\| + \|Tx\| + \|y\| + \|Ty\|.
\end{aligned} \quad (17)$$

It follows from (17) and a Bernoulli inequality argument similar to Sect. 3 in [1], that for all $\epsilon \in [0, 1]$, $x, y \in X$,

$$\begin{aligned}
\|d(Tx, Ty)\| &\leq \delta M(x, y) \\
&= \frac{(1 - \epsilon)}{\kappa} M(x, y) + \left(\delta + \frac{(\epsilon - 1)}{\kappa}\right) M(x, y) \\
&= \frac{(1 - \epsilon)}{\kappa} M(x, y) + \delta\left(1 + \frac{(\epsilon - 1)}{\kappa\delta}\right) M(x, y) \\
&\leq \frac{(1 - \epsilon)}{\kappa} M(x, y) + \delta(1 + (\epsilon - 1))^{\frac{1}{\kappa\delta}} M(x, y) \\
&\leq \frac{(1 - \epsilon)}{\kappa} M(x, y) + \delta\epsilon^{\frac{1}{\kappa\delta}}[1 + \|x\| + \|Tx\| \\
&\quad + \|y\| + \|Ty\|]. \\
&= \frac{(1 - \epsilon)}{\kappa} M(x, y) \\
&\quad + \delta\epsilon^{1+\eta}[1 + \|x\| + \|Tx\| + \|y\| + \|Ty\|],
\end{aligned} \quad (18)$$

where $\eta = \left(\frac{1}{\kappa\delta} - 1\right) > 0$. Comparing (18) with (5), one sees that T satisfies (5) with $\Lambda = \delta$, $\beta = \alpha = 1$, $\psi(\epsilon) = \epsilon^\eta$. □

Corollary 2 (The Kannan Fixed Point Theorem) Let (X, d) be a complete metric space and $\delta \in (0, 1)$. If the map $T : X \to X$ satisfies

$$d(Tx, Ty) \leq \frac{\delta}{2}(d(x, Tx) + d(y, Ty))$$

for all $x, y \in X$, then T has a unique fixed point.

Proof Observe that $\kappa = 1$ and T satisfies the hypothesis of Corollary 1. The result follows. □

We end with an example which demonstrates that Theorem 1 indeed generalizes the main result in [1]. Observe that it suffices to produce a complete metric space (X, d) and a map $T : X \to X$ which satisfies

(A) $d(Tx, Ty) \leq \delta \max\{d(x, Tx), d(y, Ty), \frac{1}{2}d(x, y)\}$ for some $\delta \in (0, 1)$ and $x, y \in X$,

(B) $d(Ta, Tb) > \frac{1}{2}\{d(a, Ta) + d(b, Tb)\}$ for some $a, b \in X$.

Example Let $X = [0, \frac{1}{2}] \cup \{1, 2\}$ with d being the usual metric. It is clear that (X, d) is a complete cone metric space with $\kappa = 1$. Define the map $T : X \to X$ by $Tx = 2$ if $x \neq 1, 2$ and $T1 = T2 = 1$. Observe that for $x \neq 1, 2$, $d(Tx, T1) = 1$ and

$$\max\left\{d(x, Tx), d(1, T1), \frac{1}{2}d(x, 1)\right\} = \max\left\{2 - x, 0, \frac{1}{2}(1 - x)\right\} = 2 - x.$$

Similarly, for $x \neq 1, 2$, $d(Tx, T2) = 1$ and

$$\max\left\{d(x, Tx), d(2, T2), \frac{1}{2}d(x, 2)\right\} = \max\left\{2 - x, 1, \frac{1}{2}(2 - x)\right\} = 2 - x.$$

Since, by choice, $x \in [0, \frac{1}{2}]$, it follows that $1 \leq \frac{2}{3}(2 - x)$. Moreover, $d(T1, T2) = d(Tx, Ty) = 0$ for all $x, y \in [0, \frac{1}{2}]$. Putting all this together implies that T satisfies condition (A) above with $\delta = \frac{2}{3}$. Since $\frac{1}{2}\{d(\frac{1}{4}, T\frac{1}{4}) + d(1, T1)\} = \frac{(2-\frac{1}{4})}{2} < 1 = d(T\frac{1}{4}, T1)$, it follows that T also satisfies condition (B) above with $a = \frac{1}{4}$ and $b = 1$. From the proof of Corollary 1, it follows that T satisfies (5). However, T does not satisfy (4), as it satisfies condition (B) above. This can be seen by setting $\epsilon = 0$ in (4). It is immediate that T has a unique fixed point.

References

1. Chakraborty, M., Samanta, S.K.: A fixed point theorem for kannan-type maps in metric spaces, pre-print (2012), arXiv:1211.7331v2 [math.GN], Nov 2012
2. Haghi, R.H., Rezapour, Sh.: Fixed points of multifunctions on regular cone metric spaces. Expositiones Mathematicae **28**(1), 71–77 (2010)
3. Huang, L.-G., Zhang, X.: Cone metric spaces and fixed point theorems of contractive mappings. J. Math. Anal. Appl. **332**, 1468–1476 (2007)
4. Pata, V.: A fixed point theorem in metric spaces. J. Fixed Point Theory Appl. **10**, 299–305 (2011)

Permissions

All chapters in this book were first published in MS, by Springer International Publishing AG.; hereby published with permission under the Creative Commons Attribution License or equivalent. Every chapter published in this book has been scrutinized by our experts. Their significance has been extensively debated. The topics covered herein carry significant findings which will fuel the growth of the discipline. They may even be implemented as practical applications or may be referred to as a beginning point for another development.

The contributors of this book come from diverse backgrounds, making this book a truly international effort. This book will bring forth new frontiers with its revolutionizing research information and detailed analysis of the nascent developments around the world.

We would like to thank all the contributing authors for lending their expertise to make the book truly unique. They have played a crucial role in the development of this book. Without their invaluable contributions this book wouldn't have been possible. They have made vital efforts to compile up to date information on the varied aspects of this subject to make this book a valuable addition to the collection of many professionals and students.

This book was conceptualized with the vision of imparting up-to-date information and advanced data in this field. To ensure the same, a matchless editorial board was set up. Every individual on the board went through rigorous rounds of assessment to prove their worth. After which they invested a large part of their time researching and compiling the most relevant data for our readers.

The editorial board has been involved in producing this book since its inception. They have spent rigorous hours researching and exploring the diverse topics which have resulted in the successful publishing of this book. They have passed on their knowledge of decades through this book. To expedite this challenging task, the publisher supported the team at every step. A small team of assistant editors was also appointed to further simplify the editing procedure and attain best results for the readers.

Apart from the editorial board, the designing team has also invested a significant amount of their time in understanding the subject and creating the most relevant covers. They scrutinized every image to scout for the most suitable representation of the subject and create an appropriate cover for the book.

The publishing team has been an ardent support to the editorial, designing and production team. Their endless efforts to recruit the best for this project, has resulted in the accomplishment of this book. They are a veteran in the field of academics and their pool of knowledge is as vast as their experience in printing. Their expertise and guidance has proved useful at every step. Their uncompromising quality standards have made this book an exceptional effort. Their encouragement from time to time has been an inspiration for everyone.

The publisher and the editorial board hope that this book will prove to be a valuable piece of knowledge for researchers, students, practitioners and scholars across the globe.

List of Contributors

Yekini Shehu
Department of Mathematics, University of Nigeria, Nsukka, Nigeria

Muhammad Arshad, Fahimuddin, Abdullah Shoaib and Aftab Hussain
Department of Mathematics, International Islamic University, Islamabad 44000, Pakistan

Mohammad Imdad
Department of Mathematics, Aligarh Muslim University, Aligarh Uttar Pradesh, India

Samir Dashputre
Department of Applied Mathematics, Shri Shankaracharya Technical Campus, Shri Shankaracharya Group of Institutions (F.E.T), Junwani, Bhilai 490020, India

Nihal Yilmaz Özgür and Nihal Taş
Department of Mathematics, Balıkesir University, 10145 Balıkesir, Turkey

W. M. Abd-Elhameed
Department of Mathematics, Faculty of Science, University of Jeddah, Jeddah, Saudi Arabia
Department of Mathematics, Faculty of Science, Cairo University, Giza, Egypt

Y. H. Youssri and E. H. Doha
Department of Mathematics, Faculty of Science, Cairo University, Giza, Egypt

Hassen Aydi
Department of Mathematics, College of Education of Jubail, University of Dammam, P.O. 12020, Industrial Jubail 31961, Saudi Arabia

Abdelbasset Felhi and Slah Sahmim
College of Sciences, Department of Mathematics, KFU University, Al-hasa, Saudi Arabia

Jerolina Fernandez and Neeraj Malviya
Department of Mathematics, NRI Institute of Information Science & Technology, Bhopal 462021, MP, India

Geeta Modi
Department of Mathematics, Government M V M College, Bhopal 462001, MP, India

Nawab Hussain
Department of Mathematics, King Abdulaziz University, Jeddah 21589, Saudi Arabia

Vahid Parvaneh
Department of Mathematics, Gilan-E-Gharb Branch, Islamic Azad University, Gilan-E-Gharb, Iran

Farhan Golkarmanesh
Department of Mathematics, Sanandaj Branch, Islamic Azad University, Sanandaj, Iran

Yaé Ulrich Gaba
Department of Mathematics and Applied Mathematics, University of Cape Town, Rondebosch 7701, South Africa

A. Kilicman and M. Wadai
Department of Mathematics and Institute for Mathematical Research, University Putra Malaysia, UPM, 43400, Serdang, Selangor, Malaysia

Ishak Altun
Department of Mathematics, Faculty of Science and Arts, Kirikkale University, Yahsihan, 71540 Kirikkale, Turkey

Kishin Sadarangani
Departamento de Matema´ticas, Universidad de Las Palmas de Gran Canaria, Campus de Tafira Baja, 35017 Las Palmas de Gran Canaria, Spain

R. A. Rashwan and S. M. Saleh
Department of Mathematics, Faculty of Science, Assiut University, Assiut, Egypt

B. N. Onagh
Department of Mathematics, Golestan University, Gorgan, Iran

Z. Kadelburg
Faculty of Mathematics, University of Belgrade, Beograd, Serbia

S. Radenović
Faculty of Mechanical Engineering, University of Belgrade, Kraljice Marije 16, 11120 Beograd, Serbia

Saeid Abbasbandy
Department of Mathematics, Imam Khomeini International University, Qazvin 34149-16818, Iran

A. Fazli and T. Allahviranloo
Department of Mathematic, Science and Research Branch, Islamic Azad University, Tehran, Iran

Sh. Javadi
Department of Mathematic, Kharazmi University, Tehran, Iran

Mohammad Imdad, Rqeeb Gubran and Mohammad Arif
Department of Mathematics, Aligarh Muslim University, Aligarh 202002, India

Dhananjay Gopal
Department of Applied Mathematics and Humanities, S.V. National Institute of Technology, Surat, Gujarat 395007, India

Zaid Mohammed Fadail and Abd Ghafur Bin Ahmad
School of Mathematical Sciences, Faculty of Science and Technology, Universiti Kebangsaan Malaysia (UKM), 43600 Bangi, Selangor Darul Ehsan, Malaysia

Stojan Radenović
Faculty of Mathematics and Information Technology, Teacher Education, Dong Thap University, Cao Lanh, Dong Thap, Viet Nam

Miloje Rajović
Faculty of Mechanical Engineering, Dositejeva19, Kraljevo 36000, Serbia

Kianoush Fathi Vajargah and Maryam Shoghi
Department of Statistics, Islamic Azad University, North Branch Tehran, Tehran, Iran

V. M. L. Hima Bindu and G. N. V. Kishore
Department of Mathematics, K L University, Vaddeswaram, Guntur 522 502, Andhra Pradesh, India

K. P. R. Rao
Department of Mathematics, Acharya Nagarjuna University, Nagarjuna Nagar, Guntur 522 510, Andhra Pradesh, India

Y. Phani
Department of Basic Sciences, Shri Vishnu Engineering College for Women, Bhimavaram, West Godavari 534 202, Andhra Pradesh, India

S. H. Rasouli and Z. Firouzjahi
Department of Mathematics, Faculty of Basic Sciences, Babol University of Technology, Babol, Iran

Rabha W. Ibrahim
Faculty of Computer Science and Information Technology, University, Malaya, 50603 Kuala Lumpur, Malaysia

M. Z. Ahmad and M. Jasim Mohammed
Institute of Engineering Mathematics, Universiti Malaysia Perlis, 02600 Arau, Perlis, Malaysia

Reza Arab
Department of Mathematics, Sari Branch, Islamic Azad University, Sari, Iran

Reza Allahyari and Ali Shole Haghighi
Department of Mathematics, Mashhad Branch, Islamic Azad University, Mashhad, Iran

Huixia Mo and Xin Sui
School of Science, Beijing University of Posts and Telecommunications, Beijing 100876, China

Sriram Balasubramanian
Department of Mathematics and Statistics, Indian Institute of Science Education and Research, Kolkata, India

Index

www.ingramcontent.com/pod-product-compliance
Lightning Source LLC
Chambersburg PA
CBHW082023190326
41458CB00010B/3252
* 9 7 8 1 6 3 2 3 8 6 6 2 5 *